Numerical

Analysis

Numerical
Analysis

Ian Jacques and **Colin Judd**

Department of Mathematics
Coventry Lanchester Polytechnic

London   New York
CHAPMAN AND HALL

First published in 1987 by
Chapman and Hall Ltd
11 New Fetter Lane, London EC4P 4EE
Published in the USA by
Chapman and Hall
29 West 35th Street, New York NY 10001

© 1987 I. Jacques and C.J. Judd

Printed in Great Britain by
J.W. Arrowsmith Ltd, Bristol

ISBN 0 412 27950 9 (hardback)
     0 412 27960 6 (paperback)

British Library Cataloguing in Publication Data

Jacques, Ian
  Numerical analysis
  1. Numerical analysis
  I. Title II. Judd, Colin
  519.4   QA297

  ISBN 0–412–27950–9
  ISBN 0–412–27960–6 Pbk

Library of Congress Cataloging in Publication Data

Jacques, Ian, 1957–
  Numerical analysis.

  Bibliography: p.
  Includes index.
  1. Numerical analysis. I. Judd, Colin, 1952–
II. Title
QA297.J26 1987   519.4   86–26887
ISBN 0–412–27950–9
ISBN 0–412–27960–6 (pbk.)

Contents

Preface

This book is primarily intended for undergraduates in mathematics, the physical sciences and engineering. It introduces students to most of the techniques forming the core component of courses in numerical analysis. The text is divided into eight chapters which are largely self-contained. However, with a subject as intricately woven as mathematics, there is inevitably some interdependence between them. The level of difficulty varies and, although emphasis is firmly placed on the methods themselves rather than their analysis, we have not hesitated to include theoretical material when we consider it to be sufficiently interesting. However, it should be possible to omit those parts that do seem daunting while still being able to follow the worked examples and to tackle the exercises accompanying each section. Familiarity with the basic results of analysis and linear algebra is assumed since these are normally taught in first courses on mathematical methods. For reference purposes a list of theorems used in the text is given in the appendix.

The main purpose of the exercises is to give the student practice in using numerical methods, but they also include simple proofs and slight digressions intended to extend the material of the text. Solutions are provided for every question in each exercise. Many of the questions are suitable for hand calculation or are theoretical in nature, although some require the use of a computer. No attempt has been made to provide listings of programs or to discuss programming style and languages. Nowadays there is a massive amount of published software available and in our opinion use should be made of this. It is not essential for students to actually program a method as a prerequisite to understanding.

We are indebted to Dr K.E. Barrett for his helpful comments on much of the text and to Mrs M. Schoales for her meticulous typing of the manuscript.

IAN JACQUES
COLIN JUDD

1

Introduction

Numerical analysis is concerned with the development and analysis of methods for the numerical solution of practical problems. Traditionally, these methods have been mainly used to solve problems in the physical sciences and engineering. However, they are finding increasing relevance in a much broader range of subjects including economics and business studies.

The first stage in the solution of a particular problem is the formulation of a mathematical model. Mathematical symbols are introduced to represent the variables involved and physical (or economic) principles are applied to derive equations which describe the behaviour of these variables. Unfortunately, it is often impossible to find the exact solution of the resulting mathematical problem using standard techniques. In fact, there are very few problems for which an analytical solution can be determined. For example, there are formulas for solving quadratic, cubic and quartic polynomial equations, but no such formula exists for polynomial equations of degree greater than four or even for a simple equation such as

$$x = \cos x$$

Similarly, we can certainly evaluate the integral

$$\int_a^b e^x \, dx$$

as $e^b - e^a$, but we cannot find the exact value of

$$\int_a^b e^{x^2} \, dx$$

since no function exists which differentiates to e^{x^2}. Even when an analytical solution can be found it may be of more theoretical than practical use. For example, if the solution of a differential equation

$$y'' = f(x, y, y')$$

is expressed as an infinite sum of Bessel functions, then it is most unsuitable for calculating the numerical value of y corresponding to some numerical value of x.

A numerical algorithm consists of a sequence of arithmetic and logical operations which produces an approximate solution to within any prescribed accuracy. There are often several different algorithms for the solution of any one problem. The particular algorithm chosen depends on the context from which the problem is taken. In economics, for example, it may be that only the general behaviour of a variable is required, in which case a simple, low accuracy method which uses only a few calculations is appropriate. On the other hand, in precision engineering, it may be essential to use a sophisticated, highly accurate method, regardless of the total amount of computational effort involved.

Once a numerical algorithm has been selected, a computer program is usually written for its implementation. This is an important stage in the solution process because there is little point in devising an efficient numerical procedure if the program is badly written. Indeed, the choice of algorithm may well be influenced by programming considerations. Finally, the program is run to obtain numerical results, although this may not be the end of the story. The computed solution could indicate that the original mathematical model needs modifying with a corresponding change in both the numerical algorithm and the program.

Although the solution of 'real problems' by numerical techniques involves the use of a digital computer or calculator, the study of numerical analysis dates back to long before such machines were invented. The basic ideas underlying many of the methods currently in use have been known for hundreds of years. However, there has been a great change over the past thirty years and the size and complexity of problems that may be tackled have increased enormously. Determination of the eigenvalues of large matrices, for example, did not become a realistic proposition until computers became available because of the amount of computation involved. Nowadays any numerical technique can at least be demonstrated on a microcomputer, although there are some problems that can only be solved using the speed and storage capacity of much larger machines.

One effect of the recent increase in computing power has been to take the tedium of repetitive arithmetic out of problem solving. However, there is a great deal more to numerical analysis than just 'number crunching'. Most, if not all, numerical methods can be made to produce erroneous results if they are given suitably mischievous data. A mathematical analysis of these methods provides a deeper insight into their behaviour, and it is for this reason that a study of numerical analysis is essential.

1.1 ROUNDING ERRORS AND INSTABILITY

The vast majority of numerical methods involve a large number of calculations which are best performed on a computer or calculator. Unfortunately, such machines are incapable of working to infinite precision and so small errors occur in nearly every arithmetic operation. Even an apparently simple number such as 2/3 cannot be represented exactly on a computer. This number has a non-terminating decimal expansion

$$0.666\,666\,666\,666\,66\ldots$$

and if, for example, the machine uses ten-digit arithmetic, then it is stored as

$$0.666\,666\,666\,7$$

(In fact, computers use binary arithmetic. However, since the substance of the argument is the same in either case, we restrict our attention to decimal arithmetic for simplicity.) The difference between the exact and stored values is called the *rounding error* which, for this example, is

$$-0.000\,000\,000\,033\,33\ldots$$

Suppose that for a given real number α the digits after the decimal point are

$$d_1 d_2 \cdots d_n d_{n+1} \cdots$$

To round α to n decimal places (abbreviated to nD) we proceed as follows. If $d_{n+1} < 5$, then α is rounded down; all digits after the nth place are removed. If $d_{n+1} \geqslant 5$, then α is rounded up; d_n is increased by one and all digits after the nth place are removed. It should be clear that in either case the magnitude of the rounding error does not exceed 0.5×10^{-n}.

In most situations the introduction of rounding errors into the calculations does not significantly affect the final results. However, in certain cases it can lead to a serious loss of accuracy so that computed results are very different from those obtained using exact arithmetic. The term *instability* is used to describe this phenomenon. There are two fundamental types of instability in numerical analysis – inherent and induced. The first of these is a fault of the problem, the second of the method of solution.

■ Definition 1.1
A problem is said to be *inherently unstable* (or *ill-conditioned*) if small changes in the data of the problem cause large changes in its solution.   □

This concept is important for two reasons. Firstly, the data may be given as a set of readings from an analogue device such as a thermometer or voltmeter and as such cannot be measured exactly. If the problem is ill-conditioned then any numerical results, irrespective of the method used to obtain them, will be highly inaccurate and may be worthless. The second reason is that even if the data is exact it will not necessarily be stored exactly on a computer. Consequently, the problem which the computer is attempting to solve may differ slightly from the one originally posed. This does not usually matter, but if the problem is ill-conditioned then the computed results may differ wildly from those expected.

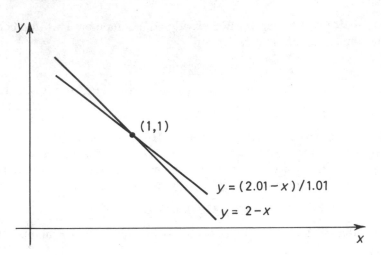

Figure 1.1

■ **Example 1.1**

Consider the simultaneous linear equations

$$x + y = 2$$
$$x + 1.01y = 2.01$$

which have solution $x = y = 1$. If the number 2.01 is changed to 2.02, the corresponding solution is $x = 0$, $y = 2$. We see that a 0.5% change in the data produces a 100% change in the solution! It is instructive to give a geometrical interpretation of this result. The solution of the system is the point of intersection of the two lines $y = 2 - x$ and $y = (2.01 - x)/1.01$. These lines are sketched in Fig. 1.1. It is clear that the point of intersection is sensitive to small movements in either of these lines since they are nearly parallel. In fact, if the coefficient of y in the second equation is 1.00, the two lines are exactly parallel and the system has no solution. This is fairly typical of ill-conditioned problems. They are often close to 'critical' problems which either possess infinitely many solutions or no solution whatsoever.   □

■ **Example 1.2**

Consider the initial value problem

$$y'' - 10y' - 11y = 0; \qquad y(0) = 1, \qquad y'(0) = -1$$

defined on $x \geqslant 0$. The corresponding auxiliary equation has roots -1 and 11, so the general solution of the differential equation is

$$y = Ae^{-x} + Be^{11x}$$

for arbitrary constants A and B. The particular solution which satisfies the given initial conditions is

$$y = e^{-x}$$

Now suppose that the initial conditions are replaced by

$$y(0) = 1 + \delta, \qquad y'(0) = -1 + \varepsilon$$

for some small numbers δ and ε. The particular solution satisfying these conditions is

$$y = \left(1 + \frac{11\delta}{12} - \frac{\varepsilon}{12}\right)e^{-x} + \left(\frac{\delta}{12} + \frac{\varepsilon}{12}\right)e^{11x}$$

and the change in the solution is therefore

$$\left(\frac{11\delta}{12} - \frac{\varepsilon}{12}\right)e^{-x} + \left(\frac{\delta}{12} + \frac{\varepsilon}{12}\right)e^{11x}$$

The term $(\delta + \varepsilon)e^{11x}/12$ is large compared with e^{-x} for $x > 0$, indicating that this problem is ill-conditioned.

To a certain extent inherent stability depends on the size of the solution to the original problem as well as on the size of any changes arising from perturbations in the data. For example, if the solution of the original problem had been $y = e^{11x}$, we might have been prepared to tolerate a change of $(\delta + \varepsilon)e^{11x}/12$, which, although large in absolute terms, is small compared with e^{11x}. Under these circumstances, one would say that the problem is absolutely ill-conditioned but relatively well-conditioned.   □

We now consider a different type of instability which is a consequence of the method of solution rather than the problem itself.

■ **Definition 1.2**
A method is said to suffer from *induced instability* if small errors present at one stage of the method adversely affect the calculations in subsequent stages to such an extent that the final results are totally inaccurate.   □

Nearly all numerical methods involve a repetitive sequence of calculations and so it is inevitable that small individual rounding errors accumulate as they proceed. However, the actual growth of these errors can occur in different ways. If, after n steps of the method, the total rounding error is approximately $Cn\varepsilon$, where C is a positive constant and ε is the size of a typical rounding error, then the growth in rounding errors is usually acceptable. For example, if $C = 1$ and $\varepsilon = 10^{-11}$, it takes about 50 000 steps before the sixth decimal place is affected. On the other hand, if the total rounding error is approximately $Ca^n\varepsilon$ or $Cn!\varepsilon$, for some number $a > 1$, then the growth in rounding errors is usually unacceptable. For example, in the first case, if $C = 1$,

$\varepsilon = 10^{-11}$ and $a = 10$, it only takes about five steps before the sixth decimal place is affected. The second case is illustrated by the following example.

■ **Example 1.3**
Consider the integral

$$\int_0^1 x^7 e^{x-1} \, dx$$

which has an exact value 0.112 38... (see Exercise 6.5, Question 5). One very simple method of evaluating this integral is to establish a reduction formula for

$$I_n = \int_0^1 x^n e^{x-1} \, dx \qquad n = 0, 1, 2, \dots \tag{1.1}$$

Integration by parts gives

$$I_n = [x^n e^{x-1}]_0^1 - n \int_0^1 x^{n-1} e^{x-1} \, dx$$

$$= 1 - n I_{n-1} \tag{1.2}$$

This formula can be used repeatedly for $n = 1, 2, \dots, 7$ in turn, with I_0 evaluated directly from

$$I_0 = \int_0^1 e^{x-1} \, dx = [e^{x-1}]_0^1 = 1 - e^{-1}$$

which, when rounded to 4D, is 0.6321.

Recurrence relation (1.2) produces the value $I_7 = 0.2160$, which is very inaccurate. Unfortunately, in rounding I_0 to 4D, an error of about 2×10^{-5} is incurred and this is multiplied by n $(n = 1, 2, \dots, 7)$ at each step. Therefore the error in the calculated value of I_7, caused by the inaccuracy in I_0, is $7! \times 2 \times 10^{-5} \approx 0.1$.

Clearly we need to devise methods which avoid repeated multiplication by numbers which are greater than one in modulus. In the present example this can be done by rearranging (1.2) as

$$I_{n-1} = \frac{1}{n}(1 - I_n) \tag{1.3}$$

Now, $x^n \to 0$ as $n \to \infty$ for values of x in the interval $(0, 1)$. It follows from (1.1) that $I_n \approx 0$ for reasonably large n. If I_{20} is taken as 0, then (1.3) can be applied repeatedly for $n = 20, 19, \dots, 8$ in turn, which gives $I_7 = 0.1124$ (4D). Even though the error in I_{20} is much larger than that of I_0 for the previous process (the exact value of I_{20} is in fact 0.045 54...), a greatly improved result has been obtained. This is because the error in I_{20} is divided by n at each step and is therefore damped down as the method proceeds. $\square$

A rather more subtle instance of induced instability is illustrated by the following example.

■ **Example 1.4**
Many successful algorithms are available for calculating individual real roots of polynomial equations of the form

$$p_n(x) = a_n x^n + a_{n-1} x^{n-1} + \cdots + a_0 = 0$$

Some of these are described in Chapter 3. An attractive idea would be to use these methods to estimate one of the real roots, α say, and then to divide $p_n(x)$ by $x - \alpha$ to produce a polynomial of degree $n - 1$ which contains the remaining roots. This process can then be repeated until all of the roots have been located. This is usually referred to as the method of *deflation*. If α were an exact root of $p_n(x) = 0$, then the remaining $n - 1$ roots would, of course, be the zeros of the deflated polynomial of degree $n - 1$. However, in practice α might only be an approximate root and in this case the zeros of the deflated polynomial can be very different from those of $p_n(x)$. For example, consider the cubic

$$p_3(x) = x^3 - 13x^2 + 32x - 20 = (x - 1)(x - 2)(x - 10)$$

and suppose that an estimate of its largest zero is taken as 10.1. If we divide $p_3(x)$ by $x - 10.1$, the quotient is $x^2 - 2.9x + 2.71$ which has zeros $1.45 \pm 0.78i$. Clearly an error of 0.1 in the largest zero of $p_3(x)$ has induced a large error into the calculation of the remaining zeros.

It is interesting to note that if we divide $p_3(x)$ by $x - 1.1$, the corresponding quadratic has zeros 1.9 and 10.0 which are perfectly acceptable. The deflation process can be applied successfully provided that certain precautions are taken. In particular, the roots should be eliminated in increasing order of magnitude. Further details of polynomial deflation can be found in Wilkinson (1963). $\square$

Of the two types of instability discussed, that of inherent instability is the most serious. Induced instability is a fault of the method and can be avoided either by modifying the existing method, as we did for the two examples given in this section, or by using a completely different solution procedure. Inherent instability, however, is a fault of the problem so there is relatively little that we can do about it. The extent to which this property is potentially disastrous depends not only on the degree of ill-conditioning involved but also on the context from which the problem is taken.

2
Linear
algebraic
equations

Many important problems in science and engineering require the solution of systems of simultaneous linear equations of the form

$$
\begin{aligned}
a_{11}x_1 + a_{12}x_2 + \cdots + a_{1n}x_n &= b_1 \\
a_{21}x_1 + a_{22}x_2 + \cdots + a_{2n}x_n &= b_2 \\
&\quad \vdots \\
a_{n1}x_1 + a_{n2}x_2 + \cdots + a_{nn}x_n &= b_n
\end{aligned}
\tag{2.1}
$$

where the coefficients a_{ij} and the right hand sides b_i are given numbers, and the quantities x_i are the unknowns which need to be determined. In matrix notation this system can be written as

$$
Ax = b \tag{2.2}
$$

where $A = (a_{ij})$, $b = (b_i)$ and $x = (x_i)$. We shall assume that the $n \times n$ matrix A is non-singular (i.e. that the determinant of A is non-zero) so that equation (2.2) possesses a unique solution.

There are two classes of method for solving systems of this type. *Direct methods* find the solution in a finite number of steps. These methods are guaranteed to succeed and are recommended for general purpose use. *Indirect* or *iterative methods* start with an arbitrary first approximation to x and then (hopefully) improve this estimate in an infinite but convergent sequence of steps. These methods are used for solving large systems of equations which have a high proportion of zero coefficients since, for these systems, direct methods require comparatively more computer storage and processor time. Systems of this type occur frequently in the numerical solution of differential equations.

Most of the techniques which are taught in introductory courses on mathematical methods are not to be recommended for practical computation.

One such method calculates $x = A^{-1}b$, where the inverse of A is found by evaluating n^2 cofactors which are determinants of order $(n-1) \times (n-1)$. Each of these determinants requires approximately $(n-1)!$ multiplications, which is prohibitively time consuming when n is large. This technique may also be subject to significant induced instability.

2.1 GAUSS ELIMINATION

Gauss elimination is used to solve a system of linear equations by transforming it to an upper triangular system (i.e. one in which all of the coefficients below the leading diagonal are zero) using elementary row operations. The solution of the upper triangular system is then found using back substitution. We shall describe the method in detail for the general 3×3 system

$$\begin{pmatrix} a_{11} & a_{12} & a_{13} \\ a_{21} & a_{22} & a_{23} \\ a_{31} & a_{32} & a_{33} \end{pmatrix} \begin{pmatrix} x_1 \\ x_2 \\ x_3 \end{pmatrix} = \begin{pmatrix} b_1 \\ b_2 \\ b_3 \end{pmatrix}$$

STEP 1   The first step eliminates the variable x_1 from the second and third equations. This can be done by subtracting multiples $m_{21} = a_{21}/a_{11}$ and $m_{31} = a_{31}/a_{11}$ of row 1 from rows 2 and 3, respectively, producing the equivalent system

$$\begin{pmatrix} a_{11} & a_{12} & a_{13} \\ 0 & a_{22}^{(2)} & a_{23}^{(2)} \\ 0 & a_{32}^{(2)} & a_{33}^{(2)} \end{pmatrix} \begin{pmatrix} x_1 \\ x_2 \\ x_3 \end{pmatrix} = \begin{pmatrix} b_1 \\ b_2^{(2)} \\ b_3^{(2)} \end{pmatrix}$$

where $a_{ij}^{(2)} = a_{ij} - m_{i1}a_{1j}$ and $b_i^{(2)} = b_i - m_{i1}b_1$ $(i, j = 2, 3)$.

STEP 2   The second step eliminates the variable x_2 from the third equation. This can be done by subtracting a multiple $m_{32} = a_{32}^{(2)}/a_{22}^{(2)}$ of row 2 from row 3, producing the equivalent upper triangular system

$$\begin{pmatrix} a_{11} & a_{12} & a_{13} \\ 0 & a_{22}^{(2)} & a_{23}^{(2)} \\ 0 & 0 & a_{33}^{(3)} \end{pmatrix} \begin{pmatrix} x_1 \\ x_2 \\ x_3 \end{pmatrix} = \begin{pmatrix} b_1 \\ b_2^{(2)} \\ b_3^{(3)} \end{pmatrix}$$

where $a_{33}^{(3)} = a_{33}^{(2)} - m_{32}a_{23}^{(2)}$ and $b_3^{(3)} = b_3^{(2)} - m_{32}b_2^{(2)}$.

Since these row operations are reversible, the original system and the upper triangular system have the same solution. The upper triangular system is solved using back substitution. The last equation implies that

$$x_3 = b_3^{(3)}/a_{33}^{(3)}$$

This number can then be substituted into the second equation and the value of x_2 obtained from

$$x_2 = (b_2^{(2)} - a_{23}^{(2)} x_3)/a_{22}^{(2)}$$

Finally, the known values of x_2 and x_3 can be substituted into the first equation and the value of x_1 obtained from

$$x_1 = (b_1 - a_{12} x_2 - a_{13} x_3)/a_{11}$$

A general system of n equations can be reduced to an upper triangular system in exactly $n - 1$ steps. Step j eliminates the variable x_j from the $(j + 1)$th, ..., nth equations. This is achieved by subtracting multiples $m_{ij} = a_{ij}^{(j)}/a_{jj}^{(j)}$ of row j from row i for $i = j + 1, ..., n$. The solution of the resulting upper triangular system is found by solving the last equation for x_n, then the $(n - 1)$th equation for x_{n-1}, and continuing in this way until x_1 is determined. These values are calculated from

$$x_n = b_n^{(n)}/a_{nn}^{(n)} \tag{2.3}$$

and

$$x_j = \left(b_j^{(j)} - \sum_{k=j+1}^{n} a_{jk}^{(j)} x_k \right) \Bigg/ a_{jj}^{(j)} \qquad j < n \tag{2.4}$$

Inspection of the previous equations indicates that the algorithm fails if any of the quantities $a_{jj}^{(j)}$ are zero, since these numbers are used as the denominators both in the multipliers m_{ij} and in the back substitution equations. These numbers are usually referred to as *pivots*. Elimination also produces poor results if any of the multipliers are greater than one in modulus. To see this, consider what happens in step 1 when, for example, $m_{21} = 20$. Any rounding errors in the coefficients $a_{12}, a_{13}, ..., a_{1n}$ and b_1 are magnified by a factor of 20 when the variable x_1 is eliminated from the second equation. If the multipliers in subsequent steps are of a similar magnitude then these rounding errors accumulate rapidly, producing an unstable method. Significant figures may also be lost in the back substitution process if any of the pivots are very small.

It is possible to prevent these difficulties by using row interchanges. At step j, the elements in column j which are on or below the diagonal are scanned. The row containing the element of largest modulus is called the pivotal row. Row j is then interchanged (if necessary) with the pivotal row. The new element on the diagonal has a modulus greater than or equal to the moduli of the elements below the diagonal in column j and so $|m_{ij}| \leqslant 1$. It can, of course, happen that all of the numbers $a_{jj}^{(j)}, a_{j+1,j}^{(j)}, ..., a_{nj}^{(j)}$ are exactly zero, in which case the coefficient matrix does not have full rank and the system fails to possess a unique solution. Furthermore, if all of these numbers are very small, then the selected pivot will also be small. It follows from equation (2.4) that the value of x_j is likely to be sensitive to small changes in the coefficients, indicating that the original system is ill-conditioned.

The process of using row interchanges in this way is called *partial pivoting*. The word partial is used to distinguish the above procedure from total pivoting which employs both row and column interchanges. Total pivoting produces an additional reduction in the effect of rounding errors and is essential for the accurate solution of certain systems. However, since for most systems this usually requires a disproportionate increase in computational effort, it will not be considered here.

■ **Example 2.1**

To illustrate the effect of partial pivoting, consider the solution of

$$\begin{pmatrix} 0.610 & 1.23 & 1.72 \\ 1.02 & 2.15 & -5.51 \\ -4.34 & 11.2 & -4.25 \end{pmatrix} \begin{pmatrix} x_1 \\ x_2 \\ x_3 \end{pmatrix} = \begin{pmatrix} 0.792 \\ 12.0 \\ 16.3 \end{pmatrix}$$

using three significant figure arithmetic with rounding. This models the more realistic case of solving a large system of equations on a computer capable of working to, say, ten significant figure accuracy. Without partial pivoting we proceed as follows:

STEP 1 The multipliers are $m_{21} = 1.02/0.610 = 1.67$ and $m_{31} = -4.34/0.610 = -7.11$, which give

$$\begin{pmatrix} 0.610 & 1.23 & 1.72 \\ 0 & 0.100 & -8.38 \\ 0 & 20.0 & 7.95 \end{pmatrix} \begin{pmatrix} x_1 \\ x_2 \\ x_3 \end{pmatrix} = \begin{pmatrix} 0.792 \\ 10.7 \\ 21.9 \end{pmatrix}$$

STEP 2 The multiplier is $m_{32} = 20.0/0.100 = 200$, which gives

$$\begin{pmatrix} 0.610 & 1.23 & 1.72 \\ 0 & 0.100 & -8.38 \\ 0 & 0 & 1690 \end{pmatrix} \begin{pmatrix} x_1 \\ x_2 \\ x_3 \end{pmatrix} = \begin{pmatrix} 0.792 \\ 10.7 \\ -2120 \end{pmatrix}$$

Solving by back substitution, we obtain

$$x_3 = -1.25, \qquad x_2 = 2.00, \qquad x_1 = 0.790$$

With partial pivoting we proceed as follows:

STEP 1 Since $|-4.34| > |0.610|$ and $|1.02|$, rows 1 and 3 are interchanged to get

$$\begin{pmatrix} -4.34 & 11.2 & -4.25 \\ 1.02 & 2.15 & -5.51 \\ 0.610 & 1.23 & 1.72 \end{pmatrix} \begin{pmatrix} x_1 \\ x_2 \\ x_3 \end{pmatrix} = \begin{pmatrix} 16.3 \\ 12.0 \\ 0.792 \end{pmatrix}$$

The multipliers are $m_{21} = 1.02/-4.34 = -0.235$ and $m_{31} = 0.610/-4.34 = -0.141$, which give

$$\begin{pmatrix} -4.34 & 11.2 & -4.25 \\ 0 & 4.78 & -6.51 \\ 0 & 2.81 & 1.12 \end{pmatrix} \begin{pmatrix} x_1 \\ x_2 \\ x_3 \end{pmatrix} = \begin{pmatrix} 16.3 \\ 15.8 \\ 3.09 \end{pmatrix}$$

STEP 2   Since $|4.78| > |2.81|$, no further interchanges are needed. The multiplier is $m_{32} = 2.81/4.78 = 0.588$, which gives

$$\begin{pmatrix} -4.34 & 11.2 & -4.25 \\ 0 & 4.78 & -6.51 \\ 0 & 0 & 4.95 \end{pmatrix} \begin{pmatrix} x_1 \\ x_2 \\ x_3 \end{pmatrix} = \begin{pmatrix} 16.3 \\ 15.8 \\ -6.20 \end{pmatrix}$$

Solving by back substitution we obtain

$$x_3 = -1.25, \qquad x_2 = 1.60, \qquad x_1 = 1.59$$

By substituting these values into the original system of equations it is easy to verify that the result obtained with partial pivoting is a reasonably accurate solution. (In fact, the exact solution, rounded to three significant figures, is given by $x_3 = -1.26$, $x_2 = 1.60$ and $x_1 = 1.61$.) However, the values obtained without partial pivoting are totally unacceptable; the value of x_1 is not even correct to one significant figure. $\square$

Gauss elimination is particularly simple to use when all of the non-zero coefficients lie along a limited number of diagonals on and on either side of the leading diagonal. Such a system is said to be *banded* and the total number of diagonals containing non-zero coefficients is called the *bandwidth*. A special, but important, case is the tridiagonal system

$$\begin{pmatrix} a_{11} & a_{12} & & & & & \mathbf{0} \\ a_{21} & a_{22} & a_{23} & & & & \\ & a_{32} & a_{33} & a_{34} & & & \\ & & a_{43} & a_{44} & a_{45} & & \\ & & & & \ddots & \ddots & a_{n-1,n} \\ \mathbf{0} & & & & & a_{n,n-1} & a_{nn} \end{pmatrix} \begin{pmatrix} x_1 \\ x_2 \\ x_3 \\ x_4 \\ \vdots \\ x_n \end{pmatrix} = \begin{pmatrix} b_1 \\ b_2 \\ b_3 \\ b_4 \\ \vdots \\ b_n \end{pmatrix}$$

which has a bandwidth of three. Such systems arise, for example, when numerical techniques are used for the construction of cubic splines and in the solution of boundary value problems. The elimination process is trivial since only one extra subdiagonal of zeros needs to be created, and back substitution is equally straightforward since the final upper triangular system takes the form

$$
\begin{pmatrix}
a_{11} & a_{12} & & & & & & \text{\Large 0} \\
 & a_{22}^{(2)} & a_{23}^{(2)} \\
 & & a_{33}^{(3)} & a_{34}^{(3)} \\
 & & & a_{44}^{(4)} & a_{45}^{(4)} \\
 & & & & \ddots & \ddots \\
 & & & & & & a_{n-1,n}^{(n-1)} \\
\text{\Large 0} & & & & & & & a_{nn}^{(n)}
\end{pmatrix}
\begin{pmatrix}
x_1 \\ x_2 \\ x_3 \\ x_4 \\ \vdots \\ \\ x_n
\end{pmatrix}
=
\begin{pmatrix}
b_1 \\ b_2^{(2)} \\ b_3^{(3)} \\ b_4^{(4)} \\ \vdots \\ \\ b_n^{(n)}
\end{pmatrix}
$$

The determinant of a general $n \times n$ matrix A can easily be found as a by-product of Gauss elimination. The operation of subtracting a multiple of one row from another row leaves the determinant unaltered, whereas the operation of interchanging two rows changes its sign. Consequently, if m interchanges are used in the elimination process, the determinant $\det A$ is $(-1)^m$ times that of the final upper triangular matrix. The determinant of a triangular matrix is simply the product of the elements on the leading diagonal, and so $\det A$ can be calculated from the formula

$$
\det A = (-1)^m \prod_{i=1}^{n} a_{ii}^{(i)}
$$

The determinant of the coefficient matrix used in Example 2.1 can, therefore, be computed as

$$
(-1)^1 \times (-4.34) \times 4.78 \times 4.95 = 103
$$

Gauss elimination also facilitates the computation of the inverse of A. The inverse $X = (x_{ij})$ is an $n \times n$ matrix satisfying the equation

$$
AX = I
$$

where I is the $n \times n$ identity matrix. Equating the jth column of both sides of this equation gives

$$
\begin{pmatrix}
a_{11} & a_{12} & a_{13} & \cdots & a_{1n} \\
a_{21} & a_{22} & a_{23} & \cdots & a_{2n} \\
a_{31} & a_{32} & a_{33} & \cdots & a_{3n} \\
\vdots & \vdots & \vdots & & \vdots \\
a_{j1} & a_{j2} & a_{j3} & \cdots & a_{jn} \\
\vdots & \vdots & \vdots & & \vdots \\
a_{n1} & a_{n2} & a_{n3} & \cdots & a_{nn}
\end{pmatrix}
\begin{pmatrix}
x_{1j} \\ x_{2j} \\ x_{3j} \\ \vdots \\ x_{jj} \\ \vdots \\ x_{nj}
\end{pmatrix}
=
\begin{pmatrix}
0 \\ 0 \\ 0 \\ \vdots \\ 1 \\ \vdots \\ 0
\end{pmatrix}
$$

The columns of X can, therefore, be found by solving n systems of linear equations with right hand sides corresponding to successive columns of the identity matrix. The total volume of work is not too large since the coefficient matrix is the same for each system, enabling the n solutions to be found simultaneously.

It is interesting to analyse the method of Gauss elimination using matrix algebra. For simplicity, we shall consider only the case $n = 3$ and assume that there are no interchanges. The system obtained in step 1 is $A_1 x = b_1$, where

$$A_1 = \begin{pmatrix} a_{11} & a_{12} & a_{13} \\ 0 & a_{22}^{(2)} & a_{23}^{(2)} \\ 0 & a_{32}^{(2)} & a_{33}^{(2)} \end{pmatrix} \qquad b_1 = \begin{pmatrix} b_1 \\ b_2^{(2)} \\ b_3^{(2)} \end{pmatrix}$$

The original matrix A and vector b can be recovered from A_1 and b_1 by adding multiples m_{21} and m_{31} of rows 2 and 3, respectively, to row 1. This is equivalent to pre-multiplication by

$$L_1 = \begin{pmatrix} 1 & 0 & 0 \\ m_{21} & 1 & 0 \\ m_{31} & 0 & 1 \end{pmatrix}$$

i.e. $A = L_1 A_1$ and $b = L_1 b_1$.

The system obtained in step 2 is $Ux = c$, where

$$U = \begin{pmatrix} a_{11} & a_{12} & a_{13} \\ 0 & a_{22}^{(2)} & a_{23}^{(2)} \\ 0 & 0 & a_{33}^{(3)} \end{pmatrix} \qquad c = \begin{pmatrix} b_1 \\ b_2^{(2)} \\ b_3^{(3)} \end{pmatrix}$$

The matrix A_1 and vector b_1 can be recovered from U and c by adding the multiple m_{32} of row 3 to row 2. This is equivalent to pre-multiplication by

$$L_2 = \begin{pmatrix} 1 & 0 & 0 \\ 0 & 1 & 0 \\ 0 & m_{32} & 1 \end{pmatrix}$$

i.e. $A_1 = L_2 U$ and $b_1 = L_2 c$.

Hence

$$A = L_1 A_1 = L_1 L_2 U$$

and

$$b = L_1 b_1 = L_1 L_2 c$$

Now

$$L_1 L_2 = \begin{pmatrix} 1 & 0 & 0 \\ m_{21} & 1 & 0 \\ m_{31} & 0 & 1 \end{pmatrix} \begin{pmatrix} 1 & 0 & 0 \\ 0 & 1 & 0 \\ 0 & m_{32} & 1 \end{pmatrix} = \begin{pmatrix} 1 & 0 & 0 \\ m_{21} & 1 & 0 \\ m_{31} & m_{32} & 1 \end{pmatrix} = L, \text{ say}$$

The matrix L is lower triangular with ones on the leading diagonal. Such a matrix is said to be unit lower triangular. We have, therefore, shown that Gauss elimination can be represented by the matrix relations

$$A = LU$$
$$Lc = b \tag{2.5}$$
$$Ux = c$$

where U is the coefficient matrix of the upper triangular system, c is the vector associated with the right hand side of the upper triangular system and L is a unit lower triangular matrix whose elements are the corresponding multipliers used in the elimination process.

EXERCISE 2.1

1. Solve the following systems of linear equations using Gauss elimination (i) without pivoting (ii) with partial pivoting.

   (a)
   $$0.005x_1 + x_2 + x_3 = 2$$
   $$x_1 + 2x_2 + x_3 = 4$$
   $$-3x_1 - x_2 + 6x_3 = 2$$

   (b)
   $$x_1 - x_2 + 2x_3 = 5$$
   $$2x_1 - 2x_2 + x_3 = 1$$
   $$3x_1 - 2x_2 + 7x_3 = 20$$

   (c)
   $$1.19x_1 + 2.37x_2 - 7.31x_3 + 1.75x_4 = 2.78$$
   $$2.15x_1 - 9.76x_2 + 1.54x_3 - 2.08x_4 = 6.27$$
   $$10.7x_1 - 1.11x_2 + 3.78x_3 + 4.49x_4 = 9.03$$
   $$2.17x_1 + 3.58x_2 + 1.70x_3 + 9.33x_4 = 5.00$$

   Work to three significant figures and comment on the results.

2. Solve the equations

   $$1.34x_1 + 7.21x_2 + 1.04x_3 = 9.60$$
   $$3.18x_1 + 4.01x_2 + 0.980x_3 = 8.17$$
   $$2.84x_1 - 24.0x_2 - 2.24x_3 = -23.4$$

   using Gauss elimination with partial pivoting. Work to three significant figures. Change the right hand side of the first equation to 9.59 and repeat the calculations. What can you deduce about this system of equations? Why could this have been deduced from the size of the pivots used in the elimination?

3. Use Gauss elimination with partial pivoting to find the determinant and inverse of the matrix

$$\begin{pmatrix} 1.723 & -1.421 & 3.784 \\ 0.113 & 4.071 & 1.213 \\ 5.131 & -4.010 & -2.176 \end{pmatrix}$$

Perform all calculations to four significant figures.

4. Use Gauss elimination without pivoting to construct a unit lower triangular matrix L and an upper triangular matrix U satisfying $A = LU$ from

$$A = \begin{pmatrix} 4 & -2 & 2 \\ 4 & -3 & -2 \\ 2 & 3 & -1 \end{pmatrix}$$

5. Solve the equations

$$x_1 + \tfrac{1}{2}x_2 + \tfrac{1}{3}x_3 + \tfrac{1}{4}x_4 = \tfrac{25}{12}$$
$$\tfrac{1}{2}x_1 + \tfrac{1}{3}x_2 + \tfrac{1}{4}x_3 + \tfrac{1}{5}x_4 = \tfrac{77}{60}$$
$$\tfrac{1}{3}x_1 + \tfrac{1}{4}x_2 + \tfrac{1}{5}x_3 + \tfrac{1}{6}x_4 = \tfrac{19}{20}$$
$$\tfrac{1}{4}x_1 + \tfrac{1}{5}x_2 + \tfrac{1}{6}x_3 + \tfrac{1}{7}x_4 = \tfrac{319}{420}$$

using Gauss elimination with partial pivoting. The fractions should be rounded to three significant figures and all subsequent calculations performed to that accuracy. Repeat using four and five significant figures throughout.

   Verify that the exact solution is $(1, 1, 1, 1)^{\mathrm{T}}$ and compare this with the computed results. What can you deduce about this system of equations?

6. Show that the total number of multiplications and divisions needed to solve an $n \times n$ tridiagonal system of linear equations is $5n - 4$.

2.2 MATRIX DECOMPOSITION METHODS

A system of linear equations can be solved by writing the coefficient matrix A as a product of two triangular matrices. The solution is then obtained using forward and back substitution. We have already seen how to decompose A into LU, where L is unit lower triangular and U is upper triangular, using Gauss elimination. In practice it is more convenient to determine L and U directly from the equation $A = LU$. The general forms of L and U are written as

$$L = \begin{pmatrix} 1 & & & & & \\ l_{21} & 1 & & & & \\ l_{31} & l_{32} & 1 & & & \\ \vdots & & & \ddots & & \\ l_{n1} & l_{n2} & l_{n3} & \cdots & l_{n,n-1} & 1 \end{pmatrix}$$

and

$$U = \begin{pmatrix} u_{11} & u_{12} & u_{13} & \cdots & u_{1n} \\ & u_{22} & u_{23} & & u_{2n} \\ & & u_{33} & & u_{3n} \\ & & & \ddots & \vdots \\ & & & & u_{nn} \end{pmatrix}$$

and the unknown elements of L and U are found by equating corresponding entries in A and LU in a systematic way.

■ **Example 2.2**

To illustrate the construction of the LU decomposition of a matrix, consider

$$\begin{pmatrix} 4 & -2 & 2 \\ 4 & -3 & -2 \\ 2 & 3 & -1 \end{pmatrix} = \begin{pmatrix} 1 & 0 & 0 \\ l_{21} & 1 & 0 \\ l_{31} & l_{32} & 1 \end{pmatrix} \begin{pmatrix} u_{11} & u_{12} & u_{13} \\ 0 & u_{22} & u_{23} \\ 0 & 0 & u_{33} \end{pmatrix}$$

STEP 1   Equate entries along the first row to get

$$4 = 1u_{11} \Rightarrow u_{11} = 4$$
$$-2 = 1u_{12} \Rightarrow u_{12} = -2$$
$$2 = 1u_{13} \Rightarrow u_{13} = 2$$

and then down the first column to get

$$4 = l_{21}u_{11} \Rightarrow l_{21} = 1$$
$$2 = l_{31}u_{11} \Rightarrow l_{31} = 1/2$$

and so

$$\begin{pmatrix} 4 & -2 & 2 \\ 4 & -3 & -2 \\ 2 & 3 & -1 \end{pmatrix} = \begin{pmatrix} 1 & 0 & 0 \\ 1 & 1 & 0 \\ 1/2 & l_{32} & 1 \end{pmatrix} \begin{pmatrix} 4 & -2 & 2 \\ 0 & u_{22} & u_{23} \\ 0 & 0 & u_{33} \end{pmatrix}$$

STEP 2   Equate entries along the second row to get

$$-3 = 1(-2) + 1u_{22} \Rightarrow u_{22} = -1$$
$$-2 = 1(2) + 1u_{23} \Rightarrow u_{23} = -4$$

and then down the second column to get

$$3 = \tfrac{1}{2}(-2) + l_{32}u_{22} \Rightarrow l_{32} = -4$$

and so

$$\begin{pmatrix} 4 & -2 & 2 \\ 4 & -3 & -2 \\ 2 & 3 & -1 \end{pmatrix} = \begin{pmatrix} 1 & 0 & 0 \\ 1 & 1 & 0 \\ 1/2 & -4 & 1 \end{pmatrix} \begin{pmatrix} 4 & -2 & 2 \\ 0 & -1 & -4 \\ 0 & 0 & u_{33} \end{pmatrix}$$

STEP 3   Equate entries along the third row to get

$$-1 = \tfrac{1}{2}(2) + (-4)(-4) + 1u_{33} \Rightarrow u_{33} = -18$$

and so

$$\begin{pmatrix} 4 & -2 & 2 \\ 4 & -3 & -2 \\ 2 & 3 & -1 \end{pmatrix} = \begin{pmatrix} 1 & 0 & 0 \\ 1 & 1 & 0 \\ 1/2 & -4 & 1 \end{pmatrix} \begin{pmatrix} 4 & -2 & 2 \\ 0 & -1 & -4 \\ 0 & 0 & -18 \end{pmatrix} \qquad \square$$

For a general $n \times n$ matrix, step j ($1 \leqslant j \leqslant n$) equates entries in row j followed by column j to obtain the jth row and column of U and L, respectively. In step 1 the values of u_{1j} and l_{i1} are calculated from

$$u_{1j} = a_{1j} \tag{2.6}$$

$$l_{i1} = a_{i1}/u_{11} \tag{2.7}$$

In subsequent steps note that the values of u_{ij} and l_{ij} are obtained by equating the (i, j) entries for $i \leqslant j$ and $i > j$, respectively. If $i \leqslant j$ this gives

$$a_{ij} = l_{i1}u_{1j} + l_{i2}u_{2j} + \cdots + l_{i,i-1}u_{i-1,j} + u_{ij}$$

and so u_{ij} is calculated from

$$u_{ij} = a_{ij} - \sum_{k=1}^{i-1} l_{ik}u_{kj} \qquad 2 \leqslant i \leqslant j \tag{2.8}$$

If $i > j$ this gives

$$a_{ij} = l_{i1}u_{1j} + l_{i2}u_{2j} + \cdots + l_{i,j-1}u_{j-1,j} + l_{ij}u_{jj}$$

and so l_{ij} is calculated from

$$l_{ij} = \left(a_{ij} - \sum_{k=1}^{j-1} l_{ik}u_{kj} \right) \Big/ u_{jj} \qquad i > j \geqslant 2 \tag{2.9}$$

Once the matrices L and U have been constructed it is a simple matter to solve $Ax = b$. Substituting LU for A in this equation gives $LUx = b$. If we let $Ux = c$, say, it follows that $Lc = b$. The components of c are obtained from $Lc = b$ using forward substitution, i.e. the first equation gives the value of the first component c_1, the second equation gives the value of the second component c_2, and so on. Back substitution is then used to calculate the components of x from $Ux = c$ as in Gauss elimination.

A comparison of these relations with (2.5) shows that the LU decomposition method and Gauss elimination are mathematically equivalent; the matrices L, U and c determined from these two methods are identical if exact arithmetic is used. However, the actual calculations used to obtain them are completely different. Equations (2.8) and (2.9) are particularly suited for practical computation on a machine capable of double precision arithmetic. The individual sums and products in (2.8) and (2.9) are formed in double length registers and they are rounded to single length precision after u_{ij} and l_{ij} have been found. The effect is that u_{ij} and l_{ij} have rounding errors normally associated with a single arithmetic operation, which ultimately results in a more accurate calculation of the solution vector x. The calculations used in Gauss elimination, which are typified by

$$a_{ij}^{(k+1)} = a_{ij}^{(k)} - m_{ik}a_{kj}^{(k)}$$

cannot be organized in this way and so the final results are less accurate.

■ Example 2.3

To illustrate the LU decomposition method, consider the system

$$4x_1 - 2x_2 + 2x_3 = 6$$
$$4x_1 - 3x_2 - 2x_3 = -8$$
$$2x_1 + 3x_2 - x_3 = 5$$

The triangular decomposition of the coefficient matrix has already been constructed in Example 2.2 as

$$\begin{pmatrix} 1 & 0 & 0 \\ 1 & 1 & 0 \\ 1/2 & -4 & 1 \end{pmatrix} \begin{pmatrix} 4 & -2 & 2 \\ 0 & -1 & -4 \\ 0 & 0 & -18 \end{pmatrix}$$

Solution of the system $Lc = b$, i.e.

$$\begin{pmatrix} 1 & 0 & 0 \\ 1 & 1 & 0 \\ 1/2 & -4 & 1 \end{pmatrix} \begin{pmatrix} c_1 \\ c_2 \\ c_3 \end{pmatrix} = \begin{pmatrix} 6 \\ -8 \\ 5 \end{pmatrix}$$

using forward substitution gives

$$c_1 = 6, \qquad c_2 = -14, \qquad c_3 = -54$$

Back substitution is then applied to $Ux = c$, i.e.

$$\begin{pmatrix} 4 & -2 & 2 \\ 0 & -1 & -4 \\ 0 & 0 & -18 \end{pmatrix} \begin{pmatrix} x_1 \\ x_2 \\ x_3 \end{pmatrix} = \begin{pmatrix} 6 \\ -14 \\ -54 \end{pmatrix}$$

to get

$$x_3 = 3, \qquad x_2 = 2, \qquad x_1 = 1 \qquad\qquad \square$$

Inspection of equations (2.7) and (2.9) indicates that the *LU* decomposition method fails if any of the quantities u_{jj} are zero. Moreover, the results may be highly inaccurate if u_{jj} is small, since rounding errors are magnified by $1/u_{jj}$. In order to prevent this, a technique analogous to partial pivoting in Gauss elimination can be introduced into the *LU* decomposition algorithm to maximize the magnitude of the diagonal elements of *U*. Row interchanges are incorporated into each step of the method as follows:

STEP 1   Row 1 is interchanged, if necessary, with the row containing the element of largest modulus in column 1 of *A*. Since $u_{11} = a_{11}$, this maximizes $|u_{11}|$.

STEP j $(1 < j < n)$.   Various options are considered. Option i ($i = 1, 2, \ldots,$ $n - j + 1$) interchanges row j and row $(j + i - 1)$ and the value of u_{jj} is calculated. The option chosen is that which produces the largest value of $|u_{jj}|$.

STEP n No interchanges are used.

If interchanges are needed, then the matrix product *LU* is equal to a row permutation A^* of *A*, i.e.

$$LU = A^* = PA \qquad\qquad (2.10)$$

where *P* is obtained by permuting rows of the $n \times n$ identity matrix corresponding to the interchanges used in the decomposition process.

■ **Example 2.4**
To illustrate the use of row interchanges, consider the solution of

$$\begin{pmatrix} 0 & 1 & 5 & 3 \\ 2 & 2 & -2 & 4 \\ 1 & 5 & 7 & -10 \\ -1 & 1 & 6 & -5 \end{pmatrix} \begin{pmatrix} x_1 \\ x_2 \\ x_3 \\ x_4 \end{pmatrix} = \begin{pmatrix} -6 \\ 4 \\ 14 \\ 0 \end{pmatrix}$$

STEP 1
The element of largest modulus in the first column occurs in the second row, so rows 1 and 2 are interchanged to give

$$\begin{pmatrix} 2 & 2 & -2 & 4 \\ 0 & 1 & 5 & 3 \\ 1 & 5 & 7 & -10 \\ -1 & 1 & 6 & -5 \end{pmatrix} = \begin{pmatrix} 1 & 0 & 0 & 0 \\ l_{21} & 1 & 0 & 0 \\ l_{31} & l_{32} & 1 & 0 \\ l_{41} & l_{42} & l_{43} & 1 \end{pmatrix} \begin{pmatrix} u_{11} & u_{12} & u_{13} & u_{14} \\ 0 & u_{22} & u_{23} & u_{24} \\ 0 & 0 & u_{33} & u_{34} \\ 0 & 0 & 0 & u_{44} \end{pmatrix}$$

Equating entries in the first row and then first column gives

$$u_{11} = 2, \qquad u_{12} = 2, \qquad u_{13} = -2, \qquad u_{14} = 4$$
$$l_{21} = 0, \qquad l_{31} = 1/2, \qquad l_{41} = -1/2$$

and so

$$
\begin{pmatrix}
2 & 2 & -2 & 4 \\
0 & 1 & 5 & 3 \\
1 & 5 & 7 & -10 \\
-1 & 1 & 6 & -5
\end{pmatrix}
=
\begin{pmatrix}
1 & 0 & 0 & 0 \\
0 & 1 & 0 & 0 \\
1/2 & l_{32} & 1 & 0 \\
-1/2 & l_{42} & l_{43} & 1
\end{pmatrix}
\begin{pmatrix}
2 & 2 & -2 & 4 \\
0 & u_{22} & u_{23} & u_{24} \\
0 & 0 & u_{33} & u_{34} \\
0 & 0 & 0 & u_{44}
\end{pmatrix}
$$

STEP 2
Option 1   Leave row 2 unchanged. Equating the $(2, 2)$ entry gives

$$u_{22} = 1$$

Option 2   Interchange rows 2 and 3 to get

$$
\begin{pmatrix}
2 & 2 & -2 & 4 \\
1 & 5 & 7 & -10 \\
0 & 1 & 5 & 3 \\
-1 & 1 & 6 & -5
\end{pmatrix}
=
\begin{pmatrix}
1 & 0 & 0 & 0 \\
1/2 & 1 & 0 & 0 \\
0 & l_{32} & 1 & 0 \\
-1/2 & l_{42} & l_{43} & 1
\end{pmatrix}
\begin{pmatrix}
2 & 2 & -2 & 4 \\
0 & u_{22} & u_{23} & u_{24} \\
0 & 0 & u_{33} & u_{34} \\
0 & 0 & 0 & u_{44}
\end{pmatrix}
$$

Equating the $(2, 2)$ entry gives

$$u_{22} = 4$$

Notice that in order to maintain the validity of the previous calculations it is necessary to interchange those elements of L which have been found in step 1.

Option 3   Interchange rows 2 and 4 to get

$$
\begin{pmatrix}
2 & 2 & -2 & 4 \\
-1 & 1 & 6 & -5 \\
1 & 5 & 7 & -10 \\
0 & 1 & 5 & 3
\end{pmatrix}
=
\begin{pmatrix}
1 & 0 & 0 & 0 \\
-1/2 & 1 & 0 & 0 \\
1/2 & l_{32} & 1 & 0 \\
0 & l_{42} & l_{43} & 1
\end{pmatrix}
\begin{pmatrix}
2 & 2 & -2 & 4 \\
0 & u_{22} & u_{23} & u_{24} \\
0 & 0 & u_{33} & u_{34} \\
0 & 0 & 0 & u_{44}
\end{pmatrix}
$$

Equating the $(2, 2)$ entry gives

$$u_{22} = 2$$

The largest value of $|u_{22}|$ is obtained with option 2. Finishing off option 2 gives

$$u_{23} = 8, \qquad u_{24} = -12$$
$$l_{32} = 1/4, \qquad l_{42} = 1/2$$

and so

$$\begin{pmatrix} 2 & 2 & -2 & 4 \\ 1 & 5 & 7 & -10 \\ 0 & 1 & 5 & 3 \\ -1 & 1 & 6 & -5 \end{pmatrix} = \begin{pmatrix} 1 & 0 & 0 & 0 \\ 1/2 & 1 & 0 & 0 \\ 0 & 1/4 & 1 & 0 \\ -1/2 & 1/2 & l_{43} & 1 \end{pmatrix} \begin{pmatrix} 2 & 2 & -2 & 4 \\ 0 & 4 & 8 & -12 \\ 0 & 0 & u_{33} & u_{34} \\ 0 & 0 & 0 & u_{44} \end{pmatrix}$$

STEP 3

Option 1 Leave row 3 unchanged. Equating the $(3,3)$ entry gives

$$u_{33} = 3$$

Option 2 Interchange rows 3 and 4 to get

$$\begin{pmatrix} 2 & 2 & -2 & 4 \\ 1 & 5 & 7 & -10 \\ -1 & 1 & 6 & -5 \\ 0 & 1 & 5 & 3 \end{pmatrix} = \begin{pmatrix} 1 & 0 & 0 & 0 \\ 1/2 & 1 & 0 & 0 \\ -1/2 & 1/2 & 1 & 0 \\ 0 & 1/4 & l_{43} & 1 \end{pmatrix} \begin{pmatrix} 2 & 2 & -2 & 4 \\ 0 & 4 & 8 & -12 \\ 0 & 0 & u_{33} & u_{34} \\ 0 & 0 & 0 & u_{44} \end{pmatrix}$$

Equating the $(3,3)$ entry gives

$$u_{33} = 1$$

The largest value of $|u_{33}|$ is obtained with option 1. Finishing off option 1 gives

$$u_{34} = 6$$
$$l_{43} = 1/3$$

and so

$$\begin{pmatrix} 2 & 2 & -2 & 4 \\ 1 & 5 & 7 & -10 \\ 0 & 1 & 5 & 3 \\ -1 & 1 & 6 & -5 \end{pmatrix} = \begin{pmatrix} 1 & 0 & 0 & 0 \\ 1/2 & 1 & 0 & 0 \\ 0 & 1/4 & 1 & 0 \\ -1/2 & 1/2 & 1/3 & 1 \end{pmatrix} \begin{pmatrix} 2 & 2 & -2 & 4 \\ 0 & 4 & 8 & -12 \\ 0 & 0 & 3 & 6 \\ 0 & 0 & 0 & u_{44} \end{pmatrix}$$

STEP 4

Equating the $(4,4)$ entry gives

$$u_{44} = 1$$

and so

$$\begin{pmatrix} 2 & 2 & -2 & 4 \\ 1 & 5 & 7 & -10 \\ 0 & 1 & 5 & 3 \\ -1 & 1 & 6 & -5 \end{pmatrix} = \begin{pmatrix} 1 & 0 & 0 & 0 \\ 1/2 & 1 & 0 & 0 \\ 0 & 1/4 & 1 & 0 \\ -1/2 & 1/2 & 1/3 & 1 \end{pmatrix} \begin{pmatrix} 2 & 2 & -2 & 4 \\ 0 & 4 & 8 & -12 \\ 0 & 0 & 3 & 6 \\ 0 & 0 & 0 & 1 \end{pmatrix}$$

which in the previous notation is $PA = LU$ with

$$P = \begin{pmatrix} 0 & 1 & 0 & 0 \\ 0 & 0 & 1 & 0 \\ 1 & 0 & 0 & 0 \\ 0 & 0 & 0 & 1 \end{pmatrix}$$

The next stage is to solve $Lc = b^*$ where $b^* = Pb$. The matrix P need not be computed explicitly; b^* is derived from b by applying the interchanges that are used in the triangular decomposition of A. In this example we interchange -6 and 4 to get $(4, -6, 14, 0)^{\mathrm{T}}$ and then -6 and 14 to get $(4, 14, -6, 0)^{\mathrm{T}}$. Hence $Lc = b^*$ is

$$\begin{pmatrix} 1 & 0 & 0 & 0 \\ 1/2 & 1 & 0 & 0 \\ 0 & 1/4 & 1 & 0 \\ -1/2 & 1/2 & 1/3 & 1 \end{pmatrix} \begin{pmatrix} c_1 \\ c_2 \\ c_3 \\ c_4 \end{pmatrix} = \begin{pmatrix} 4 \\ 14 \\ -6 \\ 0 \end{pmatrix}$$

Forward substitution gives

$$c_1 = 4, \qquad c_2 = 12, \qquad c_3 = -9, \qquad c_4 = -1$$

Finally, the system $Ux = c$, i.e.

$$\begin{pmatrix} 2 & 2 & -2 & 4 \\ 0 & 4 & 8 & -12 \\ 0 & 0 & 3 & 6 \\ 0 & 0 & 0 & 1 \end{pmatrix} \begin{pmatrix} x_1 \\ x_2 \\ x_3 \\ x_4 \end{pmatrix} = \begin{pmatrix} 4 \\ 12 \\ -9 \\ -1 \end{pmatrix}$$

is solved using back substitution to get

$$x_4 = -1, \qquad x_3 = -1, \qquad x_2 = 2, \qquad x_1 = 1 \qquad \Box$$

Elements of L calculated in this way have moduli not exceeding one, which is to be expected since these numbers are the multipliers in the equivalent elimination process. This can also be deduced from equations (2.8) and (2.9). From equation (2.8), the value of $|u_{jj}|$ selected in step j is

$$\max_{j \leqslant i \leqslant n} \left| a_{ij} - \sum_{k=1}^{j-1} l_{ik} u_{kj} \right|$$

where a_{ij} is the (i, j) entry of the matrix derived from the previous step. Substituting this expression into equation (2.9) for u_{jj} shows that $|l_{ij}| \leqslant 1$.

It might appear that the introduction of partial pivoting into the LU decomposition algorithm considerably increases the total amount of work. However, the quantities u_{jj} calculated in the rejected options are reusable;

division of these numbers by the value of u_{jj} given in the selected option produces the elements in the jth column of L. For instance, in step 2 of Example 2.3 the first and third options are rejected. The corresponding values of u_{22} are 1 and 2, respectively, and division of these numbers by the chosen diagonal element, $u_{22} = 4$, produces $l_{32} = 1/4$ and $l_{42} = 1/2$. In fact, it can be shown that the total number of arithmetic operations needed to solve an $n \times n$ system using Gauss elimination and LU decomposition, with or without partial pivoting, is the same. Each method requires $(1/3)n^3 + (1/2)n^2 - (5/6)n$ additions/subtractions and $(1/3)n^3 + (1/2)n^2 - (1/3)n$ multiplications/divisions (see Fox, 1964).

As in Gauss elimination, the number of calculations is considerably reduced if A is banded. In the case of a tridiagonal matrix, for example, the matrices L and U take the form

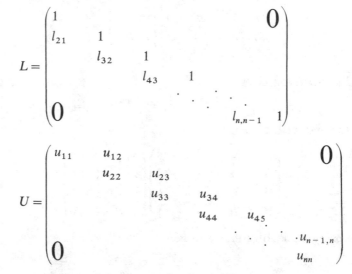

and the computation of L, U, c and x is trivial.

One useful by-product of LU decomposition is the calculation of the determinant. If $A^* = LU$, then

$$\det A^* = \det L \det U = 1^n \prod_{i=1}^{n} u_{ii} = \prod_{i=1}^{n} u_{ii}$$

making use of the matrix property that the determinant of the product of two matrices is the product of their determinants. Moreover, if A^* is obtained from A using m interchanges, then $\det A = (-1)^m \det A^*$. Hence

$$\det A = (-1)^m \prod_{i=1}^{n} u_{ii}$$

Systems of equations with multiple right hand sides are also easily solved.

The matrices L and U depend only on the coefficient matrix and so the LU decomposition need only be carried out once. The solutions are then found using repeated forward and back substitutions.

In the special case of a symmetric matrix A it is possible to construct an alternative triangular factorization which, when programmed, saves on storage space and on the number of calculations. The matrix A is written as the product LL^T where L is lower triangular. The system $Ax = b$ is solved by applying forward and back substitutions to $Lc = b$ and $L^Tx = c$, respectively. This is called *Choleski's method*. The elements of L are found by equating successive columns in the relation

$$\begin{pmatrix} a_{11} & a_{12} & & a_{1n} \\ a_{12} & a_{22} & & a_{2n} \\ & & \cdot & \\ & & & \cdot \\ a_{1n} & a_{2n} & & a_{nn} \end{pmatrix} = \begin{pmatrix} l_{11} & & & \mathbf{0} \\ l_{21} & l_{22} & & \\ & & \cdot & \\ & & & \cdot \\ l_{n1} & l_{n2} & & l_{nn} \end{pmatrix} \begin{pmatrix} l_{11} & l_{21} & & l_{n1} \\ & l_{22} & & l_{n2} \\ & & \cdot & \\ \mathbf{0} & & & \cdot \\ & & & l_{nn} \end{pmatrix}$$

■ **Example 2.5**

To illustrate the construction of the Choleski decomposition, consider

$$\begin{pmatrix} 4 & 2 & -2 \\ 2 & 10 & 2 \\ -2 & 2 & 3 \end{pmatrix} = \begin{pmatrix} l_{11} & 0 & 0 \\ l_{21} & l_{22} & 0 \\ l_{31} & l_{32} & l_{33} \end{pmatrix} \begin{pmatrix} l_{11} & l_{21} & l_{31} \\ 0 & l_{22} & l_{32} \\ 0 & 0 & l_{33} \end{pmatrix}$$

STEP 1   Equate entries in the first column to get

$$4 = l_{11}^2 \quad \Rightarrow l_{11} = 2$$
$$2 = l_{21}l_{11} \Rightarrow l_{21} = 1$$
$$-2 = l_{31}l_{11} \Rightarrow l_{31} = -1$$

i.e.

$$\begin{pmatrix} 4 & 2 & -2 \\ 2 & 10 & 2 \\ -2 & 2 & 3 \end{pmatrix} = \begin{pmatrix} 2 & 0 & 0 \\ 1 & l_{22} & 0 \\ -1 & l_{32} & l_{33} \end{pmatrix} \begin{pmatrix} 2 & 1 & -1 \\ 0 & l_{22} & l_{32} \\ 0 & 0 & l_{33} \end{pmatrix}$$

Note that l_{11} could also be -2 and so the matrix L is not (quite) unique.

STEP 2   Equate entries in the second column to get

$$10 = \quad 1 + l_{22}^2 \quad \Rightarrow l_{22} = 3$$
$$2 = -1 + l_{32}l_{22} \Rightarrow l_{32} = 1$$

i.e.

$$\begin{pmatrix} 4 & 2 & -2 \\ 2 & 10 & 2 \\ -2 & 2 & 3 \end{pmatrix} = \begin{pmatrix} 2 & 0 & 0 \\ 1 & 3 & 0 \\ -1 & 1 & l_{33} \end{pmatrix} \begin{pmatrix} 2 & 1 & -1 \\ 0 & 3 & 1 \\ 0 & 0 & l_{33} \end{pmatrix}$$

STEP 3   Equate entries in the third column to get

$$3 = 1 + 1 + l_{33}^2 \Rightarrow l_{33} = 1$$

i.e.

$$\begin{pmatrix} 4 & 2 & -2 \\ 2 & 10 & 2 \\ -2 & 2 & 3 \end{pmatrix} = \begin{pmatrix} 2 & 0 & 0 \\ 1 & 3 & 0 \\ -1 & 1 & 1 \end{pmatrix} \begin{pmatrix} 2 & 1 & -1 \\ 0 & 3 & 1 \\ 0 & 0 & 1 \end{pmatrix}$$

□

For a general matrix, the elements of L are calculated from

$$l_{11} = \sqrt{a_{11}} \tag{2.11}$$

$$l_{i1} = a_{i1}/l_{11} \qquad\qquad i > 1 \tag{2.12}$$

$$l_{jj} = \sqrt{\left(a_{jj} - \sum_{k=1}^{j-1} l_{jk}^2 \right)} \qquad j > 1 \tag{2.13}$$

$$l_{ij} = \frac{1}{l_{jj}} \left(a_{ij} - \sum_{k=1}^{j-1} l_{ik} l_{jk} \right) \qquad i > j > 1 \tag{2.14}$$

There are several difficulties associated with this decomposition. The method fails if any $l_{jj} = 0$ and is unstable if $|l_{ij}| > 1$. It is also possible for the quantity inside the square root of (2.13) to be negative, in which case all of the elements in column j are purely imaginary. There is, however, a special class of matrices for which these problems do not occur. It can be shown (see Wilkinson, 1965) that A can be factorized into LL^{T} for some real, non-singular, lower triangular matrix L, if and only if A is symmetric and positive definite. Notice that if l_{jj} is real then it follows from (2.13) that $l_{jk}^2 \leqslant a_{jj}$. Hence, if the system is scaled so that $\max |a_{jj}| = 1$, then $|l_{jk}| \leqslant 1$. This means that the Choleski decomposition of a symmetric, positive definite matrix is stable without the need for interchanges.

Choleski decomposition provides a convenient method for investigating the positive definiteness of symmetric matrices. The formal definition, $x^{\mathrm{T}} A x > 0$ for all $x \neq 0$, is not at all easy to verify in practice. However, it is relatively straightforward to attempt the construction of a Choleski decomposition of a symmetric matrix A; the factorization succeeds in real arithmetic if and only if A is positive definite. In particular, we can deduce that the matrix given in Example 2.5 is positive definite.

EXERCISE 2.2

1. Solve the following systems of linear equations using LU decomposition (i) without pivoting (ii) with partial pivoting:

(a) $x_1 + 7x_2 + 6x_3 = 2$
       $x_1 - 2x_2 - x_3 = 0$
       $2x_1 + 2x_2 - 4x_3 = 8$

(b) $x_1 - x_2 + 2x_3 = 5$
       $2x_1 - 2x_2 + x_3 = 1$
       $3x_1 - 2x_2 + 7x_3 = 20$

(c) $x_1 + 2x_2 + x_3 - x_4 = 1$
       $x_1 + x_2 + x_3 + x_4 = 2$
       $-2x_1 + x_2 - 2x_3 + 3x_4 = 1$
       $3x_1 - x_2 + 4x_3 - x_4 = 0$

2. Use the *LU* decomposition method with pivoting to calculate the determinant and inverse of the following matrices:

(a) $\begin{pmatrix} 1 & 7 & 6 \\ 1 & -2 & -1 \\ 2 & 2 & -4 \end{pmatrix}$ (b) $\begin{pmatrix} 1 & 2 & 3 \\ 5 & 7 & -1 \\ 7 & 8 & -11 \end{pmatrix}$

3. Which of the following matrices are positive definite?

(a) $\begin{pmatrix} 2 & 1 & 1 \\ 1 & 1 & 3 \\ 1 & 3 & 1 \end{pmatrix}$ (b) $\begin{pmatrix} 4 & -1 & -1 \\ -1 & 4 & -1 \\ -1 & -1 & 4 \end{pmatrix}$

(c) $\begin{pmatrix} 1 & 2 & 0 & 0 \\ 2 & 5 & -1 & -1 \\ 0 & -1 & 3 & -1 \\ 0 & -1 & -1 & 4 \end{pmatrix}$

4. If L_1, L_2 are $n \times n$ unit lower triangular and U_1, U_2 are $n \times n$ non-singular upper triangular matrices, what can you say about the nature of $L_2^{-1}L_1$ and $U_2 U_1^{-1}$?
   If $L_1 U_1 = A = L_2 U_2$, show that $L_2^{-1} L_1 = U_2 U_1^{-1}$ and hence deduce that $L_1 = L_2$ and $U_1 = U_2$, i.e. that the *LU* decomposition of A is unique.

5. Prove that if $A = LL^T$, where $L = (l_{ij})$ is a real, non-singular, lower triangular matrix with $|l_{ii}| \leqslant 1$ for all i, then $|l_{jj}| \geqslant (\det A)^{1/2}$.

2.3 ITERATIVE METHODS

In this section three iterative methods are described for the solution of

$$Ax = b$$

where A is an $n \times n$ non-singular matrix with non-zero diagonal elements. In order to provide a mathematical analysis of these techniques it is convenient to introduce some notation at this stage. The matrix A is written as $L + U + D$ where L, U and D are the lower triangular, upper triangular and diagonal parts of A respectively. (Note that the matrices L and U are not the same as those used in the previous section.) For simplicity, we shall concentrate on the 3×3 system

$$
\begin{aligned}
a_{11}x_1 + a_{12}x_2 + a_{13}x_3 &= b_1 \\
a_{21}x_1 + a_{22}x_2 + a_{23}x_3 &= b_2 \\
a_{31}x_1 + a_{32}x_2 + a_{33}x_3 &= b_3
\end{aligned}
\tag{2.15}
$$

In this case

$$
L = \begin{pmatrix} 0 & 0 & 0 \\ a_{21} & 0 & 0 \\ a_{31} & a_{32} & 0 \end{pmatrix} \qquad U = \begin{pmatrix} 0 & a_{12} & a_{13} \\ 0 & 0 & a_{23} \\ 0 & 0 & 0 \end{pmatrix}
$$

$$
D = \begin{pmatrix} a_{11} & 0 & 0 \\ 0 & a_{22} & 0 \\ 0 & 0 & a_{33} \end{pmatrix}
$$

The first stage is to rearrange the ith equation in (2.15) so that only the term involving the variable x_i occurs on the left hand side. This gives

$$
\begin{aligned}
a_{11}x_1 &= b_1 - a_{12}x_2 - a_{13}x_3 \\
a_{22}x_2 &= b_2 - a_{21}x_1 - a_{23}x_3 \\
a_{33}x_3 &= b_3 - a_{31}x_1 - a_{32}x_2
\end{aligned}
$$

or in matrix notation

$$
Dx = b - (L + U)x
\tag{2.16}
$$

The next stage is to divide both sides of the ith equation by the diagonal element a_{ii} to obtain

$$
x_1 = \frac{1}{a_{11}}(b_1 - a_{12}x_2 - a_{13}x_3)
$$

$$
x_2 = \frac{1}{a_{22}}(b_2 - a_{21}x_1 - a_{23}x_3)
$$

$$
x_3 = \frac{1}{a_{33}}(b_3 - a_{31}x_1 - a_{32}x_2)
$$

This is equivalent to pre-multiplying both sides of (2.16) by the inverse of

D, and so

$$x = D^{-1}(b - (L + U)x)$$

This rearrangement suggests the iteration

$$x_1^{[m+1]} = \frac{1}{a_{11}}(b_1 - a_{12}x_2^{[m]} - a_{13}x_3^{[m]})$$

$$x_2^{[m+1]} = \frac{1}{a_{22}}(b_2 - a_{21}x_1^{[m]} - a_{23}x_3^{[m]}) \qquad (2.17)$$

$$x_3^{[m+1]} = \frac{1}{a_{33}}(b_3 - a_{31}x_1^{[m]} - a_{32}x_2^{[m]})$$

or equivalently, in matrix form,

$$x^{[m+1]} = D^{-1}(b - (L + U)x^{[m]}) \qquad (2.18)$$

This is called the *Jacobi method*. For a general system of n equations, the Jacobi method is defined by

$$x_i^{[m+1]} = \frac{1}{a_{ii}}\left(b_i - \sum_{\substack{j=1 \\ j \neq i}}^{n} a_{ij}x_j^{[m]}\right) \qquad i = 1, 2, \ldots, n$$

As usual with iterative methods, an initial approximation $x_i^{[0]}$ must be supplied. In the absence of any prior knowledge about the exact solution, it is conventional to start with $x_i^{[0]} = 0$ for all i. These values are substituted into the right hand sides of equations (2.17) to obtain the first iterates $x_i^{[1]}$. The process can then be repeated until the condition

$$\max_i |x_i^{[m+1]} - x_i^{[m]}| \leqslant \varepsilon$$

for some prescribed tolerance ε, is satisfied.

■ **Example 2.6**

To illustrate the success (and failure) of the Jacobi method, consider the system of equations

$$\begin{aligned} 5x_1 + 2x_2 - x_3 &= 6 \\ x_1 + 6x_2 - 3x_3 &= 4 \\ 2x_1 + x_2 + 4x_3 &= 7 \end{aligned} \qquad (2.19)$$

The Jacobi iteration is given by

$$x_1^{[m+1]} = \tfrac{1}{5}(6 - 2x_2^{[m]} + x_3^{[m]})$$
$$x_2^{[m+1]} = \tfrac{1}{6}(4 - x_1^{[m]} + 3x_3^{[m]})$$
$$x_3^{[m+1]} = \tfrac{1}{4}(7 - 2x_1^{[m]} - x_2^{[m]})$$

Table 2.1

m	$x_1^{[m]}$	$x_2^{[m]}$	$x_3^{[m]}$
0	0	0	0
1	1.200	0.667	1.750
2	1.283	1.342	0.983
3	0.860	0.944	0.773
4	0.977	0.910	1.084
5	1.053	1.046	1.034
6	0.988	1.008	0.962
7	0.989	0.983	1.004
8	1.008	1.004	1.010
9	1.000	1.004	0.995
10	0.997	0.998	0.999
11	1.001	1.000	1.002

and starting with $x_1^{[0]} = x_2^{[0]} = x_3^{[0]} = 0$ we obtain

$$x_1^{[1]} = \tfrac{1}{5}[6 - (2 \times 0) + 0] = 1.200$$
$$x_2^{[1]} = \tfrac{1}{6}[4 - 0 + (3 \times 0)] = 0.667$$
$$x_3^{[1]} = \tfrac{1}{4}[7 - (2 \times 0) - 0] = 1.750$$

These and further iterates are listed in Table 2.1. The iteration is stopped when

$$\max_i |x_i^{[m+1]} - x_i^{[m]}| \leqslant \tfrac{1}{2} \times 10^{-2}$$

In this case the Jacobi method has produced the solution, correct to 2D, in 11 iterations.

On the other hand, consider the system of equations

$$
\begin{aligned}
x_1 + 6x_2 - 3x_3 &= 4 \\
5x_1 + 2x_2 - x_3 &= 6 \\
2x_1 + x_2 + 4x_3 &= 7
\end{aligned}
\tag{2.20}
$$

which is the same as the previous system but with the first and second equations interchanged. Results for this system are listed in Table 2.2. In this case the Jacobi iteration rapidly diverges and the example highlights the need for an *a priori* condition for testing the convergence (or divergence) of the method.   □

It is interesting to note from (2.17) that when the second equation is used to calculate $x_2^{[m+1]}$, the value of $x_1^{[m+1]}$ is already known, so we may as well use it in place of the old estimate $x_1^{[m]}$. Since $x_1^{[m+1]}$ is likely to be

Table 2.2

m	$x_1^{[m]}$	$x_2^{[m]}$	$x_3^{[m]}$
0	0	0	0
1	4.000	3.000	1.750
2	− 8.750	− 6.125	− 1.000
3	37.750	24.375	7.656
4	− 119.282	− 87.547	− 23.219
5	459.625	289.596	83.278

closer than $x_1^{[m]}$ to the exact value of x_1 this should speed up the convergence. Similarly, it makes sense to use $x_1^{[m+1]}$ and $x_2^{[m+1]}$ in the third equation of (2.17), since both of these numbers are known when this equation is used to update the approximation to x_3. It might, therefore, be better to try the iteration

$$x_1^{[m+1]} = \frac{1}{a_{11}}(b_1 - a_{12}x_2^{[m]} - a_{13}x_3^{[m]})$$

$$x_2^{[m+1]} = \frac{1}{a_{22}}(b_2 - a_{21}x_1^{[m+1]} - a_{23}x_3^{[m]})$$

$$x_3^{[m+1]} = \frac{1}{a_{33}}(b_3 - a_{31}x_1^{[m+1]} - a_{32}x_2^{[m+1]})$$

This is called the *Gauss–Seidel method* and, since the most recent estimates are used at each stage, we would normally expect it to converge faster than Jacobi's method. These equations can be rearranged as

$$a_{11}x_1^{[m+1]} = b_1 - a_{12}x_2^{[m]} - a_{13}x_3^{[m]}$$
$$a_{21}x_1^{[m+1]} + a_{22}x_2^{[m+1]} = b_2 \qquad - a_{23}x_3^{[m]}$$
$$a_{31}x_1^{[m+1]} + a_{32}x_2^{[m+1]} + a_{33}x_3^{[m+1]} = b_3$$

and so, in matrix form, the Gauss–Seidel method can be written as

$$(D + L)x^{[m+1]} = b - Ux^{[m]}$$

or equivalently as

$$x^{[m+1]} = (D + L)^{-1}(b - Ux^{[m]}) \qquad (2.21)$$

For a general system of n equations, the Gauss–Seidel method is defined by

$$x_i^{[m+1]} = \frac{1}{a_{ii}}\left(b_i - \sum_{j=1}^{i-1} a_{ij}x_j^{[m+1]} - \sum_{j=i+1}^{n} a_{ij}x_j^{[m]}\right) \qquad i = 1, 2, \ldots, n$$

■ **Example 2.7**

To illustrate the improved rate of convergence of the Gauss–Seidel method, consider the solution of the system (2.19) using the same stopping criterion as before. The Gauss–Seidel iteration is given by

$$x_1^{[m+1]} = \tfrac{1}{5}(6 - 2x_2^{[m]} + x_3^{[m]})$$
$$x_2^{[m+1]} = \tfrac{1}{6}(4 - x_1^{[m+1]} + 3x_3^{[m]})$$
$$x_3^{[m+1]} = \tfrac{1}{4}(7 - 2x_1^{[m+1]} - x_2^{[m+1]})$$

and starting with $x_1^{[0]} = x_2^{[0]} = x_3^{[0]} = 0$ we obtain

$$x_1^{[1]} = \tfrac{1}{5}[6 - (2 \times 0) + 0] = 1.200$$
$$x_2^{[1]} = \tfrac{1}{6}[4 - 1.200 + (3 \times 0)] = 0.467$$
$$x_3^{[1]} = \tfrac{1}{4}[7 - (2 \times 1.200) - 0.467] = 1.033$$

These and further iterates are listed in Table 2.3. Convergence is achieved in 7 iterations. This compares with 11 iterations required for convergence with the Jacobi method (see Example 2.6).

The results in Table 2.4 show that the Gauss–Seidel iteration also diverges when applied to the system (2.20).   □

We have already encountered a system of equations for which the Jacobi

Table 2.3

m	$x_1^{[m]}$	$x_2^{[m]}$	$x_3^{[m]}$
0	0	0	0
1	1.200	0.467	1.033
2	1.220	0.980	0.895
3	0.987	0.950	1.019
4	1.024	1.006	0.987
5	0.995	0.994	1.004
6	1.003	1.002	0.998
7	0.999	0.999	1.001

Table 2.4

m	$x_1^{[m]}$	$x_2^{[m]}$	$x_3^{[m]}$
0	0	0	0
1	4.000	− 7.000	1.500
2	50.500	− 122.50	7.125
3	760.375	− 1894.375	95.156

and Gauss–Seidel methods both diverge. It is also possible to construct examples (see Exercise 2.3) in which the Jacobi method converges and the Gauss–Seidel method diverges (and vice versa). It is possible to explain this behaviour by performing a simple mathematical analysis of matrix iterative procedures.

Consider the general splitting $A = M + N$ for any two $n \times n$ matrices M and N. The equation $Ax = b$ then becomes $(M + N)x = b$ which can be re-arranged as

$$x = M^{-1}(b - Nx) \tag{2.22}$$

This suggests the iteration

$$x^{[m+1]} = M^{-1}(b - Nx^{[m]}) \tag{2.23}$$

A comparison of (2.23) with equations (2.18) and (2.21) shows that the Jacobi and Gauss–Seidel methods are both special cases of the above. For the Jacobi method, $M = D$, $N = L + U$ and for the Gauss–Seidel method, $M = D + L$, $N = U$.

Subtracting (2.23) from (2.22) gives

$$e^{[m+1]} = -M^{-1}Ne^{[m]} \tag{2.24}$$

where $e^{[m]}$ is the error vector $x - x^{[m]}$. The matrix $-M^{-1}N$ is called the *iteration matrix* and it governs the change in the error from one step to the next. The iteration matrices for the Jacobi and Gauss–Seidel methods are $-D^{-1}(L + U)$ and $-(D + L)^{-1}U$, respectively. For convergence, we require the error vector $e^{[m]} \to 0$ as $m \to \infty$. A necessary and sufficient condition for this is contained in the following theorem.

■ **Theorem 2.1**
Suppose that the matrix $-M^{-1}N$ has eigenvalues λ_i $(i = 1, 2, \ldots, n)$. The iteration $x^{[m+1]} = M^{-1}(b - Nx^{[m]})$ converges if and only if the spectral radius $\rho(-M^{-1}N) < 1$, where $\rho(-M^{-1}N) = \max_i |\lambda_i|$.

■ **Proof**
We shall assume that the matrix $-M^{-1}N$ admits a set of n linearly independent eigenvectors, which will certainly be so when the eigenvalues of $-M^{-1}N$ are distinct or when the matrix $-M^{-1}N$ is symmetric. (A more general proof which does not make this assumption can be found in Varga, 1962.) Let these eigenvectors be denoted by v_i with associated eigenvalues λ_i $(i = 1, 2, \ldots, n)$. The initial error vector $e^{[0]}$ can then be written in the form

$$e^{[0]} = \alpha_1 v_1 + \alpha_2 v_2 + \cdots + \alpha_n v_n$$

for some appropriate scalars α_i.

Multiplying both sides of this equation by $-M^{-1}N$ gives

$$e^{[1]} = \alpha_1(-M^{-1}N)v_1 + \alpha_2(-M^{-1}N)v_2 + \cdots + \alpha_n(-M^{-1}N)v_n$$

since, from equation (2.24), $e^{[1]} = -M^{-1}Ne^{[0]}$. Now v_i is an eigenvector of $-M^{-1}N$ with eigenvalue λ_i, so

$$(-M^{-1}N)v_i = \lambda_i v_i$$

Hence

$$e^{[1]} = \alpha_1\lambda_1 v_1 + \alpha_2\lambda_2 v_2 + \cdots + \alpha_n\lambda_n v_n$$

Multiplying both sides of this equation by $-M^{-1}N$ gives

$$
\begin{aligned}
e^{[2]} &= (-M^{-1}N)e^{[1]} \\
&= \alpha_1\lambda_1(-M^{-1}N)v_1 + \alpha_2\lambda_2(-M^{-1}N)v_1 + \cdots \\
&\quad + \alpha_n\lambda_n(-M^{-1}N)v_n \\
&= \alpha_1\lambda_1^2 v_1 + \alpha_2\lambda_2^2 v_2 + \cdots + \alpha_n\lambda_n^2 v_n
\end{aligned}
$$

Continuing in this way gives

$$e^{[m]} = \alpha_1\lambda_1^m v_1 + \alpha_2\lambda_2^m v_2 + \cdots + \alpha_n\lambda_n^m v_n \qquad (2.25)$$

Now $\lambda_i^m \to 0$ if and only if $|\lambda_i| < 1$, and so we deduce from (2.25) that a necessary and sufficient condition for $e^{[m]} \to 0$ is $\rho(-M^{-1}N) < 1$.   □

Theorem 2.1 gives us a means of investigating the convergence of the Jacobi and Gauss–Seidel methods. We find the iteration matrices $-D^{-1}(L+U)$ and $-(D+L)^{-1}U$ and then calculate their eigenvalues. If these eigenvalues are all less than one in modulus the method converges; if not then it diverges. Notice from Theorem 2.1 that the convergence of these iterative processes depends only on the coefficient matrix and not on the right hand sides or starting vector. It is also interesting to note from equation (2.25) that the rate of convergence depends on λ_i^m, and so the smaller the spectral radius, the faster the convergence.

Now although Theorem 2.1 does provide us with a nice theoretical condition for convergence, it is not at all easy to apply in practice. For matrices of order greater than 2 it is a prohibitively time consuming task to calculate the spectral radius, even if we are prepared to use a numerical technique such as the power method to find it. Indeed, it would be just as quick to actually calculate the first few iterates for the Jacobi and Gauss–Seidel methods themselves and so discover experimentally whether the methods succeed or fail! Clearly what is required is a simple way of seeing 'at a glance' whether or not an iterative method will succeed for any particular system of equations. This can be done as follows.

■ **Definition 2.1**

An $n \times n$ matrix A is said to be (strictly) *diagonally dominant* if, for each row of the matrix, the modulus of the diagonal strictly exceeds the sum of the moduli of the off-diagonals, i.e. if

$$|a_{ii}| > \sum_{j \neq i} |a_{ij}| \qquad i = 1, 2, \ldots, n \qquad \qquad □$$

For example, the coefficient matrix of the system (2.19) is diagonally dominant since

$$|5| > |2| + |-1|$$
$$|6| > |1| + |-3|$$
$$|4| > |2| + |1|$$

The following theorem shows that diagonal dominance is a sufficient condition for the convergence of the Jacobi and Gauss–Seidel methods.

■ **Theorem 2.2**

The Jacobi and Gauss–Seidel methods both converge for a system of linear equations with a diagonally dominant coefficient matrix.

■ **Proof**

If v is an eigenvector of $-M^{-1}N$ corresponding to an eigenvalue λ, then

$$-M^{-1}Nv = \lambda v$$

which can be rearranged as

$$(\lambda M + N)v = 0 \tag{2.26}$$

From Theorem 2.1, we need only prove that $|\lambda| < 1$ whenever the coefficient matrix is diagonally dominant. Let us suppose that the largest component of v occurs in the ith position, i.e.

$$|v_j| \leqslant |v_i| \qquad j \neq i$$

The proofs of the Jacobi and Gauss–Seidel methods are considered separately.

(a) *The Jacobi method*

For the Jacobi method, $M = D$ and $N = L + U$, so equation (2.26) becomes

$$(\lambda D + L + U)v = 0$$

The ith component of this equation is

$$\lambda a_{ii} v_i + \sum_{\substack{j=1 \\ j \neq i}}^{n} a_{ij} v_j = 0$$

and so

$$\lambda = -\frac{1}{a_{ii} v_i} \sum_{\substack{j=1 \\ j \neq i}}^{n} a_{ij} v_j$$

which implies that

$$|\lambda| = \frac{1}{|a_{ii}||v_i|} \left| \sum_{\substack{j=1 \\ j \neq i}}^{n} a_{ij} v_j \right|$$

For any two numbers A and B,

$$|A + B| \leqslant |A| + |B|$$

which is known as the triangle inequality. Consequently,

$$\left| \sum_{\substack{j=1 \\ j \neq i}}^{n} a_{ij} v_j \right| \leqslant \sum_{\substack{j=1 \\ j \neq i}}^{n} |a_{ij} v_j| \leqslant |v_i| \sum_{\substack{j=1 \\ j \neq i}}^{n} |a_{ij}|$$

since $|v_j| \leqslant |v_i|$.

Hence

$$|\lambda| \leqslant \frac{|v_i|}{|a_{ii}||v_i|} \sum_{\substack{j=1 \\ j \neq i}}^{n} |a_{ij}| = \frac{1}{|a_{ii}|} \sum_{\substack{j=1 \\ j \neq i}}^{n} |a_{ij}|$$

However, from the definition of diagonal dominance,

$$|a_{ii}| > \sum_{\substack{j=1 \\ j \neq i}}^{n} |a_{ij}|$$

and so we can deduce that $|\lambda| < 1$, as required.

(b) *The Gauss–Seidel method*

For the Gauss–Seidel method, $M = L + D$ and $N = U$, so (2.26) becomes

$$(\lambda D + \lambda L + U)\mathbf{v} = \mathbf{0}$$

The ith component of this equation is

$$\lambda \sum_{j=1}^{i} a_{ij} v_j + \sum_{j=i+1}^{n} a_{ij} v_j = 0$$

and so

$$\lambda = -\left(\sum_{j=i+1}^{n} a_{ij} v_j \right) \Big/ \left(\sum_{j=1}^{i} a_{ij} v_j \right) \tag{2.27}$$

Now, for any two numbers A and B,

$$|A + B| \geqslant |A| - |B|$$

Applying this inequality to the denominator of (2.27) gives

$$\left| \sum_{j=1}^{i} a_{ij} v_j \right| \geqslant |a_{ii} v_i| - \left| \sum_{j=1}^{i-1} a_{ij} v_j \right|$$

By the triangle inequality,

$$\left| \sum_{j=1}^{i-1} a_{ij} v_j \right| \leqslant \sum_{j=1}^{i-1} |a_{ij} v_j| \leqslant |v_i| \sum_{j=1}^{i-1} |a_{ij}|$$

since $|v_j| \leqslant |v_i|$, and so

$$\left| \sum_{j=1}^{i} a_{ij} v_j \right| \geq |a_{ii} v_i| - |v_i| \sum_{j=1}^{i-1} |a_{ij}|$$

$$= |v_i| \left(|a_{ii}| - \sum_{j=1}^{i-1} |a_{ij}| \right) \tag{2.28}$$

which is positive from the diagonal dominance condition.

We can also apply the triangle inequality to the numerator of (2.27) to deduce that

$$\left| \sum_{j=i+1}^{n} a_{ij} v_j \right| \leq |v_i| \sum_{j=i+1}^{n} |a_{ij}| \tag{2.29}$$

Substituting the inequalities (2.28) and (2.29) into equation (2.27) gives

$$|\lambda| \leq \left(\sum_{j=i+1}^{n} |a_{ij}| \right) \bigg/ \left(|a_{ii}| - \sum_{j=1}^{i-1} |a_{ij}| \right)$$

However, from the definition of diagonal dominance,

$$|a_{ii}| \geq \sum_{\substack{j=1 \\ j \neq i}}^{n} |a_{ij}| = \sum_{j=1}^{i-1} |a_{ij}| + \sum_{j=i+1}^{n} |a_{ij}|$$

and so we can deduce that $|\lambda| < 1$, as required. $\qquad\qquad\square$

It is important to emphasize that diagonal dominance is only a sufficient condition for convergence; it is certainly not necessary (see Question 3 in Exercise 2.3). However, it does provide a quick check on the convergence of these methods and, although it may not seem at all obvious, the condition of diagonal dominance does occur quite frequently in practice. If a system of equations does not satisfy this condition it may even be possible to reorder the equations and unknowns to produce an equivalent system for which it does hold. Notice also from the proof of Theorem 2.2 that the greater the modulus of each diagonal compared with the sum of the moduli of the off-diagonals, the smaller the spectral radius of the iteration matrix and hence the faster the convergence.

The final iterative method to be considered is an extension of the Gauss–Seidel method. For a system of three equations the Gauss–Seidel method is defined by

$$x_1^{[m+1]} = \frac{1}{a_{11}} (b_1 - a_{12} x_2^{[m]} - a_{13} x_3^{[m]})$$

$$x_2^{[m+1]} = \frac{1}{a_{22}} (b_2 - a_{21} x_1^{[m+1]} - a_{23} x_3^{[m]})$$

$$x_3^{[m+1]} = \frac{1}{a_{33}} (b_3 - a_{31} x_1^{[m+1]} - a_{32} x_2^{[m+1]})$$

Rearranging the ith equation in the form

$$x_i^{[m+1]} = x_i^{[m]} + \cdots$$

gives

$$x_1^{[m+1]} = x_1^{[m]} + \frac{1}{a_{11}}(b_1 - a_{11}x_1^{[m]} \quad - a_{12}x_2^{[m]} \quad - a_{13}x_3^{[m]})$$

$$x_2^{[m+1]} = x_2^{[m]} + \frac{1}{a_{22}}(b_2 - a_{21}x_1^{[m+1]} - a_{22}x_2^{[m]} \quad - a_{23}x_3^{[m]})$$

$$x_3^{[m+1]} = x_3^{[m]} + \frac{1}{a_{33}}(b_3 - a_{31}x_1^{[m+1]} - a_{32}x_2^{[m+1]} - a_{33}x_3^{[m]})$$

The second term on the right hand side of each of these equations represents the change, or correction, that is made to $x_i^{[m]}$ in order to calculate $x_i^{[m+1]}$. Consider the effect of multiplying this correction by a number ω, called the relaxation parameter:

$$x_1^{[m+1]} = x_1^{[m]} + \frac{\omega}{a_{11}}(b_1 - a_{11}x_1^{[m]} \quad - a_{12}x_2^{[m]} \quad - a_{13}x_3^{[m]})$$

$$x_2^{[m+1]} = x_2^{[m]} + \frac{\omega}{a_{22}}(b_2 - a_{21}x_1^{[m+1]} - a_{22}x_2^{[m]} \quad - a_{23}x_3^{[m]}) \qquad (2.30)$$

$$x_3^{[m+1]} = x_3^{[m]} + \frac{\omega}{a_{33}}(b_3 - a_{31}x_1^{[m+1]} - a_{32}x_2^{[m+1]} - a_{33}x_3^{[m]})$$

The rate of convergence depends on ω, which is chosen so as to minimize the number of iterations. If $\omega = 1$, then this method is the same as Gauss–Seidel. If $\omega > 1$, then a larger than normal correction is taken. This is useful if the Gauss–Seidel iterates are converging monotonically; that is, if successive iterates are moving in one direction towards the exact solution. If $\omega < 1$, then a smaller than normal correction is taken. This is useful if the Gauss–Seidel iterates are oscillating; that is, if successive iterates keep overshooting the exact solution. In practice the case $\omega > 1$ is most commonly used, and because of this the method is known as *successive over-relaxation* (usually abbreviated to SOR).

For a general system of n equations, the SOR method is defined by

$$x_i^{[m+1]} = x_i^{[m]} + \frac{\omega}{a_{ii}}\left(b_i - \sum_{j=1}^{i-1} a_{ij}x_j^{[m+1]} - \sum_{j=i}^{n} a_{ij}x_j^{[m]} \right) \qquad i = 1, 2, \ldots, n$$

■ Example 2.8

To illustrate the SOR method, consider the solution of the system (2.19) using the same stopping criterion as before. The results listed in Table 2.3

indicate that the Gauss–Seidel iterates oscillate slightly. Consequently, it may be worth while to try under-relaxation using a relaxation parameter just less than one. The SOR iteration with $\omega = 0.9$ is given by

$$x_1^{[m+1]} = x_1^{[m]} + \frac{0.9}{5}(6 - 5x_1^{[m]} - 2x_2^{[m]} + x_3^{[m]})$$

$$x_2^{[m+1]} = x_2^{[m]} + \frac{0.9}{6}(4 - x_1^{[m+1]} - 6x_2^{[m]} + 3x_3^{[m]})$$

$$x_3^{[m+1]} = x_3^{[m]} + \frac{0.9}{4}(7 - 2x_1^{[m+1]} - x_2^{[m+1]} - 4x_3^{[m]})$$

and starting with $x_1^{[0]} = x_2^{[0]} = x_3^{[0]} = 0$ we obtain

$$x_1^{[1]} = 0 + \frac{0.9}{5}[6 - (5 \times 0) - (2 \times 0) + 0] = 1.080$$

$$x_2^{[1]} = 0 + \frac{0.9}{6}[4 - 1.080 - (6 \times 0) + (3 \times 0)] = 0.438$$

$$x_3^{[1]} = 0 + \frac{0.9}{4}[7 - (2 \times 1.080) - 0.438 - (4 \times 0)] = 0.990$$

The results in Table 2.5 show that convergence is achieved in six iterations, which is only one iteration less than that obtained using the Gauss–Seidel method. Only a marginal improvement is possible for this particular system since the Gauss–Seidel iteration itself converges quite rapidly. □

For a more noticeable improvement in the rate of convergence, consider the following example.

■ Example 2.9
The Gauss–Seidel iterates for the linear system

$$
\begin{aligned}
-3x_1 + x_2 \qquad\quad + 3x_4 &= 1 \\
x_1 + 6x_2 + x_3 \qquad\quad &= 1 \\
x_2 + 6x_3 + x_4 &= 1 \\
3x_1 \qquad\quad + x_3 - 3x_4 &= 1
\end{aligned}
$$

are listed in Table 2.6, using the stopping criterion

$$\max_i |x_i^{[m+1]} - x_i^{[m]}| \leqslant \tfrac{1}{2} \times 10^{-2}$$

The coefficient matrix is not diagonally dominant and, although the method converges, it does so extremely slowly. Successive iterates converge monotonically, so it makes sense to try over-relaxation. Table 2.7 lists the results

Table 2.5

m	$x_1^{[m]}$	$x_2^{[m]}$	$x_3^{[m]}$
0	0	0	0
1	1.080	0.438	0.990
2	1.209	0.908	0.926
3	1.041	0.951	0.985
4	1.019	0.986	0.993
5	1.006	0.995	0.998
6	1.002	0.998	0.999

Table 2.6

m	$x_1^{[m]}$	$x_2^{[m]}$	$x_3^{[m]}$	$x_4^{[m]}$
0	0	0	0	0
1	-0.333	0.222	0.130	-0.623
2	-0.883	0.292	0.222	-1.142
3	-1.378	0.359	0.297	-1.612
4	-1.826	0.421	0.365	-2.037
5	-2.230	0.478	0.427	-2.421
$\vdots$	$\vdots$	$\vdots$	$\vdots$	$\vdots$
48	-5.953	0.993	0.993	-5.955
49	-5.957	0.994	0.994	-5.960

Table 2.7

m	$x_1^{[m]}$	$x_2^{[m]}$	$x_3^{[m]}$	$x_4^{[m]}$
0	0	0	0	0
1	-0.533	0.409	0.158	-1.303
2	-2.079	0.534	0.377	-2.878
3	-3.605	0.807	0.593	-4.259
4	-4.754	0.892	0.809	-5.153
5	-5.450	0.969	0.897	-5.683
$\vdots$	$\vdots$	$\vdots$	$\vdots$	$\vdots$
14	-6.003	1.000	1.000	-6.001
15	-6.000	1.000	1.000	-6.000

calculated with $\omega = 1.6$ and shows that the SOR method can be used to obtain a considerable reduction in the number of iterations.   □

In Examples 2.8 and 2.9, the value of the relaxation parameter was chosen more or less on a trial and error basis. However, it is possible to prove various theoretical results which may assist in the choice of ω. For example, it can be shown that a necessary condition for convergence is that ω should lie in the interval $(0, 2)$. Moreover, this condition is both necessary and sufficient if the coefficient matrix is symmetric and positive definite. However, these proofs are beyond the scope of this text. The interested reader is advised to consult the books by Varga (1962) and Young (1971) for these and many other theorems concerning this method.

EXERCISE 2.3

1. For which of the following systems of equations does the (i) Jacobi method (ii) Gauss–Seidel method converge? Start with the zero vector and use a tolerance of 0.001.

   (a) $\begin{aligned} -4x_1 + x_2 &= 1 \\ x_1 - 4x_2 + x_3 &= 1 \\ x_2 - 4x_3 + x_4 &= 1 \\ x_3 - 4x_4 &= 1 \end{aligned}$

   (b) $\begin{aligned} 2x_1 + x_2 + x_3 &= 4 \\ x_1 + 2x_2 + x_3 &= 4 \\ x_1 + x_2 + 2x_3 &= 4 \end{aligned}$

   (c) $\begin{aligned} x_1 + 2x_2 - 2x_3 &= 3 \\ x_1 + x_2 + x_3 &= 0 \\ 2x_1 + 2x_2 + x_3 &= 1 \end{aligned}$

2. Find the spectral radii of the Jacobi and Gauss–Seidel iteration matrices corresponding to each of the systems given in Question 1. Hence comment on the convergence or divergence in each case.

3. Which, if any, of the coefficient matrices used in Question 1 are diagonally dominant? Is diagonal dominance a necessary condition for convergence of the Jacobi and Gauss–Seidel methods?

4. Rearrange the following equations into a form suitable for solution using an iterative method. Hence solve this system, correct to 2D, using the Gauss–Seidel method.

$$3x_1 + x_2 + 5x_3 = 1$$
$$4x_1 - x_2 - x_3 = 2$$
$$x_1 - 6x_2 + 2x_3 = 3$$

5. The SOR method converges if and only if $a < \omega < b$ for some numbers a and b which depend on the system of equations being solved. Find the values of a and b, correct to 1D, for the following system of equations:

$$5x_1 + 2x_2 - x_3 = 6$$
$$x_1 + 6x_2 - 3x_3 = 4$$
$$2x_1 + x_2 + 4x_3 = 7$$

Find the value of ω, correct to 1D, which produces the fastest rate of convergence for this system.

6. Use (2.30) to show that the SOR iteration matrix is

$$(D + \omega L)^{-1}[(1 - \omega)D - \omega U]$$

3
Non-linear
algebraic
equations

One of the fundamental problems of mathematics is that of solving equations of the form

$$f(x) = 0 \qquad (3.1)$$

where f is a real valued function of a real variable. Any number α satisfying (3.1) is called a *root* of the equation or a *zero* of f. If

$$f(x) = mx + d$$

for some given constants m and d, then f is said to be linear; otherwise it is said to be non-linear. Most equations arising in practice are non-linear and are rarely of a form which allows the roots to be determined exactly. Consequently, numerical techniques must be used to find them. These techniques can be divided into two categories; two-point and one-point methods. At each stage of a two-point method, a pair of numbers is calculated defining an interval within which a root is known to lie. The length of this interval generally decreases as the method proceeds, enabling the root to be calculated to any prescribed accuracy. Examples are the bisection and false position methods. One-point methods, on the other hand, generate a single sequence of numbers which converge to one of the roots. Examples are the fixed point, Newton and secant methods.

All of the methods described in this chapter are applicable to general non-linear functions. Special techniques do, however, exist for certain restricted classes of functions, such as polynomials. Some of these are described in the books by Householder (1970) and Ralston (1965).

3.1 BRACKETING METHODS

All of the numerical techniques described in this chapter can be used to find a root α of equation (3.1) to any specified accuracy. They start with an initial estimate of α and iteratively improve this approximation until the required accuracy is obtained. Information about the precise number of roots and their approximate location can be obtained by sketching a graph of f and finding the points where it crosses the x-axis. However, it is sometimes more convenient to rearrange equation (3.1) as

$$f_1(x) = f_2(x)$$

for two functions f_1 and f_2 whose graphs are easier to sketch than that of f. The roots of the original equation are then given by the points where the graphs of f_1 and f_2 intersect. For example, the equation

$$(x+1)^2 e^{(x^2-2)} - 1 = 0 \tag{3.2}$$

can be rearranged as

$$(x+1)^2 = e^{(2-x^2)}$$

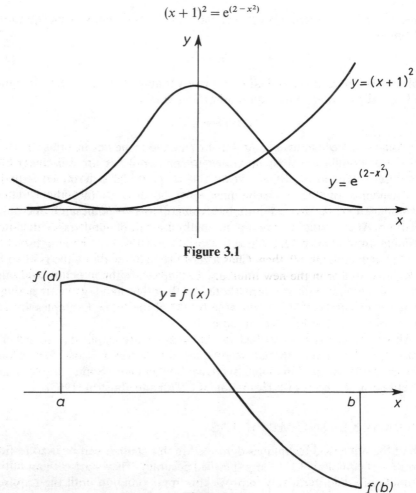

Figure 3.1

Figure 3.2

Table 3.1

x	-3	-2	-1	0	1	2	3
$f(x)$	4385.5	6.4	-1.0	-0.9	0.5	65.5	17 545.1

It is clear from Fig. 3.1 that the graphs of $y = (x + 1)^2$ and $y = e^{(2-x^2)}$ intersect twice. Therefore, equation (3.2) has exactly two roots.

Suppose that f is defined and continuous on an interval $[a, b]$. If the values of $f(x)$ at $x = a$ and $x = b$ are of opposite sign then it is clear from Fig. 3.2 that f must have at least one zero between a and b. This result, known as the intermediate value theorem, provides a simple and effective way of finding the approximate location of the zeros of f. As an example, consider the function

$$f(x) = (x + 1)^2 e^{(x^2-2)} - 1$$

Values of $f(x)$ for $x = -3, -2, \ldots, 3$ are given in Table 3.1, rounded to 1D. In this case, $f(-2) > 0$, $f(-1) < 0$, $f(0) < 0$ and $f(1) > 0$. The function therefore has zeros in the intervals $[-2, -1]$ and $[0, 1]$.

Any interval known to contain one or more roots of equation (3.1) is referred to as a bracketing interval. We now describe two methods, known as bisection and false position, which reduce the length of the bracketing interval in a systematic way.

Consider a bracketing interval $[a, b]$ for which $f(a)f(b) < 0$. In the *bisection method* the value of f at the mid-point, $c = (a + b)/2$, is calculated. There are then three possibilities. Firstly, it could happen (although it is very unlikely) that $f(c) = 0$. In this case, c is a zero of f and no further action is necessary. Secondly, if $f(a)f(c) < 0$, then f has a zero between a and c. The process can then be repeated on the new interval $[a, c]$. Finally, if $f(a)f(c) > 0$ it follows that $f(b)f(c) < 0$ since it is known that $f(a)$ and $f(b)$ have opposite signs. Hence, f has a zero between c and b and the process can be repeated with $[c, b]$. After one step of the algorithm, we have found either a zero or a new bracketing interval which is precisely half the length of the original one. The process continues until the length of the bracketing interval is less than a prescribed tolerance. This method is shown graphically in Fig. 3.3.

■ **Example 3.1**
To illustrate the bisection method, consider the calculation of the positive root of the equation

$$f(x) = (x + 1)^2 e^{(x^2-2)} - 1 = 0$$

starting with the bracketing interval $[0, 1]$. For this equation, $f(0) < 0$ and $f(1) > 0$. The mid-point of the interval is $c = 0.5$ with $f(c) = -0.609\,009$ (6D). Since $f(0.5)f(1) < 0$, we deduce that the root lies in $[0.5, 1]$.

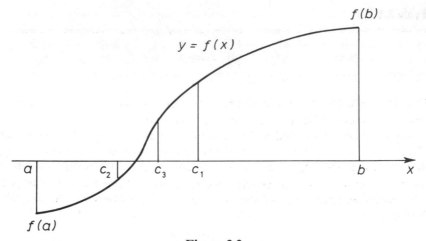

Figure 3.3

Table 3.2

n	a	b	c	$f(c)$
1	0	1	0.5	− 0.609 009
2	0.5	1	0.75	− 0.272 592
3	0.75	1	0.875	0.023 105
4	0.75	0.875	0.812 5	− 0.139 662
5	0.812 5	0.875	0.843 75	− 0.062 448
6	0.843 75	0.875	0.859 375	− 0.020 775
7	0.859 375	0.875	0.867 188	0.000 883
8	0.859 375	0.867 188	0.863 282	− 0.010 015
9	0.863 282	0.867 188	0.865 235	− 0.004 584
10	0.865 235	0.867 188	0.866 212	− 0.001 854
11	0.866 212	0.867 188	0.866 700	− 0.000 487
12	0.866 700	0.867 188	0.866 944	0.000 198
13	0.866 700	0.866 944	0.866 822	− 0.000 145
14	0.866 822	0.866 944	0.866 883	0.000 027
15	0.866 822	0.866 883	0.866 853	− 0.000 058
16	0.866 853	0.866 883	0.866 868	− 0.000 016
17	0.866 868	0.866 883	0.866 876	0.000 007

This and further steps of the bisection method are shown in Table 3.2. After 17 bisections the bracketing interval is [0.866 868, 0.866 876]. The value of α can, therefore, be quoted as

$$0.866\,872 \pm 0.000\,004 \qquad \square$$

It is interesting to notice that the number of bisections needed to obtain any given accuracy can be predicted in advance. Suppose that we wish to calculate a root to within $\pm(1/2) \times 10^{-k}$. Taking as our approximation the mid-point of a bracketing interval of length 10^{-k} ensures that the magnitude of the error is at most $(1/2) \times 10^{-k}$. If the initial interval is $[a, b]$ then after n steps the bracketing interval has length $(b - a)/2^n$. Hence we require

$$(b - a)/2^n < 10^{-k}$$

i.e.

$$2^n > 10^k(b - a)$$

Taking logs gives

$$n \log_{10} 2 > \log_{10}[10^k(b - a)]$$

i.e.

$$n > \frac{\log_{10}[10^k(b - a)]}{\log_{10} 2}$$

For example, when $a = 0$, $b = 1$ and $k = 5$ this inequality becomes

$$n > \frac{\log_{10}[10^5(1 - 0)]}{\log_{10} 2} = \frac{5}{\log_{10} 2} = 16.6$$

confirming that 17 iterations are required to estimate the root in Example 3.1 to within $\pm(1/2) \times 10^{-5}$.

One drawback of the bisection method is that it finds successive values of c using only the signs of $f(a)$ and $f(b)$ and does not reflect their relative magnitudes. For example, if $f(a) = 5000$ and $f(b) = -0.1$, it clearly makes sense to choose a value of c closer to b than to a. The *false position method* provides a way of doing this. Instead of simply halving the interval $[a, b]$,

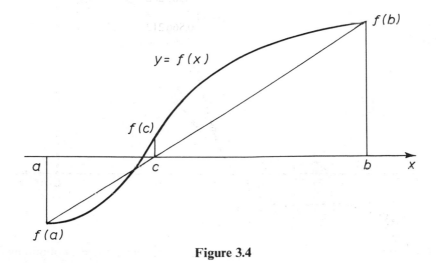

Figure 3.4

we take the point c to be the intersection of the chord joining $(a, f(a))$ and $(b, f(b))$ with the x-axis. This is illustrated in Fig. 3.4. The general equation of a straight line is

$$y = mx + d$$

and the condition that it passes through the points $(a, f(a))$ and $(b, f(b))$ leads to

$$m = \frac{f(b) - f(a)}{b - a}, \qquad d = \frac{af(b) - bf(a)}{a - b}$$

The value of y is zero at the point

$$c = -\frac{d}{m} = \frac{af(b) - bf(a)}{f(b) - f(a)}$$

Once c has been calculated from this formula, the process is repeated with the interval $[a, c]$ if $f(a)f(c) < 0$ and $[c, b]$ if $f(b)f(c) < 0$. The method is shown graphically in Fig. 3.5.

■ Example 3.2

We illustrate the false position method by solving the problem given in Example 3.1. If $a = 0$ and $b = 1$ then

$$c = \frac{0f(1) - 1f(0)}{f(1) - f(0)} = 0.647\,116$$

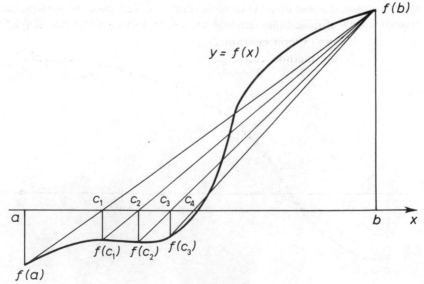

Figure 3.5

Table 3.3

n	a	b	c	$f(c)$
1	0	1	0.647 116	− 0.441 884
2	0.647 116	1	0.817 834	− 0.127 032
3	0.817 834	1	0.856 496	− 0.028 616
4	0.856 496	1	0.864 707	− 0.006 056
5	0.864 707	1	0.866 423	− 0.001 263
6	0.866 422	1	0.866 780	− 0.000 262
7	0.866 780	1	0.866 854	− 0.000 055
8	0.866 854	1	0.866 870	− 0.000 010
9	0.866 870	1	0.866 873	− 0.000 002

Now $f(c) = -0.441 884$, so $f(1)f(0.647 116) < 0$. Therefore, the root is bracketed by [0.647 116, 1]. This and further steps of the false position method are shown in Table 3.3. □

Although this method is ostensibly very similar to bisection, there are significant differences. In particular, the results of Example 3.2 show that the length of the bracketing interval does not necessarily tend to the limit zero. For this problem, the right hand end of the interval is fixed at $x = 1$ and the left hand end converges to the root $\alpha = 0.866 87$ (5D). A general stopping criterion which guarantees that a zero will be found to a prescribed number of decimal places is difficult to state. The method may be stopped when the magnitude of either the function value $f(c)$ or the difference between two successive values of c is sufficiently small. However, it must be emphasized that neither of these criteria guarantees the accuracy of the calculated zero. In Example 3.2 the method converged in only 9 steps using the stopping criterion $|f(c)| < (1/2) \times 10^{-5}$. Therefore, for this particular problem, the false position method converges much faster than bisection. However, this is not always the case; the relative performance of the two methods depends crucially on the shape of the graph of f. An example for which the false position method is actually slower than bisection can be found in Exercise 3.1.

EXERCISE 3.1

1. Show that the following equations have exactly one root and that in each case the root lies in the interval $[1/2, 1]$:

   (a) $x - \cos x = 0$
   (b) $x^2 + \ln x = 0$
   (c) $xe^x - 1 = 0$

Hence find the roots, to within $\pm(1/2) \times 10^{-2}$, using the bisection method.

2. Find the roots of the equations given in Question 1 using the false position method. Start with the bracketing interval $[1/2, 1]$ and stop when $|f(c)| < (1/2) \times 10^{-2}$.

3. Consider the bracketing interval $[a, b]$ for which $f(a) f(b) < 0$. If a number c is calculated using the false position method, show that the length of the new bracketing interval is

$$\frac{f(a)}{f(b) - f(a)}(a - b) \qquad \text{if} \quad f(a) f(c) < 0$$

and

$$\frac{f(b)}{f(b) - f(a)}(b - a) \qquad \text{if} \quad f(b) f(c) < 0$$

4. Find the root of the equation

$$f(x) = xe^x - 1 = 0$$

using the false position method. Start with the bracketing intervals (a) $[0, 2]$ (b) $[0, 3]$, and stop when $|f(c)| < (1/2) \times 10^{-1}$.

Write down (without performing any detailed calculations) the number of bisections required to find this root to within $\pm(1/2) \times 10^{-1}$ using the same starting intervals.

By sketching a graph of f, or otherwise, explain the relatively slow convergence of the false position method.

3.2 FIXED POINT ITERATION

The techniques described in the previous section depend on our ability to find an initial bracketing interval. This can be difficult for some equations and in these circumstances it is desirable to use a method which does not require any starting information. One such method is based on fixed point iteration.

Suppose that the equation $f(x) = 0$ can be rearranged as

$$x = g(x) \tag{3.3}$$

Any solution of this equation is called a fixed point of g. An obvious iteration to try for the calculation of fixed points is

$$x_{n+1} = g(x_n) \qquad n = 0, 1, 2, \ldots \tag{3.4}$$

The value of x_0 is chosen arbitrarily and the hope is that the sequence $x_0, x_1, x_2, \ldots$ converges to a number α which will automatically satisfy (3.3).

Moreover, since (3.3) is a rearrangement of (3.1), α is guaranteed to be a zero of f.

In general, there are many different ways of rearranging $f(x) = 0$ in the form (3.3). However, only some of these are likely to give rise to successful iterations, as the following example demonstrates.

■ **Example 3.3**

Consider the quadratic equation

$$x^2 - 2x - 8 = 0$$

with roots -2 and 4. Three possible rearrangements of this equation are

(a)
$$x = \sqrt{(2x + 8)}$$

(b)
$$x = \frac{2x + 8}{x}$$

(c)
$$x = \frac{x^2 - 8}{2}$$

Numerical results for the corresponding iterations, starting with $x_0 = 5$, are given in Table 3.4. The sequence converges for (a) and (b), but diverges for (c).   □

This example highlights the need for a mathematical analysis of the method. Sufficient conditions for the convergence of the fixed point iteration are given in the following theorem.

Table 3.4

n	(a) $x_{n+1} = \sqrt{(2x_n + 8)}$	(b) $x_{n+1} = (2x_n + 8)/x_n$	(c) $x_{n+1} = (x_n^2 - 8)/2$
0	5	5	5
1	4.243	3.600	8.5
2	4.060	4.222	32.125
3	4.015	3.895	512.008
4	4.004	4.054	
5	4.001	3.973	
6	4.000	4.014	
7		3.993	
8		4.004	
9		3 998	
10		4.001	
11		4.000	

■ Theorem 3.1

If g' exists on an interval $I = [\alpha - A, \alpha + A]$ containing the starting value x_0 and fixed point α, then x_n converges to α provided

$$|g'(x)| \leqslant K \qquad \text{on } I$$

for some constant $K < 1$.

■ Proof

Suppose that $x_n \in I$. Since α is a fixed point, $\alpha = g(\alpha)$. Subtraction of this equation from (3.4) gives

$$x_{n+1} - \alpha = g(x_n) - g(\alpha)$$

By the mean value theorem, there exists a number ξ_n between α and x_n such that

$$\frac{g(x_n) - g(\alpha)}{x_n - \alpha} = g'(\xi_n)$$

and so

$$x_{n+1} - \alpha = g'(\xi_n)(x_n - \alpha) \tag{3.5}$$

Now $\xi_n \in I$, since α and x_n are both in I. Hence

$$|x_{n+1} - \alpha| \leqslant K|x_n - \alpha| \tag{3.6}$$

since, by hypothesis, $|g'(x)| \leqslant K$ for any $x \in I$. It follows that $x_{n+1} \in I$ and is closer to α than x_n, because $K < 1$.

Now we are given that $x_0 \in I$, and so all subsequent iterates are contained in I with

$$|x_n - \alpha| \leqslant K|x_{n-1} - \alpha| \leqslant K^2|x_{n-2} - \alpha| \leqslant \cdots \leqslant K^n|x_0 - \alpha|$$

Finally, note that $K^n \to 0$ as $n \to \infty$, since $K < 1$. Hence $x_n \to \alpha$ as required.   □

If the iteration produces a convergent sequence then $\xi_n \to \alpha$, since ξ_n is squashed between α and x_n. From equation (3.5) it follows that

$$x_{n+1} - \alpha \approx g'(\alpha)(x_n - \alpha) \tag{3.7}$$

for sufficiently large n. This shows that, when $g'(\alpha) \neq 0$, the error in x_{n+1} is ultimately proportional to the error in the previous iterate x_n. Since the constant of proportionality is $g'(\alpha)$ we deduce that the rate of convergence depends on the value of $|g'(\alpha)|$; the smaller it is, the faster the convergence. Such a sequence is said to converge linearly, and the corresponding numerical method is said to be first order. In general, if a numerical method produces

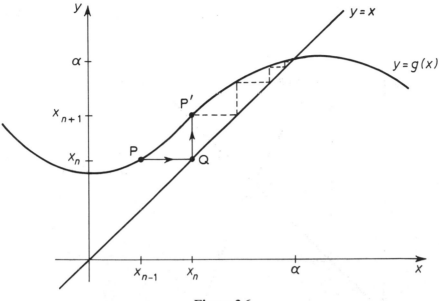

Figure 3.6

a sequence $x_0, x_1, \ldots$ converging to α, it is called pth order whenever

$$x_{n+1} - \alpha \approx C(x_n - \alpha)^p \qquad (3.8)$$

for some constant C and sufficiently large n. More rapid convergence is obtained for methods of higher order. We shall derive two such methods in the next section.

A graphical interpretation of the fixed point iteration is given in Fig. 3.6, which shows the graphs of $y = x$ and $y = g(x)$ intersecting at a fixed point α. The point P has coordinates (x_{n-1}, x_n) and lies on the curve $y = g(x)$. To calculate x_{n+1} it is necessary to evaluate $g(x)$ at $x = x_n$. A point Q which has an x-coordinate x_n can easily be constructed by sliding P along a line parallel to the x-axis until it intersects the line $y = x$. This point has co-ordinates (x_n, x_n). Finally, to determine $g(x_n)$ we slide Q along a line parallel to the y-axis until it intersects the curve $y = g(x)$. This gives a new point P′ with coordinates $(x_n, g(x_n)) = (x_n, x_{n+1})$. The whole process is then repeated. It is possible to use this graphical approach to prove Theorem 3.1. Figure 3.7(a)–(d) shows the behaviour of x_n in four cases: $g'(x) < -1$; $-1 < g'(x) \leqslant 0$; $0 \leqslant g'(x) < 1$; and $g'(x) > 1$. It is clear that x_n converges in graphs (b) and (c) for which $|g'(x)| < 1$, but diverges in graphs (a) and (d) for which $|g'(x)| > 1$.

■ Example 3.4
We are now in a position to explain the results of Example 3.3.

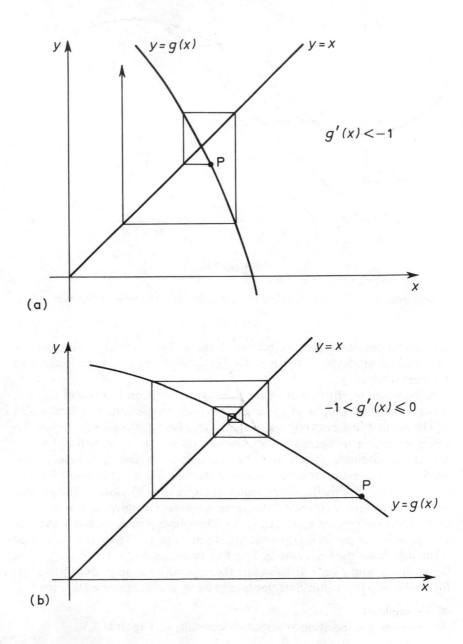

(a)

$$y = g(x)$$

$$y = x$$

$$g'(x) < -1$$

P

(b)

$$y = x$$

$$-1 < g'(x) \leqslant 0$$

P

$$y = g(x)$$

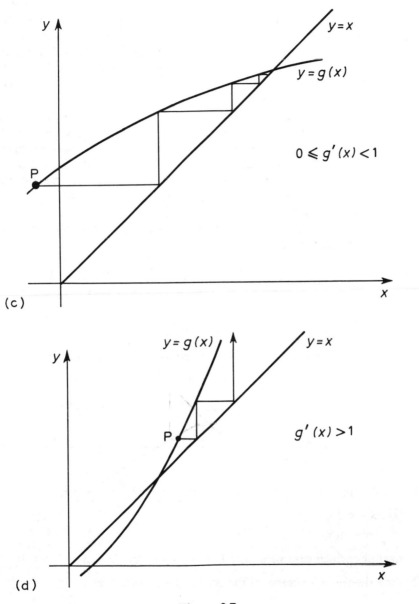

$y = x$

$y = g(x)$

$0 \leqslant g'(x) < 1$

P

(c)

x

$y = g(x)$

$y = x$

$g'(x) > 1$

P

(d)

x

Figure 3.7

(a) If $g(x) = (2x + 8)^{1/2}$ then $g'(x) = (2x + 8)^{-1/2}$

Theorem 3.1 guarantees convergence to the positive root $\alpha = 4$, because $|g'(x)| < 1$ on the interval $I = [3, 5] = [\alpha - 1, \alpha + 1]$ containing the starting value $x_0 = 5$. Moreover, since $g'(x) > 0$ on I, equation (3.5) shows that the errors in two consecutive iterates are of the same sign and so convergence is monotonic. This is also apparent from Fig. 3.7(c). Finally, note that since $|g'(4)| = 1/4$ the errors are ultimately reduced by a factor of 1/4 at each stage, which is in agreement with the results of Table 3.4(a).

(b) If $g(x) = (2x + 8)/x$ then $g'(x) = -8/x^2$

Theorem 3.1 guarantees convergence to $\alpha = 4$ because $|g'(x)| < 1$ on $I = [3, 5]$. Moreover, since $g'(x) < 0$ on I, equation (3.5) shows that the errors in two consecutive iterates are of opposite sign and so convergence is oscillatory. This is also apparent from Fig. 3.7(b). Finally, note that since $|g'(4)| = 1/2$ the errors are ultimately reduced by a factor of 1/2 at each stage, which is in agreement with the results of Table 3.4(b).

(c) If $g(x) = (x^2 - 8)/2$ then $g'(x) = x$

Since $|g'(4)| > 1$, Theorem 3.1 cannot be used to guarantee convergence. Moreover, from equation (3.5)

$$x_1 - \alpha = \xi_0(x_0 - \alpha) \geqslant 4(x_0 - \alpha)$$

because $\alpha = 4$, $x_0 = 5$ and ξ_0 lies between α and x_0. Hence $x_1 \geqslant 8$. Similarly,

$$x_2 - \alpha = \xi_1(x_1 - \alpha) \geqslant 4(x_1 - \alpha) \geqslant 4^2(x_0 - \alpha)$$

because ξ_1 lies between α and x_1. Continuing in this way we obtain

$$x_n - \alpha \geqslant 4^n(x_0 - \alpha)$$

showing the rapid divergence of x_n. This is also apparent from Fig. 3.7(d).   □

EXERCISE 3.2

1. Use an appropriate fixed point iteration to find the root of

(a) $x - \cos x = 0$

(b) $x^2 + \ln x = 0$

starting in each case with $x_0 = 1$. Stop when $|x_{n+1} - x_n| < (1/2) \times 10^{-2}$.

2. Find the first nine terms of the sequence generated by

$$x_{n+1} = e^{-x_n}$$

starting with $x_0 = 1$.
Explain why this iteration converges for any starting value x_0.

3. Assuming that the sequences generated by

(a) $x_{n+1} = \dfrac{x_n^2 + 2}{3}$

(b) $x_{n+1} = 3 - \dfrac{2}{x_n}$

both converge, show that they do so to different roots of the same equation.

4. Show by curve sketching that the equation

$$5x - e^x = 0$$

has two roots.

By constructing appropriate fixed point iterations, calculate these roots starting with $x_0 = 2$. Stop when $|x_{n+1} - x_n| < (1/2) \times 10^{-2}$.

5. Suppose that the iteration $x_{n+1} = g(x_n)$ produces a sequence of numbers converging to a fixed point α. If g'' exists, use Taylor's theorem to show that

$$g(x_n) = g(\alpha) + (x_n - \alpha)g'(\alpha) + \frac{(x_n - \alpha)^2}{2} g''(\xi_n)$$

for some number ξ_n between α and x_n.

Deduce that the fixed point iteration is a second order process whenever $g'(\alpha) = 0$ and $g''(\alpha) \neq 0$.

6. Suppose that the equation $f(x) = 0$ can be rearranged as $x = h(x)$ for some function h. Show that if, for some non-zero value of a parameter ω, the sequence generated by

$$x_{n+1} = x_n + \omega[h(x_n) - x_n]$$

converges to a number α, then α must be a fixed point of h.

Taking $g(x) = x + \omega[h(x) - x]$, use the result of Question 5 to deduce that the above iteration is second order if

$$\omega = \frac{1}{1 - h'(\alpha)} \qquad h'(\alpha) \neq 1$$

Approximating ω by $1/[1 - h'(x_0)]$, use this method to find the root of $x - \cos x = 0$, starting with $x_0 = 1$. Stop when $|x_{n+1} - x_n| < (1/2) \times 10^{-2}$, and compare the number of iterations with those of Question 1.

3.3 NEWTON'S METHOD

Newton's method, also called the Newton–Raphson method, is one of the most popular techniques for finding roots of non-linear equations. Suppose

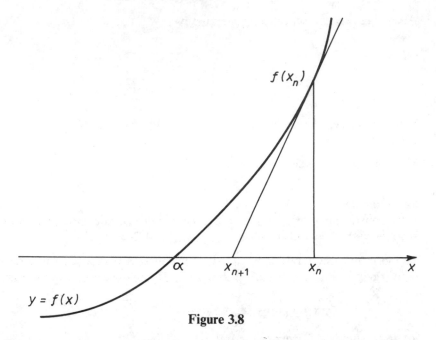

$f(x_n)$

α

x_{n+1}

x_n

x

$y = f(x)$

Figure 3.8

that x_n is a known approximation to a root α of $f(x) = 0$, shown in Fig. 3.8. The next approximation, x_{n+1}, is taken to be the point where the tangent to the graph of $y = f(x)$ at $x = x_n$ intersects the x-axis. A tangent is a straight line, so its equation is of the form $y = mx + d$. The condition that it has slope $f'(x_n)$ and passes through the point $(x_n, f(x_n))$ leads to

$$m = f'(x_n), \qquad d = f(x_n) - x_n f'(x_n)$$

The value of y is zero when x is $-d/m$, so Newton's method is defined by

$$x_{n+1} = x_n - \frac{f(x_n)}{f'(x_n)} \tag{3.9}$$

Note that when $f'(x_n) = 0$ the calculation of x_{n+1} fails. This is because the tangent at x_n is horizontal.

■ **Example 3.5**
Newton's method for calculating the zeros of

$$f(x) = e^x - x - 2$$

is given by

$$x_{n+1} = x_n - \frac{e^{x_n} - x_n - 2}{e^{x_n} - 1}$$

$$= \frac{e^{x_n}(x_n - 1) + 2}{e^{x_n} - 1}$$

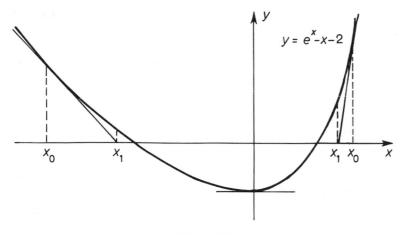

Figure 3.9

The graph of f, sketched in Fig. 3.9, shows that it has two zeros. It is clear from this graph that x_n converges to the negative root if $x_0 < 0$ and to the positive root if $x_0 > 0$, and that it breaks down if $x_0 = 0$. The results obtained with $x_0 = -10$ and $x_0 = 10$ are listed in Table 3.5.

If x_0 is a large negative number, then the first few iterates converge quickly; in this region $f(x) \approx -x - 2$ and the tangents are very good approximations to f. On the other hand, if x_0 is a large positive number, then the first few iterates converge slowly; in this region $f(x) \approx e^x$, which has a large curvature,

Table 3.5

n	x_n	x_n
0	-10	10
1	$-1.999\,591$	9.000\,499
2	$-1.843\,472$	8.001\,733
3	$-1.841\,406$	7.004\,749
4	$-1.841\,406$	6.012\,020
5		5.029\,236
6		4.068\,950
7		3.157\,113
8		2.341\,852
9		1.697\,350
10		1.302\,210
11		1.162\,073
12		1.146\,376
13		1.146\,193
14		1.146\,193

and the tangents are very poor approximations to f. In both cases, once x_n is reasonably close to a root, convergence is extremely fast.   □

Sufficient conditions for the convergence of Newton's method are given in the following theorem.

■ **Theorem 3.2**

If f'' is continuous on an interval $[\alpha - A, \alpha + A]$, then x_n converges to α provided $f'(\alpha) \neq 0$ and x_0 is sufficiently close to α.

■ **Proof**

Comparison of equations (3.4) and (3.9) shows that Newton's method is a fixed point iteration with

$$g(x) = x - \frac{f(x)}{f'(x)}$$

By the quotient rule,

$$g'(x) = 1 - \frac{f'(x)f'(x) - f(x)f''(x)}{[f'(x)]^2} = \frac{f(x)f''(x)}{[f'(x)]^2}$$

This implies that $g'(\alpha) = 0$, because $f(\alpha) = 0$ and $f'(\alpha) \neq 0$. Hence by the continuity of f'', there exists an interval $I = [\alpha - \delta, \alpha + \delta]$, for some $\delta > 0$, on which $|g'(x)| < 1$. Theorem 3.1 then guarantees convergence provided x_0 is contained in I, i.e. provided x_0 is sufficiently close to α.   □

Although Theorem 3.2 guarantees convergence, it gives us no information about the rate of convergence. (Equation (3.7) is of little use because $g'(\alpha) = 0$.) This can, however, be obtained using Taylor's theorem. Now $f(\alpha) = 0$, so writing $\alpha = x_n + (\alpha - x_n)$ we have

$$0 = f(x_n + (\alpha - x_n)) = f(x_n) + (\alpha - x_n)f'(x_n) + \frac{(\alpha - x_n)^2}{2!}f''(\xi_n)$$

for some ξ_n between α and x_n. This equation can be rearranged as

$$\left(x_n - \frac{f(x_n)}{f'(x_n)} \right) - \alpha = \frac{(\alpha - x_n)^2}{2!}\frac{f''(\xi_n)}{f'(x_n)}$$

i.e.

$$x_{n+1} - \alpha = \frac{(\alpha - x_n)^2}{2}\frac{f''(\xi_n)}{f'(x_n)}$$

If Newton's method converges then $\xi_n \to \alpha$, since ξ_n lies between α and x_n. Hence, provided $f'(\alpha) \neq 0$,

$$x_{n+1} - \alpha \approx C(x_n - \alpha)^2$$

for sufficiently large n, where $C = f''(\alpha)/2f'(\alpha)$. This shows that when $f''(\alpha) \neq 0$, the error in x_{n+1} is ultimately proportional to the square of the error in the previous iterate x_n. Newton's method is therefore a second order process and x_n is said to converge quadratically. A consequence of this property is that once x_n is reasonably close to α, the number of decimal places of accuracy roughly doubles at each step. For example, if $C = 1$ and the error in some iterate is 10^{-1}, then the errors in the next few iterates are approximately $(10^{-1})^2 = 10^{-2}$, $(10^{-2})^2 = 10^{-4}$, $(10^{-4})^2 = 10^{-8}$ and so on.

The high order convergence of Newton's method is the main reason why it is commonly used in practice. However, it does have the drawback that the derivative f' needs to be determined. This can be difficult if f is complicated and impossible if f is only defined implicitly. The calculation of $f'(x_n)$ may be avoided by approximating the slope of the tangent at $x = x_n$ by that of the chord joining the two points $(x_{n-1}, f(x_{n-1}))$ and $(x_n, f(x_n))$. This is illustrated in Fig. 3.10. The slope of the chord (or secant as it is sometimes called) is

$$\frac{f(x_n) - f(x_{n-1})}{x_n - x_{n-1}}$$

and if this expression is substituted for $f'(x_n)$ in equation (3.9), then

$$x_{n+1} = x_n - \frac{(x_n - x_{n-1})f(x_n)}{f(x_n) - f(x_{n-1})}$$

$$= \frac{x_{n-1}f(x_n) - x_n f(x_{n-1})}{f(x_n) - f(x_{n-1})} \tag{3.10}$$

Note that when $f(x_n) = f(x_{n-1})$ the calculation of x_{n+1} fails. This is because the chord is horizontal.

The iteration defined by (3.10), known as the *secant method*, needs two starting values x_0 and x_1. This method is very similar to the false position method described in Section 3.1. However, for the secant method it is not necessary for the interval $[x_0, x_1]$ to contain a root and no account is taken of the signs of the numbers $f(x_n)$. It can be shown (see Ostrowski, 1966) that the secant method has order $(1 + \sqrt{5})/2 = 1.618$ (3D), so its ultimate convergence is not quite as fast as Newton's method. In spite of this it does provide a viable alternative, since it only requires one additional function evaluation per step compared with the two evaluations ($f(x_n)$ and $f'(x_n)$) needed for Newton's method.

■ Example 3.6

We illustrate the secant method by finding the zeros of the function given in Example 3.5. The results obtained with $x_0 = -10$, $x_1 = -9$ and $x_0 = 10$, $x_1 = 9$ are listed in Table 3.6.

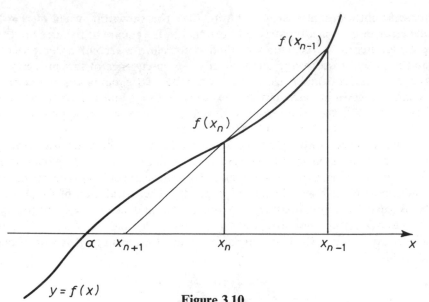

Figure 3.10

Table 3.6

n	x_n	x_n
0	-10	10
1	-9	9
2	$-1.999\,331$	8.418\,772
3	$-1.861\,918$	7.682\,967
4	$-1.841\,689$	7.008\,962
5	$-1.841\,407$	6.313\,655
6	$-1.841\,406$	5.630\,911
7	$-1.841\,406$	4.951\,111
8		4.283\,842
9		3.635\,290
10		3.018\,816
11		2.452\,933
12		1.962\,876
13		1.577\,321
14		1.319\,644
15		1.190\,394
16		1.151\,363
17		1.146\,357
18		1.146\,194
19		1.146\,193
20		1.146\,193

These results enable us to confirm the order of the secant method. Suppose that

$$x_{n+1} - \alpha = C(x_n - \alpha)^p$$

with $\alpha = 1.146\,193$. The third column of Table 3.6 gives

$$0.005\,170 = C(0.044\,201)^p$$
$$0.000\,164 = C(0.005\,170)^p$$

for $n = 15$ and 16, respectively. The constant C can be eliminated by dividing these two equations to get

$$31.524\,39 = (8.549\,52)^p$$

Taking logs gives

$$p = \frac{\log(31.524\,39)}{\log(8.549\,52)} = 1.608 \qquad (3D)$$

which is roughly in agreement with the exact value, 1.618 (3D). $\qquad\square$

EXERCISE 3.3

1. Sketch a graph of the function $f(x) = x^2 - 2$. For which ranges of x_0 will Newton's method converge to (a) the positive root (b) the negative root (c) not at all?
   Find $\sqrt{2}$, starting with $x_0 = 1$. Stop when $|x_{n+1} - x_n| < 10^{-6}$.

2. (a) Use Newton's method to find the roots of

   (i) $x - \cos x = 0$
   (ii) $x^2 + \ln x = 0$
   (iii) $x^3 + 4x^2 + 4x + 3 = 0$

   starting in each case with $x_0 = 1$. Stop when $|x_{n+1} - x_n| < 10^{-6}$.
   (b) Use the secant method to find the roots of the equations given in part (a) Start with $x_0 = 2$ and $x_1 = 1$ and stop when $|x_{n+1} - x_n| < 10^{-6}$.

3. Show that Newton's method for the solution of

   $$x^k e^x = 0$$

   is given by

   $$x_{n+1} = \frac{(k-1)x_n + x_n^2}{k + x_n}$$

   Starting with $x_0 = 1$, calculate x_5 in the case when (a) $k = 1$ (b) $k = 2$. Explain why one of the sequences converges much faster than the other.

4. One way of avoiding the calculation of $f'(x_n)$ at every step of Newton's method is to approximate $f'(x_n)$ by $f'(x_0)$ and so use the iteration defined by

$$x_{n+1} = x_n - \frac{f(x_n)}{f'(x_0)}$$

Give a graphical interpretation of this method and use it to find the root of

$$x - \cos x = 0$$

starting with $x_0 = 1$. Use the stopping criterion $|x_{n+1} - x_n| < 10^{-6}$, and compare the number of iterations with that of Question 2(a) (i).

5. Suppose that f''' exists, with $f(\alpha) = 0$, $f'(\alpha) \neq 0$ and $f''(\alpha) \neq 0$. Show that if

$$g(x) = x - \frac{f(x)}{f'(x)}$$

then (a) $g(\alpha) = \alpha$ (b) $g'(\alpha) = 0$ (c) $g''(\alpha) \neq 0$. Use the result of Exercise 3.2, Question 5 to deduce that Newton's method is second order.

6. Apply Newton's method to determine a zero, correct to 3D, of the polynomial $f(x) = -x^3 + 3x^2 - 2x$, starting from (a) $x_0 = 1.4$ (b) $x_0 = 1 + 1/\sqrt{5}$ (c) $x_0 = 1.5$ (d) $x_0 = 1.6$. Explain your results by examining the graph of f.

3.4 SYSTEMS OF NON-LINEAR EQUATIONS

In this section we consider the solution of systems of non-linear equations of the form

$$f_1(x_1, x_2, \ldots, x_m) = 0$$
$$f_2(x_1, x_2, \ldots, x_m) = 0$$
$$\vdots$$
$$f_m(x_1, x_2, \ldots, x_m) = 0$$

where each f_i $(i = 1, 2, \ldots, m)$ is a real valued function of m real variables. Although this topic is of considerable importance, we can do no more than give a brief introduction here. For an extensive treatment the interested reader is advised to consult the books by Ortega and Rheinboldt (1970) and Traub (1964).

The one-point iterative methods discussed in the previous two sections for the solution of a single equation may all be extended to systems of equations. However, we shall only consider the generalization of Newton's method, since this and methods based on it are most commonly used. In order to motivate the general case, consider a system of two non-linear simultaneous equations in two unknowns given by

$$f(x, y) = 0$$
$$g(x, y) = 0 \qquad (3.11)$$

Geometrically, the roots of this system are the points in the (x, y) plane where the curves defined by f and g intersect. For example, the curves represented by

$$f(x, y) = x^2 + y^2 - 4 = 0$$
$$g(x, y) = 2x - y^2 = 0$$

are shown in Fig. 3.11. The roots of this system are then (α_1, β_1) and (α_2, β_2).

Suppose that (x_n, y_n) is an approximation to a root (α, β). Writing $\alpha = x_n + (\alpha - x_n)$ and $\beta = y_n + (\beta - y_n)$ we can use Taylor's theorem for functions of two variables to deduce that

$$0 = f[x_n + (\alpha - x_n), y_n + (\beta - y_n)]$$
$$= f(x_n, y_n) + (\alpha - x_n)f_x(x_n, y_n) + (\beta - y_n)f_y(x_n, y_n) + \cdots$$

and

$$0 = g[x_n + (\alpha - x_n), y_n + (\beta - y_n)]$$
$$= g(x_n, y_n) + (\alpha - x_n)g_x(x_n, y_n) + (\beta - y_n)g_y(x_n, y_n) + \cdots$$

The notation f_x, f_y is used as an abbreviation for $\partial f/\partial x$, $\partial f/\partial y$, etc. If (x_n, y_n) is sufficiently close to (α, β) then higher order terms may be neglected to obtain

$$0 = f(x_n, y_n) + (\alpha - x_n)f_x(x_n, y_n) + (\beta - y_n)f_y(x_n, y_n)$$
$$0 = g(x_n, y_n) + (\alpha - x_n)g_x(x_n, y_n) + (\beta - y_n)g_y(x_n, y_n) \qquad (3.12)$$

This represents a system of two linear algebraic equations for α and β. Of course, since higher order terms are omitted in the derivation of these

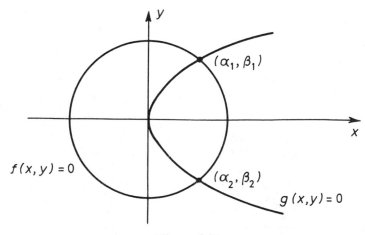

Figure 3.11

equations, their solution (α, β) is no longer an exact root of (3.11). However, it will usually be a better approximation than (x_n, y_n), so replacing (α, β) by (x_{n+1}, y_{n+1}) in (3.12) gives the iterative scheme

$$0 = f(x_n, y_n) + (x_{n+1} - x_n)f_x(x_n, y_n) + (y_{n+1} - y_n)f_y(x_n, y_n)$$
$$0 = g(x_n, y_n) + (x_{n+1} - x_n)g_x(x_n, y_n) + (y_{n+1} - y_n)g_y(x_n, y_n)$$
(3.13)

At a starting approximation (x_0, y_0), the functions f, f_x, f_y, g, g_x and g_y are evaluated. The linear equations are then solved for (x_1, y_1) and the whole process is repeated until convergence is obtained.

■ **Example 3.7**

As an illustration of the above, consider the solution of

$$x^2 + y^2 - 4 = 0$$
$$2x - y^2 = 0$$

starting with $x_0 = y_0 = 1$. In this case

$$f = x^2 + y^2 - 4, \qquad f_x = 2x, \qquad f_y = 2y$$
$$g = 2x - y^2, \qquad g_x = 2, \qquad g_y = -2y$$

At the point $(1, 1)$, equations (3.13) are given by

$$2(x_1 - 1) + 2(y_1 - 1) = 2$$
$$2(x_1 - 1) - 2(y_1 - 1) = -1$$

which have solution $x_1 = 1.25$ and $y_1 = 1.75$. This and further steps of the method are listed in Table 3.7. ☐

In matrix notation, (3.13) may be written as

$$\begin{pmatrix} f_x & f_y \\ g_x & g_y \end{pmatrix} \begin{pmatrix} x_{n+1} - x_n \\ y_{n+1} - y_n \end{pmatrix} = - \begin{pmatrix} f \\ g \end{pmatrix}$$

Table 3.7

n	x_n	y_n
0	1.000 000	1.000 000
1	1.250 000	1.750 000
2	1.236 111	1.581 349
3	1.236 068	1.572 329
4	1.236 068	1.572 303
5	1.236 068	1.572 303

where f, g and their partial derivatives are evaluated at (x_n, y_n). Hence

$$\begin{pmatrix} x_{n+1} \\ y_{n+1} \end{pmatrix} = \begin{pmatrix} x_n \\ y_n \end{pmatrix} - \begin{pmatrix} f_x & f_y \\ g_x & g_y \end{pmatrix}^{-1} \begin{pmatrix} f \\ g \end{pmatrix} \tag{3.14}$$

The matrix

$$J = \begin{pmatrix} f_x & f_y \\ g_x & g_y \end{pmatrix}$$

is called the *Jacobian matrix*. Comparison of equations (3.9) and (3.14) shows that the above procedure is indeed an extension of Newton's method in one variable, where division by f' generalizes to pre-multiplication by J^{-1}.

For a larger system of equations it is convenient to use vector notation. Consider a system

$$f(x) = 0$$

where $f = (f_1, f_2, \ldots, f_m)^{\mathrm{T}}$ and $x = (x_1, x_2, \ldots, x_m)^{\mathrm{T}}$. Denoting the nth iterate by $x^{[n]} = (x_1^{[n]}, x_2^{[n]}, \ldots, x_m^{[n]})^{\mathrm{T}}$, Newton's method is defined by

$$x^{[n+1]} = x^{[n]} - [J(x^{[n]})]^{-1} f(x^{[n]}) \tag{3.15}$$

where the Jacobian matrix J is

$$\begin{pmatrix} \dfrac{\partial f_1}{\partial x_1} & \dfrac{\partial f_1}{\partial x_2} & \cdots & \dfrac{\partial f_1}{\partial x_m} \\[2ex] \dfrac{\partial f_2}{\partial x_1} & \dfrac{\partial f_2}{\partial x_2} & \cdots & \dfrac{\partial f_2}{\partial x_m} \\[2ex] \vdots & \vdots & & \vdots \\[2ex] \dfrac{\partial f_m}{\partial x_1} & \dfrac{\partial f_m}{\partial x_2} & \cdots & \dfrac{\partial f_m}{\partial x_m} \end{pmatrix}$$

Although the iteration defined by (3.15) involves the inverse of J, in practice we do not attempt to find this explicitly. Instead, equation (3.15) is rearranged as

$$J(x^{[n]}) u^{[n]} = -f(x^{[n]}) \tag{3.16}$$

where $u^{[n]} = x^{[n+1]} - x^{[n]}$. This represents a system of linear equations for $u^{[n]}$ and can be solved using any of the methods described in the previous chapter, such as LU decomposition. Once $u^{[n]}$ has been found, the next iterate is calculated from

$$x^{[n+1]} = u^{[n]} + x^{[n]}$$

■ **Example 3.8**

As an illustration of the general case, consider the solution of

$$-2.0625x_1 + 2x_2 - 0.0625x_1^3 + 0.5 = 0$$
$$x_3 - 2x_2 + x_1 - 0.0625x_2^3 + 0.125x_2(x_3 - x_1) = 0$$
$$x_4 - 2x_3 + x_2 - 0.0625x_3^3 + 0.125x_3(x_4 - x_2) = 0$$
$$-1.9375x_4 + x_3 - 0.0625x_4^3 - 0.125x_3x_4 + 0.5 = 0$$

starting with $x^{[0]} = (1.5, 1.25, 1.0, 0.75)^T$. The Jacobian matrix for this system is

$$\begin{pmatrix} -2.0625 - 0.1875x_1^2 & 2 & 0 & 0 \\ 1 - 0.125x_2 & -2 - 0.1875x_2^2 + 0.125(x_3 - x_1) & 1 + 0.125x_2 & 0 \\ 0 & 1 - 0.125x_3 & -2 - 0.1875x_3^2 + 0.125(x_4 - x_2) & 1 + 0.125x_3 \\ 0 & 0 & 1 - 0.125x_4 & -1.9375 - 0.1875x_4^2 - 0.125x_3 \end{pmatrix}$$

Taking $n = 0$ in (3.16) we have

$$\begin{pmatrix} -2.484\,375 & 2 & 0 & 0 \\ 0.843\,750 & -2.355\,469 & 1.156\,250 & 0 \\ 0 & 0.875\,000 & -2.250\,000 & 1.125\,000 \\ 0 & 0 & 0.906\,250 & -2.167\,969 \end{pmatrix} \begin{pmatrix} u_1^{[0]} \\ u_2^{[0]} \\ u_3^{[0]} \\ u_4^{[0]} \end{pmatrix}$$

$$= \begin{pmatrix} 0.304\,688 \\ 0.200\,195 \\ 0.125\,000 \\ 0.073\,242 \end{pmatrix}$$

which has solution

$$u^{[0]} = (-0.420\,290, -0.369\,735, -0.273\,370, -0.148\,057)^T$$

Hence

$$x^{[1]} = u^{[0]} + x^{[0]} = (1.079\,710, 0.880\,265, 0.726\,630, 0.601\,943)^T$$

This and further steps of the method are given in Table 3.8.   □

Note the rapid convergence in Example 3.8. It can be shown (see Ortega, 1972) that provided $x^{[0]}$ is sufficiently close to a root α and $J(\alpha)$ is non-singular, then $x^{[n]}$ converges to α quadratically.

However, there are significant disadvantages associated with Newton's method for several variables. All of the m^2 partial derivatives of f have to be found, which may be difficult if f is complicated. Moreover, it can be costly to implement, since it is necessary to perform $m^2 + m$ function evaluations and to

Table 3.8

n	$x_1^{[n]}$	$x_2^{[n]}$	$x_3^{[n]}$	$x_4^{[n]}$
0	1.5	1.25	1	0.75
1	1.079 710	0.880 265	0.726 630	0.601 943
2	0.983 468	0.793 018	0.663 840	0.570 456
3	0.978 921	0.788 825	0.660 774	0.568 916
4	0.978 911	0.788 816	0.660 767	0.568 913
5	0.978 911	0.788 816	0.660 767	0.568 913

solve a system of linear equations at each step. One modification that is often used is to update the Jacobian matrix every kth iteration, for some $k > 1$. The extreme case of this is to evaluate J once only at $x^{[0]}$ and to use the iteration

$$x^{[n+1]} = x^{[n]} - [J(x^{[0]})]^{-1}f(x^{[n]})$$

but more usually a value of, say, $k = 3$ is used. There is a considerable saving in computation with methods of this type since the number of function evaluations is greatly reduced. Furthermore, if the LU decomposition method (see Section 2.2) is used to solve (3.16) then the triangular decomposition itself need only be calculated every kth step. It is possible to show (see Isaacson and Keller, 1966) that convergence is linear for the case given above in which J is evaluated once only. Other values of k define methods which are pth order for some $p \in [1, 2]$.

EXERCISE 3.4

1. Show that the system

$$3x^2 + y^2 + 9x - y - 12 = 0$$
$$x^2 + 36y^2 - 36 = 0$$

has exactly four roots. Find these roots using Newton's method starting with $(1, 1)$, $(1, -1)$, $(-4, 1)$ and $(-4, -1)$. Stop when successive iterates differ by less than 10^{-7}.

2. Show that the system

$$4x^3 + y - 6 = 0$$
$$x^2 y - 1 = 0$$

has exactly three roots. Find these roots using Newton's method starting with $(1, 1)$, $(0.5, 5)$ and $(-1, 5)$. Stop when successive iterates differ by less than 10^{-7}.

3. Solve the equations given in Example 3.8 by modifying Newton's method so that the Jacobian matrix is

(a) Updated every second iteration
(b) Updated every fifth iteration
(c) Evaluated once only at $x = x^{[0]}$.

Take $x^{[0]} = (1.5, 1.25, 1.0, 0.75)^T$ and stop when $|x_i^{[n+1]} - x_i^{[n]}| < 10^{-7}$ for all i. Compare the number of iterations required for convergence.

4. Show that Newton's method converges in exactly one iteration when applied to a system of linear equations

$$a_{11}x_1 + a_{12}x_2 + \cdots + a_{1m}x_m - b_1 = 0$$
$$a_{21}x_1 + a_{22}x_2 + \cdots + a_{2m}x_m - b_2 = 0$$
$$\vdots \qquad\qquad\qquad\qquad \vdots$$
$$a_{m1}x_1 + a_{m2}x_2 + \cdots + a_{mm}x_m - b_m = 0$$

5. The stationary points of a function of two variables, $F(x, y)$, are the roots of the equations

$$\frac{\partial F}{\partial x} = 0, \qquad \frac{\partial F}{\partial y} = 0$$

Show that, when applied to this system, Newton's method is given by

$$x^{[n+1]} = x^{[n]} - [G(x^{[n]})]^{-1} g(x^{[n]})$$

where G is the (Hessian) matrix

$$\begin{pmatrix} F_{xx} & F_{xy} \\ F_{xy} & F_{yy} \end{pmatrix}$$

and g is the (gradient) vector $(F_x, F_y)^T$.

Use Newton's method to find the stationary points of the function

$$F = 2x - y - \frac{x^2}{2} + \frac{x^3}{y}$$

Experiment with different starting values.

Show that, for this particular problem, Newton's method converges in exactly one iteration whenever $x^{[0]} = y^{[0]}$.

4

Eigenvalues

and

eigenvectors

In this chapter we describe numerical techniques for the calculation of a scalar λ and non-zero vector x in the equation

$$Ax = \lambda x \tag{4.1}$$

where A is a given $n \times n$ matrix. The quantities λ and x are usually referred to as an eigenvalue and an eigenvector of A. They arise in many different branches of mathematics including quadratic forms, differential systems and non-linear optimization, and can be used to solve problems from such diverse fields as economics, information theory, structural analysis, electronics and control theory.

The classical method of finding the eigenvalues of A is to estimate the roots of the characteristic equation

$$\det(A - \lambda I) = 0$$

where I is the $n \times n$ identity matrix. The eigenvectors are then determined by setting one of the non-zero elements of x to unity and calculating the remaining elements by equating coefficients in relation (4.1). Unfortunately, if $n > 3$ it is very time consuming to expand the determinant and the process may well suffer from induced instability. Once the characteristic equation has been found we still need to estimate its roots, which is a non-trivial task. Also, even if λ is a highly accurate approximation to an eigenvalue, the vector x obtained via (4.1) can be very different from the exact eigenvector.

Some numerical techniques for the calculation of eigenvalues can only be applied to symmetric matrices. Within this category are the methods of Jacobi, Givens and Householder which reduce a given symmetric matrix to a special form whose eigenvalues are readily computed. The analogous methods for unsymmetric matrices are the LR and QR methods, which are amongst the most widely used techniques for solving problems of this type. All of these procedures make use of a series of similarity transformations and, as such, are usually referred to as transformation or direct methods.

Another class of techniques, called iterative methods, can be used to find some or all of the eigenvalues and eigenvectors of a matrix. They start with an arbitrary approximation to one of the eigenvectors and successively improve this until the required accuracy is obtained. Included in this class is the powerful method of inverse iteration, which is used to find all of the eigenvectors of a matrix from known approximations to its eigenvalues.

An extensive coverage of these techniques can be found in the books by Wilkinson (1965) and Wilkinson and Reinsch (1971). They provide a description and analysis of most of the methods which form the basis of sophisticated library routines in use today. For a more recent, theoretical treatment of the general topic of numerical linear algebra, the reader is advised to consult the book by Golub and Van Loan (1983).

4.1 THE POWER METHOD

In its simplest form the power method starts with an arbitrary vector u_0 and produces a sequence of vectors u_s defined by

$$u_{s+1} = Au_s \qquad s = 0, 1, 2, \ldots \tag{4.2}$$

Under certain conditions it is possible to prove that this sequence converges to the eigenvector corresponding to the eigenvalue of A which has the largest modulus.

■ Theorem 4.1

Let the $n \times n$ matrix A have eigenvalues λ_i and linearly independent eigenvectors x_i. If

$$|\lambda_1| > |\lambda_2| \geqslant |\lambda_3| \geqslant \cdots \geqslant |\lambda_n| \qquad \text{and} \qquad u_0 = \sum_{j=1}^{n} \alpha_j x_j$$

for some scalars α_j with $\alpha_1 \neq 0$, then

$$u_s \to (\lambda_1^s \alpha_1) x_1$$

■ Proof

$$
\begin{aligned}
u_1 = Au_0 &= A(\alpha_1 x_1 + \alpha_2 x_2 \quad + \cdots + \alpha_n x_n) \\
&= \alpha_1 A x_1 + \alpha_2 A x_2 \quad + \cdots + \alpha_n A x_n \\
&= \alpha_1 \lambda_1 x_1 + \alpha_2 \lambda_2 x_2 \quad + \cdots + \alpha_n \lambda_n x_n
\end{aligned}
$$

since, from the definition of an eigenvector, $Ax_i = \lambda_i x_i$.
Similarly,

$$
\begin{aligned}
u_2 = Au_1 &= A(\alpha_1 \lambda_1 x_1 + \alpha_2 \lambda_2 x_2 \quad + \cdots + \alpha_n \lambda_n x_n) \\
&= \alpha_1 \lambda_1 A x_1 + \alpha_2 \lambda_2 A x_2 + \cdots + \alpha_n \lambda_n A x_n \\
&= \alpha_1 \lambda_1^2 x_1 \quad + \alpha_2 \lambda_2^2 x_2 \quad + \cdots + \alpha_n \lambda_n^2 x_n
\end{aligned}
$$

Continuing in this way gives

$$u_s = \alpha_1 \lambda_1^s x_1 \quad + \alpha_2 \lambda_2^s x_2 \quad + \cdots + \alpha_n \lambda_n^s x_n$$

which may be written as

$$u_s = \lambda_1^s \left[\alpha_1 x_1 + \alpha_2 \left(\frac{\lambda_2}{\lambda_1}\right)^s x_2 + \cdots + \alpha_n \left(\frac{\lambda_n}{\lambda_1}\right)^s x_n \right] \qquad (4.3)$$

All of the terms except the first in the square brackets converge to the zero vector as $s \to \infty$, since $|\lambda_1| > |\lambda_j|$ for $j \neq 1$. Hence $u_s \to (\lambda_1^s \alpha_1) x_1$ as required. □

The quantities λ_1 and x_1 are usually referred to as the *dominant* eigenvalue and eigenvector, respectively. Notice from the proof of Theorem 4.1 that although u_s converges to a multiple of x_1, the actual multiple itself, namely $\lambda_1^s \alpha_1$, changes with s. If $|\lambda_1| > 1$, the elements of u_s increase without bound as $s \to \infty$, eventually leading to computer overflow. Similarly, if $|\lambda_1| < 1$, the elements of u_s decrease to zero, resulting in a loss of significant figures. For computational purposes it is more convenient to rescale at each stage, so that the largest element of u_s is unity, since we are only interested in the ratio of the elements of an eigenvector. In practice, therefore, it is better to compute the sequence of vectors defined by

$$v_{s+1} = A u_s$$

$$s = 0, 1, 2, \ldots$$

$$u_{s+1} = v_{s+1}/\max(v_{s+1})$$

where $\max(v_{s+1})$ denotes the element of v_{s+1} of largest modulus. Theorem 4.1 is unaffected by this modification; equation (4.3) is replaced by

$$u_s = k_s \lambda_1^s \left[\alpha_1 x_1 + \alpha_2 \left(\frac{\lambda_2}{\lambda_1}\right)^s x_2 + \cdots + \alpha_n \left(\frac{\lambda_n}{\lambda_1}\right)^s x_n \right] \qquad (4.4)$$

where

$$k_s = 1/[\max(v_1)\max(v_2)\cdots\max(v_s)]$$

We see that u_s converges to a multiple of x_1 as before. The vector x_1 is, of course, the eigenvector corresponding to the eigenvalue λ_1, and so $A u_s \to \lambda_1 u_s$, i.e. $v_{s+1} \to \lambda_1 u_s$. It follows that the scaling factors $\max(v_{s+1})$ converge to the dominant eigenvalue λ_1 because the largest component of u_s is unity.

■ **Example 4.1**
As an illustration of the power method, consider the matrix

$$A = \begin{pmatrix} 3 & 0 & 1 \\ 2 & 2 & 2 \\ 4 & 2 & 5 \end{pmatrix}$$

which has exact eigenvalues $1, 2$ and 7 with eigenvectors $(1, 2, -2)^T$, $(-2, 1, 2)^T$ and $(1, 2, 4)^T$, respectively. The results obtained with $u_0 = (1, 1, 1)^T$ are listed in Table 4.1.

Table 4.1

s	u_s^T	v_{s+1}^T	$\max(v_{s+1})$
0	$(1,1,1)$	$(4,6,11)$	11
1	$(0.36, 0.55, 1)$	$(2.08, 3.82, 7.54)$	7.54
2	$(0.28, 0.51, 1)$	$(1.84, 3.58, 7.14)$	7.14
3	$(0.26, 0.50, 1)$	$(1.78, 3.52, 7.04)$	7.04
4	$(0.25, 0.50, 1)$	$(1.75, 3.50, 7.00)$	7.00
5	$(0.25, 0.50, 1)$		

It is interesting to notice from equation (4.3) that the rate of convergence depends on $(\lambda_2/\lambda_1)^s$, and so the larger $|\lambda_1|$ is compared with $|\lambda_2|$, the faster the convergence. In the present example $|\lambda_2/\lambda_1| = 2/7$, which accounts for the rapid convergence in this case.   □

We now take a closer look at the hypotheses of Theorem 4.1. One of the assumptions made is that α_1 is non-zero. If $\alpha_1 = 0$, then the proof of Theorem 4.1 indicates that in theory u_s converges to x_2 rather than x_1. However, in practice, division by $\max(v_{s+1})$ introduces rounding errors so there will be a small non-zero component in the x_1 direction after one or two iterations. This component increases as the method proceeds. Hence, even if we unwittingly choose a starting vector which is deficient in the x_1 direction, we still get convergence to x_1 as before.

Another assumption made in Theorem 4.1 is the uniqueness of the dominant eigenvalue. Suppose instead that there are exactly two eigenvalues of largest modulus. This situation can arise in three different ways; either $\lambda_1 = \lambda_2, \lambda_1 = -\lambda_2$ or $\lambda_1 = \bar{\lambda}_2$. If $\lambda_1 = \lambda_2$ then equation (4.4) shows that, in the limit,

$$u_s = k_s \lambda_1^s [\alpha_1 x_1 + \alpha_2 x_2]$$

and so u_s converges to a linear combination of x_1 and x_2 (which is itself an eigenvector corresponding to the dominant eigenvalue). If $\lambda_1 = -\lambda_2$, then u_s converges to

$$k_s \lambda_1^s [\alpha_1 x_1 + (-1)^s \alpha_2 x_2]$$

and ultimately u_s oscillates between two linear combinations of x_1 and x_2. Finally, if λ_1 and λ_2 are a pair of complex conjugates, then u_s converges to

$$k_s \lambda_1^s \left[\alpha_1 x_1 + \left(\frac{\bar{\lambda}_1}{\lambda_1} \right)^s \alpha_2 x_2 \right]$$

The complex number $\bar{\lambda}_1/\lambda_1$ has a modulus of unity, and so $(\bar{\lambda}_1/\lambda_1)^s$ lies on the boundary of the unit circle in the complex plane. The vectors u_s do not display any recognizable pattern for changing values of s. It is, however, possible to find the dominant eigenvalues and eigenvectors by considering three consecutive terms, u_s, u_{s+1} and u_{s+2}, for sufficiently large values of s. Details of this can be found in Wilkinson (1965).

Suppose now that A has real eigenvalues λ_i with

$$\lambda_1 > \lambda_2 \geqslant \lambda_3 \geqslant \cdots \geqslant \lambda_{n-1} > \lambda_n$$

and consider the sequence of vectors defined by

$$
\begin{aligned}
v_{s+1} &= (A - pI)u_s \\
u_{s+1} &= v_{s+1}/\max(v_{s+1})
\end{aligned}
\qquad s = 0, 1, 2, \ldots
$$

where I is the $n \times n$ identity matrix and p is a real parameter. This is referred to as the *power method with shifting* because $A - pI$ has eigenvalues $\lambda_i - p$, i.e. the eigenvalues of A are shifted p units along the real line. To see this, simply note that

$$(A - pI)x_i = Ax_i - px_i = \lambda_i x_i - px_i = (\lambda_i - p)x_i \qquad (4.5)$$

Theorem 4.1 can be applied to the matrix $A - pI$, and we deduce that u_s converges to the eigenvector corresponding to the eigenvalue which maximizes $|\lambda_i - p|$ (i.e. to the one furthest away from p). Hence if $p < (\lambda_1 + \lambda_n)/2$ then $u_s \to x_1$ and $\max(v_s) \to \lambda_1 - p$, and if $p > (\lambda_1 + \lambda_n)/2$ then $u_s \to x_n$ and $\max(v_s) \to \lambda_n - p$. A suitable choice of p can therefore be used to produce convergence to the two extreme eigenvectors corresponding to the algebraically largest and algebraically smallest eigenvalue. Incidentally, if $p = (\lambda_1 + \lambda_n)/2$ then $\lambda_1 - p = -(\lambda_n - p)$, and so $A - pI$ has two eigenvalues of the same modulus but of opposite signs. In this case, the sequence of vectors oscillates in the limit between two linear combinations of x_1 and x_n.

The extreme eigenvalues and eigenvectors are usually calculated by taking a zero shift to get the dominant solution, λ_1 and x_1, say. The matrix can then be shifted by the estimated eigenvalue λ_1 to get the solution, λ_n and x_n, at the other extreme.

■ Example 4.2

The dominant eigenvalue of

$$A = \begin{pmatrix} 3 & 0 & 1 \\ 2 & 2 & 2 \\ 4 & 2 & 5 \end{pmatrix}$$

has been estimated as 7.00 in Example 4.1. Successive iterates for the matrix

$$A - 7I = \begin{pmatrix} -4 & 0 & 1 \\ 2 & -5 & 2 \\ 4 & 2 & -2 \end{pmatrix}$$

with $u_0 = (1, 1, 1)^{\mathrm{T}}$ are listed in Table 4.2.

Table 4.2

s	$\mathbf{u}_s^T$	$\mathbf{v}_{s+1}^T$	$\max(\mathbf{v}_{s+1})$
0	$(1, 1, 1)$	$(-3, -1, 4)$	4
1	$(-0.75, -0.25, 1)$	$(4.00, 1.75, -5.50)$	-5.50
2	$(-0.73, -0.32, 1)$	$(3.92, 2.14, -5.56)$	-5.56
3	$(-0.71, -0.38, 1)$	$(3.84, 2.48, -5.60)$	-5.60
4	$(-0.69, -0.44, 1)$	$(3.76, 2.82, -5.64)$	-5.64
5	$(-0.67, -0.50, 1)$	$(3.68, 3.16, -5.68)$	-5.68
6	$(-0.65, -0.56, 1)$	$(3.60, 3.50, -5.72)$	-5.72
7	$(-0.63, -0.61, 1)$	$(3.52, 3.79, -5.74)$	-5.74
8	$(-0.61, -0.66, 1)$	$(3.44, 4.08, -5.76)$	-5.76
9	$(-0.60, -0.71, 1)$	$(3.40, 4.35, -5.82)$	-5.82
10	$(-0.58, -0.75, 1)$	$(3.32, 4.59, -5.82)$	-5.82
11	$(-0.57, -0.79, 1)$	$(3.28, 4.81, -5.86)$	-5.86
12	$(-0.56, -0.82, 1)$	$(3.24, 4.98, -5.88)$	-5.88
13	$(-0.55, -0.85, 1)$	$(3.20, 5.15, -5.90)$	-5.90
14	$(-0.54, -0.87, 1)$	$(3.16, 5.27, -5.90)$	-5.90
15	$(-0.54, -0.89, 1)$	$(3.16, 5.37, -5.94)$	-5.94
16	$(-0.53, -0.90, 1)$	$(3.12, 5.44, -5.92)$	-5.92
17	$(-0.53, -0.92, 1)$	$(3.12, 5.54, -5.96)$	-5.96
18	$(-0.52, -0.93, 1)$	$(3.08, 5.61, -5.94)$	-5.94
19	$(-0.52, -0.94, 1)$	$(3.08, 5.66, -5.96)$	-5.96
20	$(-0.52, -0.95, 1)$	$(3.08, 5.71, -5.98)$	-5.98
21	$(-0.52, -0.95, 1)$		

These results indicate that $A - 7I$ has an approximate eigenvector $(-0.52, -0.95, 1)^T$ with approximate eigenvalue -5.98. Therefore the original matrix A has the same eigenvector but with eigenvalue $-5.98 + 7 = 1.02$. The exact values of x_3 and λ_3 are $(-0.5, -1, 1)^T$ and 1, respectively, so these results are correct to 1D. The slow convergence of the present example can easily be accounted for; the matrix $A - 7I$ has eigenvalues -6, -5 and 0 and so the rate of convergence is governed by $(5/6)^s$. This compares with a convergence rate of $(2/7)^s$ in Example 4.1. $\qquad\square$

In general, if $\mathbf{u}_s \to \mathbf{x}_1$, then in the presence of a shift p the rate of convergence depends on

$$\left(\frac{\lambda_i - p}{\lambda_1 - p}\right)^s$$

and so a suitably chosen shift can accelerate convergence. For example, if a 3×3 matrix A has eigenvalues 15, 17 and 20, then without shifting the convergence rate depends on $(17/20)^s$, but with a shift of 16 it depends on

$(1/4)^s$ because $A - 16I$ has eigenvalues $-1, 1$ and 4. However, in practice it is non-trivial to find the best value of p unless some of the eigenvalues are known in advance.

The power method with shifting can only be used to calculate two of the eigenvalues and eigenvectors of a matrix. Whilst this may be sufficient for some applications, there are occasions when the complete eigensystem is required. This can be achieved by iterating with the inverse of $A - pI$ rather than $A - pI$ itself, i.e. by computing the sequence of vectors defined by

$$v_{s+1} = (A - pI)^{-1} u_s \qquad s = 0, 1, 2, \ldots \qquad (4.6)$$

$$u_{s+1} = v_{s+1} / \max(v_{s+1})$$

This method is referred to as *inverse iteration*. From equation (4.5),

$$(A - pI)x_i = (\lambda_i - p)x_i$$

If both sides of this equation are pre-multiplied by $(A - pI)^{-1}$ and $1/(\lambda_i - p)$, it follows that

$$(A - pI)^{-1} x_i = \frac{1}{(\lambda_i - p)} x_i$$

and so $(A - pI)^{-1}$ has the same eigenvectors as A but with eigenvalues $1/(\lambda_i - p)$. Theorem 4.1 can be applied to $(A - pI)^{-1}$ and we deduce that u_s converges to the eigenvector corresponding to the eigenvalue which maximizes $1/|\lambda_i - p|$ (i.e. to the one closest to p). Suitably chosen values of p enable us to find all of the eigenvectors, not just those corresponding to the extreme eigenvalues.

In practice, the inverse of $A - pI$ is never actually computed. It is computationally more efficient to rearrange (4.6) as

$$(A - pI)v_{s+1} = u_s \qquad (4.7)$$

and solve for v_{s+1} using the LU decomposition method with partial pivoting. This is particularly convenient since the matrices L and U are independent of s and so need only be computed once (for each value of p). If the eigenvalue closest to p is λ_j, then the value of λ_j can be calculated from

$$\max(v_s) \rightarrow \frac{1}{\lambda_j - p}$$

The rate of convergence of inverse iteration depends on

$$\left(\frac{\lambda_j - p}{\lambda_i - p} \right)^s$$

This shows that the nearer p is to λ_j, the faster u_s converges to x_j. However, there is a danger that if p is too close to λ_j, then $A - pI$ will be nearly singular. In this case $u_{nn} \approx 0$ and the system of equations (4.7) might be ill-conditioned, with the resulting difficulty of obtaining an accurate solution. Indeed, if p is exactly equal to one of the eigenvalues of A, the matrix $A - pI$ is singular, $u_{nn} = 0$ and the back substitution fails completely. However, it can be shown (see Wilkinson, 1965) that provided the starting vector u_0 is not severely deficient in the x_j component (i.e. $\alpha_j \neq 0$), inverse iteration does give good results after only one or two iterations, even when p is close to λ_j. In the unlikely event of $p = \lambda_j$, we can take u_{nn} to be an arbitrarily small number and still get an accurate approximation to x_j.

■ Example 4.3

To illustrate the robustness and fast convergence of inverse iteration, consider the matrix

$$A = \begin{pmatrix} 3 & 0 & 1 \\ 2 & 2 & 2 \\ 4 & 2 & 5 \end{pmatrix}$$

We have already estimated two of the eigenvalues of this matrix as 7.00 and 1.02 (see Examples 4.1 and 4.2). Now, the trace of a matrix is equal to the sum of the eigenvalues. In the present example the trace is 10 and so the third eigenvalue is approximately 1.98, which is taken as the value of p in the inverse iteration with $u_0 = (1, 1, 1)^T$. Now

$$A - 1.98I = \begin{pmatrix} 1.02 & 0 & 1 \\ 2 & 0.02 & 2 \\ 4 & 2 & 3.02 \end{pmatrix}$$

Decomposing this matrix into the product LU with partial pivoting gives

$$L = \begin{pmatrix} 1 & 0 & 0 \\ 0.50 & 1 & 0 \\ 0.26 & 0.52 & 1 \end{pmatrix}$$

$$U = \begin{pmatrix} 4 & 2 & 3.02 \\ 0 & -0.98 & 0.49 \\ 0 & 0 & -0.03 \end{pmatrix}$$

Results of successive forward and back substitutions are given in Table 4.3. The matrix A therefore has an approximate eigenvector $x_2 = (-1.00, -0.50, 1)^T$ with eigenvalue

$$\lambda_2 = p + \frac{1}{\max(v_3)} = 1.98 + \frac{1}{42.00} = 2.00 \qquad \square$$

Table 4.3

s	u_s^T	v_{s+1}^T	$\max(v_{s+1})$
0	$(1, 1, 1)$	$(16.59, -8.51, -16.00)$	16.59
1	$(1, -0.51, -0.96)$	$(42.28, -21.13, -42.33)$	-42.33
2	$(-1.00, 0.50, 1)$	$(-41.96, 21.00, 42.00)$	42.00
3	$(-1.00, 0.50, 1)$		

The success of inverse iteration depends on our ability to obtain accurate estimates of the eigenvalues to use as appropriate shifts. In Example 4.3 an estimate of λ_2 was made using the trace check. Unfortunately, for matrices of order greater than three this property is of little use and we need to have an independent means of calculating the eigenvalues. This can be done using numerical methods such as Householder or QR which are described later in this chapter.

EXERCISE 4.1

1. Use the power method with shifting to find the two extreme eigenvalues and eigenvectors of the following matrices. Start with $(1, 1, 1)^T$ and stop when the components of two successive iterates differ by less than 10^{-6}.

(a) $\begin{pmatrix} 10 & -6 & -4 \\ -6 & 11 & 2 \\ -4 & 2 & 6 \end{pmatrix}$ (b) $\begin{pmatrix} 1 & 0 & -1 \\ 1 & 2 & 1 \\ 2 & 2 & 3 \end{pmatrix}$

Use the trace check to write down the remaining eigenvalue of each matrix. For which matrix does the power method (with zero shift) converge fastest, and why was this to be expected?

2. Find, correct to 1D, the two optimal values of p for calculating the two extreme eigenvectors of the matrix given in Question 1(b). Suggest a formula for the optimal values of p in terms of λ_1, λ_2 and λ_3, where $\lambda_1 > \lambda_2 > \lambda_3$.

3. Attempt to use the power method (without shifting) to find the dominant eigenvalue and eigenvector of

(a) $\begin{pmatrix} 1 & -2 \\ 3 & 1 \end{pmatrix}$ (b) $\begin{pmatrix} 2 & 5 \\ 1 & -2 \end{pmatrix}$

What can you deduce about the nature of the eigenvalues in each case? Does a shift help you to find them?

4. Perform a single step of the method of inverse iteration to find the middle

eigenvector of the matrix given in Question 1(b) using $p = 2.01$ and starting vector $(1, 1, 1)^T$.

5. Use the power method with shifting to find as many eigenvalues and eigenvectors as you can of the matrices

(a) $\begin{pmatrix} 5 & -2 \\ -1 & 4 \end{pmatrix}$ (b) $\begin{pmatrix} 12 & 3 & 1 \\ -9 & -2 & -3 \\ 14 & 6 & 2 \end{pmatrix}$

(c) $\begin{pmatrix} 2 & 4 & -2 \\ 4 & 2 & 2 \\ -2 & 2 & 5 \end{pmatrix}$ (d) $\begin{pmatrix} 8 & -2 & -3 & 1 \\ 7 & -1 & -3 & 1 \\ 6 & -2 & -1 & 1 \\ 5 & -2 & -3 & 4 \end{pmatrix}$

Experiment with different starting vectors, shifts and stopping criteria.

6. Calculate the first two iterates of the power method in the case when

$$A = \begin{pmatrix} 2 & -2 & 10 \\ -2 & 11 & 8 \\ 10 & 8 & 5 \end{pmatrix} \qquad u_0 = \begin{pmatrix} 1 \\ -1/4 \\ -1/4 \end{pmatrix}$$

Use exact arithmetic. Explain why it follows from these calculations that

$$A^2 u_0 = 81 u_0$$

Deduce that two of the eigenvalues of A are ± 9, and by taking appropriate linear combinations of u_0 and u_1 find the corresponding eigenvectors.

4.2 DEFLATION

Suppose that one of the eigenvalues and its corresponding eigenvector, λ_1 and x_1 say, of an $n \times n$ matrix A are known. They might, for example, have been calculated using the power method described in the previous section. The deflation process uses x_1 to construct an $(n-1) \times (n-1)$ matrix C which contains, in a sense to be made precise later, the remaining eigenvalues and eigenvectors of A. This process can then be repeated, reducing the order of the matrix by one at each stage, until a 2×2 matrix is obtained. Before this method is described in detail it is convenient to establish some preliminary results.

■ Lemma 4.1

If a matrix A has eigenvalues λ_i corresponding to eigenvectors x_i, then $P^{-1}AP$ has the same eigenvalues as A but with eigenvectors $P^{-1}x_i$ for any non-singular matrix P.

■ Proof

We are given that $Ax_i = \lambda_i x_i$. Hence

$$(P^{-1}AP)P^{-1}x_i = P^{-1}A(PP^{-1})x_i = P^{-1}Ax_i = P^{-1}(\lambda_i x_i) = \lambda_i(P^{-1}x_i)$$

The result then follows from the definition of an eigenvalue and an eigenvector. $\square$

As a consequence of Lemma 4.1, the matrices A and $P^{-1}AP$ are said to be similar. Matrix transformations of this type are known as similarity transformations.

■ Lemma 4.2

Let

$$B = \begin{pmatrix} \lambda_1 & a_{12} & a_{13} & \cdots & a_{1n} \\ 0 & c_{22} & c_{23} & & c_{2n} \\ 0 & c_{32} & c_{33} & & c_{3n} \\ \vdots & & & & \vdots \\ 0 & c_{n2} & c_{n3} & \cdots & c_{nn} \end{pmatrix}$$

and let C be the $(n-1) \times (n-1)$ matrix obtained by deleting the first row and column of B.

The matrix B has eigenvalues λ_1 together with the $n-1$ eigenvalues of C. Moreover, if $(\beta_2, \beta_3, \ldots, \beta_n)^{\mathrm{T}}$ is an eigenvector of C with eigenvalue $\mu \neq \lambda_1$, then the corresponding eigenvector of B is $(\beta_1, \beta_2, \ldots, \beta_n)^{\mathrm{t}}$ with

$$\beta_1 = \frac{\displaystyle\sum_{j=2}^{n} a_{1j}\beta_j}{\mu - \lambda_1} \tag{4.8}$$

■ Proof

The characteristic polynomial of B is

$$\begin{vmatrix} \lambda_1 - \lambda & a_{12} & a_{13} & \cdots & a_{1n} \\ 0 & c_{22} - \lambda & c_{23} & & c_{2n} \\ 0 & c_{32} & c_{33} - \lambda & & c_{3n} \\ \vdots & & & & \vdots \\ 0 & c_{n2} & c_{n3} & \cdots & c_{nn} - \lambda \end{vmatrix}$$

Expanding this determinant down the first column gives

$$(\lambda_1 - \lambda)\begin{vmatrix} c_{22} - \lambda & c_{23} & \cdots & c_{2n} \\ c_{32} & c_{33} - \lambda & & c_{3n} \\ \vdots & & & \vdots \\ c_{n2} & c_{n3} & \cdots & c_{nn} - \lambda \end{vmatrix} = (\lambda_1 - \lambda)\det(C - \lambda I)$$

This polynomial therefore has zeros λ_1 together with the zeros of $\det(C - \lambda I)$ which are the eigenvalues of C.

If $(\beta_1, \beta_2, \ldots, \beta_n)^{\mathrm{T}}$ is an eigenvector of B with eigenvalue μ, then

$$\begin{pmatrix} \lambda_1 & a_{12} & a_{13} & \cdots & a_{1n} \\ 0 & c_{22} & c_{23} & & c_{2n} \\ 0 & c_{32} & c_{33} & & c_{3n} \\ \vdots & & & & \vdots \\ 0 & c_{n2} & c_{n3} & \cdots & c_{nn} \end{pmatrix} \begin{pmatrix} \beta_1 \\ \beta_2 \\ \beta_3 \\ \vdots \\ \beta_n \end{pmatrix} = \mu \begin{pmatrix} \beta_1 \\ \beta_2 \\ \beta_3 \\ \vdots \\ \beta_n \end{pmatrix}$$

The last $n-1$ equations of this system are automatically satisfied since $(\beta_2, \beta_3, \ldots, \beta_n)^{\mathrm{T}}$ is given to be an eigenvector of C with eigenvalue μ. It only remains to choose β_1 to ensure that the first equation, namely

$$\lambda_1 \beta_1 + a_{12}\beta_2 + a_{13}\beta_3 + \cdots + a_{1n}\beta_n = \mu\beta_1$$

is true. This will be so if

$$\beta_1 = \left(\sum_{j=2}^{n} a_{ij}\beta_j \right) \bigg/ (\mu - \lambda_1)$$

as required. $\square$

Let the known eigenvector of A be scaled so that its first element is unity, i.e. $x_1 = (1, \xi_2, \xi_3, \ldots, \xi_n)^{\mathrm{T}}$, for some known numbers ξ_i. The aim of deflation is to construct a non-singular matrix P such that $P^{-1}AP = B$, where B has the special form given in Lemma 4.2. This is convenient since it follows from Lemmas 4.1 and 4.2 that the remaining eigenvalues of A are those of the deflated $(n-1) \times (n-1)$ matrix C. Once the eigenvectors of C have been calculated, it is a simple matter to write down the eigenvectors of B using equation (4.8). From Lemma 4.1 these are $P^{-1}x_i$, and so the eigenvectors x_i of A can be recovered by pre-multiplication by P.

The matrix P which does the trick is

$$\begin{pmatrix} 1 & 0 & 0 & \cdots & 0 \\ \xi_2 & 1 & 0 & & 0 \\ \xi_3 & 0 & 1 & & 0 \\ \vdots & & & \ddots & \vdots \\ \xi_n & & & & 1 \end{pmatrix}$$

To see this, note that the first column of AP is Ax_1, which is $\lambda_1 x_1$ since x_1 is the eigenvector of A with eigenvalue λ_1, and so

$$AP = \begin{pmatrix} \lambda_1 & a_{12} & a_{13} & \cdots & a_{1n} \\ \lambda_1\xi_2 & a_{22} & a_{23} & & a_{2n} \\ \lambda_1\xi_3 & a_{32} & a_{33} & & a_{3n} \\ \vdots & & & & \vdots \\ \lambda_1\xi_n & a_{n2} & a_{n3} & \cdots & a_{nn} \end{pmatrix}$$

The matrix P can be derived from the identity matrix by adding multiples ξ_i of row 1 to row i. The inverse operation is to subtract these multiples from row i, and so

$$P^{-1} = \begin{pmatrix} 1 & 0 & 0 & \cdots & 0 \\ -\xi_2 & 1 & 0 & & 0 \\ -\xi_3 & 0 & 1 & & 0 \\ \vdots & & & \ddots & \\ -\xi_n & & & & 1 \end{pmatrix}$$

In fact it is easily verified that $P^{-1}P = I$ by direct multiplication. Therefore

$$P^{-1}AP = \begin{pmatrix} 1 & 0 & 0 & \cdots & 0 \\ -\xi_2 & 1 & 0 & & 0 \\ -\xi_3 & 0 & 1 & & 0 \\ \vdots & & & \ddots & \\ -\xi_n & & & & 1 \end{pmatrix} \begin{pmatrix} \lambda_1 & a_{12} & a_{13} & \cdots & a_{1n} \\ \lambda_1\xi_2 & a_{22} & a_{23} & \cdots & a_{2n} \\ \lambda_1\xi_3 & a_{32} & a_{33} & \cdots & a_{3n} \\ \vdots & & & & \vdots \\ \lambda_1\xi_n & a_{n2} & a_{n3} & \cdots & a_{nn} \end{pmatrix}$$

$$= \begin{pmatrix} \lambda_1 & a_{12} & a_{13} & \cdots & a_{1n} \\ 0 & c_{22} & c_{23} & & c_{2n} \\ 0 & c_{32} & c_{33} & & c_{3n} \\ \vdots & & & & \vdots \\ 0 & c_{n2} & c_{n3} & \cdots & c_{nn} \end{pmatrix}$$

which is of the required form with

$$c_{ij} = a_{ij} - \xi_i a_{1j} \tag{4.9}$$

■ **Example 4.4**

As an illustration of the foregoing, consider the matrix

$$A = \begin{pmatrix} 10 & -6 & -4 \\ -6 & 11 & 2 \\ -4 & 2 & 6 \end{pmatrix}$$

for which $x_1 = (1, -1, -1/2)^T$ and $\lambda_1 = 18$. The transformation matrix P is

$$\begin{pmatrix} 1 & 0 & 0 \\ -1 & 1 & 0 \\ -1/2 & 0 & 1 \end{pmatrix}$$

Hence

$B = P^{-1}AP =$

$$\begin{pmatrix} 1 & 0 & 0 \\ 1 & 1 & 0 \\ 1/2 & 0 & 1 \end{pmatrix} \begin{pmatrix} 10 & -6 & -4 \\ -6 & 11 & 2 \\ -4 & 2 & 6 \end{pmatrix} \begin{pmatrix} 1 & 0 & 0 \\ -1 & 1 & 0 \\ -1/2 & 0 & 1 \end{pmatrix} = \begin{pmatrix} 18 & -6 & -4 \\ 0 & 5 & -2 \\ 0 & -1 & 4 \end{pmatrix}$$

The deflated matrix

$$C = \begin{pmatrix} 5 & -2 \\ -1 & 4 \end{pmatrix}$$

has eigenvectors $(1, -1/2)^T$ and $(1,1)^T$ with eigenvalues 6 and 3, respectively. The matrix A therefore has eigenvalues $\lambda_2 = 6$ and $\lambda_3 = 3$. The eigenvectors of B can be calculated from the relations

$$\begin{pmatrix} 18 & -6 & -4 \\ 0 & 5 & -2 \\ 0 & -1 & 4 \end{pmatrix} \begin{pmatrix} \beta_1 \\ 1 \\ -1/2 \end{pmatrix} = 6 \begin{pmatrix} \beta_1 \\ 1 \\ -1/2 \end{pmatrix} \qquad \text{i.e. } 18\beta_1 - 4 = 6\beta_1$$

and

$$\begin{pmatrix} 18 & -6 & -4 \\ 0 & 5 & -2 \\ 0 & -1 & 4 \end{pmatrix} \begin{pmatrix} \beta_1 \\ 1 \\ 1 \end{pmatrix} = 3 \begin{pmatrix} \beta_1 \\ 1 \\ 1 \end{pmatrix} \qquad \text{i.e. } 18\beta_1 - 10 = 3\beta_1$$

The eigenvectors of B are therefore $(1/3, 1, -1/2)^T$ and $(2/3, 1, 1)^T$.

Finally, the eigenvectors of the original matrix A can be found by pre-multiplying these vectors by P, i.e.

$$x_2 = \begin{pmatrix} 1 & 0 & 0 \\ -1 & 1 & 0 \\ -1/2 & 0 & 1 \end{pmatrix} \begin{pmatrix} 1/3 \\ 1 \\ -1/2 \end{pmatrix} = \begin{pmatrix} 1/3 \\ 2/3 \\ -2/3 \end{pmatrix} \qquad \text{or equivalently} \begin{pmatrix} 1/2 \\ 1 \\ -1 \end{pmatrix}$$

and

$$x_3 = \begin{pmatrix} 1 & 0 & 0 \\ -1 & 1 & 0 \\ -1/2 & 0 & 1 \end{pmatrix} \begin{pmatrix} 2/3 \\ 1 \\ 1 \end{pmatrix} = \begin{pmatrix} 2/3 \\ 1/3 \\ 2/3 \end{pmatrix} \qquad \text{or equivalently} \begin{pmatrix} 1 \\ 1/2 \\ 1 \end{pmatrix}$$

It is interesting to notice in this example that the deflated matrix C is

$(5-\lambda)(4-\lambda) - 2$ $(\lambda-6)(\lambda-3)$

$+\lambda^2 \ -9\lambda +18$

unsymmetric even though the original matrix A is symmetric. We deduce that the property of symmetry is not preserved in the deflation process. This is also apparent from equation (4.9), since even if $a_{ij} = a_{ji}$ there is no reason to suppose that $c_{ij} = c_{ji}$. □

The method of deflation fails whenever the first element of x_1 is zero, since x_1 cannot then be scaled so that this number is one. Furthermore, this technique may exhibit induced instability if any of the quantities $|\xi_i|$ are greater than one, since from equation (4.9) the coefficients a_{1j} are multiplied by ξ_i in the calculation of c_{ij}. These difficulties can be prevented by performing a preliminary similarity transformation of the current matrix A. If the element of largest modulus of x_1 occurs in the ith position then the deflation process is applied to the transformed matrix QAQ, where Q is obtained from the identity matrix by interchanging rows 1 and i. It is easy to see that $Q^{-1} = Q$ because the operation of interchanging rows is self-inverse. Therefore we can use Lemma 4.2 to deduce that the eigenvalues are preserved and that the first and ith elements of the eigenvector are interchanged. Hence the largest element of the eigenvector of QAQ occurs as the first element and, when this vector is scaled, all of the other elements satisfy $|\xi_i| \leqslant 1$. As an example, consider

$$A = \begin{pmatrix} 3 & 0 & 1 \\ 2 & 2 & 2 \\ 4 & 2 & 5 \end{pmatrix}$$

which has a dominant eigenvector $(1, 2, 4)^{\mathrm{T}}$. The matrix Q is constructed by interchanging the first and third rows of the 3×3 identity matrix to get

$$Q = \begin{pmatrix} 0 & 0 & 1 \\ 0 & 1 & 0 \\ 1 & 0 & 0 \end{pmatrix}$$

and so

$$QAQ = \begin{pmatrix} 5 & 2 & 4 \\ 2 & 2 & 2 \\ 1 & 0 & 3 \end{pmatrix}$$

The usual deflation procedure can now be applied to the new matrix, since this has a dominant eigenvector $(4, 2, 1)^{\mathrm{T}}$ which is scaled to $(1, 1/2, 1/4)^{\mathrm{T}}$.

Even if these precautions are taken, a loss of accuracy may still occur in practice, although this need not be disastrous. The eigenvector x_1 is usually computed using a numerical technique such as the power method and so will not be known exactly. This means that none of the eigenvalues of the deflated matrix C will be exactly the same as the eigenvalues of A. If, in its turn, the matrix C is deflated, additional errors will be introduced.

Consequently, the final eigenvalues (and eigenvectors) to be calculated may be slightly less accurate than those found during the early stages of the method.

EXERCISE 4.2

1. The matrix

$$A = \begin{pmatrix} 1 & 1 & -2 \\ -1 & 2 & 1 \\ 0 & 1 & -1 \end{pmatrix}$$

has an eigenvector

$$\begin{pmatrix} 1 \\ 3 \\ 1 \end{pmatrix}$$

with eigenvalue 2. Construct matrices Q and E such that

$$Q^{-1}AQ = E$$

where the eigenvector of E corresponding to the eigenvalue 2 is $(3, 1, 1)^{\mathrm{T}}$.

   Use the method of deflation to find the remaining eigenvalues and eigenvectors of E. Use exact arithmetic and calculate the eigenvalues and eigenvectors of the deflated matrix via the characteristic equation and relation (4.1).

   Hence write down the complete eigensystem of A.

2. The matrix

$$A = \begin{pmatrix} 8 & -2 & -3 & 1 \\ 7 & -1 & -3 & 1 \\ 6 & -2 & -1 & 1 \\ 5 & -2 & -3 & 4 \end{pmatrix}$$

has an eigenvector

$$\begin{pmatrix} 1 \\ 1 \\ 1 \\ 1 \end{pmatrix}$$

with eigenvalue 4. Use the deflation process to find the remaining eigenvalues and eigenvectors of A. Use exact arithmetic.

3. Use the power method with deflation to find as many eigenvalues as you

can of the following matrices:

(a) $\begin{pmatrix} 3 & 0 & 1 \\ 2 & 2 & 2 \\ 4 & 2 & 5 \end{pmatrix}$ (b) $\begin{pmatrix} 12 & 3 & 1 \\ -9 & -2 & -3 \\ 14 & 6 & 2 \end{pmatrix}$ (c) $\begin{pmatrix} 2 & 4 & -2 \\ 4 & 2 & 2 \\ -2 & 2 & 5 \end{pmatrix}$

(d) $\begin{pmatrix} 3 & 2 & 2 & -4 \\ 2 & 3 & 2 & -1 \\ 1 & 1 & 2 & -1 \\ 2 & 2 & 2 & -1 \end{pmatrix}$

4. Use the power method with deflation to determine the eigenvalues of the following matrices:

(a) $\begin{pmatrix} -2.0 & -1.0 & 4.0 \\ 2.0 & 1.0 & -2.0 \\ -1.0 & -1.0 & 3.0 \end{pmatrix}$ (b) $\begin{pmatrix} -2.1 & -0.9 & 3.9 \\ 1.9 & 1.1 & -2.1 \\ -0.9 & -1.1 & 3.1 \end{pmatrix}$

Start with $(1, 1, 1)^T$ and $(1, 1)^T$ at each stage and stop when the components of two successive iterates differ by less than 0.001. What can you deduce about the inherent stability of these eigenvalues?

5. The matrix

$$\begin{pmatrix} 10 & -6 & -4 \\ -6 & 11 & 2 \\ -4 & 2 & 6 \end{pmatrix}$$

has an exact eigenvector

$$\begin{pmatrix} 1 \\ -1 \\ -1/2 \end{pmatrix}$$

with eigenvalue 18. Starting with the approximation $(1, -0.99, -0.49)^T$, use the deflation process to estimate the remaining eigenvalues and eigenvectors. Use the characteristic equation and relation (4.1) to calculate the eigenvalues and eigenvectors of the deflated matrix. Compare these results with the exact eigensystem obtained in Example 4.4. Comment on how the inaccuracy of the dominant eigenvector affects the subsequent calculation of the other eigenvalues and eigenvectors in the deflation process.

4.3 JACOBI'S METHOD

In this and the following two sections we restrict our attention to $n \times n$ symmetric matrices. Matrices of this type are known to have real eigenvalues with a complete set of n linearly independent eigenvectors. Moreover, the eigenvectors corresponding to distinct eigenvalues are orthogonal. (See Goult *et al.*, 1974 for a proof of these theoretical results.)

Given a symmetric matrix A, Jacobi's method may be used to find its eigenvalues and eigenvectors by performing a series of similarity transformations

$$A_{r+1} = P_r^{-1} A_r P_r \quad r = 1, 2, \ldots \tag{4.10}$$

where $A_1 = A$. The matrices $A_1, A_2, \ldots$ converge in an infinite number of steps to a diagonal matrix. The eigenvalues and eigenvectors can then be estimated by virtue of the following lemma (which applies to both symmetric and unsymmetric matrices).

■ **Lemma 4.3**

Let A be an $n \times n$ matrix with eigenvalues λ_i corresponding to linearly independent eigenvectors x_i $(i = 1, 2, \ldots, n)$, and let

$$D = \begin{pmatrix} \lambda_1 & & & & 0 \\ & \lambda_2 & & & \\ & & \lambda_3 & & \\ & & & \ddots & \\ 0 & & & & \lambda_n \end{pmatrix}$$

Then $D = X^{-1}AX$ if and only if the ith column of X is x_i.

■ **Proof**

If the ith column of an $n \times n$ matrix X is denoted by x_i, then the ith columns of AX and XD are Ax_i and $\lambda_i x_i$, respectively. Therefore the vectors x_i are the eigenvectors of A if and only if $AX = XD$. This equation can be rearranged as $X^{-1}AX = D$ provided X is invertible, which is the case whenever the columns of X are linearly independent.   □

After m steps of Jacobi's method we have

$$A_{m+1} = P_m^{-1} \cdots P_2^{-1} P_1^{-1} A_1 P_1 P_2 \cdots P_m$$

Hence, if $A_{m+1} \approx D$, it follows that the diagonal elements of A_{m+1} are approximate eigenvalues and the columns of $P_1 P_2 \cdots P_m$ are approximate eigenvectors.

The transformation matrices P_r are taken to be of the form

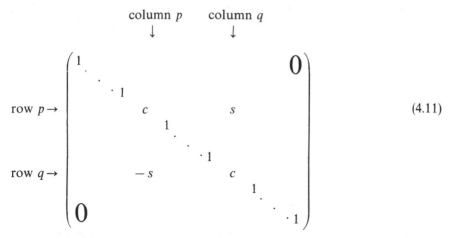

$$(4.11)$$

where $c = \cos \theta$ and $s = \sin \theta$ for some suitably chosen angle θ and integers p, q. The elements of P_r are the same as those of the $n \times n$ identity matrix with the exception of the (p, p), (p, q), (q, p) and (q, q) elements, which are $\cos \theta$, $\sin \theta$, $-\sin \theta$ and $\cos \theta$, respectively.

In the 2×2 case, the transformation

$$\begin{pmatrix} x_1 \\ x_2 \end{pmatrix} \to \begin{pmatrix} \cos \theta & \sin \theta \\ -\sin \theta & \cos \theta \end{pmatrix} \begin{pmatrix} x_1 \\ x_2 \end{pmatrix}$$

is a clockwise rotation of the Cartesian plane through an angle θ. In general, $x \to P_r x$ is a rotation of $\mathscr{R}^n$ in the (x_p, x_q) plane, and because of this P_r is usually called a (plane) *rotation matrix*.

It is easy to check, using the rules of matrix multiplication, that P_r satisfies the relation

$$P_r^T P_r = I$$

which is the defining property of an orthogonal matrix. Orthogonal matrices are particularly advantageous for use in (4.10) for several reasons. Firstly, the inverse of P_r is trivially computed as P_r^T. Secondly, symmetry is preserved at each stage, since if $A_r^T = A_r$ then

$$A_{r+1}^T = (P_r^T A_r P_r)^T = P_r^T A_r^T (P_r^T)^T = P_r^T A_r P_r = A_{r+1}$$

This is useful because the storage requirements of a symmetric matrix are only about half those of an unsymmetric matrix. Finally, since the elements of an orthogonal matrix have a modulus not exceeding unity, any rounding errors present in A_r are damped down in the evaluation of $P_r^T A_r P_r$, leading to a stable method.

In order to describe the technical details of Jacobi's method it is necessary to derive expressions for the elements of A_{r+1} in terms of those of A_r. If the (i, j) elements of A_r and A_{r+1} are denoted by a_{ij} and b_{ij}, respectively, then

$$\left.\begin{aligned} b_{ij} &= a_{ij} \\ b_{ip} &= ca_{ip} - sa_{iq} \\ b_{iq} &= sa_{ip} + ca_{iq} \\ b_{pj} &= ca_{pj} - sa_{qj} \\ b_{qj} &= sa_{pj} + ca_{qj} \end{aligned}\right\} \quad i, j \neq p, q$$

$$b_{pp} = c^2 a_{pp} - 2sc a_{pq} + s^2 a_{qq}$$
$$b_{pq} = b_{qp} = cs(a_{pp} - a_{qq}) + (c^2 - s^2)a_{pq} \qquad (4.12)$$
$$b_{qq} = s^2 a_{pp} + 2sc a_{pq} + c^2 a_{qq}$$

To see this, note that post-multiplication by P_r only changes columns p and q. If $Z = A_r P_r$, then column p of Z is

$$c(\text{column } p \text{ of } A_r) - s(\text{column } q \text{ of } A_r)$$

and column q of Z is

$$s(\text{column } p \text{ of } A_r) + c(\text{column } q \text{ of } A_r)$$

Hence, if $Z = (z_{ij})$, this gives

$$\begin{aligned} z_{ij} &= a_{ij} \qquad j \neq p, q \\ z_{ip} &= ca_{ip} - sa_{iq} \qquad (4.13) \\ z_{iq} &= sa_{ip} + ca_{iq} \end{aligned}$$

The matrix P_r^T is the same as P_r but with the elements s and $-s$ interchanged. Pre-multiplication by P_r^T only changes rows p and q. If $A_{r+1} = P_r^T Z$, then row p of A_{r+1} is

$$c(\text{row } p \text{ of } Z) - s(\text{row } q \text{ of } Z)$$

and row q of A_{r+1} is

$$s(\text{row } p \text{ of } Z) + c(\text{row } q \text{ of } Z)$$

Hence, if $A_{r+1} = (b_{ij})$, this gives

$$\begin{aligned} b_{ij} &= z_{ij} \qquad i \neq p, q \\ b_{pj} &= cz_{pj} - sz_{qj} \qquad (4.14) \\ b_{qj} &= sz_{pj} + cz_{qj} \end{aligned}$$

The first five equations in (4.12) follow immediately from equations (4.13) and (4.14). The remaining equations for b_{pp}, b_{pq} and b_{qq} are obtained by substituting the expressions for z_{ip} and z_{iq} into (4.14), which gives

$$b_{pp} = cz_{pp} - sz_{qp}$$
$$= c(ca_{pp} - sa_{pq}) - s(ca_{qp} - sa_{qq})$$
$$= c^2 a_{pp} - 2sca_{pq} + s^2 a_{qq}$$

since, by symmetry of A_r, $a_{pq} = a_{qp}$.

We also get

$$b_{pq} = cz_{pq} - sz_{qq}$$
$$= c(sa_{pp} + ca_{pq}) - s(sa_{qp} + ca_{qq})$$
$$= cs(a_{pp} - a_{qq}) + (c^2 - s^2)a_{pq}$$

and

$$b_{qq} = sz_{pq} + cz_{qq}$$
$$= s(sa_{pp} + ca_{pq}) + c(sa_{qp} + ca_{qq})$$
$$= s^2 a_{pp} + 2sca_{pq} + c^2 a_{qq}$$

This completes the derivation of equations (4.12).

The value of θ is chosen to create a zero in the (p, q) and (q, p) positions of A_{r+1}. The value of b_{pq} (and by symmetry also of b_{qp}) is

$$(\cos \theta \sin \theta)(a_{pp} - a_{qq}) + (\cos^2 \theta - \sin^2 \theta)a_{pq}$$

which can be written as

$$(\tfrac{1}{2}\sin 2\theta)(a_{pp} - a_{qq}) + (\cos 2\theta)a_{pq}$$

using the double angle formulas for sines and cosines.

Hence b_{pq} and b_{qp} are both zero if θ is chosen so that

$$\tan 2\theta = \frac{2a_{pq}}{a_{qq} - a_{pp}}$$

i.e.

$$\theta = \tfrac{1}{2}\tan^{-1}\left(\frac{2a_{pq}}{a_{qq} - a_{pp}}\right) \tag{4.15}$$

This means that at each step of Jacobi's method we can create zeros in two off-diagonal positions. It might, therefore, appear that a diagonal matrix can be obtained after a finite number of steps by working through the matrix annihilating two off-diagonal elements at a time. Unfortunately, this is not possible because subsequent orthogonal transformations usually destroy zeros previously created. However, it is possible to show that when a zero is created in the (p, q) and (q, p) positions, the sum of the squares of the off-diagonal elements decreases by $2a_{pq}^2$. This can be readily deduced from equations (4.12) since, if $i, j \neq p, q$,

$$b_{ij}^2 = a_{ij}^2$$
$$b_{ip}^2 + b_{iq}^2 = (ca_{ip} - sa_{iq})^2 + (sa_{ip} + ca_{iq})^2 = a_{ip}^2 + a_{iq}^2$$
$$b_{pj}^2 + b_{qj}^2 = (ca_{pj} - sa_{qj})^2 + (sa_{pj} + ca_{qj})^2 = a_{pj}^2 + a_{qj}^2$$

using the relation $\cos^2 \theta + \sin^2 \theta = 1$.

Moreover, by choice of θ, $b_{pq}^2 = b_{qp}^2 = 0$. Therefore

$$\sum_{\substack{j=1 \\ i \neq j}}^{n} \sum_{\substack{i=1 \\ i \neq j}}^{n} b_{ij}^2 = \sum_{\substack{j=1 \\ i \neq j}}^{n} \sum_{\substack{i=1 \\ i \neq j}}^{n} a_{ij}^2 - 2a_{pq}^2$$

Hence the overall effect, as we cycle through the matrix creating and destroying zeros, is to convert it into a diagonal matrix.

There are three standard strategies for annihilating off-diagonals. The *serial method* selects the elements in row order, that is, in positions $(1, 2), (1, 3), \ldots,$ $(1, n); (2, 3), (2, 4), \ldots, (2, n); \ldots; (n-1, n)$ in turn, which is then repeated. The *natural method* searches through all of the off-diagonals and annihilates the element of largest modulus at each stage. Although this method converges faster than the serial method it is not recommended for large values of n, since the actual search procedure itself can be extremely time consuming. In the third strategy, known as the *threshold serial method*, the off-diagonals are cycled in row order as in the serial method, omitting transformations on any element whose magnitude is below some threshold value. This value is usually decreased after each cycle. The advantage of this approach is that zeros are only created in positions where it is worth while to do so, without the need for a lengthy search.

■ Example 4.5
As an illustration of the threshold serial Jacobi method, consider the matrix

$$A = \begin{pmatrix} 3 & 0.4 & 5 \\ 0.4 & 4 & 0.1 \\ 5 & 0.1 & -2 \end{pmatrix}$$

taking the threshold values for the first and second cycles to be 0.5 and 0.05 respectively.

For the first cycle the $(1, 2)$ transformation is omitted because $|0.4| < 0.5$. However, since $|5| > 0.5$, a zero is created in position $(1, 3)$. The rotation matrix is

$$P_1 = \begin{pmatrix} c & 0 & s \\ 0 & 1 & 0 \\ -s & 0 & c \end{pmatrix} \quad \text{with } \theta = \tfrac{1}{2}\tan^{-1}\left(\frac{2 \times 5}{(-2) - 3}\right)$$

which gives

$$P_1 = \begin{pmatrix} 0.8507 & 0 & -0.5257 \\ 0 & 1 & 0 \\ 0.5257 & 0 & 0.8507 \end{pmatrix} \qquad A_2 = \begin{pmatrix} 6.0902 & 0.3928 & 0 \\ 0.3928 & 4 & -0.1252 \\ 0 & -0.1252 & -5.0902 \end{pmatrix}$$

The (2, 3) transformation is omitted because $|-0.1252| < 0.5$. This completes the first cycle.

For the second cycle a zero is created in position (1, 2) because $|0.3928| > 0.05$. This gives

$$P_2 = \begin{pmatrix} 0.9839 & -0.1788 & 0 \\ 0.1788 & 0.9839 & 0 \\ 0 & 0 & 1 \end{pmatrix} \qquad A_3 = \begin{pmatrix} 6.1616 & 0 & -0.0224 \\ 0 & 3.9286 & -0.1232 \\ -0.0224 & -0.1232 & -5.0902 \end{pmatrix}$$

The (1, 3) transformation is omitted because $|-0.0224| < 0.05$. Finally, a zero is created in position (2, 3) with

$$P_3 = \begin{pmatrix} 1 & 0 & 0 \\ 0 & 0.9999 & 0.0137 \\ 0 & -0.0137 & 0.9999 \end{pmatrix} \qquad A_4 = \begin{pmatrix} 6.1616 & 0.0003 & -0.0224 \\ 0.0003 & 3.9303 & 0 \\ -0.0224 & 0 & -5.0919 \end{pmatrix}$$

Hence

$$P_1 P_2 P_3 = \begin{pmatrix} 0.8370 & -0.1449 & -0.5277 \\ 0.1788 & 0.9838 & 0.0135 \\ 0.5172 & -0.1056 & 0.8493 \end{pmatrix}$$

Therefore A has approximate eigenvalues 6.1616, 3.9303 and -5.0919 corresponding to approximate eigenvectors

$$\begin{pmatrix} 0.8370 \\ 0.1788 \\ 0.5172 \end{pmatrix} \qquad \begin{pmatrix} -0.1449 \\ 0.9838 \\ -0.1056 \end{pmatrix} \qquad \begin{pmatrix} -0.5277 \\ 0.0135 \\ 0.8493 \end{pmatrix}$$

respectively. □

We conclude this section by considering an elegant theorem concerning the location of the eigenvalues of both symmetric and unsymmetric matrices. The eigenvalues of an unsymmetric matrix could, of course, be complex, in which case the theorem gives us a means of locating these numbers in the complex plane. There are occasions when it is not necessary to obtain accurate approximations to the eigenvalues of a matrix. All that may be required, for example, is to decide whether they are all positive or whether they lie inside the unit circle. The following theorem can often be used to provide a quick answer to these questions, avoiding the need for detailed calculations.

■ **Theorem 4.2**

(a) *Gerschgorin's first theorem*   Every eigenvalue of a matrix $A = (a_{ij})$ lies in at least one of the circles in the complex plane with centre a_{ii} and radius

$$\sum_{\substack{j \neq i \\ j=1}}^{n} |a_{ij}|$$

(b) *Gerschgorin's second theorem*   If the union of q of these circles forms a connected region isolated from the remaining circles, then there are exactly q eigenvalues in this region.

■ **Proof**

(a) Let $Av = \lambda v$ with v scaled so that its largest element is one, and suppose that

$$v = (v_1, \ldots, v_{i-1}, 1, v_{i+1}, \ldots, v_n)^{\mathrm{T}}$$

Equating the ith components of Av and λv gives

$$\sum_{j=1}^{n} a_{ij} v_j = \lambda v_i = \lambda$$

Therefore

$$|\lambda - a_{ii}| = \left| \sum_{j=1}^{n} a_{ij} v_j - a_{ii} \right| = \left| \sum_{\substack{j \neq i \\ j=1}}^{n} a_{ij} v_j \right|$$

$$\leqslant \sum_{\substack{j \neq i \\ j=1}}^{n} |a_{ij}| |v_j| \leqslant \sum_{\substack{j \neq i \\ j=1}}^{n} |a_{ij}|$$

where we have used the triangle inequality and the fact that $|v_j| \leqslant 1$.

(b) The proof of Gerschgorin's second theorem is fairly sophisticated and can be found in Noble and Daniel (1977).   □

■ **Example 4.6**

The Gerschgorin circles associated with the matrix

$$A = \begin{pmatrix} 5 & 1 & 0 \\ -1 & 3 & 1 \\ -2 & 1 & 10 \end{pmatrix}$$

are given by:

Centre	Radius
C_1: 5	$|1| + |0| = 1$
C_2: 3	$|-1| + |1| = 2$
C_3: 10	$|-2| + |1| = 3$

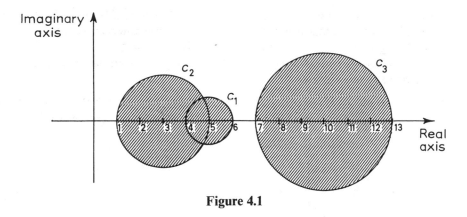

Figure 4.1

These circles are illustrated in Fig. 4.1, and Gerschgorin's first theorem indicates that the eigenvalues of A lie inside the shaded region.

Furthermore, since $C_1 \cup C_2$ does not intersect C_3, Gerschgorin's second theorem shows that two of these eigenvalues lie in $C_1 \cup C_2$, with the remaining eigenvalue in C_3.   □

Gerschgorin's theorems provide rigorous bounds on the errors of the eigenvalues calculated via Jacobi's method. In Example 4.5, the Gerschgorin circles of the transformed matrix A_4 are given by:

Centre	Radius
6.1616	0.0227
3.9303	0.0003
−5.0919	0.0224

These circles are isolated and so there is exactly one eigenvalue in each. The eigenvalues can therefore be quoted as

$$6.1616 \pm 0.0227 \qquad 3.9303 \pm 0.0003 \qquad -5.0919 \pm 0.0224$$

Quite generally, if the off-diagonals of an $n \times n$ symmetric matrix have a modulus not exceeding ε, then, provided that the Gerschgorin circles are isolated, the eigenvalues differ from the diagonals by at most $(n-1)\varepsilon$.

EXERCISE 4.3

1. If

$$P = \begin{pmatrix} \cos\theta & 0 & \sin\theta \\ 0 & 1 & 0 \\ -\sin\theta & 0 & \cos\theta \end{pmatrix} \qquad A = \begin{pmatrix} 5 & 0 & 1 \\ 0 & -3 & 0.1 \\ 1 & 0.1 & 2 \end{pmatrix}$$

calculate $P^{\mathrm{T}}AP$, and deduce that if $\theta = -(1/2)\tan^{-1}(2/3)$ then the $(1,3)$ and $(3,1)$ entries of this product are zero. Hence write down approximations to the eigenvalues and eigenvectors of A. Work to 4D.

Use Gerschgorin's theorems to obtain an upper bound on the errors of the eigenvalue estimates.

2. Use equations (4.12) to prove that, when a zero is created in the (p,q) and (q,p) positions, the sum of the squares of the diagonals increases by $2a_{pq}^2$.

3. Use Jacobi's method to estimate the eigenvalues of

(a) $\begin{pmatrix} 10 & -6 & -4 \\ -6 & 11 & 2 \\ -4 & 2 & 6 \end{pmatrix}$ (b) $\begin{pmatrix} 337 & 304 & 176 \\ 304 & 361 & 128 \\ 176 & 128 & 121 \end{pmatrix}$

(c) $\begin{pmatrix} 2 & 4 & -2 \\ 4 & 2 & 2 \\ -2 & 2 & 5 \end{pmatrix}$ (d) $\begin{pmatrix} 123 & 1 & 1 & 901 \\ 1 & 234 & 1 & 1 \\ 1 & 1 & 3 & 678 \\ 901 & 1 & 678 & 1 \end{pmatrix}$

using the

(i) serial method
(ii) natural method
(iii) threshold serial method with threshold values 10^{2-i} for the ith cycle.

Each method should be stopped when all of the off-diagonals are less than 10^{-4} in modulus.

4. If

$$P = \begin{pmatrix} 1 & 0 & 0 & 0 \\ 0 & 1 & 0 & 0 \\ 0 & 0 & k & 0 \\ 0 & 0 & 0 & 1 \end{pmatrix} \qquad A = \begin{pmatrix} 2.30 & 0.04 & 0.02 & -0.05 \\ 0.02 & 7.60 & 0.01 & 0.02 \\ -0.03 & 0.02 & 0.80 & 0.02 \\ 0.01 & -0.01 & -0.01 & -5.20 \end{pmatrix}$$

calculate $P^{-1}AP$. Find the value of k which minimizes the radius of the Gerschgorin disc centred on 0.80, whilst still isolating it from the other discs. Deduce that A has a real eigenvalue which can be quoted as

$$0.8 \pm 0.001$$

5. If

$$P = \begin{pmatrix} \cos\theta & 0 & \sin\theta & 0 \\ 0 & 1 & 0 & 0 \\ -\sin\theta & 0 & \cos\theta & 0 \\ 0 & 0 & 0 & 1 \end{pmatrix} \qquad A = \begin{pmatrix} 2 & 0.1 & 2 & 0.1 \\ 0.1 & 3 & 0.1 & 0.2 \\ 2 & 0.1 & 6 & 0.05 \\ 0.1 & 0.2 & 0.05 & 1 \end{pmatrix}$$

calculate $P^{\mathrm{T}}AP$ and deduce that A is non-singular.

4.4 STURM SEQUENCE ITERATION

When a symmetric matrix is tridiagonal, the eigenvalues can be estimated to any specified precision using a simple method based on Sturm sequences. In the next section we show that a symmetric matrix can always be converted to this special form using similarity transformations. The iteration described below can, therefore, be used in the calculation of the eigenvalues of any symmetric matrix.

Suppose that a symmetric tridiagonal matrix C is given by

$$
C = \begin{pmatrix}
a_1 & b_2 & & & & & & \\
b_2 & a_2 & b_3 & & & & & \\
& b_3 & a_3 & b_4 & & & & \\
& & & \cdot & \cdot & \cdot & & \\
& & & & \cdot & \cdot & \cdot & \\
& & & & & \cdot & \cdot & \cdot \\
& & & & & b_{n-1} & a_{n-1} & b_n \\
& & & & & & b_n & a_n
\end{pmatrix}
$$

and assume that $b_i \neq 0$ $(i = 2, 3, \ldots, n)$. This assumption is not at all restrictive in practice since, if any $b_i = 0$, the characteristic polynomial of C factorizes into the product of characteristic polynomials of two or more symmetric tridiagonal submatrices of C. For example, the eigenvalues of

$$
\begin{pmatrix}
1 & 2 & 0 & 0 & 0 \\
2 & 3 & -1 & 0 & 0 \\
0 & -1 & 5 & 0 & 0 \\
0 & 0 & 0 & 1 & 4 \\
0 & 0 & 0 & 4 & 2
\end{pmatrix}
$$

in which $b_4 = 0$ are the same as the eigenvalues of

$$
\begin{pmatrix}
1 & 2 & 0 \\
2 & 3 & -1 \\
0 & -1 & 5
\end{pmatrix}
$$

together with those of

$$
\begin{pmatrix}
1 & 4 \\
4 & 2
\end{pmatrix}
$$

For any numerical value of p, let $f_r(p)$ denote the determinant of the $r \times r$ matrix obtained by deleting the last $n - r$ rows and columns of $C - pI$, so that

$$
f_1(p) = \det (a_1 - p) = a_1 - p
$$

$$
f_2(p) = \det \begin{pmatrix} a_1 - p & b_2 \\ b_2 & a_2 - p \end{pmatrix} = (a_2 - p)(a_1 - p) - b_2^2 \tag{4.16}
$$

and, in general,

$$f_r(p) = \det \begin{pmatrix} a_1 - p & b_2 & & & & \\ b_2 & a_2 - p & b_3 & & & \\ & & \ddots & \ddots & \ddots & \\ & & & b_{r-1} & a_{r-1} - p & b_r \\ & & & & b_r & a_r - p \end{pmatrix}$$

It is very tedious to expand each determinant using cofactors, particularly if r is large. However, it is possible to show that successive terms of the sequence $\{f_r(p)\}$ can be calculated from the recurrence relation

$$f_r(p) = (a_r - p)f_{r-1}(p) - b_r^2 f_{r-2}(p) \qquad 2 \leqslant r \leqslant n \qquad (4.17)$$

provided that we define $f_0(p) = 1$ and evaluate $f_1(p)$ separately as $a_1 - p$. The case $r = 2$ follows immediately from equation (4.16). For general r, the result can be proved by expanding $f_r(p)$ along its bottom row, to get

$$f_r(p) = -b_r \begin{vmatrix} a_1 - p & b_2 & & & \\ b_2 & a_2 - p & b_3 & & \\ & & \ddots & \ddots & \ddots \\ & & b_{r-2} & a_{r-2} - p & 0 \\ & & & b_{r-1} & b_r \end{vmatrix} + (a_r - p)f_{r-1}(p)$$

observing the sign convention for cofactors. Expanding this determinant down its last column gives

$$f_r(p) = -b_r^2 f_{r-2}(p) + (a_r - p)f_{r-1}(p)$$

as required. The practical importance of the sequence $f_r(p)$ is contained in the following theorem.

■ **Theorem 4.3** *Sturm sequence property*
For any value of p, the number of agreements in sign of successive terms of the sequence $\{f_r(p)\}_{r=0}^n$ is equal to the number of eigenvalues of C which are strictly greater than p. The sign of a zero is taken to be opposite to that of the previous term.

■ **Proof**
A readable proof of this result is given in Goult *et al* (1974).   □

This theorem can be used to determine the number of eigenvalues lying in any particular interval of the real axis, as the following example demonstrates.

■ **Example 4.7**

To find the number of eigenvalues of the matrix

$$C = \begin{pmatrix} 2 & 4 & 0 & 0 \\ 4 & 10 & 3 & 0 \\ 0 & 3 & 9 & -1 \\ 0 & 0 & -1 & 5 \end{pmatrix}$$

lying in the interval $(0, 5)$, we need to compute the Sturm sequences $f_r(0)$ and $f_r(5)$. Taking $p = 0$,

$$f_0(0) = 1$$
$$f_1(0) = a_1 - 0 = 2 - 0 = 2$$
$$f_2(0) = (a_2 - 0)f_1(0) \quad b_2^2 f_0(0) = \quad [(10 - 0) \times 2] \quad -[4^2 \times 1] \quad\quad = 4$$
$$f_3(0) = (a_3 - 0)f_2(0) - b_3^2 f_1(0) = \quad [(9 - 0) \times 4] \quad -[3^2 \times 2] \quad\quad = 18$$
$$f_4(0) = (a_4 - 0)f_3(0) - b_4^2 f_2(0) = \quad [(5 - 0) \times 18] \quad -[(-1)^2 \times 4] \quad = 86$$

which have signs $+ + + + +$, with four agreements.

Similar calculations for $p = 5$ give

$$f_0(5) = 1, \quad f_1(5) = -3, \quad f_2(5) = -31, \quad f_3(5) = -97, \quad f_4(5) = 31$$

which have signs $+ - - - +$, with two agreements.

Theorem 4.3 shows that there are four eigenvalues strictly greater than 0 and two eigenvalues strictly greater than 5. Hence, there are exactly two eigenvalues in $(0, 5]$. Furthermore, since $f_4(5) \neq 0$, we deduce that no eigenvalue is exactly equal to 5, because $f_4(p) = \det(C - pI)$, the characteristic polynomial of C. Therefore, there are two eigenvalues in the open interval $(0, 5)$.

Incidentally, we can also deduce that C is positive definite, since we have shown that all of the eigenvalues of C are strictly positive. □

Sturm sequences can be used to estimate any individual eigenvalue by repeatedly bisecting any interval within which the desired eigenvalue is known to lie. This is illustrated in the following example.

■ **Example 4.8**

Gerschgorin's first theorem shows that all of the eigenvalues of

$$C = \begin{pmatrix} 2 & -1 & 0 \\ -1 & 2 & -1 \\ 0 & -1 & 1 \end{pmatrix}$$

lie in the interval $[0, 4]$.

Let $0 \leqslant \lambda_3 < \lambda_2 < \lambda_1 \leqslant 4$ and suppose, for example, that we seek an estimate of λ_2, with an error at most $1/4$.

Evaluation of the Sturm sequence at the mid-point of this interval gives

$$f_0(2) = 1, \quad f_1(2) = 0, \quad f_2(2) = -1, \quad f_3(2) = 1$$

which have signs $+ - - +$ (adopting the convention that a zero has the opposite sign to its predecessor), with one agreement. Hence there is only one eigenvalue greater than 2, which must be λ_1, since this is the largest. Therefore $\lambda_2 \in [0, 2]$.

Evaluation of the Sturm sequence at the mid-point of this new interval gives

$$f_0(1) = 1, \quad f_1(1) = 1, \quad f_2(1) = 0, \quad f_3(1) = -1$$

which have signs $+ + - -$, with two agreements. Hence there are two eigenvalues greater than 1, which must be λ_1 and λ_2. Therefore $\lambda_2 \in [1, 2]$.

The process can be repeated, halving the length of the interval containing λ_2 at each stage. The next step gives

$$f_0(1.5) = 1, \quad f_1(1.5) = 0.5, \quad f_2(1.5) = -0.75, \quad f_3(1.5) = -0.125$$

which have signs $+ + - -$, with two agreements. So $\lambda_2 \in [1.5, 2]$, i.e.

$$\lambda_2 = 1.75 \pm 0.25 \qquad\qquad \Box$$

For the general $n \times n$ case, let $s(p)$ denote the number of agreements in sign of $\{f_r(p)\}_{r=0}^n$, and suppose that the ith eigenvalue λ_i is known to lie in an interval $[a, b]$, where $\lambda_1 > \lambda_2 > \cdots > \lambda_n$. (The distinctness of the eigenvalues follows from the assumption that the off-diagonal elements of C are non-zero. A proof of this can be found in Goult *et al.*, 1974.) Suitable starting values for a and b can always be found via Gerschgorin's theorems.

The bisection algorithm can be summarized as follows:

1. The value of $s(p)$ is calculated for $p = (a + b)/2$.
2. If $s(p) < i$ then $\lambda_i \in [a, p]$ and the procedure is repeated, setting $a = a$ and $b = p$.
3. If $s(p) \geqslant i$ then $\lambda_i \in [p, b]$ and the procedure is repeated, setting $a = p$ and $b = b$.
4. The bisection is stopped once the desired accuracy has been attained.

EXERCISE 4.4

1. Find the number of eigenvalues of

$$\begin{pmatrix} 1 & 2 & 0 & 0 \\ 2 & -1 & 3 & 0 \\ 0 & 3 & 6 & 4 \\ 0 & 0 & 4 & -3 \end{pmatrix}$$

which lie in the interval $[1, 2]$.

2. Show that the matrix

$$
\begin{pmatrix}
1 & -1 & 0 & 0 & 0 \\
-1 & 2 & -1 & 0 & 0 \\
0 & -1 & 2 & -1 & 0 \\
0 & 0 & -1 & 2 & -1 \\
0 & 0 & 0 & -1 & 2
\end{pmatrix}
$$

is positive definite.

3. Find, to within $\pm 1/4$, all of the eigenvalues of

$$
\begin{pmatrix}
2 & 1 & 0 \\
1 & -1 & -1 \\
0 & -1 & 2
\end{pmatrix}
$$

starting with the interval $[-3, 3]$.

4. Find, to within ± 0.0001, all of the eigenvalues of

(a)
$$
\begin{pmatrix}
1 & 1 & 0 & 0 \\
1 & 2 & -2 & 0 \\
0 & -2 & 3 & 1 \\
0 & 0 & 1 & 4
\end{pmatrix}
$$

(b)
$$
\begin{pmatrix}
1 & -1 & 0 & 0 & 0 & 0 \\
-1 & 2 & 4 & 0 & 0 & 0 \\
0 & 4 & 5 & 0 & 0 & 0 \\
0 & 0 & 0 & 1 & 2 & 0 \\
0 & 0 & 0 & 2 & 10 & -3 \\
0 & 0 & 0 & 0 & -3 & -4
\end{pmatrix}
$$

5. Show that the sequence $\{f'_r(p)\}^n_{r=0}$ satisfies

$$f'_0(p) = 0$$
$$f'_1(p) = -1$$
$$f'_r(p) = (a_r - p)f'_{r-1}(p) - f_{r-1}(p) - b_r^2 f'_{r-2}(p) \qquad r \geqslant 2$$

Apply Newton's method to the characteristic equation

$$f_n(p) = \det(C - pI) = 0$$

to find, correct to 2D, the largest eigenvalue of

$$
\begin{pmatrix}
0 & 1 & 0 \\
1 & -1 & -1 \\
0 & -1 & 1
\end{pmatrix}
$$

starting with the estimate $p_0 = 1.5$. Work to 3D.

6. A seemingly attractive method for the computation of the eigenvectors of the tridiagonal matrix C is to solve the linear equations

$$(C - \lambda I)x = 0$$

i.e.

$$
\begin{aligned}
(a_1 - \lambda)x_1 && + b_2 x_2 &= 0 \\
b_2 x_1 + && (a_2 - \lambda)x_2 && + b_3 x_3 &= 0 \\
&& \vdots \\
b_{n-1} x_{n-2} + (a_{n-1} - \lambda)x_{n-1} + b_n x_n &= 0 \\
b_n x_{n-1} + && (a_n - \lambda)x_n && &= 0
\end{aligned}
$$

in a recursive manner. The value of x_1 can be chosen to be unity and the first equation solved for x_2. These numbers can then be substituted into the second equation to calculate x_3, and so on. The process continues until the $(n-1)$th equation is used to calculate x_n. If λ is an exact eigenvalue and all calculations are performed in exact arithmetic, the last equation will be automatically satisfied.

Illustrate this method by computing the eigenvector of the 21×21 matrix C for which

$$
\begin{aligned}
a_i &= 11 - i && i = 1, \ldots, 21 \\
b_i &= 1 && i = 2, \ldots, 21
\end{aligned}
$$

corresponding to the approximate eigenvalue 10.7462. (The exact value of λ is 10.746 194 2. ...) Work to 4D.

Compare these results with the exact eigenvector given in the following table (components rounded to 4D). Is this method to be recommended as a practical means of calculating eigenvectors of tridiagonal matrices from known approximations to eigenvalues?

i	x_i
1	1.0000
2	0.7462
3	0.3030
4	0.0859
5	0.0188
6	0.0034
7	0.0005
8	0.0001
>8	0.0000

4.5 GIVENS' AND HOUSEHOLDER'S METHODS

We now describe two algorithms for reducing a given symmetric matrix A to a symmetric tridiagonal matrix C using a finite number of orthogonal similarity transformations. The eigenvalues of A are the same as those of C, which can be readily computed using Sturm sequences.

The first algorithm, due to W. J. Givens, uses plane rotation matrices. Let X_{pq} denote the matrix representing a rotation in the (x_p, x_q) plane given by (4.11). For Givens' method the angle θ is chosen to create zeros, not in the (p, q) and (q, p) positions as in Jacobi's method, but in the $(p - 1, q)$ and $(q, p - 1)$ positions. This subtle change of tactic means that zeros can be created in row order without destroying those previously obtained. The $(p - 1, q)$ element of $X_{pq}^{T} A X_{pq}$ is given from equations (4.12) as

$$sa_{p-1,p} + ca_{p-1,q}$$

which is zero if θ is chosen to be

$$-\tan^{-1}\left(\frac{a_{p-1,q}}{a_{p-1,p}}\right)$$

The first stage of Givens' method annihilates elements along the first row (and, by symmetry, down the first column) in positions $(1, 3), (1, 4), \ldots, (1, n)$ using rotation matrices $X_{23}, X_{24}, \ldots, X_{2n}$ in turn. Once a zero has been created in position $(1, j)$, subsequent transformations use matrices X_{pq} with $p, q \neq 1, j$ and so zeros are not destroyed. The second stage annihilates elements in positions $(2, 4), (2, 5), \ldots, (2, n)$ using $X_{34}, X_{35}, \ldots, X_{3n}$. Again, any zeros produced from these transformations are not destroyed as subsequent zeros are created along the second row. Furthermore, zeros previously obtained in the first row are also preserved. To see this consider, for example, what happens to the $(1, 3)$ element when performing the X_{34} transformation. From (4.12), the new element in this position is

$$ca_{13} - sa_{14}$$

where a_{13} and a_{14} denote the elements in the $(1, 3)$ and $(1, 4)$ positions resulting from the previous step, which are known to be zeros. Consequently, the element in the $(1, 3)$ position remains zero. The algorithm continues until a zero is created in position $(n - 2, n)$ using $X_{n-1,n}$. The original matrix can, therefore, be converted into a tridiagonal matrix in exactly

$$(n - 2) + (n - 3) + \cdots + 1 = \tfrac{1}{2}(n - 1)(n - 2)$$

steps.

■ **Example 4.9**
Givens' method reduces

$$A_1 = \begin{pmatrix} 2 & -1 & 1 & 4 \\ -1 & 3 & 1 & 2 \\ 1 & 1 & 5 & -3 \\ 4 & 2 & -3 & 6 \end{pmatrix}$$

to tridiagonal form as follows:

STEP 1   Create a zero in the $(1, 3)$ position using

$$X_{23} = \begin{pmatrix} 1 & 0 & 0 & 0 \\ 0 & c & s & 0 \\ 0 & -s & c & 0 \\ 0 & 0 & 0 & 1 \end{pmatrix} \quad \text{with } \theta = -\tan^{-1}\left(\frac{1}{-1}\right)$$

which gives

$$A_2 = X_{23}^{\mathrm{T}} A_1 X_{23} = \begin{pmatrix} 2.0000 & -1.4142 & 0 & 4.0000 \\ \cdot & 3.0000 & -1.0000 & 3.5355 \\ & & 5.0000 & -0.7071 \\ & & & 6.0000 \end{pmatrix}$$

(By symmetry, the lower triangular part of A_2 is just a mirror image of its upper triangular part.)

STEP 2   Create a zero in the $(1, 4)$ position using

$$X_{24} = \begin{pmatrix} 1 & 0 & 0 & 0 \\ 0 & c & 0 & s \\ 0 & 0 & 1 & 0 \\ 0 & -s & 0 & c \end{pmatrix} \quad \text{with } \theta = -\tan^{-1}\left(\frac{4.0000}{-1.4142}\right)$$

which gives

$$A_3 = X_{24}^{\mathrm{T}} A_2 X_{24} = \begin{pmatrix} 2.0000 & -4.2426 & 0 & 0 \\ \cdot & 3.4444 & 0.3333 & -3.6927 \\ & & 5.0000 & -1.1785 \\ & & & 5.5556 \end{pmatrix}$$

STEP 3   Create a zero in the $(2, 4)$ position using

$$X_{34} = \begin{pmatrix} 1 & 0 & 0 & 0 \\ 0 & 1 & 0 & 0 \\ 0 & 0 & c & s \\ 0 & 0 & -s & c \end{pmatrix} \quad \text{with } \theta = -\tan^{-1}\left(\frac{-3.6927}{0.3333}\right)$$

which gives

$$C = X_{34}^{\mathrm{T}} A_3 X_{34} = \begin{pmatrix} 2.0000 & -4.2426 & 0 & 0 \\ \cdot & 3.4444 & 3.7077 & 0 \\ & & 5.7621 & 1.1097 \\ & & \cdot & 4.7934 \end{pmatrix}$$

The eigenvalues of C (and hence of A) are found, using the Sturm sequence iteration described in Section 4.4, to be -2.3846, 3.5601, 5.2035 and 9.6210 (4D).   □

The second algorithm, due to A. S. Householder, uses matrices of the form

$$Q = I - 2ww^T$$

where I is the $n \times n$ identity and w is some $n \times 1$ vector satisfying $w^Tw = 1$ (i.e. w has unit length). It is easy to verify that Q is both symmetric and orthogonal because

$$Q^T = I^T - 2(ww^T)^T = I - 2(w^T)^Tw^T = I - 2ww^T = Q$$

(making use of the general matrix properties $(A + B)^T = A^T + B^T$, $(AB)^T = B^TA^T$ and $(A^T)^T = A$) and because

$$\begin{aligned} Q^TQ = Q^2 &= (I - 2ww^T)(I - 2ww^T) \\ &= I - 4ww^T + 4w(w^Tw)w^T \\ &= I \end{aligned}$$

(making use of the fact that $w^Tw = 1$).

Given an $n \times 1$ vector x, we now show how to choose the vector w so as to ensure that

$$Qx = ke \qquad (4.18)$$

for some scalar k, where e is the first column of I. Although it is not at all obvious at this stage, it turns out that such a choice of w enables us to convert an $n \times n$ symmetric matrix A to tridiagonal form. Firstly, note that if equation (4.18) is to hold, then k must be given by

$$k = \pm (x^Tx)^{1/2}$$

To see this, we pre-multiply both sides of (4.18) by the transposed equation

$$x^TQ^T = ke^T$$

to get

$$x^TQ^TQx = k^2e^Te$$

The result then follows since $Q^TQ = I$ and $e^Te = 1$.

Now, from equation (4.18),

$$(I - 2ww^T)x = ke$$

which can be rearranged as

$$2w(w^Tx) = x - ke$$

Hence, if we define

$$u = x - ke \qquad (4.19)$$

then

$$w = \alpha u$$

where $\alpha = 1/2(w^T x)$. However, $w^T w = 1$ and so $\alpha^2 u^T u = 1$, i.e.

$$\alpha^2 = \frac{1}{u^T u}$$

The matrix Q is therefore given by

$$Q = I - 2ww^T = I - 2\alpha^2 uu^T = I - 2\frac{uu^T}{u^T u} \qquad (4.20)$$

where u can be trivially computed from equation (4.19) with $k = \pm(x^T x)^{1/2}$. The sign of k is chosen to be opposite to that of the first component of x. This maximizes the magnitude of the first component of $x - ke$. Matrices Q obtained in this way are referred to as Householder matrices.

■ **Example 4.10**

Consider the construction of the Householder matrix satisfying

$$Qx = k\begin{pmatrix} 1 \\ 0 \\ 0 \end{pmatrix} \qquad \text{where } x = \begin{pmatrix} -1 \\ 1 \\ 4 \end{pmatrix}$$

The value of k is $\pm[(-1)^2 + (1)^2 + (4)^2]^{1/2} = \pm 3\sqrt{2}$, which is taken to be $+3\sqrt{2}$ adopting the sign convention. Hence

$$u = \begin{pmatrix} -1 \\ 1 \\ 4 \end{pmatrix} - 3\sqrt{2}\begin{pmatrix} 1 \\ 0 \\ 0 \end{pmatrix} = \begin{pmatrix} -(1+3\sqrt{2}) \\ 1 \\ 4 \end{pmatrix}$$

with $u^T u = [-(1 + 3\sqrt{2})]^2 + 1^2 + 4^2 = 36 + 6\sqrt{2}$.

Finally,

$$Q = \begin{pmatrix} 1 & 0 & 0 \\ 0 & 1 & 0 \\ 0 & 0 & 1 \end{pmatrix} - \frac{2}{36 + 6\sqrt{2}}\begin{pmatrix} -(1+3\sqrt{2}) \\ 1 \\ 4 \end{pmatrix}\big(-(1+3\sqrt{2}), 1, 4\big)$$

$$= \begin{pmatrix} 1 & 0 & 0 \\ 0 & 1 & 0 \\ 0 & 0 & 1 \end{pmatrix} - \frac{18 - 3\sqrt{2}}{306}\begin{pmatrix} 19 + 6\sqrt{2} & -1 - 3\sqrt{2} & -4 - 12\sqrt{2} \\ \cdot & 1 & 4 \\ \cdot & \cdot & 16 \end{pmatrix}$$

$$= \frac{1}{102}\begin{pmatrix} -17\sqrt{2} & 17\sqrt{2} & 68\sqrt{2} \\ \cdot & 96 + \sqrt{2} & -24 + 4\sqrt{2} \\ \cdot & \cdot & 6 + 16\sqrt{2} \end{pmatrix} \qquad \square$$

Householder's algorithm transforms a given $n \times n$ symmetric matrix A_1 to a symmetric tridiagonal matrix C in exactly $n-2$ steps. Each step of the algorithm creates zeros in a complete row and column. The first step annihilates elements in positions $(1,3),(1,4),\ldots,(1,n)$ simultaneously. To see how this can be done, consider the partitioning

$$A_1 = \left(\begin{array}{cccc} a_{11} & a_{12} & \cdots & a_{1n} \\ \hline a_{12} & a_{22} & \cdots & a_{2n} \\ \vdots & \vdots & & \vdots \\ a_{1n} & a_{2n} & \cdots & a_{nn} \end{array}\right) = \left(\begin{array}{c|c} a_{11} & x_1^{\mathrm{T}} \\ \hline x_1 & B_1 \end{array}\right) \text{ say}$$

and let

$$P_1 = \left(\begin{array}{c|c} 1 & 0^{\mathrm{T}} \\ \hline 0 & Q_1 \end{array}\right)$$

where 0 is the $(n-1) \times 1$ zero matrix and Q_1 is a Householder matrix of order $n-1$, constructed to satisfy

$$Q_1 x_1 = k_1 e$$

The matrix P_1 is symmetric and orthogonal since Q_1 has these properties. Hence

$$A_2 = P_1^{-1} A_1 P_1 = P_1^{\mathrm{T}} A_1 P_1 = P_1 A_1 P_1$$

$$= \left(\begin{array}{c|c} 1 & 0^{\mathrm{T}} \\ \hline 0 & Q_1 \end{array}\right)\left(\begin{array}{c|c} a_{11} & x_1^{\mathrm{T}} \\ \hline x_1 & B_1 \end{array}\right)\left(\begin{array}{c|c} 1 & 0^{\mathrm{T}} \\ \hline 0 & Q_1 \end{array}\right)$$

$$= \left(\begin{array}{c|c} 1 & 0^{\mathrm{T}} \\ \hline 0 & Q_1 \end{array}\right)\left(\begin{array}{c|c} a_{11} & (Q_1 x_1)^{\mathrm{T}} \\ \hline x_1 & B_1 Q_1 \end{array}\right)$$

$$= \left(\begin{array}{c|c} a_{11} & (Q_1 x_1)^{\mathrm{T}} \\ \hline Q_1 x_1 & Q_1 B_1 Q_1 \end{array}\right)$$

$$= \left(\begin{array}{c|c} a_{11} & k_1 e^{\mathrm{T}} \\ \hline k_1 e & Q_1 B_1 Q_1 \end{array}\right)$$

which has the required form because the first row of A_2 is $(a_{11}, k_1, 0, 0, \ldots, 0)$.

Similarly, step r annihilates elements in positions $(r, r+2), (r, r+3), \ldots, (r, n)$ simultaneously, using

$$A_{r+1} = P_r A_r P_r$$

with

$$P_r = \left(\begin{array}{c|c} I_r & 0^{\mathrm{T}} \\ \hline 0 & Q_r \end{array}\right)$$

The matrix I_r is the rth order identity, 0 is the $(n-r) \times r$ zero matrix, and Q_r is an $(n-r) \times (n-r)$ Householder matrix satisfying

$$Q_r x_r = k_r e$$

where e is the first column of I_{n-r}, and x_r is the $(n-r) \times 1$ vector comprising the elements below the diagonal in column r of A_r. Note that this transformation does not alter the first $r-1$ rows and columns of A_r, and so no zeros are lost.

■ **Example 4.11**

Householder's method reduces

$$A_1 = \begin{pmatrix} 2 & -1 & 1 & 4 \\ -1 & 3 & 1 & 2 \\ 1 & 1 & 5 & -3 \\ 4 & 2 & -3 & 6 \end{pmatrix}$$

to tridiagonal form as follows:

STEP 1   Create zeros in positions $(1, 3)$ and $(1, 4)$ using

$$P_1 = \left(\begin{array}{c|c} 1 & 0 \\ \hline 0 & Q_1 \end{array} \right)$$

where Q_1 satisfies

$$Q_1 \begin{pmatrix} -1 \\ 1 \\ 4 \end{pmatrix} = k_1 \begin{pmatrix} 1 \\ 0 \\ 0 \end{pmatrix}$$

From Example 4.10, Q_1 is given by

$$\begin{pmatrix} -0.2357 & 0.2357 & 0.9428 \\ \cdot & 0.9550 & -0.1798 \\ \cdot & \cdot & 0.2807 \end{pmatrix}$$

which gives

$$A_2 = P_1 A_1 P_1 = \begin{pmatrix} 2.0000 & 4.2426 & 0 & 0 \\ \cdot & 3.4444 & -2.2729 & 2.9293 \\ \cdot & \cdot & 6.2324 & -0.7448 \\ \cdot & \cdot & \cdot & 4.3232 \end{pmatrix}$$

STEP 2   Create a zero in position $(2, 4)$ using

$$P_2 = \left(\begin{array}{c|c} I_2 & 0 \\ \hline 0 & Q_2 \end{array} \right)$$

where Q_2 satisfies

$$Q_2 \begin{pmatrix} -2.2729 \\ 2.9293 \end{pmatrix} = k_2 \begin{pmatrix} 1 \\ 0 \end{pmatrix}$$

Using equations (4.19) and (4.20), Q_2 is calculated as

$$\begin{pmatrix} -0.6130 & 0.7901 \\ 0.7901 & 0.6130 \end{pmatrix}$$

which gives

$$C = P_2 A_2 P_2 = \begin{pmatrix} 2.0000 & 4.2426 & 0 & 0 \\ \cdot & 3.4444 & 3.7077 & 0 \\ \cdot & \cdot & 5.7621 & -1.1097 \\ \cdot & \cdot & \cdot & 4.7934 \end{pmatrix} \qquad \square$$

It is interesting to compare the results of Examples 4.9 and 4.11; the final matrices are identical apart from a change of sign in some of the off-diagonal elements. In fact it is possible to prove that, whenever a symmetric matrix is reduced to a tridiagonal matrix C using orthogonal transformations, the elements of C are unique, except for the signs of its off-diagonal elements. A simple proof of this is given in Fox (1964). Although the final reduced matrices are almost identical for the two methods described in this section, the calculations used to obtain them are completely different. In fact, Givens' method requires approximately twice as many arithmetic operations as Householder's method to compute the tridiagonal matrix. Consequently, Householder's method is the one most widely used in practice. The only time that Givens' algorithm is preferred is when the original matrix already has a large number of zero off-diagonal elements, since it is easy to omit the transformation X_{pq} if the current element in position $(p-1, q)$ is zero.

Inverse iteration can easily be used to calculate the eigenvectors of the tridiagonal matrix C and, from Lemma 4.1, those of A can be found by pre-multiplying these eigenvectors by the product of successive transformation matrices.

EXERCISE 4.5

1. If

$$\tan \theta = -\frac{a_{p-1,q}}{a_{p-1,p}}$$

show that

$$\sin \theta = \pm \frac{a_{p-1,q}}{\sqrt{(a_{p-1,q}^2 + a_{p-1,p}^2)}}$$

$$\cos \theta = \mp \frac{a_{p-1,p}}{\sqrt{(a_{p-1,q}^2 + a_{p-1,p}^2)}}$$

2. If $w = (1/3, 2/3, 2/3)^T$ show that $w^T w = 1$. Calculate $Q = I - 2ww^T$ and verify that Q is symmetric and orthogonal.

3. Transform the matrix

$$
\begin{pmatrix}
3 & 2 & 1 & 2 \\
2 & -1 & 1 & 2 \\
1 & 1 & 4 & 3 \\
2 & 2 & 3 & 1
\end{pmatrix}
$$

to tridiagonal form using Givens' method.

4. Construct Householder matrices Q_1 and Q_2 such that

$$
Q_1 \begin{pmatrix} 2 \\ 1 \\ 2 \end{pmatrix} = k_1 \begin{pmatrix} 1 \\ 0 \\ 0 \end{pmatrix}
$$

and

$$
Q_2 \begin{pmatrix} -3 \\ -1 \end{pmatrix} = k_2 \begin{pmatrix} 1 \\ 0 \end{pmatrix}
$$

Hence reduce the 4×4 matrix given in Question 3 to tridiagonal form.

5. Use Givens' method to convert

$$
A = \begin{pmatrix}
10 & -6 & -4 \\
-6 & 11 & 2 \\
-4 & 2 & 6
\end{pmatrix}
$$

into a tridiagonal matrix C, and calculate its eigenvalues to within ± 0.05 using Sturm sequences.

Find the approximate eigenvectors of C by performing two steps of inverse iteration, starting with $(1, 1, 1)^T$ and using shifts 18.0001, 6.0001 and 3.0001. Hence write down approximations to the eigenvectors of A.

4.6 THE LR AND QR METHODS

We have already seen in Section 4.3 how Jacobi's method may be used to find the eigenvalues of a symmetric matrix by computing a sequence of matrices $A_1, A_2, \ldots$ which converge to a diagonal matrix. The analogous methods for unsymmetric matrices are the LR and QR methods in which the limiting form of A_s is upper triangular.

The LR method is based upon the triangular decomposition of a matrix.

The first stage is to carry out the LU decomposition of a given matrix A_1, i.e.

$$A_1 = L_1 R_1$$

(In the present context it is customary to denote the upper triangular matrix by R instead of U. This notation is used because L and R are the first letters of the words 'left' and 'right'.) A new matrix A_2 is then formed by multiplying these matrices in reverse order, i.e.

$$A_2 = R_1 L_1$$

The whole process is repeated, and the general equations defining the method are

$$\begin{aligned} A_s &= L_s R_s \\ A_{s+1} &= R_s L_s \end{aligned} \qquad s = 1, 2, \ldots$$

Each complete step is a similarity transformation because

$$A_{s+1} = R_s L_s = L_s^{-1} A_s L_s$$

and so all of the matrices A_s have the same eigenvalues.

Suppose now that A_1 has real eigenvalues λ_i with

$$|\lambda_1| > |\lambda_2| > \cdots > |\lambda_n|$$

It is possible to prove (see Wilkinson, 1965) that, under certain not very restrictive conditions, A_s converges to an upper triangular matrix with

$$\lim_{s \to \infty} A_s = \begin{pmatrix} \lambda_1 & & & X \\ & \lambda_2 & & \\ & & \ddots & \\ 0 & & & \lambda_n \end{pmatrix}$$

The numbers on the leading diagonal are the eigenvalues of A, in decreasing order of magnitude. Moreover, the rate at which the lower triangular element $a_{ij}^{(s)}$ of A_s converges to zero is of order

$$\left(\frac{\lambda_i}{\lambda_j} \right)^s \qquad i > j$$

This implies, in particular, that the order of convergence of the elements along the first subdiagonal is

$$\left(\frac{\lambda_{i+1}}{\lambda_i} \right)^s$$

and so convergence will be slow whenever two or more eigenvalues are close together. If any of the eigenvalues are complex then the situation is rather more complicated. In this case,

$$
\lim_{s \to \infty} A_s =
\begin{pmatrix}
\lambda_1 & & & & & & X \\
& \lambda_2 & & & & & \\
& & \ddots & & & & \\
& & & \lambda_{r-1} & & & \\
& & & & a_{r,r}^{(s)} & a_{r,r+1}^{(s)} & \\
& & & & a_{r+1,r}^{(s)} & a_{r+1,r+1}^{(s)} & \\
& & & & & & \lambda_{r+2} \\
& 0 & & & & & \ddots \\
& & & & & & \lambda_n
\end{pmatrix}
$$

The real eigenvalues $\lambda_1, \ldots, \lambda_{r-1}, \lambda_{r+2}, \ldots, \lambda_n$ are obtained as before. The elements of the 2×2 submatrices

$$
\begin{pmatrix}
a_{r,r}^{(s)} & a_{r,r+1}^{(s)} \\
a_{r+1,r}^{(s)} & a_{r+1,r+1}^{(s)}
\end{pmatrix}
$$

jutting out of the leading diagonal of A_s do not converge, but their eigenvalues converge to a complex conjugate pair of eigenvalues of A_1. These can be trivially computed as the roots of the characteristic polynomials of each 2×2 submatrix.

Unfortunately, there are several computational difficulties associated with this method which severely limit its practical use. The actual triangular decomposition itself may not exist at each stage and, even if it does, the numerical calculations used to obtain it may not be stable. However, we have already seen in Section 2.2 that this problem may be resolved by the use of partial pivoting. In the presence of interchanges,

$$
P_s A_s = L_s R_s
$$

where P_s represents the row permutations used in the decomposition (see equation (2.10)). In order to preserve eigenvalues it is necessary to calculate A_{s+1} from

$$
A_{s+1} = (R_s P_s) L_s
$$

It is easy to see that this is a similarity transformation because

$$
A_{s+1} = (R_s P_s) L_s = L_s^{-1} P_s A_s P_s L_s
$$

and $P_s^{-1} = P_s$.

The matrix P_s does not have to be computed explicitly; $R_s P_s$ is just a column permutation of R_s using interchanges corresponding to the row interchanges used in the decomposition of A_s'.

■ **Example 4.12**

To illustrate the success of the stable *LR* method, consider

$$A_1 = \begin{pmatrix} 2 & 1 & 3 & 1 \\ -1 & 2 & 2 & 1 \\ 1 & 0 & 1 & 0 \\ -1 & -1 & -1 & 1 \end{pmatrix}$$

which has exact eigenvalues $(3 \pm \sqrt{5})/2$ and $[3 \pm (\sqrt{3})i]/2$. The triangular decomposition of A_1 produces

$$L_1 = \begin{pmatrix} 1 & 0 & 0 & 0 \\ -0.500 & 1 & 0 & 0 \\ -0.500 & -0.200 & 1 & 0 \\ 0.500 & -0.200 & 0.167 & 1 \end{pmatrix}$$

$$R_1 = \begin{pmatrix} 2.000 & 1.000 & 3.000 & 1.000 \\ 0 & 2.500 & 3.500 & 1.500 \\ 0 & 0 & 1.200 & 1.800 \\ 0 & 0 & 0 & -0.500 \end{pmatrix}$$

where rows 3 and 4 are interchanged when equating the third columns of A_1 and $L_1 R_1$. The matrix $R_1 P_1$ is therefore obtained by interchanging columns 3 and 4 of R_1, i.e.

$$R_1 P_1 = \begin{pmatrix} 2.000 & 1.000 & 1.000 & 3.000 \\ 0 & 2.500 & 1.500 & 3.500 \\ 0 & 0 & 1.800 & 1.200 \\ 0 & 0 & -0.500 & 0 \end{pmatrix}$$

Post-multiplying $R_1 P_1$ by L_1 gives

$$A_2 = (R_1 P_1)L_1 = \begin{pmatrix} 2.500 & 0.200 & 1.500 & 3.000 \\ -0.250 & 1.500 & 2.083 & 3.500 \\ -0.300 & -0.600 & 2.000 & 1.200 \\ 0.250 & 0.100 & -0.500 & 0.000 \end{pmatrix}$$

The next few matrices in the sequence are

$$A_3 = \begin{pmatrix} 2.600 & -0.211 & 0.739 & 3.000 \\ -0.040 & 0.874 & 1.270 & 3.800 \\ -0.063 & -0.989 & 2.265 & 3.000 \\ 0.026 & 0.014 & -0.066 & 0.261 \end{pmatrix}$$

$$A_4 = \begin{pmatrix} 2.615 & 0.876 & -0.245 & 3.000 \\ -0.009 & 3.105 & -1.029 & 3.073 \\ -0.014 & 3.176 & -0.074 & 6.537 \\ 0.004 & -0.006 & -0.004 & 0.354 \end{pmatrix}$$

$$A_5 = \begin{pmatrix} 2.618 & 0.605 & 0.888 & 3.000 \\ -0.002 & 3.019 & 3.207 & 6.553 \\ 0.001 & -0.949 & -0.013 & -3.319 \\ 0.001 & -0.001 & 0.002 & 0.377 \end{pmatrix}$$

$$A_6 = \begin{pmatrix} 2.618 & 0.325 & 0.895 & 3.000 \\ 0.000 & 2.009 & 3.223 & 6.556 \\ 0.000 & -0.312 & 0.992 & -1.259 \\ 0.000 & 0.000 & 0.001 & 0.382 \end{pmatrix}$$

The $(1, 1)$ and $(4, 4)$ elements have converged to the real eigenvalues 2.618 and 0.382 (3D), respectively. Furthermore, the appropriate 2×2 submatrices of A_5 and A_6 which are given by

$$\begin{pmatrix} 3.019 & 3.207 \\ -0.949 & -0.013 \end{pmatrix} \qquad \begin{pmatrix} 2.009 & 3.223 \\ -0.312 & 0.992 \end{pmatrix}$$

have eigenvalues $[3.006 \pm (\sqrt{2.980})\mathrm{i}]/2$ and $[3.001 \pm (\sqrt{2.990})\mathrm{i}]/2$. Although the submatrices themselves are not converging, their eigenvalues converge to the conjugate pair $[3 \pm (\sqrt{3})\mathrm{i}]/2$.   □

Another technique, which is very similar to the *LR* method, is based upon the decomposition of a matrix into the product of an orthogonal matrix Q and an upper triangular matrix R. Such a decomposition always exists and can be found using plane rotation matrices. As remarked in Section 4.3, methods involving the products of orthogonal matrices are automatically stable without the need for interchanges, and because of this the so-called *QR* method is usually preferred to the *LR* method. The general equations defining this method are

$$\begin{aligned} A_s &= Q_s R_s \\ A_{s+1} &= R_s Q_s \end{aligned} \qquad s = 1, 2, \ldots$$

The sequence $A_1, A_2, \ldots$ exhibits the same limiting behaviour as that of the *LR* method, enabling all of the eigenvalues, both real and complex, to be determined. It remains to show how the matrices Q_s and R_s can be constructed at each stage.

The simplest way of calculating the *QR* decomposition of an $n \times n$ matrix A is to pre-multiply A by a series of rotation matrices X_{pq}^T, where X_{pq} is given by (4.11) and the values of p, q and θ are chosen to annihilate one of the lower triangular elements. From equation (4.14), the element in the (q, p) position of $X_{pq}^T A$ is

$$sa_{pp} + ca_{qp}$$

which is zero if $\theta = -\tan^{-1}(a_{qp}/a_{pp})$. The first stage of the decomposition annihilates the element in position $(2, 1)$ using X_{12}^T. The next two stages annihilate elements in positions $(3, 1)$ and $(3, 2)$ using X_{13}^T and X_{23}^T. The process continues in this way, creating zeros in row order, until $X_{n-1,n}^T$ is used to annihilate the element in position $(n, n-1)$. Applying similar reasoning to that which was used in the Givens method, it is easy to see that no zeros are destroyed and an upper triangular matrix R is produced after $n(n-1)/2$ pre-multiplications, i.e.

$$X_{n-1,n}^T \cdots X_{13}^T X_{12}^T A = R$$

which can be rearranged as

$$A = (X_{12} X_{13} \cdots X_{n-1,n}) R$$

since $X_{pq}^T = X_{pq}^{-1}$. The matrix $Q = X_{12} X_{13} \cdots X_{n-1,n}$ is readily checked to be orthogonal because

$$Q^T = (X_{12} X_{13} \cdots X_{n-1,n})^T = X_{n-1,n}^T \cdots X_{13}^T X_{12}^T$$
$$= X_{n-1,n}^{-1} \cdots X_{13}^{-1} X_{12}^{-1} = (X_{12} X_{13} \cdots X_{n-1,n})^{-1} = Q^{-1}$$

■ **Example 4.13**
Consider the orthogonal decomposition of

$$A = \begin{pmatrix} 1 & 4 & 2 \\ -1 & 2 & 0 \\ 1 & 3 & -1 \end{pmatrix}$$

STEP 1   Create a zero in the $(2, 1)$ position using

$$X_{12} = \begin{pmatrix} c & s & 0 \\ -s & c & 0 \\ 0 & 0 & 1 \end{pmatrix} \quad \text{with } \theta = -\tan^{-1}\left(\frac{-1}{1}\right)$$

which gives

$$X_{12}^T A = \begin{pmatrix} 1.4142 & 1.4142 & 1.4142 \\ 0 & 4.2426 & 1.4142 \\ 1.0000 & 3.0000 & -1.0000 \end{pmatrix}$$

STEP 2   Create a zero in the $(3, 1)$ position using

$$X_{13} = \begin{pmatrix} c & 0 & s \\ 0 & 1 & 0 \\ -s & 0 & c \end{pmatrix} \quad \text{with } \theta = -\tan^{-1}\left(\frac{1.0000}{1.4142}\right)$$

which gives

$$X_{13}^{T}(X_{12}^{T}A) = \begin{pmatrix} 1.7321 & 2.8868 & 0.5774 \\ 0 & 4.2426 & 1.4142 \\ 0 & 1.6330 & -1.6330 \end{pmatrix}$$

STEP 3   Create a zero in the $(3,2)$ position using

$$X_{23} = \begin{pmatrix} 1 & 0 & 0 \\ 0 & c & s \\ 0 & -s & c \end{pmatrix} \quad \text{with } \theta = -\tan^{-1}\left(\frac{1.6330}{4.2426}\right)$$

which gives

$$R = X_{23}^{T}(X_{13}^{T}X_{12}^{T}A) = \begin{pmatrix} 1.7321 & 2.8868 & 0.5774 \\ 0 & 4.5461 & 0.7332 \\ 0 & 0 & -2.0320 \end{pmatrix}$$

Therefore

$$Q = X_{12}X_{13}X_{23}$$

$$= \begin{pmatrix} 0.7071 & 0.7071 & 0 \\ -0.7071 & 0.7071 & 0 \\ 0 & 0 & 1 \end{pmatrix} \begin{pmatrix} 0.8165 & 0 & -0.5774 \\ 0 & 1 & 0 \\ 0.5774 & 0 & 0.8165 \end{pmatrix} \begin{pmatrix} 1 & 0 & 0 \\ 0 & 0.9332 & -0.3592 \\ 0 & 0.3592 & 0.9332 \end{pmatrix}$$

$$= \begin{pmatrix} 0.5774 & 0.5133 & -0.6350 \\ -0.5774 & 0.8066 & 0.1270 \\ 0.5774 & 0.2933 & 0.7620 \end{pmatrix}$$

Hence the original matrix can be decomposed as

$$\begin{pmatrix} 0.5774 & 0.5133 & -0.6350 \\ -0.5774 & 0.8066 & 0.1270 \\ 0.5774 & 0.2933 & 0.7620 \end{pmatrix} \begin{pmatrix} 1.7321 & 2.8868 & 0.5774 \\ 0 & 4.5461 & 0.7332 \\ 0 & 0 & -2.0320 \end{pmatrix} \qquad \square$$

Once the eigenvalues have been determined, the corresponding eigenvectors are computed via inverse iteration. This is particularly easy to apply in conjunction with the *LR* and *QR* methods since the limiting form is upper triangular and only one or two steps of back substitution are needed. From Lemma 4.1, the eigenvectors of the original matrix A_1 are found from the computed eigenvectors of A_s by pre-multiplying by $L_1 L_2 \cdots L_{s-1}$ and $Q_1 Q_2 \cdots Q_{s-1}$ in the case of the *LR* and *QR* methods, respectively. However, this can be costly if only a selected number of eigenvectors are required. Under these circumstances it is better to apply inverse iteration directly to A_1.

EXERCISE 4.6

1. If $s = a_{qp}/(a_{qp}^2 + a_{pp}^2)^{1/2}$ and $c = -a_{pp}/(a_{qp}^2 + a_{pp}^2)^{1/2}$, show that

   (a) $c^2 + s^2 = 1$
   (b) $sa_{pp} + ca_{pq} = 0$

   Hence, or otherwise, construct an orthogonal decomposition of

   $$\begin{pmatrix} 1 & 1 & 3 \\ 2 & 0 & 1 \\ 2 & 1 & -1 \end{pmatrix}$$

   Use exact arithmetic.

2. Use the (i) stable *LR* method (ii) *QR* method to compute the eigenvalues of the following matrices. The algorithms should be stopped when all of the (appropriate) lower triangular elements of R_s are less than 0.0001.

   (a) $\begin{pmatrix} 3 & 0 & 1 \\ 2 & 2 & 2 \\ 4 & 2 & 5 \end{pmatrix}$
   (b) $\begin{pmatrix} 337 & 304 & 176 \\ 304 & 361 & 128 \\ 176 & 128 & 121 \end{pmatrix}$

   (c) $\begin{pmatrix} 12 & 3 & 1 \\ -9 & -2 & -3 \\ 14 & 6 & 2 \end{pmatrix}$
   (d) $\begin{pmatrix} 2 & -2 & 10 \\ -2 & 11 & 8 \\ 10 & 8 & 5 \end{pmatrix}$

   (e) $\begin{pmatrix} 1 & -1 & 0 & 0 \\ -1 & 2 & -1 & 0 \\ 0 & -1 & 3 & -1 \\ 0 & 0 & -1 & 4 \end{pmatrix}$
   (f) $\begin{pmatrix} -1 & -2 & 1 & 3 \\ -4 & -1 & 5 & 4 \\ -4 & -2 & 2 & 5 \\ 0 & -2 & -1 & 4 \end{pmatrix}$

3. Use the

   (a) *LR* method without interchanges
   (b) *LR* method with interchanges

   to find the eigenvalues of

   $$\begin{pmatrix} 1 & 3 \\ 2 & 0 \end{pmatrix}$$

4. If Q_1, Q_2 are $n \times n$ orthogonal matrices and R_1, R_2 are $n \times n$ non-singular upper triangular matrices, comment on the nature of $Q_1^T Q_2$ and $R_1 R_2^{-1}$.
   If $Q_1 R_1 = Q_2 R_2$, show that $R_1 R_2^{-1} = Q_1^T Q_2$. Deduce that there exists a diagonal matrix D with ± 1s on its leading diagonal such that $R_1 = DR_2$ and $Q_2 = Q_1 D$ (i.e. that the *QR* decomposition is unique to within a change of sign in the rows of R and the columns of Q).

5. Prove that (a) eigenvalues (b) symmetry are preserved in the QR algorithm.
   If Q_sR_s denotes an orthogonal decomposition of a 4×4 matrix of the form

$$A_s = \begin{pmatrix} x & x & 0 & 0 \\ x & x & x & 0 \\ 0 & x & x & x \\ 0 & 0 & x & x \end{pmatrix}$$

indicate which elements of Q_s and R_s are (possibly) non-zero and deduce that R_sQ_s has the form

$$A_{s+1} = \begin{pmatrix} x & x & x & x \\ x & x & x & x \\ 0 & x & x & x \\ 0 & 0 & x & x \end{pmatrix}$$

Hence show that if A_s is a 4×4 symmetric tridiagonal matrix, then so is A_{s+1}.

6. If $A = QR$, $Ax = b$ and $Rx = c$, show that

$$c = Q^{\mathrm{T}}b$$

Use these relations, together with your answer to Question 1, to find the solution of the linear system

$$\begin{pmatrix} 1 & 1 & 3 \\ 2 & 0 & 1 \\ 2 & 1 & -1 \end{pmatrix} \begin{pmatrix} x_1 \\ x_2 \\ x_3 \end{pmatrix} = \begin{pmatrix} 3 \\ 3 \\ 0 \end{pmatrix}$$

4.7 HESSENBERG FORM

The major drawback of the LR and QR algorithms is the high volume of work involved in using them. For example, the LR method requires approximately $(1/3)n^3$ multiplications for the triangular decomposition and an additional $(1/3)n^3$ multiplications for the matrix product – all for just one step of the algorithm. However, the number of calculations is considerably reduced if the original matrix is first transformed into a special form containing a high proportion of zero elements which are retained at each step. One type of matrix which satisfies these conditions is an upper Hessenberg matrix whose lower triangular elements are all zero, with the possible exception of those elements on the first subdiagonal. More formally, $H = (h_{ij})$ is upper Hessenberg if $h_{ij} = 0$ for $i > j + 1$.

We first show how to convert a general $n \times n$ matrix into this form and then go on to demonstrate its usefulness in the LR and QR algorithms. For simplicity, we begin with the case $n = 4$ and consider the reduction to upper Hessenberg form using transformation matrices resembling those used in the analysis of Gauss elimination. Let

$$A_1 = \begin{pmatrix} a_{11} & a_{12} & a_{13} & a_{14} \\ a_{21} & a_{22} & a_{23} & a_{24} \\ a_{31} & a_{32} & a_{33} & a_{34} \\ a_{41} & a_{42} & a_{43} & a_{44} \end{pmatrix}$$

STEP 1   The first step annihilates the elements in the (3, 1) and (4, 1) positions. This can be done by subtracting multiples $m_{31} = a_{31}/a_{21}$ and $m_{41} = a_{41}/a_{21}$ of row 2 from rows 3 and 4, respectively, which is equivalent to pre-multiplying A_1 by

$$\begin{pmatrix} 1 & 0 & 0 & 0 \\ 0 & 1 & 0 & 0 \\ 0 & -m_{31} & 1 & 0 \\ 0 & -m_{41} & 0 & 1 \end{pmatrix}$$

Since we wish to carry out a similarity transformation to preserve eigenvalues, it is also necessary to post-multiply by the inverse of this matrix, which, from Section 4.2, is

$$\begin{pmatrix} 1 & 0 & 0 & 0 \\ 0 & 1 & 0 & 0 \\ 0 & m_{31} & 1 & 0 \\ 0 & m_{41} & 0 & 1 \end{pmatrix}$$

In step 1, we therefore calculate

$$\begin{pmatrix} 1 & 0 & 0 & 0 \\ 0 & 1 & 0 & 0 \\ 0 & -m_{31} & 1 & 0 \\ 0 & -m_{41} & 0 & 1 \end{pmatrix} \begin{pmatrix} a_{11} & a_{12} & a_{13} & a_{14} \\ a_{21} & a_{22} & a_{23} & a_{24} \\ a_{31} & a_{32} & a_{33} & a_{34} \\ a_{41} & a_{42} & a_{43} & a_{44} \end{pmatrix} \begin{pmatrix} 1 & 0 & 0 & 0 \\ 0 & 1 & 0 & 0 \\ 0 & m_{31} & 1 & 0 \\ 0 & m_{41} & 0 & 1 \end{pmatrix}$$

which is conveniently written as $A_2 = M_1^{-1} A_1 M_1$. Now, pre-multiplication by M_1^{-1} alters the last two rows and post-multiplication by M_1 alters the second column. Hence A_2 takes the form

$$\begin{pmatrix} a_{11} & a_{12}^{(2)} & a_{13} & a_{14} \\ a_{21} & a_{22}^{(2)} & a_{23} & a_{24} \\ 0 & a_{32}^{(2)} & a_{33}^{(2)} & a_{34}^{(2)} \\ 0 & a_{42}^{(2)} & a_{43}^{(2)} & a_{44}^{(2)} \end{pmatrix}$$

where $a_{ij}^{(2)}$ denotes the new element in position (i, j).

STEP 2   The second step annihilates the element in the $(4, 2)$ position. This can be done by subtracting a multiple $m_{42} = a_{42}^{(2)}/a_{32}^{(2)}$ of row 3 from row 4. We therefore perform the transformation

$$
\begin{pmatrix}
1 & 0 & 0 & 0 \\
0 & 1 & 0 & 0 \\
0 & 0 & 1 & 0 \\
0 & 0 & -m_{42} & 1
\end{pmatrix}
\begin{pmatrix}
a_{11} & a_{12}^{(2)} & a_{13} & a_{14} \\
a_{21} & a_{22}^{(2)} & a_{23} & a_{24} \\
0 & a_{32}^{(2)} & a_{33}^{(2)} & a_{34}^{(2)} \\
0 & a_{42}^{(2)} & a_{43}^{(2)} & a_{44}^{(2)}
\end{pmatrix}
\begin{pmatrix}
1 & 0 & 0 & 0 \\
0 & 1 & 0 & 0 \\
0 & 0 & 1 & 0 \\
0 & 0 & m_{42} & 1
\end{pmatrix}
$$

which we write as $A_3 = M_2^{-1} A_2 M_2$. This produces a matrix

$$
A_3 =
\begin{pmatrix}
a_{11} & a_{12}^{(2)} & a_{13}^{(3)} & a_{14} \\
a_{12} & a_{22}^{(2)} & a_{23}^{(3)} & a_{24} \\
0 & a_{32}^{(2)} & a_{33}^{(3)} & a_{34}^{(2)} \\
0 & 0 & a_{43}^{(3)} & a_{44}^{(3)}
\end{pmatrix}
$$

which is upper Hessenberg.

A general $n \times n$ matrix A_1 can be reduced to upper Hessenberg form in exactly $n - 2$ steps. Step j annihilates elements in positions $(j + 2, j)$, $(j + 3, j), \ldots, (n, j)$, which is achieved by performing the similarity transformation

$$
A_{j+1} = M_j^{-1} A_j M_j
$$

The columns of M_j are those of the $n \times n$ identity matrix with the exception of the $(j + 1)$th column, which is

$$
(0, 0, \ldots, 1, m_{j+2, j}, m_{j+3, j}, \ldots, m_{nj})^{\mathrm{T}}
$$

where $m_{ij} = a_{ij}^{(j)}/a_{j+1, j}^{(j)}$.

The reduction fails if any $a_{j+1, j}^{(j)} = 0$ and, as in Gauss elimination, is unstable whenever $|m_{ij}| > 1$. Row and column interchanges are used to avoid these difficulties. At step j, the elements below the diagonal in column j are scanned. If the element of largest modulus occurs in row r_j, say, then rows $j + 1$ and r_j are interchanged. This is equivalent to pre-multiplication by

$$
I_{j+1, r_j}
$$

where I_{ab} denotes the matrix obtained from the identity matrix by interchanging rows a and b. Preservation of the eigenvalues is assured provided we also post-multiply by

$$
I_{j+1, r_j}^{-1} = I_{j+1, r_j}
$$

which is equivalent to interchanging columns $j+1$ and r_j. Hence step j is given by

$$A_{j+1} = M_j^{-1}(I_{j+1,r_j}A_jI_{j+1,r_j})M_j$$

where the elements of M_j are all less than or equal to one in modulus.

■ **Example 4.14**

Consider the reduction of

$$A_1 = \begin{pmatrix} 4 & 2 & 1 & -3 \\ 2 & 4 & 1 & -3 \\ 3 & 2 & 2 & -3 \\ 1 & 2 & 1 & 0 \end{pmatrix}$$

to upper Hessenberg form using elementary transformations with interchanges.

STEP 1 The element of largest modulus below the diagonal in the first column occurs in the third row, so we need to interchange rows 2 and 3 and columns 2 and 3 to get

$$I_{23}A_1I_{23} = \begin{pmatrix} 4 & 1 & 2 & -3 \\ 3 & 2 & 2 & -3 \\ 2 & 1 & 4 & -3 \\ 1 & 1 & 2 & 0 \end{pmatrix}$$

Hence

$$M_1 = \begin{pmatrix} 1 & 0 & 0 & 0 \\ 0 & 1 & 0 & 0 \\ 0 & 2/3 & 1 & 0 \\ 0 & 1/3 & 0 & 1 \end{pmatrix}$$

which gives

$$A_2 = M_1^{-1}(I_{23}A_1I_{23})M_1 = \begin{pmatrix} 4 & 4/3 & 2 & -3 \\ 3 & 7/3 & 2 & -3 \\ 0 & 10/9 & 8/3 & -1 \\ 0 & 14/9 & 4/3 & 1 \end{pmatrix}$$

STEP 2 The element of largest modulus below the diagonal in the second column occurs in the fourth row, and so we need to interchange rows 3 and

4 and columns 3 and 4 to get

$$I_{34}A_2I_{34} = \begin{pmatrix} 4 & 4/3 & -3 & 2 \\ 3 & 7/3 & -3 & 2 \\ 0 & 14/9 & 1 & 4/3 \\ 0 & 10/9 & -1 & 8/3 \end{pmatrix}$$

Hence

$$M_2 = \begin{pmatrix} 1 & 0 & 0 & 0 \\ 0 & 1 & 0 & 0 \\ 0 & 0 & 1 & 0 \\ 0 & 0 & 5/7 & 1 \end{pmatrix}$$

which gives

$$A_3 = M_2^{-1}(I_{34}A_2I_{34})M_2 = \begin{pmatrix} 4 & 4/3 & -11/7 & 2 \\ 3 & 7/3 & -11/7 & 2 \\ 0 & 14/9 & 41/21 & 4/3 \\ 0 & 0 & -24/49 & 12/7 \end{pmatrix} \qquad \Box$$

We now show how a preliminary similarity transformation to an upper Hessenberg matrix H facilitates the computation of eigenvalues via the LR and QR algorithms. To avoid unnecessary repetition we concentrate on the QR algorithm only. (All that follows has an obvious counterpart for the LR algorithm.) The important thing to notice is that the calculation of the QR decomposition itself is greatly simplified if the matrix is upper Hessenberg, because only $n - 1$ rotation matrices are required (one for each of the non-zero subdiagonal elements) compared with the $n(n - 1)/2$ rotation matrices used for a full matrix. This saving is offset by the extra calculations needed to produce H, which would be particularly damaging if it were not for the fact that upper Hessenberg form is preserved in the QR algorithm and so need only be obtained once. To see this, consider the 4×4 matrix

$$H_s = \begin{pmatrix} x & x & x & x \\ x & x & x & x \\ 0 & x & x & x \\ 0 & 0 & x & x \end{pmatrix}$$

where an x denotes some (possibly) non-zero element of H_s. The rotation matrices used in the orthogonal decomposition of H_s are X_{12}, X_{23} and X_{34}.

Hence the orthogonal matrix Q_s is calculated from

$$Q_s = X_{12}X_{23}X_{34} = \begin{pmatrix} x & x & 0 & 0 \\ x & x & 0 & 0 \\ 0 & 0 & x & 0 \\ 0 & 0 & 0 & x \end{pmatrix}\begin{pmatrix} x & 0 & 0 & 0 \\ 0 & x & x & 0 \\ 0 & x & x & 0 \\ 0 & 0 & 0 & x \end{pmatrix}\begin{pmatrix} x & 0 & 0 & 0 \\ 0 & x & 0 & 0 \\ 0 & 0 & x & x \\ 0 & 0 & x & x \end{pmatrix}$$

$$= \begin{pmatrix} x & x & x & x \\ x & x & x & x \\ 0 & x & x & x \\ 0 & 0 & x & x \end{pmatrix}$$

and so

$$H_{s+1} = R_sQ_s = \begin{pmatrix} x & x & x & x \\ 0 & x & x & x \\ 0 & 0 & x & x \\ 0 & 0 & 0 & x \end{pmatrix}\begin{pmatrix} x & x & x & x \\ x & x & x & x \\ 0 & x & x & x \\ 0 & 0 & x & x \end{pmatrix} = \begin{pmatrix} x & x & x & x \\ x & x & x & x \\ 0 & x & x & x \\ 0 & 0 & x & x \end{pmatrix}$$

which is of the required form.

Unfortunately, although transformation to upper Hessenberg form reduces the number of calculations at each step, the method may still prove to be computationally inefficient if the number of steps required for convergence is too large. It was noted in the previous section that the rate of convergence of the elements in positions $(i+1, i)$ is $(\lambda_{i+1}/\lambda_i)^s$, similar to that of the power method. The successful use of an origin shift for that method suggests a similar modification to the QR algorithm. The equations defining the QR *method with shifting* are

$$\begin{aligned} H_s - p_sI &= Q_sR_s \\ H_{s+1} &= R_sQ_s + p_sI \end{aligned} \qquad s = 1, 2, \ldots$$

and since

$$H_{s+1} = Q_s^T(Q_sR_s + p_sI)Q_s = Q_s^TH_sQ_s$$

the eigenvalues are still preserved. In Section 4.1 the matrix $H_s - p_sI$ was shown to have eigenvalues $\lambda_i - p_s$ and so, in particular, the element $h_{n,n-1}^{(s)}$ converges to zero as $(\lambda_n - p_s)^s/(\lambda_{n-1} - p_s)^s$. Now, if the eigenvalues are all real, the element $h_{nn}^{(s)}$ converges to λ_n. Consequently, if p_s is chosen to be $h_{nn}^{(s)}$ at each stage, the $(n, n-1)$ element of H_s converges rapidly to zero.

After a few steps,

$$H_s = \begin{pmatrix} x & x & x & \cdots & & & x \\ x & x & x & \cdots & & & x \\ 0 & x & x & \cdots & & & x \\ 0 & 0 & x & \cdots & & & x \\ & & & & & & \vdots \\ & & & & x & x & x \\ & & & & 0 & \varepsilon & x \end{pmatrix}$$

where ε is small. If the $(n, n-1)$ element is exactly zero, the eigenvalues are $h_{nn}^{(s)}$, together with those of the $(n-1) \times (n-1)$ upper Hessenberg matrix shown above. In practice, once $|h_{n,n-1}^{(s)}|$ falls below some threshold value, $h_{nn}^{(s)}$ is taken as the approximation to λ_n and the whole process is repeated with the deflated matrix of order $(n-1) \times (n-1)$. In the case of complex eigenvalues a modification based on double shifts can be used to deflate the matrix, and details can be found in Goult *et al.* (1974).

■ **Example 4.15**
In Example 4.14, the matrix

$$A_1 = \begin{pmatrix} 4 & 2 & 1 & -3 \\ 2 & 4 & 1 & -3 \\ 3 & 2 & 2 & -3 \\ 1 & 2 & 1 & 0 \end{pmatrix}$$

was converted into the upper Hessenberg matrix

$$H_1 = \begin{pmatrix} 4.0000 & 1.3333 & -1.5714 & 2.0000 \\ 3.0000 & 2.3333 & -1.5714 & 2.0000 \\ 0 & 1.5556 & 1.9524 & 1.3333 \\ 0 & 0 & -0.4898 & 1.7143 \end{pmatrix}$$

We now consider the calculation of the eigenvalues via the QR method with origin shift, deflating when the magnitude of the $(n, n-1)$ element of the current upper Hessenberg matrix is less than 0.0005.

STEP 1   The shift p_1 is taken to be the (4, 4) element of H_1, which is 1.7143. The matrix $H_1 - 1.7143I$ has an orthogonal decomposition $Q_1 R_1$ where

$$Q_1 = \begin{pmatrix} 0.6060 & -0.3207 & -0.4375 & -0.5818 \\ 0.7954 & 0.2444 & 0.3333 & 0.4432 \\ 0 & -0.9151 & 0.2423 & 0.3222 \\ 0 & 0 & -0.7992 & 0.6010 \end{pmatrix}$$

$$R_1 = \begin{pmatrix} 3.7715 & 1.3004 & -2.2023 & 2.8029 \\ 0 & -1.6999 & -0.0979 & -1.3728 \\ 0 & 0 & 0.6129 & 0.1148 \\ 0 & 0 & 0 & 0.1526 \end{pmatrix}$$

and so

$$H_2 = R_1 Q_1 + 1.7143I = \begin{pmatrix} 5.0344 & 1.1235 & -3.9905 & -0.6427 \\ -1.3522 & 1.3885 & 0.5068 & -1.6102 \\ 0 & -0.5608 & 1.7711 & 0.2665 \\ 0 & 0 & -0.1220 & 1.8060 \end{pmatrix}$$

STEP 2 The shift p_2 is 1.8060, which gives

$$H_3 = \begin{pmatrix} 4.5717 & 4.0754 & 2.1002 & -0.2093 \\ -0.2175 & 1.9096 & 0.4619 & -0.4714 \\ 0 & 1.0760 & 1.5215 & -1.6714 \\ 0 & 0 & 0.0217 & 1.9972 \end{pmatrix}$$

STEP 3 The shift p_3 is 1.9972, which gives

$$H_4 = \begin{pmatrix} 4.2292 & 2.9860 & -3.6739 & -0.2782 \\ -0.0931 & 1.9452 & -1.0930 & -1.7721 \\ 0 & -0.7103 & 1.8258 & 0.0825 \\ 0 & 0 & -0.0001 & 1.9998 \end{pmatrix}$$

Note the very rapid convergence of $h_{43}^{(s)}$. Since $|h_{43}^{(4)}| < 0.0005$, we can quote one of the eigenvalues as 1.9998 and deflate to get

$$K_1 = \begin{pmatrix} 4.2292 & 2.9860 & -3.6739 \\ -0.0931 & 1.9452 & -1.0930 \\ 0 & -0.7103 & 1.8258 \end{pmatrix}$$

The process can now be repeated for the upper Hessenberg matrix K_1.

STEP 1 The shift p_1 is 1.8258, which gives

$$K_2 = \begin{pmatrix} 4.1139 & 4.4092 & 1.7754 \\ -0.0290 & 2.4285 & 0.5881 \\ 0 & 1.1127 & 1.4578 \end{pmatrix}$$

STEP 2   The shift p_2 is 1.4578, which gives

$$K_3 = \begin{pmatrix} 4.0659 & 4.2942 & -2.0707 \\ -0.0165 & 2.7791 & -0.8356 \\ 0 & -0.3304 & 1.1553 \end{pmatrix}$$

STEP 3   The shift p_3 is 1.1553, which gives

$$K_4 = \begin{pmatrix} 4.0416 & 4.6235 & -1.1802 \\ -0.0095 & 2.9667 & -0.4842 \\ 0 & 0.0327 & 0.9920 \end{pmatrix}$$

STEP 4   The shift p_4 is 0.9920, which gives

$$K_5 = \begin{pmatrix} 4.0272 & 4.6068 & -1.2546 \\ -0.0062 & 2.9731 & -0.5204 \\ 0 & 0.0001 & 1.0000 \end{pmatrix}$$

Since $|k_{32}^{(5)}| < 0.0005$, we can quote one of the eigenvalues as 1.0000 and deflate to get

$$L_1 = \begin{pmatrix} 4.0272 & 4.6068 \\ -0.0062 & 2.9731 \end{pmatrix}$$

The process can now be repeated for the upper Hessenberg matrix L_1.

STEP 1   The shift p_1 is 2.9731, which gives

$$L_2 = \begin{pmatrix} 4.0001 & 4.6128 \\ -0.0002 & 3.0002 \end{pmatrix}$$

Since $|l_{21}^{(2)}| < 0.0005$, the remaining two eigenvalues can be quoted as 3.0002 and 4.0001.

Estimates of all four eigenvalues have been found in only eight iterations. In fact, without shifting, the QR method takes 28 iterations (each involving a 4×4 matrix) to attain the same accuracy. Notice that in this example the first eigenvalue to be calculated is not the one of smallest magnitude; the element $h_{44}^{(s)}$ converges to the eigenvalue closest to the initial shift. If desired, the eigenvalues can be computed in increasing order of magnitude by using a zero shift for the first few steps until $h_{nn}^{(s)}$ is closest to λ_n.   □

The QR method with shifting provides a very efficient and stable means of calculating eigenvalues and eigenvectors. A further improvement is possible in the case of symmetric matrices, since they can be converted into tridiagonal form via Givens or Householder transformations. It is easy to see that this form is invariant with respect to the *QR* decomposition (see Exercise 4.6, Question 5), and so can be used in place of upper Hessenberg form throughout.

EXERCISE 4.7

1. Transform the matrix

$$A = \begin{pmatrix} 3 & -4 & 1 & -4 \\ 4 & 1 & -6 & 5 \\ -4 & 1 & 7 & -4 \\ -4 & -2 & 3 & -3 \end{pmatrix}$$

to upper Hessenberg form. Hence write down its eigenvalues.

2. Transform the matrices

$$\text{(a)} \begin{pmatrix} 3 & 0 & 1 \\ 2 & 2 & 2 \\ 4 & 2 & 5 \end{pmatrix} \qquad \text{(b)} \begin{pmatrix} 8 & -2 & -3 & 1 \\ 7 & -1 & -3 & 1 \\ 6 & -2 & -1 & 1 \\ 5 & -2 & -3 & 4 \end{pmatrix}$$

to upper Hessenberg form. Use the *QR* method with shifting to find their eigenvalues, deflating when the magnitude of the $(n, n-1)$ element is less than 0.0001.

3. Use the *QR* method with shifting to calculate the eigenvalues of

$$\begin{pmatrix} 1 & -1 & 0 & 0 \\ -1 & 2 & -1 & 0 \\ 0 & -1 & 3 & -1 \\ 0 & 0 & -1 & 4 \end{pmatrix}$$

Deflate when the magnitude of the $(n, n-1)$ clement is less than 0.0001.

4. Show that the reduction of a 4×4 matrix A to upper Hessenberg form H, using elementary matrices without interchanges, is equivalent to the matrix equation $AL = LH$, where L is of the form

$$\begin{pmatrix} 1 & 0 & 0 & 0 \\ 0 & 1 & 0 & 0 \\ 0 & l_{32} & 1 & 0 \\ 0 & l_{42} & l_{43} & 1 \end{pmatrix}$$

Transform the matrix given in Question 2(b) to upper Hessenberg form by equating entries in successive columns of AL and LH.

5. Show that the characteristic equation of

$$\begin{pmatrix} 0 & 0 & 0 & d \\ 1 & 0 & 0 & c \\ 0 & 1 & 0 & b \\ 0 & 0 & 1 & a \end{pmatrix}$$

is

$$\lambda^4 - a\lambda^3 - b\lambda^2 - c\lambda - d = 0$$

Use the *QR* method with shifting to calculate the roots of

$$\lambda^4 + 7\lambda^3 - 7\lambda^2 - 43\lambda + 42 = 0$$

correct to 1D.

5
Methods of
approximation
theory

In this chapter we describe some of the numerical methods that are used in approximation theory. The main purpose of these techniques is to replace a complicated function by one which is simpler and more manageable. The classes of approximating functions that are employed include polynomials, trigonometric functions and rational functions. However, we shall restrict our attention to polynomials. These are widely used in practice, since they are easy to evaluate, differentiate and integrate, and they are well behaved–all derivatives exist and are continuous. Whilst most of the techniques described in this chapter are of considerable practical value, some are also studied for their theoretical importance. In particular, the material on polynomial interpolation will be used in subsequent chapters to derive numerical methods for differentiation, integration and the solution of ordinary differential equations.

For the most part, we shall be concerned with the problem of constructing a polynomial approximation to a set of discrete data points. These points might, for example, be the tabulated values of a given function or the results of a physical experiment. The particular method of approximation chosen depends on the context from which the data is taken, as well as on the form of the data itself.

Consider the data listed in Table 5.1. If it is known that this consists of the exact values of a function, then the problem becomes one of interpolation; we would like the approximating polynomial to reproduce the given points exactly, as well as supplying accurate intermediate values. On the other hand, suppose that it consists of a set of readings arising from a physical experiment. Data resulting from experiments is invariably subject to error, and in this situation the problem is one of smoothing; we try to pick up the underlying relationship and are not interested in minor irregularities or wild points. Inspection of Table 5.1 indicates that in this case the relationship is likely to be linear, and it makes sense to attempt to find a straight line passing close to all five points.

A more theoretical approach to the methods described in this chapter can

Table 5.1

x	0.0	0.1	0.4	0.7	1.0
y	0.00	0.15	0.35	0.8	1.05

be found in the texts by Powell (1981) and Cheney (1966). These books also contain a description of some of the more sophisticated techniques of approximation theory.

5.1 POLYNOMIAL INTERPOLATION: LAGRANGE FORM

We consider the problem of constructing a polynomial which passes through a given set of $n+1$ distinct data points (x_i, y_i) $(i = 0, 1, \ldots, n)$. The numbers y_i might, for example, be the values of a given function f at $x = x_i$. It is not immediately obvious that such a polynomial can be found. We cannot, for

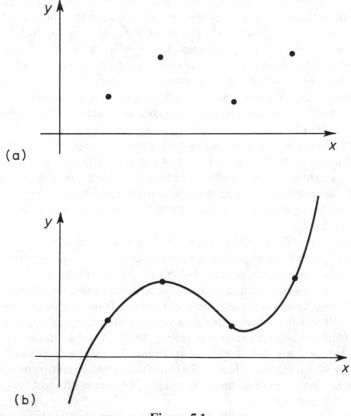

(a)

(b)

Figure 5.1

instance, fit a straight line (i.e. a polynomial of degree one) through the data shown in Fig. 5.1(a); neither can we fit a quadratic. It is, however, possible to fit a cubic, as shown in Fig. 5.1(b).

A polynomial p_n of degree n has a total of $n+1$ coefficients, so one might expect to be able to choose these to satisfy the $n+1$ conditions

$$p_n(x_i) = y_i \qquad i = 0, 1, \ldots, n \qquad (5.1)$$

However, we have no guarantee at this stage that such coefficients exist, or that they are unique.

In order to motivate the general case, consider the construction of a linear polynomial p_1 passing through (x_0, y_0) and (x_1, y_1). If $p_1(x) = a_0 + a_1 x$, then we require

$$a_0 + a_1 x_0 = y_0$$

and

$$a_0 + a_1 x_1 = y_1$$

which have a unique solution

$$a_0 = \frac{x_0 y_1 - x_1 y_0}{x_0 - x_1}, \qquad a_1 = \frac{y_1 - y_0}{x_1 - x_0}$$

Hence

$$p_1(x) = \frac{x_0 y_1 - x_1 y_0}{x_0 - x_1} + \frac{y_1 - y_0}{x_1 - x_0} x$$

$$= \frac{x - x_1}{x_0 - x_1} y_0 + \frac{x - x_0}{x_1 - x_0} y_1$$

This is usually denoted by $L_0(x) y_0 + L_1(x) y_1$, where

$$L_0(x) = \frac{x - x_1}{x_0 - x_1}, \qquad L_1(x) = \frac{x - x_0}{x_1 - x_0}$$

are referred to as Lagrange polynomials. This result will now be generalized to $n+1$ data points. We show that, given a set of $n+1$ data points (x_i, y_i) $(i = 0, 1, \ldots, n)$, where the numbers x_i are distinct, a unique polynomial p_n of degree at most n can always be found to satisfy (5.1).

Assume for the moment that there exist polynomials $L_i (i = 0, 1, \ldots, n)$ of degree n having the property

$$L_i(x_j) = 0 \qquad j \neq i \qquad (5.2)$$

$$L_i(x_i) = 1 \qquad (5.3)$$

We may then write

$$p_n(x) = \sum_{i=0}^{n} L_i(x) y_i \qquad (5.4)$$

which is clearly a polynomial of degree at most n and satisfies (5.1) because

$$p_n(x_i) = L_0(x_i)y_0 + \cdots + L_{i-1}(x_i)y_{i-1} + L_i(x_i)y_i + L_{i+1}(x_i)y_{i+1}$$
$$+ \cdots + L_n(x_i)y_n$$
$$= 0y_0 + \cdots + 0y_{i-1} + 1y_i + 0y_{i+1} + \cdots + 0y_n$$
$$= y_i$$

It remains to be shown how the polynomials L_i can be constructed so that they satisfy (5.2) and (5.3).

If L_i is to satisfy (5.2), then it must contain a factor

$$(x - x_0)\cdots(x - x_{i-1})(x - x_{i+1})\cdots(x - x_n)$$

Furthermore, since this expression has exactly n terms and L_i is a polynomial of degree n, we can deduce that

$$L_i(x) = A_i(x - x_0)\cdots(x - x_{i-1})(x - x_{i+1})\cdots(x - x_n)$$

for some multiplicative constant A_i. The value of A_i is chosen so that L_i satisfies (5.3), i.e.

$$A_i = 1/[(x_i - x_0)\cdots(x_i - x_{i-1})(x_i - x_{i+1})\cdots(x_i - x_n)]$$

where none of the terms in the denominator can be zero from the assumption of distinct points. Hence

$$L_i(x) = \prod_{\substack{j=0 \\ j \neq i}}^{n} \frac{x - x_j}{x_i - x_j} \tag{5.5}$$

The numerator of $L_i(x)$ consists of the product of all possible factors $(x - x_j)$ with the exception of $(x - x_i)$, and the denominator is the same as the numerator but with x replaced by x_i.

We have shown that an interpolating polynomial always exists and can be written in the so-called Lagrange form defined by (5.4) and (5.5).

To prove uniqueness, suppose that there is another polynomial q_n, of degree at most n, such that $q_n(x_i) = y_i$. Now $p_n - q_n$ is itself a polynomial of degree at most n and vanishes at the $n + 1$ points x_i because

$$p_n(x_i) - q_n(x_i) = y_i - y_i = 0$$

However, by the fundamental theorem of algebra, a polynomial of degree n has at most n roots unless it is identically zero. Hence $p_n = q_n$ as required.

■ Example 5.1

Consider the construction of the quadratic interpolating polynomial to the function $f(x) = \ln x$ using the following data:

x_i	2	2.5	3
$f(x_i)$	0.693 15	0.916 29	1.098 61

From (5.5),

$$L_0(x) = \frac{(x - 2.5)(x - 3)}{(2 - 2.5)(2 - 3)} = \frac{x^2 - 5.5x + 7.5}{0.5}$$

$$L_1(x) = \frac{(x - 2)(x - 3)}{(2.5 - 2)(2.5 - 3)} = \frac{x^2 - 5x + 6}{-0.25}$$

$$L_2(x) = \frac{(x - 2)(x - 2.5)}{(3 - 2)(3 - 2.5)} = \frac{x^2 - 4.5x + 5}{0.5}$$

Hence, from (5.4),

$$p_2(x) = \frac{x^2 - 5.5x + 7.5}{0.5} 0.693\,15 + \frac{x^2 - 5x + 6}{-0.25} 0.916\,29$$

$$+ \frac{x^2 - 4.5x + 5}{0.5} 1.098\,61$$

$$= -0.081\,64\,x^2 + 0.813\,66\,x - 0.607\,61$$

Intermediate estimates of $\ln x$ can be calculated by evaluating $p_2(x)$. For example, an approximation to $\ln 2.7$ is $0.994\,12$. In fact, the true value of $\ln 2.7$ is $0.993\,25\,(5D)$, and so the error in this approximation is $-0.000\,87$.

□

It is not possible, in general, to say how accurately the interpolating polynomial p_n approximates a given function f. All that can be said with any certainty is that $f(x) - p_n(x) = 0$ at $x = x_0, x_1, \dots, x_n$. However, it is sometimes possible to obtain a bound on the error $f(x) - p_n(x)$ at an intermediate point x using the following theorem.

■ **Theorem 5.1**

Let $[a, b]$ be any interval containing the distinct points $x_i (i = 0, 1, \dots, n)$. If all derivatives of f up to and including $f^{(n+1)}$ exist and are continuous on $[a, b]$ then, corresponding to each value of x in $[a, b]$, there exists a number ξ in (a, b) such that

$$f(x) - p_n(x) = (x - x_0)(x - x_1) \cdots (x - x_n) \frac{f^{(n+1)}(\xi)}{(n+1)!}$$

■ **Proof**

Consider the function g defined by

$$g(t) = f(t) - p_n(t) - C(t - x_0)(t - x_1) \cdots (t - x_n)$$

for $t \in [a, b]$, where C is a constant to be determined. It is easy to see that $g(t) = 0$ for $t = x_i$ since $f(x_i) = p_n(x_i)$. Let the value of C be chosen so that

$g(x) = 0$ for some intermediate point x in $[a, b]$, i.e.

$$C = \frac{f(x) - p_n(x)}{(x - x_0) \cdots (x - x_n)} \tag{5.6}$$

Now f has continuous derivatives up to and including the $(n + 1)$th order. Therefore g also has continuous derivatives of order up to and including $n + 1$, since p_n and $(t - x_0) \cdots (t - x_n)$ are polynomials. We have arranged that g vanishes at the $n + 2$ distinct points $x_0, x_1, \ldots, x_n, x$. By repeated application of Rolle's theorem, we deduce that g' vanishes at $n + 1$ points, g'' vanishes at n points, etc., until finally $g^{(n+1)}$ vanishes at one point in (a, b) which we denote by ξ. Evaluating $g^{(n+1)}$ at ξ gives

$$0 = g^{(n+1)}(\xi) = f^{(n+1)}(\xi) - p_n^{(n+1)}(\xi) - C \frac{d^{(n+1)}}{dt^{(n+1)}} \left[(t - x_0) \cdots (t - x_n) \right] \Bigg|_{t=\xi}$$

Now p_n is a polynomial of degree n, so its $(n + 1)$th derivative must be zero. Also, $(t - x_0) \cdots (t - x_n)$ is a polynomial of degree $n + 1$ with leading term t^{n+1}, so its $(n + 1)$th derivative is simply $(n + 1)!$. Hence

$$0 = f^{(n+1)}(\xi) - 0 - (n + 1)! C$$

If the value of C given by (5.6) is substituted into this equation, it can be rearranged as

$$f(x) - p_n(x) = (x - x_0) \cdots (x - x_n) \frac{f^{(n+1)}(\xi)}{(n + 1)!} \qquad \qquad \square$$

In fact, Theorem 5.1 is of more theoretical than practical importance because it is only of use when we can specify bounds on the derivatives of f. However, in simple cases it can be used to obtain realistic error bounds, as the following example shows.

■ Example 5.2
To illustrate the use of Theorem 5.1, consider the problem defined in Example 5.1, namely

$$f(x) = \ln x, \qquad n = 2, \qquad x_0 = 2, \qquad x_1 = 2.5, \qquad x_2 = 3, \qquad x = 2.7$$

In this case,

$$f'(x) = 1/x, \qquad f''(x) = -1/x^2, \qquad f'''(x) = 2/x^3$$

and so f satisfies the continuity conditions on $[2, 3]$. Therefore

$$\ln x - p_2(x) = (x - 2)(x - 2.5)(x - 3) \frac{2}{3! \, \xi^3}$$

for some $\xi \in (2, 3)$. On this interval, $|1/\xi^3| \leqslant 1/8$. Therefore

$$|\ln x - p_2(x)| \leqslant |(x - 2)(x - 2.5)(x - 3)|/24$$

and so the error in $p_2(2.7)$ does not exceed

$$|(2.7 - 2)(2.7 - 2.5)(2.7 - 3)|/24 = 0.001\,75$$

This theoretical bound is in agreement with the actual error, which was found to be $-0.000\,87$. □

Serious problems can arise as a result of the shape of the polynomials L_i. At each of the points x_i there is only one non-zero Lagrange polynomial L_i, and so one might have expected that L_i would have most influence on the shape of the curve near to x_i. This is certainly the case if L_i takes its

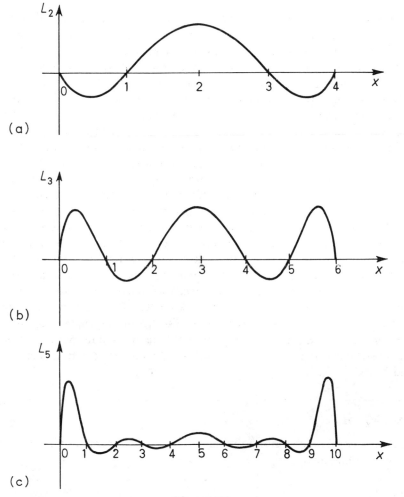

(a)

(b)

(c)

Figure 5.2

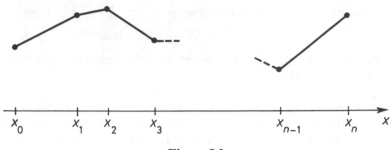

Figure 5.3

maximum value at x_i and decays in magnitude further away from x_i. However, Fig. 5.2(a)–(c) show that this is not necessarily true, particularly if n is large. Figure 5.2(a) is a graph of L_2 with $x_i = i$ ($i = 0, 1, \ldots, 4$). This polynomial does take its maximum value at $x = 2$. However, if we consider Fig. 5.2(b), which is a graph of L_3 with $x_i = i$ ($i = 0, 1, \ldots, 6$), we see that this is no longer the case and the function does not reduce in magnitude further away from $x = 3$. In Fig. 5.2(c) the situation is even worse. This is a graph of L_5 with $x_i = i$ ($i = 0, 1, \ldots, 10$). Near the ends of the interval, L_5 exhibits considerable oscillation. The interpolating polynomial p_n is a weighted linear combination of the polynomials L_i, and so any oscillations in the L_i will also be present in p_n. These examples indicate that considerable care must be exercised when constructing interpolating polynomials of high degree for data which is equally spaced or nearly equally spaced.

The easiest way of avoiding these spurious oscillations is to divide the set of data points into subsets, each containing a small number of consecutive data points, and to construct low order interpolating polynomials on each subset. This is called *piecewise polynomial interpolation*. The simplest type of piecewise approximation, known as piecewise linear interpolation, involves joining a set of data points (x_i, y_i) by a sequence of straight lines, as shown in Fig. 5.3. Although easy to construct, this type of approximating function has the disadvantage of not being differentiable at the data points. Methods for constructing smooth piecewise interpolants are considered in Sections 5.4 and 5.5.

EXERCISE 5.1

1. Construct the cubic interpolating polynomial to the following data and hence estimate $f(1)$:

x_i	-2	0	3	4
$f(x_i)$	5	1	55	209

2. Construct the linear interpolating polynomial to the function $f(x) = \sin x$ using the following data:

x_i	0.5	0.6
$f(x_i)$	0.479 425 5	0.564 642 4

Hence estimate $\sin 0.55$. Use Theorem 5.1 to derive a theoretical bound on the error and compare this with the actual error. ($\sin 0.55 = 0.522\,687\,2$ (7D))

3. Estimate $f(1.2)$ for the function $f(x) = e^{-2x}$ by constructing interpolating polynomials to each of the following sets of data:

(a)

x_i	0	1	2
$f(x_i)$	1	0.135 335	0.018 316

(b)

x_i	0	0.5	1	1.5	2
$f(x_i)$	1	0.367 879	0.135 335	0.049 787	0.018 316

(c)

x_i	0	0.25	0.5	0.75	1
$f(x_i)$	1	0.606 531	0.367 879	0.223 130	0.135 335
x_i	1.25	1.5	1.75	2	
$f(x_i)$	0.082 085	0.049 787	0.030 197	0.018 316	

Compare these results with the exact value, 0.090 718 (6D).

4. Sketch a graph of L_6 with $x_i = i$ $(i = 0, 1, \ldots, 12)$ and compare with Fig. 5.2.

5. By considering the error term given in Theorem 5.1, show that the interpolating polynomial $p_n(x)$ will reproduce any polynomial of degree at most n exactly.
   Deduce that $\sum_{i=0}^{n} L_i(x) = 1$ for all x.

6. Construct the interpolating polynomial of degree five to the function

$$f(x) = \frac{(x+5)(x-2)}{(x+3)(x+1)}$$

using the following data, and hence estimate $f(7)$:

x_i	0	1	1.5	3	5	10
$f(x_i)$	$-3.333\,333$	-0.75	$-0.288\,889$	$0.333\,333$	0.625	$0.839\,161$

Construct a piecewise linear approximation to the same data and hence estimate $f(7)$.
   Compare the accuracy of these two results.

5.2 POLYNOMIAL INTERPOLATION: DIVIDED DIFFERENCE FORM

It is a simple matter to write a computer program for the Lagrange form of the interpolating polynomial. However, for a large number of data points there will be many multiplications – enough for the plotting of a smooth curve on a microcomputer to take a second or two. More significantly, whenever a new data point is added to an existing set, the interpolating polynomial has to be completely recalculated. In this section, we describe an efficient way of organizing the calculations so as to overcome these disadvantages.

It is possible to express the interpolating polynomial in terms of *divided differences*, which are defined as follows. The ratio

$$f[x_p, x_{p+1}] = \frac{f(x_{p+1}) - f(x_p)}{x_{p+1} - x_p}$$

is referred to as the first divided difference of f relative to x_p and x_{p+1}. Geometrically, it represents the slope of the straight line joining $(x_p, f(x_p))$ to $(x_{p+1}, f(x_{p+1}))$. The second divided difference of f relative to x_p, x_{p+1} and x_{p+2} is defined by

$$f[x_p, x_{p+1}, x_{p+2}] = \frac{f[x_{p+1}, x_{p+2}] - f[x_p, x_{p+1}]}{x_{p+2} - x_p}$$

which is a divided difference of two divided differences. In general, the mth divided difference of f relative to $x_p, x_{p+1}, \ldots, x_{p+m}$ is defined recursively by

$$f[x_p, x_{p+1}, \ldots, x_{p+m}] = \frac{f[x_{p+1}, x_{p+2}, \ldots, x_{p+m}] - f[x_p, x_{p+1}, \ldots, x_{p+m-1}]}{x_{p+m} - x_p}$$

Finally, for consistency, we write $f[x_p] = f(x_p)$ and so define the zeroth divided difference to be the function value itself. For hand calculations, it is convenient to set up a table of divided differences of the following form:

x_i	$f[\;\;]$	$f[\;,\;]$	$f[\;,\;,\;]$	$f[\;,\;,\;,\;]$	$\cdots$
x_0	$f[x_0]$				
		$f[x_0, x_1]$			
x_1	$f[x_1]$		$f[x_0, x_1, x_2]$		
		$f[x_1, x_2]$		$f[x_0, x_1, x_2, x_3]$	
x_2	$f[x_2]$		$f[x_1, x_2, x_3]$		
		$f[x_2, x_3]$			
x_3	$f[x_3]$				
$\vdots$					

A new element $(x_4, f(x_4))$ can be added to the bottom of the table and $f[x_3, x_4]$, $f[x_2, x_3, x_4]$, $f[x_1, x_2, x_3, x_4]$ and $f[x_0, x_1, x_2, x_3, x_4]$ calculated without affecting existing entries.

■ Example 5.3

Consider the construction of a divided difference table for the function $f(x) = \ln x$ using the following data:

x_i	1	1.5	1.75	2
$f(x_i)$	0.000 00	0.405 47	0.559 62	0.693 15

The results are listed in Table 5.2.

If additional data becomes available the table can be trivially extended. For example, since $f(1.1) = 0.095\,31$, we have Table 5.3. (We have redrawn the table for clarity; existing entries, however, have not changed.) □

The interpolating polynomial $p_n(x)$, passing through the points $(x_i, f(x_i))$ $(i = 0, 1, \ldots, n)$, can be written in terms of divided differences as

$$p_n(x) = f[x_0] + f[x_0, x_1](x - x_0) + f[x_0, x_1, x_2](x - x_0)(x - x_1)$$
$$+ \cdots + f[x_0, x_1, \ldots, x_n](x - x_0)(x - x_1) \cdots (x - x_{n-1})$$
$$(5.7)$$

It is possible to prove that (5.7) is simply a rearrangement of the Lagrange form defined by (5.4) and (5.5). For example, putting $n = 1$ gives

$$p_1(x) = f[x_0] + f[x_0, x_1](x - x_0)$$
$$= f(x_0) + \left(\frac{f(x_1) - f(x_0)}{x_1 - x_0} \right)(x - x_0)$$
$$= \frac{(x_1 - x_0)f(x_0) + f(x_1)(x - x_0) - f(x_0)(x - x_0)}{x_1 - x_0}$$
$$= \left(\frac{x - x_1}{x_0 - x_1} \right)f(x_0) + \left(\frac{x - x_0}{x_1 - x_0} \right)f(x_1)$$
$$= L_0(x)f(x_0) + L_1(x)f(x_1)$$

A proof of the general case is presented at the end of this section.

The main advantage of the divided difference form over the Lagrange form is that p_n can be calculated from p_{n-1} by adding just one extra term, since it follows from (5.7) that

$$p_n(x) = p_{n-1}(x) + f[x_0, x_1, \ldots, x_n](x - x_0)(x - x_1) \cdots (x - x_{n-1})$$

■ Example 5.4

Consider the evaluation of $p_3(1.3)$ and $p_4(1.3)$, where p_3 and p_4 are the

Table 5.2

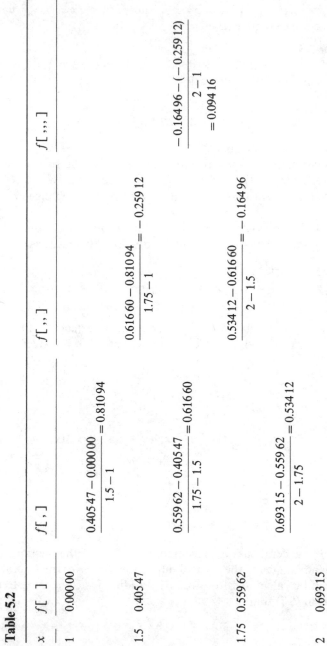

x	$f[\ ]$	$f[\ ,\ ]$	$f[\ ,\ ,\ ]$	$f[\ ,\ ,\ ,\ ]$
1	0.00000			
		$\dfrac{0.40547 - 0.00000}{1.5 - 1} = 0.81094$		
1.5	0.40547		$\dfrac{0.61660 - 0.81094}{1.75 - 1} = -0.25912$	
		$\dfrac{0.55962 - 0.40547}{1.75 - 1.5} = 0.61660$		$\dfrac{-0.16496 - (-0.25912)}{2 - 1}$ $= 0.09416$
1.75	0.55962		$\dfrac{0.53412 - 0.61660}{2 - 1.5} = -0.16496$	
		$\dfrac{0.69315 - 0.55962}{2 - 1.75} = 0.53412$		
2	0.69315			

Table 5.3

x	$f[\ ]$	$f[\ ,\ ]$	$f[\ ,\ ,]$	$f[\ ,\ ,\ ]$	$f[\ ,\ ,\ ,\ ]$
1	0.000 00				
		0.810 94			
1.5	0.405 47		− 0.259 12		
		0.616 60		0.094 16	
1.75	0.559 62		− 0.164 96		− 0.059 80
		0.534 12		0.088 18	
2	0.693 15		− 0.200 23		
		0.664 27			
1.1	0.095 31				

interpolating polynomials based upon the data supplied in Tables 5.2 and 5.3 respectively.

From equation (5.7),

$$p_3(x) = f[x_0] + f[x_0, x_1](x - x_0) + f[x_0, x_1, x_2](x - x_0)(x - x_1)$$
$$+ f[x_0, x_1, x_2, x_3](x - x_0)(x - x_1)(x - x_2)$$

and substitution of the tabulated divided differences gives

$$p_3(x) = 0 + 0.810\,94(x - 1) - 0.259\,12(x - 1)(x - 1.5)$$
$$+ 0.094\,16(x - 1)(x - 1.5)(x - 1.75)$$

Therefore

$$p_3(1.3) = 0.261\,37$$

The polynomial p_4 can be determined from p_3 by the addition of

$$f[x_0, x_1, x_2, x_3, x_4](x - x_0)(x - x_1)(x - x_2)(x - x_3)$$

i.e.

$$p_4(x) = p_3(x) - 0.059\,80(x - 1)(x - 1.5)(x - 1.75)(x - 2)$$

and so

$$p_4(1.3) = 0.261\,37 + 0.001\,13 = 0.262\,50 \qquad \square$$

In order to prove the validity of (5.7), it is necessary to establish the following technical lemma.

■ **Lemma 5.1**

Given a set of $n + 1$ distinct data points (x_i, y_i) $(i = p, p + 1, \ldots, p + n)$ with $y_i = f(x_i)$, the nth divided difference of f relative to $x_p, x_{p+1}, \ldots, x_{p+n}$ may be expressed in the form

$$f[x_p, x_{p+1}, \ldots, x_{p+n}] = \sum_{i=p}^{p+n} \left[y_i \bigg/ \prod_{\substack{j=p \\ j \neq i}}^{p+n} (x_i - x_j) \right]$$

■ **Proof**

The proof is by induction on the order of the divided difference. If $n = 1$ then

$$f[x_p, x_{p+1}] = \frac{y_{p+1} - y_p}{x_{p+1} - x_p} = \frac{y_p}{x_p - x_{p+1}} + \frac{y_{p+1}}{x_{p+1} - x_p}$$

$$= \sum_{i=p}^{p+1} \left[y_i \bigg/ \prod_{\substack{j=p \\ j \neq i}}^{p+1} (x_i - x_j) \right]$$

For the inductive stage, assume that

$$f[x_p, x_{p+1}, \ldots, x_{p+r}] = \sum_{i=p}^{p+r} \left[y_i \bigg/ \prod_{\substack{j=p \\ j \neq i}}^{p+r} (x_i - x_j) \right]$$

and consider

$$f[x_p, x_{p+1}, \ldots, x_{p+r+1}]$$

$$= \frac{1}{x_{p+r+1} - x_p} \{ f[x_{p+1}, x_{p+2}, \ldots, x_{p+r+1}] - f[x_p, x_{p+1}, \ldots, x_{p+r}] \}$$

$$= \frac{1}{x_{p+r+1} - x_p} \left\{ \sum_{i=p+1}^{p+r+1} \left[y_i \bigg/ \prod_{\substack{j=p+1 \\ j \neq i}}^{p+r+1} (x_i - x_j) \right] - \sum_{i=p}^{p+r} \left[y_i \bigg/ \prod_{\substack{j=p \\ j \neq i}}^{p+r} (x_i - x_j) \right] \right\}$$

$$= \frac{1}{x_{p+r+1} - x_p} \left\{ \left[\frac{y_{p+1}}{(x_{p+1} - x_{p+2}) \cdots (x_{p+1} - x_{p+r+1})} \right. \right.$$

$$+ \frac{y_{p+2}}{(x_{p+2} - x_{p+1})(x_{p+2} - x_{p+3}) \cdots (x_{p+2} - x_{p+r+1})}$$

$$\left. + \cdots + \frac{y_{p+r+1}}{(x_{p+r+1} - x_{p+1}) \cdots (x_{p+r+1} - x_{p+r})} \right]$$

$$- \left[\frac{y_p}{(x_p - x_{p+1}) \cdots (x_p - x_{p+r})} + \frac{y_{p+1}}{(x_{p+1} - x_p)(x_{p+1} - x_{p+2}) \cdots (x_{p+1} - x_{p+r})} \right.$$

$$\left. \left. + \cdots + \frac{y_{p+r}}{(x_{p+r} - x_p) \cdots (x_{p+r} - x_{p+r-1})} \right] \right\}$$

Collecting terms with numerator y_{p+k} in $\pm$ pairs we see that this expression can be written as

$$\frac{1}{x_{p+r+1} - x_p} \left\{ \frac{-y_p}{(x_p - x_{p+1}) \cdots (x_p - x_{p+r})} \right.$$

$$+ \frac{(x_{p+1} - x_p)y_{p+1} - (x_{p+1} - x_{p+r+1})y_{p+1}}{(x_{p+1} - x_p)(x_{p+1} - x_{p+2}) \cdots (x_{p+1} - x_{p+r+1})}$$

$$+ \cdots + \frac{(x_{p+r} - x_p)y_{p+r} - (x_{p+r} - x_{p+r+1})y_{p+r}}{(x_{p+r} - x_p) \cdots (x_{p+r} - x_{p+r-1})(x_{p+r} - x_{p+r+1})}$$

$$\left. + \frac{y_{p+r+1}}{(x_{p+r+1} - x_{p+1}) \cdots (x_{p+r+1} - x_{p+r})} \right\}$$

$$= \frac{y_p}{(x_p - x_{p+1}) \cdots (x_p - x_{p+r+1})}$$

$$+ \frac{y_{p+1}}{(x_{p+1} - x_p)(x_{p+1} - x_{p+2}) \cdots (x_{p+1} - x_{p+r+1})}$$

$$+ \cdots + \frac{y_{p+r}}{(x_{p+r} - x_p) \cdots (x_{p+r} - x_{p+r-1})(x_{p+r} - x_{p+r+1})}$$

$$+ \frac{y_{p+r+1}}{(x_{p+r+1} - x_p) \cdots (x_{p+r+1} - x_{p+r})}$$

$$= \sum_{i=p}^{p+r+1} \left[y_i \left/ \prod_{\substack{j=p \\ j \neq i}}^{p+r+1} (x_i - x_j) \right. \right]$$

as required. □

We are now in a position to prove the main result.

■ Theorem 5.2

Given a set of $n + 1$ distinct data points (x_i, y_i) $(i = 0, 1, \ldots, n)$ with $y_i = f(x_i)$, the interpolating polynomial p_n may be expressed in the form

$$p_n(x) = p_{n-1}(x) + f[x_0, x_1, \ldots, x_n](x - x_0)(x - x_1) \cdots (x - x_{n-1})$$

where p_{n-1} interpolates (x_i, y_i) for $i = 0, 1, \ldots, n - 1$.

■ Proof

Let $g = p_n - p_{n-1}$, which is a polynomial of degree at most n, with roots $x_i (i = 0, 1, \ldots, n - 1)$, because

$$g(x_i) = p_n(x_i) - p_{n-1}(x_i) = y_i - y_i = 0$$

Hence

$$g(x) = A_n(x - x_0)(x - x_1) \cdots (x - x_{n-1})$$

for some multiplicative constant A_n. Therefore

$$p_n(x) = p_{n-1}(x) + A_n(x - x_0)(x - x_1) \cdots (x - x_{n-1})$$

which has a leading term of $A_n x^n$ since p_{n-1} is of degree at most $n - 1$.
 Now, in Lagrange form,

$$p_n(x) = \sum_{i=0}^{n} y_i \prod_{\substack{j=0 \\ j \neq i}}^{n} \frac{x - x_j}{x_i - x_j}$$

which has a leading term of

$$\sum_{i=0}^{n}\left[y_i \Big/ \prod_{\substack{j=0 \\ j \neq i}}^{n} (x_i - x_j) \right] x^n$$

We therefore deduce that

$$A_n = \sum_{i=0}^{n}\left[y_i \Big/ \prod_{\substack{j=0 \\ j \neq i}}^{n} (x_i - x_j) \right]$$

The result then follows by taking $p = 0$ in Lemma 5.1. $\square$

EXERCISE 5.2

1. Construct a divided difference table for the data given in Exercise 5.1, Question 1. Hence calculate $p_3(1)$. Add the data points $(-1, -1)$ and $(2, 5)$ to this set one at a time and calculate the corresponding values of $p_4(1)$ and $p_5(1)$. Comment on what you find.

2. Values of the function $f(x) = \ln x$ are given in the following table. Use as many of these data points as you need to estimate $f(2.6)$ to within 0.000 05. Experiment with different orderings of these points to try to minimize the total number of points required.

x_i	1.5	1.7	1.8	1.9	2.2	2.3
$f(x_i)$	0.405 47	0.530 63	0.587 79	0.641 85	0.788 46	0.832 91

x_i	2.5	2.7	2.9	3.1	3.2	3.4
$f(x_i)$	0.916 29	0.993 25	1.064 71	1.131 40	1.163 15	1.223 78

x_i	3.5	3.6	3.7	4.0
$f(x_i)$	1.252 76	1.280 93	1.308 33	1.386 29

3. Show that if x_0, x_1 and x_2 are distinct then

$$f[x_0, x_1, x_2] = f[x_1, x_2, x_0] = f[x_2, x_0, x_1]$$

4. The divided difference form of the interpolating polynomial p_2 is

$$p_2(x) = f[x_0] + f[x_0, x_1](x - x_0) + f[x_0, x_1, x_2](x - x_0)(x - x_1)$$

By expressing these divided differences in terms of the function values $f(x_i)$ ($i = 0, 1, 2$), verify that p_2 does pass through the points $(x_i, f(x_i))$ ($i = 0, 1, 2$).

5.3 POLYNOMIAL INTERPOLATION: FINITE DIFFERENCE FORM

In deriving the Lagrange and divided difference forms of the interpolating polynomial, the only restriction imposed on the points x_i $(i = 0, 1, \ldots, n)$ was that they should be distinct. We now consider the simplifications that may be made when the data points are equally spaced. The study of interpolation to equally spaced data used to be an extremely important part of numerical analysis before computers and calculators became available, and virtually all function evaluations were performed using tables. These tables usually gave data at equally spaced values of x and special techniques were developed for polynomial interpolation to such data. Although less use is made of tables nowadays, methods based on equally spaced data can be used to derive formulas for numerical differentiation and integration and for the solution of ordinary differential equations.

Let (x_i, f_i) be a given set of data points tabulated at equal intervals h so that $x_i = x_0 + ih$, with $f_i = f(x_i)$. It is possible to derive an alternative expression for the interpolating polynomial in terms of *forward differences*, which are defined as follows. The number $f_{i+1} - f_i$ is referred to as the first forward difference at $x = x_i$ and is denoted by Δf_i. Higher order differences are defined recursively by

$$\Delta^m f_i = \Delta(\Delta^{m-1} f_i)$$

For example, the second forward difference at $x = x_i$ is

$$\Delta^2 f_i = \Delta(\Delta f_i) = \Delta(f_{i+1} - f_i)$$
$$= \Delta f_{i+1} - \Delta f_i$$
$$= (f_{i+2} - f_{i+1}) - (f_{i+1} - f_i)$$
$$= f_{i+2} - 2f_{i+1} + f_i$$

Successive differences can be listed in a table of the following form:

x_i	f_i	Δf_i	$\Delta^2 f_i$	$\Delta^3 f_i$	$\cdots$
x_0	f_0				
		$\Delta f_0 = f_1 - f_0$			
x_1	f_1		$\Delta^2 f_0 = \Delta f_1 - \Delta f_0$		
		$\Delta f_1 = f_2 - f_1$		$\Delta^3 f_0 = \Delta^2 f_1 - \Delta^2 f_0$	
x_2	f_2		$\Delta^2 f_1 = \Delta f_2 - \Delta f_1$		
		$\Delta f_2 = f_3 - f_2$			
x_3	f_3				
$\vdots$					

■ **Example 5.5**

Consider the construction of a forward difference table for the function

$$f(x) = \frac{(x-1)(x+5)}{(x+2)(x+1)}$$

with $x_0 = 0$, $h = 0.1$ and $i = 0, 1, \ldots, 8$. The forward differences, calculated as far as $\Delta^4 f_i$, are shown in Table 5.4. (For convenience, the decimal point has been suppressed and only significant figures quoted.)   □

In the special case of equally spaced data we can rewrite the divided difference form of the interpolating polynomial in terms of forward differences, making use of the relation

$$f[x_p, x_{p+1}, \ldots, x_{p+m}] = \frac{1}{m! h^m} \Delta^m f_p \qquad m \geqslant 0 \tag{5.8}$$

with $\Delta^0 f_p = f_p$. This formula is easily proved by induction since, if $m = 0$,

$$f[x_p] = f(x_p) = \frac{1}{0! h^0} \Delta^0 f_p$$

Table 5.4

x_i	f_i	Δf_i	$\Delta^2 f_i$	$\Delta^3 f_i$	$\Delta^4 f_i$
0.0	− 2.500 000				
		512 987			
0.1	− 1.987 013		− 101 732		
		411 255		25 432	
0.2	− 1.575 758		− 76 300		− 7 570
		334 955		17 862	
0.3	− 1.240 803		− 58 438		− 4 988
		276 517		12 874	
0.4	− 0.964 286		− 45 564		− 3 392
		230 953		9 482	
0.5	− 0.733 333		− 36 082		− 2 358
		194 871		7 124	
0.6	− 0.538 462		− 28 958		− 1 689
		165 913		5 435	
0.7	− 0.372 549		− 23 523		
		142 390			
0.8	− 0.230 159				

and, on the assumption that (5.8) is true for $m = r$, we have

$$f[x_p, x_{p+1}, \ldots, x_{p+r+1}] = \frac{f[x_{p+1}, x_{p+2}, \ldots, x_{p+r+1}] - f[x_p, x_{p+1}, \ldots, x_{p+r}]}{x_{p+r+1} - x_p}$$

$$= \frac{\frac{1}{r!h^r} \Delta^r f_{p+1} - \frac{1}{r!h^r} \Delta^r f_p}{(r+1)h} = \frac{1}{(r+1)!h^{r+1}} (\Delta^r f_{p+1} - \Delta^r f_p)$$

$$= \frac{1}{(r+1)!h^{r+1}} \Delta^{r+1} f_p$$

If the divided differences in equation (5.7) are replaced by their equivalent forward difference expressions given in (5.8), then

$$p_n(x) = f_0 + \frac{1}{h} \Delta f_0 (x - x_0) + \frac{1}{2!h^2} \Delta^2 f_0 (x - x_0)(x - x_1)$$

$$+ \cdots + \frac{1}{n!h^n} \Delta^n f_0 (x - x_0)(x - x_1) \cdots (x - x_{n-1}) \qquad (5.9)$$

This can be further simplified by the change of variable $x = x_0 + sh$. The effect is to convert a polynomial of degree at most n in x into a polynomial of degree at most n in s. Note that

$$x - x_i = (x_0 + sh) - (x_0 + ih) = (s - i)h$$

and so equation (5.9) becomes

$$p_n(x_0 + sh) = f_0 + \Delta f_0 s + \frac{1}{2!} \Delta^2 f_0 s(s-1) + \cdots$$

$$+ \frac{1}{n!} \Delta^n f_0 s(s-1) \cdots (s-n+1)$$

$$= \sum_{i=0}^{n} \Delta^i f_0 \binom{s}{i} \qquad (5.10)$$

where

$$\binom{s}{i}$$

denotes the binomial coefficient $s!/[i!(s-i)!]$.

Equation (5.10) is called *Newton's forward difference formula*. With the change of variable $x = x_0 + sh$ the error term in the interpolating polynomial, given in Theorem 5.1, becomes

$$s(s-1)(s-2) \cdots (s-n) \frac{h^{n+1}}{(n+1)!} f^{(n+1)}(x_0 + \theta h) \qquad (5.11)$$

for some $\theta \in (0, n)$ which depends on s.

■ **Example 5.6**

As an illustration of Newton's forward difference formula, consider the calculation of $p_4(0.12)$ where p_4 interpolates the first five data points listed in Table 5.4. From equation (5.10),

$$p_4(x_0 + sh) = f_0 + \Delta f_0 s + \Delta^2 f_0 \frac{s(s-1)}{2!}$$

$$+ \Delta^3 f_0 \frac{s(s-1)(s-2)}{3!} + \Delta^4 f_0 \frac{s(s-1)(s-2)(s-3)}{4!}$$

In this case $x_0 = 0$, $h = 0.1$ and $x_0 + sh = 0.12$. Hence $s = 1.2$, and substitution of the tabulated forward differences gives

$$
\begin{aligned}
p_4(0.12) = & -2.5 + (0.512\,987)(1.2) + \tfrac{1}{2}(-0.101\,732)(1.2)(0.2) \\
& + \tfrac{1}{6}(0.025\,432)(1.2)(0.2)(-0.8) \\
& + \tfrac{1}{24}(-0.007\,570)(1.2)(0.2)(-0.8)(-1.8) \\
= & -1.897\,546
\end{aligned}
$$

The exact value of $f(0.12)$ is $-1.897\,574$ (6D), and so this estimate has an error $0.000\,028$.   □

There are occasions when it is convenient to use an alternative notation for the differences constructed in this section. The first *backward difference* at $x = x_i$ is defined by

$$\nabla f_i = f_i - f_{i-1}$$

and higher order differences by

$$\nabla^m f_i = \nabla(\nabla^{m-1} f_i)$$

The layout of the backward difference table is as follows:

x_i	f_i	∇f_i	$\nabla^2 f_i$	$\nabla^3 f_i$	$\cdots$
x_0	f_0				
		∇f_1			
x_1	f_1		$\nabla^2 f_2$		
		∇f_2		$\nabla^3 f_3$	
x_2	f_2		$\nabla^2 f_3$		
		∇f_3			
x_3	f_3				
$\vdots$					

It is easy to show, using an inductive argument analogous to the one used to prove equation (5.8), that

$$f[x_p, x_{p-1}, \ldots, x_{p-m}] = \frac{1}{m! h^m} \nabla^m f_p$$

If the points x_i are reordered as $x_n, x_{n-1}, \ldots, x_0$, then equation (5.7) becomes

$$
\begin{aligned}
p_n(x) = f[x_n] &+ f[x_n, x_{n-1}](x - x_n) + f[x_n, x_{n-1}, x_{n-2}](x - x_n)(x - x_{n-1}) \\
&+ \cdots + f[x_n, x_{n-1}, \ldots, x_0](x - x_n)(x - x_{n-1}) \cdots (x - x_1)
\end{aligned}
$$

Replacing these divided differences by their equivalent backward difference expressions and making the change of variable $x = x_n + sh$ produces

$$p_n(x_n + sh) = f_n + \nabla f_n s + \frac{1}{2!} \nabla^2 f_n s(s + 1) + \cdots + \frac{1}{n!} \nabla^n f_n s(s + 1) \cdots (s + n - 1)$$

$$(5.12)$$

This is called *Newton's backward difference formula*.

Given the values of a function f tabulated at $n + 1$ equally spaced points x_i, Newton's forward and backward difference formulas provide an efficient means of estimating $f(x)$ at some intermediate point x. In particular, we can calculate $p_j(x)$ from $p_{j-1}(x)$ by adding just one extra term. This property can be exploited since it may be possible to obtain an accurate estimate of $f(x)$ using only a subset of the data points supplied; we can calculate $p_1(x), p_2(x), \ldots,$ in turn and stop when successive estimates differ by less than a prescribed tolerance. Applied in this way, Newton's forward difference formula is suitable for interpolation when x is close to x_0 because p_j passes through the points $(x_0, f_0), (x_1, f_1), \ldots, (x_j, f_j)$. Newton's backward difference formula is suitable for interpolation at the other end of the table because in this case p_j passes through $(x_n, f_n), (x_{n-1}, f_{n-1}), \ldots, (x_{n-j}, f_{n-j})$. If x lies in the middle of the table then a *central difference formula* is appropriate. The first and second central differences are defined by

$$\delta f_{i+1/2} = f_{i+1} - f_i$$
$$\delta^2 f_i = \delta f_{i+1/2} - \delta f_{i-1/2}$$

and higher order differences by

$$\delta^{2n-1} f_{i+1/2} = \delta^{2n-2} f_{i+1} - \delta^{2n-2} f_i$$
$$\delta^{2n} f_i = \delta^{2n-1} f_{i+1/2} - \delta^{2n-1} f_{i-1/2}$$

The layout of the central difference table is as follows:

$$x_i \quad f_i \quad \delta f_{i+1/2} \quad \delta^2 f_i \quad \delta^3 f_{i+1/2} \quad \cdots$$

$$x_0 \quad f_0$$
$$\delta f_{1/2}$$
$$x_1 \quad f_1 \qquad\qquad \delta^2 f_1$$
$$\delta f_{3/2} \qquad\qquad \delta^3 f_{3/2}$$
$$x_2 \quad f_2 \qquad\qquad \delta^2 f_2$$
$$\delta f_{5/2}$$
$$x_3 \quad f_3$$
$$\vdots$$

Notice that, for a given set of data points, the actual numbers appearing in the forward, backward and central difference tables are identical. The three operators Δ, ∇ and δ simply allow us to label the elements of the difference table in different ways. For instance, $\Delta f_0, \nabla f_1$ and $\delta f_{1/2}$ are all equal to $f_1 - f_0$.

We shall not derive the central difference formula here, but simply quote one form of it which is applicable to an odd number of data points. If $n = 2m$, then

$$p_n(x_m + sh) = f_m + \left\{ \frac{s}{2(1!)}(\delta f_{m+1/2} + \delta f_{m-1/2}) + \frac{s^2}{2!}\delta^2 f_m \right\}$$

$$+ \left\{ \frac{s(s^2 - 1^2)}{2(3!)}(\delta^3 f_{m+1/2} + \delta^3 f_{m-1/2}) + \frac{s^2(s^2 - 1^2)}{4!}\delta^4 f_m \right\} + \cdots$$

$$+ \left\{ \frac{s(s^2 - 1^2)\cdots[s^2 - (m-1)^2]}{2[(2m-1)!]}(\delta^{2m-1} f_{m+1/2} + \delta^{2m-1} f_{m-1/2}) \right.$$

$$+ \left. \frac{s^2(s^2 - 1^2)\cdots[s^2 - (m-1)^2]}{(2m)!}\delta^{2m} f_m \right\} \tag{5.13}$$

This is known as *Stirling's formula*. A proof of the fact that this polynomial does interpolate the points (x_i, f_i) $(i = 0, 1, \ldots, 2m)$ is presented in Hildebrand (1974), which also contains an extensive treatment of other formulas of this type. We shall content ourselves with an illustrative example of its use.

■ Example 5.7
The central difference table for the function $\sin x$ based on the points $x_i = x_0 + ih$, with $x_0 = 0$, $h = 0.2$, for $i = 0, 1, \ldots, 6$, is shown in Table 5.5.

If an approximation to $\sin (0.67)$ is required, then we can use (5.13) with $x_m = 0.6$ and $s = (0.67 - 0.6)/0.2 = 0.35$. The central differences used in this

Table 5.5

x_i	f_i	$\delta f_{i+1/2}$	$\delta^2 f_i$	$\delta^3 f_{i+1/2}$	$\delta^4 f_i$	$\delta^5 f_{i+1/2}$	$\delta^6 f_i$
0	0						
		198 67					
0.2	0.198 67		$-7\,92$				
		190 75		$-7\,61$			
0.4	0.389 42		$-15\,53$		64		
		175 22		$-6\,97$		22	
0.6	0.564 64		$-22\,50$		86		10
		152 72		$-6\,11$		32	
0.8	0.717 36		$-28\,61$		1 18		
		124 11		$-4\,93$			
1.0	0.841 47		$-33\,54$				
		90 57					
1.2	0.932 04						

formula are underlined in Table 5.5. This gives

$$p_2(0.67) = 0.564\,64 + \left\{ \frac{0.35}{2}(0.152\,72 + 0.175\,22) + \frac{(0.35)^2}{2}(-0.022\,50) \right\}$$

$$= 0.564\,64 + 0.057\,39 - 0.001\,38$$

$$= 0.620\,65$$

which is the value of the quadratic interpolating polynomial based on the points $x = 0.4, 0.6$ and 0.8.

Adding on the next two terms,

$$p_4(0.27) = p_2(0.67) + \left\{ \frac{(0.35)(0.35^2 - 1^2)}{12}(-0.006\,11 - 0.006\,97) \right.$$

$$\left. + \frac{(0.35)^2(0.35^2 - 1^2)}{24}(0.000\,86) \right\}$$

$$= 0.620\,65 + 0.000\,33 + 0.000\,00$$

$$= 0.620\,98$$

which is the value of the quartic interpolating polynomial based on the points $x = 0.2, 0.4, 0.6, 0.8$ and 1.0.

Notice that the second term in the brackets is zero to 5D. Consequently, there is little point in continuing the calculations since the data has only been supplied to 5D. In fact the exact value of $\sin(0.67)$ is $0.620\,99$ (5D), and so the estimated $p_4(0.67)$ has an error of only $0.000\,01$. ☐

EXERCISE 5.3

1. Construct the difference table for the following data:

x_i	1.0	1.1	1.2	1.3	1.4
f_i	0.183 939 7	0.166 435 5	0.150 597 1	0.136 265 9	0.123 298 5

Write down the corresponding quartic interpolation polynomial p_4 in forward and backward difference forms, and confirm that both produce the same estimate for $f(1.25)$.

2. The exact values of a polynomial q are given in the following table. Construct a difference table for this data and hence identify q.

x_i	0.0	0.1	0.2	0.3	0.4	0.5	0.6	0.7
q_i	6.000 00	5.909 98	5.879 36	5.965 14	6.219 52	6.687 50	7.404 48	8.393 86

3. Construct a difference table for the values of $\log_{10} x$, rounded to 6D, for $x = 1, 1.1, \ldots, 1.6$. Use the forward difference formula to estimate $\log_{10}(1.05)$, the backward difference formula to estimate $\log_{10}(1.55)$ and the central difference formula to estimate $\log_{10}(1.35)$, correct to 4D.

4. Construct a difference table for the values of $\sin x$, rounded to 3D, for $x = 0, 0.2, \ldots, 1.2$. Comment on what you find.

5. Express the following differences in terms of function values f_j:

   (a) $\Delta^4 f_i$ (b) $\nabla^3 f_i$ (c) $\delta^2 f_i$

6. Verify that the quartic central difference interpolating polynomial

$$p_4(x_2 + sh) = f_2 + \left\{ \frac{s}{2(1!)}(\delta f_{5/2} + \delta f_{3/2}) + \frac{s^2}{2!}\delta^2 f_2 \right\}$$

$$+ \left\{ \frac{s(s^2 - 1^2)}{2(3!)}(\delta^3 f_{5/2} + \delta^3 f_{3/2}) + \frac{s^2(s^2 - 1^2)}{4!}\delta^4 f_2 \right\}$$

passes through the points (x_i, f_i) $(i = 0, 1, \ldots, 4)$.

5.4 HERMITE INTERPOLATION

So far, we have only considered the interpolation of data of the form $(x_i, y_i = f(x_i))$. In certain situations the values of the first derivative f' may also be available at these points. In these circumstances it is sometimes useful to construct a polynomial which takes not only the same function value but also the same first derivative value as f at appropriate points. This is referred to as Hermite interpolation.

Let x_i ($i = 0, 1, \ldots, n$) denote a set of $n + 1$ distinct points, and suppose that y_i and y_i' denote the numerical values of f and f' at x_i. A polynomial H_{2n+1} of degree $2n + 1$ has a total of $2n + 2$ coefficients, so one might expect to be able to choose these to satisfy the $2n + 2$ conditions

$$H_{2n+1}(x_j) = y_j, \qquad H_{2n+1}'(x_j) = y_j' \qquad j - 0, 1, \ldots, n \tag{5.14}$$

We shall prove the existence of H_{2n+1} using a constructive proof analogous to the one given in Section 5.1 for Lagrange interpolation. We express the interpolating polynomial H_{2n+1} in the form

$$H_{2n+1}(x) = \sum_{i=0}^{n} r_i(x) y_i + \sum_{i=0}^{n} s_i(x) y_i' \tag{5.15}$$

where r_i and s_i ($i = 0, 1, \ldots, n$) are polynomials of degree at most $2n + 1$.
Now the condition $H_{2n+1}(x_j) = y_j$ is satisfied if

$$r_i(x_j) = \begin{cases} 1 & i = j \\ 0 & i \neq j \end{cases} \tag{5.16a}$$

and

$$s_i(x_j) = 0 \qquad \text{for all } i \tag{5.16b}$$

since, from (5.15),

$$H_{2n+1}(x_j) = \sum_{i=0}^{n} r_i(x_j) y_i + \sum_{i=0}^{n} s_i(x_j) y_i'$$

$$= r_j(x_j) y_j = y_j$$

Also, the condition $H_{2n+1}'(x_j) = y_j'$ is satisfied if

$$r_i'(x_j) = 0 \qquad \text{for all } i \tag{5.17a}$$

and

$$s_i'(x_j) = \begin{cases} 1 & i = j \\ 0 & i \neq j \end{cases} \tag{5.17b}$$

since, from (5.15),

$$H_{2n+1}'(x_j) = \sum_{i=0}^{n} r_i'(x_j) y_i + \sum_{i=0}^{n} s_i'(x_j) y_i'$$

$$= s_j'(x_j) y_j' = y_j'$$

It remains to be shown how the polynomials r_i and s_i can be constructed.
Consider first the calculation of r_i. Conditions (5.16a) and (5.17a) imply that $r_i(x_j) = r_i'(x_j) = 0$ for all $j \neq i$, and so r_i must have a double root at x_j. Hence $r_i(x)$ has a factor

$$(x - x_0)^2 (x - x_1)^2 \cdots (x - x_{i-1})^2 (x - x_{i+1})^2 \cdots (x - x_n)^2$$

Furthermore, since this expression has exactly $2n$ terms and r_i is a polynomial of degree $2n + 1$, we can deduce that

$$r_i(x) = (\alpha_i x + \beta_i)(x - x_0)^2(x - x_1)^2 \cdots (x - x_{i-1})^2(x - x_{i+1})^2 \cdots (x - x_n)^2$$

for some linear factor $\alpha_i x + \beta_i$. More succinctly, we can write

$$r_i(x) = (a_i x + b_i)[L_i(x)]^2$$

where

$$a_i = \alpha_i \prod_{\substack{j=0 \\ j \neq i}}^{n} (x_i - x_j)^2, \qquad b_i = \beta_i \prod_{\substack{j=0 \\ j \neq i}}^{n} (x_i - x_j)^2$$

and L_i is the Lagrange polynomial of degree n defined in Section 5.1.

The unknown values of a_i and b_i can be determined from the remaining two conditions of (5.16a) and (5.17a), which are $r_i(x_i) = 1$ and $r_i'(x_i) = 0$. These give

$$(a_i x_i + b_i)[L_i(x_i)]^2 = 1$$

and, by applying the product rule for differentiation,

$$2(a_i x_i + b_i)L_i(x_i)L_i'(x_i) + a_i[L_i(x_i)]^2 = 0$$

Now $L_i(x_i) = 1$, so that

$$a_i x_i + b_i = 1$$

and

$$2(a_i x_i + b_i)L_i'(x_i) + a_i = 0$$

which have solution

$$a_i = -2L_i'(x_i), \qquad b_i = 1 + 2x_i L_i'(x_i)$$

Hence

$$r_i(x) = [1 - 2(x - x_i)L_i'(x_i)][L_i(x)]^2 \qquad (5.18)$$

The calculation of s_i is similar to that of r_i. Equations (5.16b) and (5.17b), for $i \neq j$, lead to

$$s_i(x) = (c_i x + d_i)[L_i(x)]^2$$

for some linear factor $c_i x + d_i$. The remaining two equations are $s_i(x_i) = 0$ and $s_i'(x_i) = 1$, which give $c_i = 1$ and $d_i = -x_i$. Hence

$$s_i(x) = (x - x_i)[L_i(x)]^2 \qquad (5.19)$$

This completes the proof of the existence of Hermite polynomials.

To prove uniqueness, suppose that there is another polynomial G_{2n+1}, of degree at most $2n + 1$, such that $G_{2n+1}(x_i) = y_i$ and $G_{2n+1}'(x_i) = y_i'$. Now $H_{2n+1} - G_{2n+1}$ is itself a polynomial of degree at most $2n + 1$, and has a double root at the $n+1$ points x_i because

$$H_{2n+1}(x_i) - G_{2n+1}(x_i) = y_i - y_i = 0$$

and

$$H'_{2n+1}(x_i) - G'_{2n+1}(x_i) = y'_i - y'_i = 0$$

Therefore $H_{2n+1} - G_{2n+1}$ has $2n+2$ roots, including multiplicities. However, by the fundamental theorem of algebra, a polynomial of degree $2n+1$ has at most $2n+1$ roots unless it is identically zero. Hence $H_{2n+1} = G_{2n+1}$ as required.

■ **Example 5.8**

Consider the estimation of $\ln(2.7)$ from the following data:

x_i	2	2.5	3
$\ln x_i$	0.693 147	0.916 291	1.098 612
$1/x_i$	0.500 000	0.400 000	0.333 333

In this case $n = 2$, so we can interpolate using the Hermite polynomial of degree 5. The first step is to find the Lagrange polynomials and their derivatives. From Example 5.1,

$$L_0(x) = 2x^2 - 11x + 15, \quad L_1(x) = -4x^2 + 20x - 24, \quad L_2(x) = 2x^2 - 9x + 10$$

and so

$$L'_0(x) = 4x - 11, \qquad L'_1(x) = -8x + 20, \qquad L'_2(x) = 4x - 9$$

Hence

$$L'_0(x_0) = L'_0(2) = -3, \qquad L'_1(x_1) = L'_1(2.5) = 0, \qquad L'_2(x_2) = L'_2(3) = 3$$

We are now in a position to calculate r_i and s_i from equations (5.18) and (5.19), i.e.

$$r_0(x) = [1 - 2(x-2)(-3)][L_0(x)]^2 = (6x-11)(2x^2 - 11x + 15)^2$$
$$r_1(x) = [1 - 2(x-2.5)(0)][L_1(x)]^2 = (-4x^2 + 20x - 24)^2$$
$$r_2(x) = [1 - 2(x-3)(3)][L_2(x)]^2 = (-6x + 19)(2x^2 - 9x + 10)^2$$

and

$$s_0(x) = (x-2)[L_0(x)]^2 = (x-2)(2x^2 - 11x + 15)^2$$
$$s_1(x) = (x-2.5)[L_1(x)]^2 = (x-2.5)(-4x^2 + 20x - 24)^2$$
$$s_2(x) = (x-3)[L_2(x)]^2 = (x-3)(2x^2 - 9x + 10)^2$$

Finally,

$$H_5(x) = \sum_{i=0}^{2} r_i(x)y_i + \sum_{i=0}^{2} s_i(x)y'_i$$

with

$$H_5(2.7) = (0.074\,88)(0.693\,147) + (0.7056)(0.916\,291) + (0.219\,52)(1.098\,612)$$
$$+ (0.010\,08)(0.5) + (0.141\,12)(0.4) + (-0.023\,52)(0.333\,333)$$
$$= 0.993\,253$$

which has an error $-0.000\,001$.   □

In Section 5.1 the disadvantages of high order polynomial interpolation to equally spaced data were briefly discussed. It was suggested that simple piecewise linear interpolation often produces results which are as good as, if not better than, the more elaborate technique of Lagrange interpolation. Hermite interpolation is subject to the same problems, and again a piecewise strategy may be desirable.

The simplest case is piecewise Hermite cubic interpolation in which each of the subintervals $[x_i, x_{i+1}]$ is treated separately. The four pieces of information (x_i, y_i), (x_i, y_i'), (x_{i+1}, y_{i+1}) and (x_{i+1}, y_{i+1}') are used to construct the unique Hermite cubic on each subinterval. By matching function and derivative values, the cubic segments are joined with continuity of both position and slope. This is illustrated in Fig. 5.4, in which the notation $H_{3,i}$ is used to represent the cubic on $[x_i, x_{i+1}]$. Notice that the graph of the piecewise Hermite interpolant is smooth, which contrasts with the rather jagged curve associated with piecewise Lagrange interpolation (see Fig. 5.3).

Let L_i and L_{i+1} denote the linear Lagrange polynomials associated with the points $\{x_i, x_{i+1}\}$, so that

$$L_i(x) = \frac{x - x_{i+1}}{x_i - x_{i+1}}, \qquad L_{i+1}(x) = \frac{x - x_i}{x_{i+1} - x_i}$$

$$L_i'(x) = \frac{1}{x_i - x_{i+1}}, \qquad L_{i+1}'(x) = \frac{1}{x_{i+1} - x_i}$$

If $h_i = x_{i+1} - x_i$ is the length of the ith subinterval, it follows from (5.15),

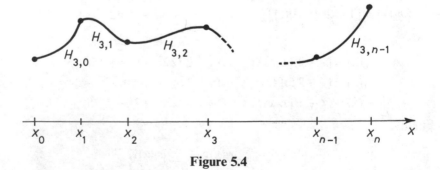

Figure 5.4

(5.18) and (5.19) that

$$H_{3,i}(x) = y_i \left\{ \frac{(x - x_{i+1})^2}{h_i^2} + \frac{2(x - x_i)(x - x_{i+1})^2}{h_i^3} \right\}$$

$$+ y_{i+1} \left\{ \frac{(x - x_i)^2}{h_i^2} - \frac{2(x - x_{i+1})(x - x_i)^2}{h_i^3} \right\}$$

$$+ y_i' \left\{ \frac{(x - x_i)(x - x_{i+1})^2}{h_i^2} \right\} + y_{i+1}' \left\{ \frac{(x - x_{i+1})(x - x_i)^2}{h_i^2} \right\} \quad (5.20)$$

defined on $[x_i, x_{i+1}]$. This formula can be used to find the piecewise Hermite cubic interpolant of any given set of data points.

■ **Example 5.9**
Consider the data

x_i	−2	0	1
y_i	3	1	−2
y_i'	−1	0	1

In this case, $x_0 = -2$, $x_1 = 0$, $x_2 = 1$ and so $h_0 = 2$ and $h_1 = 1$. From (5.20),

$$H_{3,0}(x) = 3 \left\{ \frac{(x - 0)^2}{2^2} + \frac{2[x - (-2)](x - 0)^2}{2^3} \right\}$$

$$+ 1 \left\{ \frac{[x - (-2)]^2}{2^2} - \frac{2(x - 0)[x - (-2)]^2}{2^3} \right\}$$

$$+ (-1) \left\{ \frac{[x - (-2)](x - 0)^2}{2^2} \right\} + 0 \left\{ \frac{(x - 0)[x - (-2)]^2}{2^2} \right\}$$

$$= \frac{x^3}{4} + x^2 + 1$$

and

$$H_{3,1}(x) = 1 \left\{ \frac{(x - 1)^2}{1^2} + \frac{2(x - 0)(x - 1)^2}{1^3} \right\}$$

$$+ (-2) \left\{ \frac{(x - 0)^2}{1^2} - \frac{2(x - 1)(x - 0)^2}{1^3} \right\}$$

$$+ 0 \left\{ \frac{(x - 0)(x - 1)^2}{1^2} \right\} + 1 \left\{ \frac{(x - 1)(x - 0)^2}{1^2} \right\}$$

$$= 7x^3 - 10x^2 + 1$$

Therefore the piecewise Hermite cubic interpolant, $H(x)$ say, is defined by

$$H(x) = \begin{cases} x^3/4 + x^2 + 1 & \text{on } -2 \leqslant x \leqslant 0 \\ 7x^3 - 10x^2 + 1 & \text{on } \quad 0 \leqslant x \leqslant 1 \end{cases} \qquad \square$$

EXERCISE 5.4

1. Construct the Hermite interpolating polynomial of degree 7 to the following data. Hence estimate $f(1)$.

x_i	-2	0	3	4
$f(x_i)$	-63	1	82	513
$f'(x_i)$	16	0	189	768

2. Estimate $f(1.2)$ for the function $f(x) = e^{-2x}$ by constructing the Hermite interpolating polynomial corresponding to the following data:

x_i	0	1	2
$f(x_i)$	1	0.135 335	0.018 316
$f'(x_i)$	-2	$-0.270 671$	$-0.036 631$

Compare this estimate with those of Exercise 5.1, Question 3.

3. Let $[a, b]$ be any interval containing the distinct points x_i $(i = 0, 1, \ldots, n)$. If all derivatives of f up to and including $f^{(2n+2)}$ exist and are continuous on $[a, b]$, prove that corresponding to each value of x in $[a, b]$ there exists a number ξ in (a, b) such that

$$f(x) - H_{2n+1}(x) = (x - x_0)^2 (x - x_1)^2 \cdots (x - x_n)^2 \frac{f^{(2n+2)}(\xi)}{(2n+2)!}$$

4. Values of the function $f(x) = e^x$ are given in the following table at $x = 0, 0.5$ and 1. Use the result of Question 3 to obtain a theoretical bound on the error of the estimate $H_5(0.75)$ and compare this with the actual error. ($f(0.75) = 2.117\,000\,0\,(7D)$)

x_i	0	0.5	1
$f(x_i)$	1	1.648 721 3	2.718 281 8

5. Sketch a graph of the Hermite interpolating polynomial to the following data:

x_i	0	1	2	3	4	5	6	7	8	9	10	11	12
y_i	0	0	0	0	0	0	1	0	0	0	0	0	0
y_i'	0	0	0	0	0	0	0	0	0	0	0	0	0

Comment on what you find.

6. Find the piecewise cubic Hermite approximation to the following data:

x_i	0	1	3
y_i	2	1	0
y_i'	1	0	-1

5.5 CUBIC SPLINE INTERPOLATION

Polynomial interpolation, in either Lagrange or Hermite form, often produces unacceptable results. A polynomial interpolant to a large number of data points will, of necessity, be of high degree. These polynomials can be highly oscillatory in nature and small localized changes to some of the data can cause large fluctuations over the entire interval.

An alternative approach is to divide the interval into a series of subintervals, and to construct a different polynomial interpolant on each subinterval. This strategy is known as piecewise polynomial interpolation. There are, however, significant disadvantages associated with both the piecewise Lagrange approach discussed in Section 5.1 and with the piecewise Hermite approach discussed in Section 5.4. For piecewise Lagrange interpolation only continuity of position can be guaranteed, whereas for piecewise Hermite interpolation we can only guarantee continuity of position and slope. A further problem is that although function values will generally be available, derivative values may not be.

In this section we construct piecewise cubic interpolating polynomials which are continuous in position, slope and curvature, i.e. the function, first derivative and second derivative values are matched at the ends of each adjoining subinterval. Piecewise polynomials possessing these properties are known as cubic splines. It is possible to construct splines of other orders, such as quadratic splines (with continuity of function and first derivative values only) and quintic splines (with continuity of derivatives up to and including those of fourth order). However, we shall restrict our attention to cubic splines, since these are most widely used in practice.

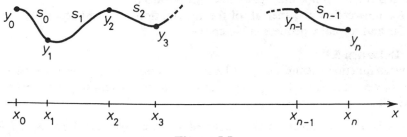

Figure 5.5

Let s_i denote a cubic polynomial defined on the subinterval $[x_i, x_{i+1}]$. This is illustrated in Fig. 5.5. Each cubic polynomial s_i has four unknown coefficients. Therefore, if there are $n + 1$ points $x_i (i = 0, 1, \ldots, n)$, there will be a total of $4n$ potentially free parameters or unknowns to be determined.

If each s_i interpolates the data supplied at the ends of its subinterval $[x_i, x_{i+1}]$, then

$$s_i(x_i) = y_i \quad \text{and} \quad s_i(x_{i+1}) = y_{i+1} \qquad i = 0, 1, \ldots, n - 1$$

The imposition of these conditions gives rise to $2n$ equations. Moreover, if the first and second derivatives are to be continuous at each internal data point, then

$$s'_{i-1}(x_i) = s'_i(x_i) \quad \text{and} \quad s''_{i-1}(x_i) = s''_i(x_i) \qquad i = 1, 2, \ldots, n - 1$$

which provide a further $2(n - 1)$ equations. This gives a total of $2n + 2(n - 1) = 4n - 2$ equations for the $4n$ unknowns. Two more equations can be obtained by imposing boundary conditions at the end points, x_0 and x_n. These may be defined in a variety of ways. The most commonly used boundary conditions are

$$s''_0(x_0) = s''_{n-1}(x_n) = 0$$

which are called *free* boundary conditions, and

$$s'_0(x_0) = y'_0 \quad \text{and} \quad s'_{n-1}(x_n) = y'_n$$

which are called *clamped* boundary conditions. The latter case has the disadvantage that additional data y'_0 and y'_n needs to be supplied.

The word 'spline' originates from the name of a draughtman's tool used for drawing smooth curves. It consists of a long, flexible piece of wood with moveable weights attached to it. The weights are sufficiently heavy to keep the wood in place. As the weights (called knots) are moved, the shape of the spline changes. The ends of the spline are usually free of knots, so it will naturally assume zero curvature there. This corresponds to the free boundary conditions and such splines are often called *natural* splines. Clamped boundary conditions, on the other hand, correspond to the situation where the ends of the spline are rigidly fixed in position.

We now collect together all of the assumptions and requirements made so far and formally define a cubic spline.

■ Definition 5.1
Given a function f defined on $[a, b]$ and a set of numbers x_i, called *knots*, with $a = x_0 < x_1 < \cdots < x_{n-1} < x_n = b$, a *cubic spline* interpolant to f is a function s such that

(i) s is a cubic polynomial s_i defined on the subinterval $[x_i, x_{i+1}]$ for $i = 0, 1, \ldots, n - 1$.

(ii) $s_i(x_i) = y_i$ and $s_i(x_{i+1}) = y_{i+1}$ $i = 0, 1, \ldots, n-1$
where $y_i = f(x_i)$.
(iii) $s'_{i-1}(x_i) = s'_i(x_i)$ $i = 1, 2, \ldots, n-1$
(iv) $s''_{i-1}(x_i) = s''_i(x_i)$ $i = 1, 2, \ldots, n-1$
(v) Either

$$s''_0(x_0) = s''_{n-1}(x_n) = 0 \quad \text{(free boundary conditions)}$$

or

$$s'_0(x_0) = y'_0 \quad \text{and} \quad s'_{n-1}(x_n) = y'_n \quad \text{(clamped boundary conditions)}$$

where $y'_0 = f'(x_0)$ and $y'_n = f'(x_n)$.

$\square$

■ Example 5.10

To illustrate this definition, consider the construction of the cubic spline interpolant to $f(x) = x^4$ with knots $-1, 0, 1$, subject to clamped boundary conditions. In this case $n = 2$, $x_0 = -1$, $x_1 = 0$, $x_2 = 1$, $y_0 = 1$, $y_1 = 0$ and $y_2 = 1$. Moreover, since $f'(x) = 4x^3$, $y'_0 = -4$ and $y'_2 = 4$. Let

$$s_0(x) = a_0 + b_0 x + c_0 x^2 + d_0 x^3, \qquad s_1(x) = a_1 + b_1 x + c_1 x^2 + d_1 x^3$$

Hence

$$s'_0(x) = b_0 + 2c_0 x + 3d_0 x^2, \qquad\qquad s'_1(x) = b_1 + 2c_1 x + 3d_1 x^2$$
$$s''_0(x) = 2c_0 + 6d_0 x, \qquad\qquad\qquad s''_1(x) = 2c_1 + 6d_1 x$$

From condition (ii),

$$s_0(-1) = 1 = a_0 - b_0 + c_0 - d_0$$
$$s_0(0) = 0 = a_0$$
$$s_1(0) = 0 = a_1$$
$$s_1(1) = 1 = a_1 + b_1 + c_1 + d_1$$

From condition (iii),

$$s'_0(0) = s'_1(0), \quad \text{i.e. } b_0 = b_1$$

From condition (iv),

$$s''_0(0) = s''_1(0), \quad \text{i.e. } 2c_0 = 2c_1$$

Finally, applying the clamped boundary conditions in condition (v),

$$s'_0(-1) = -4 = b_0 - 2c_0 + 3d_0$$
$$s'_1(1) = 4 = b_1 + 2c_1 + 3d_1$$

Altogether there are eight equations for the eight unknown coefficients of s_0

and s_1. In matrix form the equations are

$$\begin{pmatrix} 1 & -1 & 1 & -1 & 0 & 0 & 0 & 0 \\ 1 & 0 & 0 & 0 & 0 & 0 & 0 & 0 \\ 0 & 0 & 0 & 0 & 1 & 0 & 0 & 0 \\ 0 & 0 & 0 & 0 & 1 & 1 & 1 & 1 \\ 0 & 1 & 0 & 0 & 0 & -1 & 0 & 0 \\ 0 & 0 & 1 & 0 & 0 & 0 & -1 & 0 \\ 0 & 1 & -2 & 3 & 0 & 0 & 0 & 0 \\ 0 & 0 & 0 & 0 & 0 & 1 & 2 & 3 \end{pmatrix} \begin{pmatrix} a_0 \\ b_0 \\ c_0 \\ d_0 \\ a_1 \\ b_1 \\ c_1 \\ d_1 \end{pmatrix} = \begin{pmatrix} 1 \\ 0 \\ 0 \\ 1 \\ 0 \\ 0 \\ -4 \\ 4 \end{pmatrix}$$

with solution $a_0 = b_0 = a_1 = b_1 = 0$, $c_0 = c_1 = -1$, $d_0 = -2$ and $d_1 = 2$.
Hence

$$s(x) = \begin{cases} -2x^3 - x^2 & -1 \leqslant x \leqslant 0 \\ 2x^3 - x^2 & 0 \leqslant x \leqslant 1 \end{cases}$$

A graph of s is sketched in Fig. 5.6.   □

For even a moderately large number of data points (10–15 say), this method of constructing cubic splines directly from Definition 5.1 is extremely tedious and involves the solution of a large system of linear equations. Although the coefficient matrix is sparse, it lacks any well-defined structure which might assist in the solution procedure. Fortunately, there are a number of techniques for constructing cubic splines which are computationally more efficient. We shall investigate two quite different approaches. Both give rise to systems of

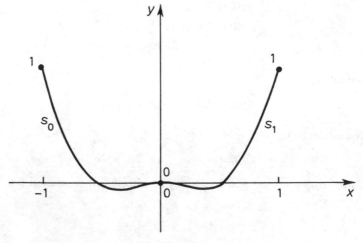

Figure 5.6

$n + 1$ linear equations (compared with $4n$ equations in the previous approach) with symmetric, tridiagonal coefficient matrices.

The first method is based on the piecewise Hermite cubic interpolant discussed at the end of Section 5.4, which automatically satisfies conditions (i)–(iii) in Definition 5.1. It is uniquely determined once the values of y_i and y_i' are known. In the present context only the values of y_i are given, and we shall attempt to use the remaining conditions to set up a system of $n + 1$ equations for the unknowns y_i' $(i = 0, 1, \ldots, n)$. If the expression for $H_{3,i}$ given by equation (5.20) is differentiated with respect to x, then

$$
\begin{aligned}
H'_{3,i}(x) = 2y_i &\left\{ \frac{(x - x_{i+1})}{h_i^2} + \frac{1}{h_i^3}[2(x - x_i)(x - x_{i+1}) + (x - x_{i+1})^2] \right\} \\
+ 2y_{i+1} &\left\{ \frac{(x - x_i)}{h_i^2} - \frac{1}{h_i^3}[2(x - x_{i+1})(x - x_i) + (x - x_i)^2] \right\} \\
+ y_i' &\left\{ \frac{2(x - x_i)(x - x_{i+1})}{h_i^2} + \frac{(x - x_{i+1})^2}{h_i^2} \right\} \\
+ y_{i+1}' &\left\{ \frac{2(x - x_{i+1})(x - x_i)}{h_i^2} + \frac{(x - x_i)^2}{h_i^2} \right\} \tag{5.21}
\end{aligned}
$$

(where $h_i = x_{i+1} - x_i$), and

$$
\begin{aligned}
H''_{3,i}(x) = 2y_i &\left\{ \frac{1}{h_i^2} + \frac{2}{h_i^3}[(x - x_i) + 2(x - x_{i+1})] \right\} \\
+ 2y_{i+1} &\left\{ \frac{1}{h_i^2} - \frac{2}{h_i^3}[(x - x_{i+1}) + 2(x - x_i)] \right\} \\
+ 2y_i' &\left\{ \frac{(x - x_i)}{h_i^2} + \frac{2(x - x_{i+1})}{h_i^2} \right\} + 2y_{i+1}' \left\{ \frac{(x - x_{i+1})}{h_i^2} + \frac{2(x - x_i)}{h_i^2} \right\}
\end{aligned}
$$

$$\tag{5.22}$$

If the subscript i is replaced by $i - 1$ in (5.22), then

$$
\begin{aligned}
H''_{3,i-1}(x) = 2y_{i-1} &\left\{ \frac{1}{h_{i-1}^2} + \frac{2}{h_{i-1}^3}[(x - x_{i-1}) + 2(x - x_i)] \right\} \\
+ 2y_i &\left\{ \frac{1}{h_{i-1}^2} - \frac{2}{h_{i-1}^3}[(x - x_i) + 2(x - x_{i-1})] \right\} \\
+ 2y_{i-1}' &\left\{ \frac{(x - x_{i-1})}{h_{i-1}^2} + \frac{2(x - x_i)}{h_{i-1}^2} \right\} + 2y_i' \left\{ \frac{(x - x_i)}{h_{i-1}^2} + \frac{2(x - x_{i-1})}{h_{i-1}^2} \right\}
\end{aligned}
$$

$$\tag{5.23}$$

Denoting s_i by $H_{3,i}$, condition (iv) becomes

$$H''_{3,i-1}(x_i) = H''_{3,i}(x_i) \qquad i = 1, 2, \ldots, n-1$$

From (5.22) and (5.23), this gives

$$2y_{i-1}\left\{\frac{1}{h_{i-1}^2} + \frac{2}{h_{i-1}^3}(x_i - x_{i-1})\right\} + 2y_i\left\{\frac{1}{h_{i-1}^2} - \frac{4}{h_{i-1}^3}(x_i - x_{i-1})\right\}$$

$$+ 2y'_{i-1}\left\{\frac{(x_i - x_{i-1})}{h_{i-1}^2}\right\} + 4y'_i\left\{\frac{(x_i - x_{i-1})}{h_{i-1}^2}\right\}$$

$$= 2y_i\left\{\frac{1}{h_i^2} + \frac{4}{h_i^3}(x_i - x_{i+1})\right\} + 2y_{i+1}\left\{\frac{1}{h_i^2} - \frac{2}{h_i^3}(x_i - x_{i+1})\right\}$$

$$+ 4y'_i\left\{\frac{(x_i - x_{i+1})}{h_i^2}\right\} + 2y'_{i+1}\left\{\frac{(x_i - x_{i+1})}{h_i^2}\right\}$$

Now $x_{i+1} - x_i = h_i$ and $x_i - x_{i-1} = h_{i-1}$, so this equation can be simplified to give

$$\frac{y'_{i-1}}{h_{i-1}} + 2\left(\frac{1}{h_{i-1}} + \frac{1}{h_i}\right)y'_i + \frac{y'_{i+1}}{h_i} = 3\left[\frac{-y_{i-1}}{h_{i-1}^2} + \left(\frac{1}{h_{i-1}^2} - \frac{1}{h_i^2}\right)y_i + \frac{y_{i+1}}{h_i^2}\right] \quad (5.24)$$

for $i = 1, 2, \ldots, n-1$. The terms on the right hand side of equation (5.24) are easily calculated, to produce a system of $n-1$ linear equations for the $n+1$ unknowns y'_i $(i = 0, 1, \ldots, n)$.

If free boundary conditions are used, then we require

$$H''_{3,0}(x_0) = H''_{3,n-1}(x_n) = 0$$

From equation (5.22), the first of these gives

$$2y_0\left\{\frac{1}{h_0^2} + \frac{4}{h_0^3}(x_0 - x_1)\right\} + 2y_1\left\{\frac{1}{h_0^2} - \frac{2}{h_0^3}(x_0 - x_1)\right\}$$

$$+ 4y'_0\left\{\frac{(x_0 - x_1)}{h_0^2}\right\} + 2y'_1\left\{\frac{(x_0 - x_1)}{h_0^2}\right\} = 0$$

and, since $x_1 - x_0 = h_0$, this can be simplified to

$$\frac{2y'_0}{h_0} + \frac{y'_1}{h_0} = \frac{3}{h_0^2}(-y_0 + y_1) \quad (5.25)$$

Similarly, the condition $H''_{3,n-1}(x_n) = 0$ gives

$$\frac{y'_{n-1}}{h_{n-1}} + \frac{2y'_n}{h_{n-1}} = \frac{3}{h_{n-1}^2}(-y_{n-1} + y_n) \quad (5.26)$$

The complete tridiagonal system can then be written in matrix form as in equation (5.27).

$$
\begin{pmatrix}
\dfrac{2}{h_0} & \dfrac{1}{h_0} & & & & & & \\[6pt]
\dfrac{1}{h_0} & \dfrac{2}{h_0}+\dfrac{2}{h_1} & \dfrac{1}{h_1} & & & & \text{\Large 0} & \\[6pt]
& \dfrac{1}{h_1} & \dfrac{2}{h_1}+\dfrac{2}{h_2} & \dfrac{1}{h_2} & & & & \\[6pt]
& & \ddots & \ddots & \ddots & & & \\[6pt]
& & \dfrac{1}{h_{i-1}} & \dfrac{2}{h_{i-1}}+\dfrac{2}{h_i} & \dfrac{1}{h_i} & & & \\[6pt]
& & & \ddots & \ddots & \ddots & & \\[6pt]
& & & & \dfrac{1}{h_{n-3}} & \dfrac{2}{h_{n-3}}+\dfrac{2}{h_{n-2}} & \dfrac{1}{h_{n-2}} & \\[6pt]
& \text{\Large 0} & & & & \dfrac{1}{h_{n-2}} & \dfrac{2}{h_{n-2}}+\dfrac{2}{h_{n-1}} & \dfrac{1}{h_{n-1}} \\[6pt]
& & & & & & \dfrac{1}{h_{n-1}} & \dfrac{2}{h_{n-1}}
\end{pmatrix}
\begin{pmatrix}
y'_0 \\[6pt] y'_1 \\[6pt] y'_2 \\[6pt] \vdots \\[6pt] y'_i \\[6pt] \vdots \\[6pt] y'_{n-2} \\[6pt] y'_{n-1} \\[6pt] y'_n
\end{pmatrix}
= 3
\begin{pmatrix}
-\dfrac{y_0}{h_0^2}+\dfrac{y_1}{h_0^2} \\[8pt]
-\dfrac{y_0}{h_0^2}+\dfrac{y_1}{h_0^2}-\dfrac{y_1}{h_1^2}+\dfrac{y_2}{h_1^2} \\[8pt]
-\dfrac{y_1}{h_1^2}+\dfrac{y_2}{h_1^2}-\dfrac{y_2}{h_2^2}+\dfrac{y_3}{h_2^2} \\[8pt]
\vdots \\[8pt]
-\dfrac{y_{i-1}}{h_{i-1}^2}+\dfrac{y_i}{h_{i-1}^2}-\dfrac{y_i}{h_i^2}+\dfrac{y_{i+1}}{h_i^2} \\[8pt]
\vdots \\[8pt]
-\dfrac{y_{n-3}}{h_{n-3}^2}+\dfrac{y_{n-2}}{h_{n-3}^2}-\dfrac{y_{n-2}}{h_{n-2}^2}+\dfrac{y_{n-1}}{h_{n-2}^2} \\[8pt]
-\dfrac{y_{n-2}}{h_{n-2}^2}+\dfrac{y_{n-1}}{h_{n-2}^2}-\dfrac{y_{n-1}}{h_{n-1}^2}+\dfrac{y_n}{h_{n-1}^2} \\[8pt]
-\dfrac{y_{n-1}}{h_{n-1}^2}+\dfrac{y_n}{h_{n-1}^2}
\end{pmatrix}
\tag{5.27}
$$

It can be shown (see Exercise 5.5, Question 2) that the coefficient matrix is symmetric and positive definite, and so this system may be solved using the Choleski decomposition method described in Section 2.2.

In the case of clamped boundary conditions the values of y_0' and y_n' are known explicitly. These numbers can be substituted into (5.24) to obtain a system of $n-1$ equations for the $n-1$ unknowns y_i' $(i = 1, 2, \ldots, n-1)$. The corresponding coefficient matrix is the same as that of (5.27) but with the first and last rows and columns deleted.

■ **Example 5.11**

Consider the estimation of $\cos(3.141\,59)$ from the following data:

x_i	0	1	3	3.5	5
y_i	1.000 00	0.540 30	$-0.989\,99$	$-0.936\,46$	0.283 66

In this case $n = 4$, $h_0 = 1$, $h_1 = 2$, $h_2 = 0.5$ and $h_3 = 1.5$.

If free boundary conditions are imposed we may use (5.27) to obtain

$$\begin{pmatrix} 2 & 1 & 0 & 0 & 0 \\ 1 & 3 & 1/2 & 0 & 0 \\ 0 & 1/2 & 5 & 2 & 0 \\ 0 & 0 & 2 & 16/3 & 2/3 \\ 0 & 0 & 0 & 2/3 & 4/3 \end{pmatrix} \begin{pmatrix} y_0' \\ y_1' \\ y_2' \\ y_3' \\ y_4' \end{pmatrix} = \begin{pmatrix} -1.379\,10 \\ -2.526\,82 \\ -0.505\,36 \\ 2.269\,19 \\ 1.626\,83 \end{pmatrix} \tag{5.28}$$

which has solution

$$y_0' = -0.339\,66, \qquad y_1' = -0.699\,78, \qquad y_2' = -0.175\,66$$
$$y_3' = 0.361\,42, \qquad y_4' = 1.039\,41$$

The cubic polynomial defined on $3 \leqslant x \leqslant 3.5$ is s_2, which is given from (5.20) as

$$s_2(x) = -0.989\,99 \left\{ \frac{(x-3.5)^2}{(0.5)^2} + \frac{2(x-3)(x-3.5)^2}{(0.5)^3} \right\}$$

$$-0.936\,46 \left\{ \frac{(x-3)^2}{(0.5)^2} - \frac{2(x-3.5)(x-3)^2}{(0.5)^3} \right\}$$

$$-0.175\,66 \left\{ \frac{(x-3)(x-3.5)^2}{(0.5)^2} \right\} + 0.361\,42 \left\{ \frac{(x-3.5)(x-3)^2}{(0.5)^2} \right\}$$

Hence $s_2(3.141\,59) = -1.002\,71$.

If clamped boundary conditions are imposed then we require $y_0' = -\sin(0) = 0$ and $y_4' = -\sin(5) = 0.958\,92$. Deleting the first and last equations of (5.28) and substituting the known values of y_0' and y_4' gives

$$\begin{pmatrix} 3 & 1/2 & 0 \\ 1/2 & 5 & 2 \\ 0 & 2 & 16/3 \end{pmatrix} \begin{pmatrix} y_1' \\ y_2' \\ y_3' \end{pmatrix} = \begin{pmatrix} -2.526\,82 \\ -0.505\,36 \\ 1.629\,91 \end{pmatrix}$$

which has solution

$$y_1' = -0.814\,46, \qquad y_2' = -0.166\,91, \qquad y_3' = 0.368\,20$$

The corresponding value of $s_2(3.141\,59)$ is $-1.002\,27$. □

We now consider an alternative way of constructing cubic splines without reference to Hermite polynomials. If s_i is a cubic polynomial on $[x_i, x_{i+1}]$, then s_i'' is a linear polynomial on $[x_i, x_{i+1}]$. Let us write $s_i''(x_i) = c_i$ and $s_i''(x_{i+1}) = c_{i+1}$ for $i = 0, 1, \ldots, n-1$. It follows from this that the second derivative of s is continuous at the internal knots because, if we replace i by $i-1$ in the second of these, then $s_{i-1}''(x_i) = c_i = s_i''(x_i)$. In Lagrange form,

$$s_i''(x) = L_i(x)c_i + L_{i+1}(x)c_{i+1} = \left(\frac{x - x_{i+1}}{x_i - x_{i+1}} \right)c_i + \left(\frac{x - x_i}{x_{i+1} - x_i} \right)c_{i+1}$$

$$= \frac{(x - x_{i+1})}{-h_i}c_i + \frac{(x - x_i)}{h_i}c_{i+1}$$

Integrating twice with respect to x gives

$$s_i(x) = \frac{(x - x_{i+1})^3}{-h_i} \frac{c_i}{6} + \frac{(x - x_i)^3}{h_i} \frac{c_{i+1}}{6} + Ax + B \qquad (5.29)$$

for some constants of integration A and B. Now, from the interpolation conditions $s_i(x_i) = y_i$ and $s_i(x_{i+1}) = y_{i+1}$, it follows that

$$h_i^2 \frac{c_i}{6} + Ax_i + B = y_i \qquad (5.30)$$

and

$$h_i^2 \frac{c_{i+1}}{6} + Ax_{i+1} + B = y_{i+1} \qquad (5.31)$$

These equations represent a system of equations for the unknowns A and B. The constant B can be eliminated by subtracting (5.30) from (5.31), which gives

$$A = \frac{y_{i+1} - y_i}{h_i} + \frac{h_i}{6}(c_i - c_{i+1})$$

The constant A can be eliminated by multiplying both sides of equations (5.30) and (5.31) by x_{i+1} and x_i, respectively, and subtracting, which gives

$$B = \frac{x_{i+1}y_i - x_i y_{i+1}}{h_i} + \frac{h_i}{6}(x_i c_{i+1} - x_{i+1} c_i)$$

If these expressions are substituted into (5.29), then

$$s_i(x) = \frac{c_i}{6h_i}(x_{i+1} - x)^3 + \frac{c_{i+1}}{6h_i}(x - x_i)^3$$

$$+ \left(\frac{y_i}{h_i} - \frac{h_i c_i}{6}\right)(x_{i+1} - x) + \left(\frac{y_{i+1}}{h_i} - \frac{h_i c_{i+1}}{6}\right)(x - x_i) \qquad (5.32)$$

Equation (5.32) gives us a formula for $s_i(x)$ in terms of the known values of x_i, y_i and h_i and the unknown values of c_i. Furthermore, the derivation of (5.32) shows that conditions (i), (ii) and (iv) in Definition 5.1 are automatically satisfied. The remaining conditions can be used to set up a system of $n+1$ equations for the unknowns c_i ($i = 0, 1, \ldots, n$). If the expression for s_i given by equation (5.32) is differentiated with respect to x, then

$$s_i'(x) = -\frac{c_i}{2h_i}(x_{i+1} - x)^2 + \frac{c_{i+1}}{2h_i}(x - x_i)^2$$

$$- \left(\frac{y_i}{h_i} - \frac{h_i c_i}{6}\right) + \left(\frac{y_{i+1}}{h_i} - \frac{h_i c_{i+1}}{6}\right) \qquad (5.33)$$

and replacing the subscript i by $i-1$ gives

$$s_{i-1}'(x) = -\frac{c_{i-1}}{2h_{i-1}}(x_i - x)^2 + \frac{c_i}{2h_{i-1}}(x - x_{i-1})^2$$

$$- \left(\frac{y_{i-1}}{h_{i-1}} - \frac{h_{i-1} c_{i-1}}{6}\right) + \left(\frac{y_i}{h_{i-1}} - \frac{h_{i-1} c_i}{6}\right)$$

From condition (iii), $s_{i-1}'(x_i) = s_i'(x_i)$, and so

$$\frac{c_i h_{i-1}}{2} - \left(\frac{y_{i-1}}{h_{i-1}} - \frac{h_{i-1} c_{i-1}}{6}\right) + \left(\frac{y_i}{h_{i-1}} - \frac{h_{i-1} c_i}{6}\right)$$

$$= -\frac{c_i h_i}{2} - \left(\frac{y_i}{h_i} - \frac{h_i c_i}{6}\right) + \left(\frac{y_{i+1}}{h_i} - \frac{h_i c_{i+1}}{6}\right)$$

which can be rearranged as

$$h_{i-1} c_{i-1} + 2c_i(h_{i-1} + h_i) + h_i c_{i+1}$$

$$= 6\left[\frac{y_{i-1}}{h_{i-1}} - y_i\left(\frac{1}{h_{i-1}} + \frac{1}{h_i}\right) + \frac{y_{i+1}}{h_i}\right] \qquad (5.34)$$

The terms on the right hand side of equation (5.34) are easily calculated to produce a system of $n-1$ linear equations for the $n+1$ unknowns c_i ($i = 0, 1, \ldots, n$).

In the case of free boundary conditions, $c_0 = s_0''(x_0) = 0$ and $c_n = s_{n-1}''(x_n) = 0$. These numbers can be substituted into (5.34) to obtain the system shown in equation (5.35).

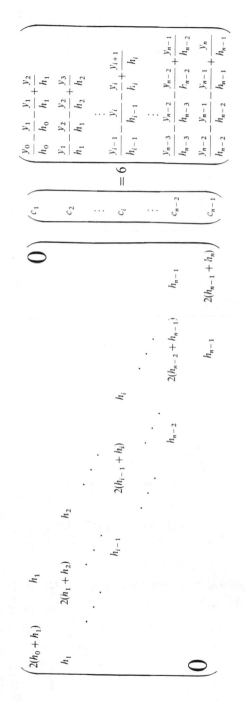

$$
\begin{pmatrix}
2(h_0+h_1) & h_1 & & & & & \\
h_1 & 2(h_1+h_2) & h_2 & & & & \\
& h_{i-1} & 2(h_{i-1}+h_i) & h_i & & & \\
& & & \ddots & & & \\
& & & h_{n-2} & 2(h_{n-2}+h_{n-1}) & h_{n-1} \\
& & & & h_{n-1} & 2(h_{n-1}+h_n)
\end{pmatrix}
\begin{pmatrix}
c_1 \\ c_2 \\ \vdots \\ c_i \\ \vdots \\ c_{n-2} \\ c_{n-1}
\end{pmatrix}
= 6
\begin{pmatrix}
\dfrac{y_0}{h_0} - \dfrac{y_1}{h_1} + \dfrac{y_2}{h_1} \\[4pt]
\dfrac{y_1}{h_1} - \dfrac{y_2}{h_2} + \dfrac{y_3}{h_2} \\[4pt]
\vdots \\[4pt]
\dfrac{y_{i-1}}{h_{i-1}} - \dfrac{y_i}{h_i} + \dfrac{y_{i+1}}{h_i} \\[4pt]
\vdots \\[4pt]
\dfrac{y_{n-3}}{h_{n-3}} - \dfrac{y_{n-2}}{h_{n-2}} + \dfrac{y_{n-1}}{h_{n-2}} \\[4pt]
\dfrac{y_{n-2}}{h_{n-2}} - \dfrac{y_{n-1}}{h_{n-1}} + \dfrac{y_n}{h_{n-1}}
\end{pmatrix}
\tag{5.35}
$$

The tridiagonal coefficient matrix is symmetric and positive definite, and so (5.35) can be solved via Choleski's decomposition.

If clamped boundary conditions are used, we require $s_0'(x_0) = y_0'$ and $s_{n-1}'(x_n) = y_n'$. From equation (5.33) these give

$$2h_0 c_0 + h_0 c_1 = 6\left(-\frac{y_0}{h_0} + \frac{y_1}{h_0} - y_0'\right) \tag{5.36}$$

and

$$h_{n-1} c_{n-1} + 2h_{n-1} c_n = 6\left(\frac{y_{n-1}}{h_{n-1}} - \frac{y_n}{h_{n-1}} + y_n'\right) \tag{5.37}$$

Equations (5.36) and (5.37), together with those of (5.34), furnish a system of $n + 1$ equations for the $n + 1$ unknowns c_i ($i = 0, 1, \ldots, n$). Moreover, it is easy to check that provided (5.36) is taken to be the first equation of the system and (5.37) as the last, the corresponding coefficient matrix is also tridiagonal, symmetric and positive definite.

■ Example 5.12

To illustrate this approach we repeat the calculations of Example 5.11.

In the case of free boundary conditions we can use (5.35) to get

$$\begin{pmatrix} 6 & 2 & 0 \\ 2 & 5 & 1/2 \\ 0 & 1/2 & 4 \end{pmatrix} \begin{pmatrix} c_1 \\ c_2 \\ c_3 \end{pmatrix} = 6 \begin{pmatrix} -0.305\,45 \\ 0.872\,21 \\ 0.706\,35 \end{pmatrix}$$

which has solution

$$c_1 = -0.720\,23, \qquad c_2 = 1.244\,35, \qquad c_3 = 0.903\,98$$

From (5.32),

$$s_2(x) = \frac{1.244\,35}{6(0.5)}(3.5 - x)^3 + \frac{0.903\,98}{6(0.5)}(x - 3)^3$$

$$+ \left(-\frac{0.989\,99}{0.5} - \frac{0.5(1.244\,35)}{6}\right)(3.5 - x)$$

$$+ \left(-\frac{0.936\,46}{0.5} - \frac{0.5(0.903\,98)}{6}\right)(x - 3)$$

Hence $s_2(3.141\,59) = -1.002\,71$.

In the case of clamped conditions we use (5.34), (5.36) and (5.37) to get

$$\begin{pmatrix} 2 & 1 & 0 & 0 & 0 \\ 1 & 6 & 2 & 0 & 0 \\ 0 & 2 & 5 & 1/2 & 0 \\ 0 & 0 & 1/2 & 4 & 3/2 \\ 0 & 0 & 0 & 3/2 & 3 \end{pmatrix} \begin{pmatrix} c_0 \\ c_1 \\ c_2 \\ c_3 \\ c_4 \end{pmatrix} = 6 \begin{pmatrix} -0.459\,70 \\ -0.305\,45 \\ 0.872\,21 \\ 0.706\,35 \\ 0.145\,51 \end{pmatrix}$$

which has solution

$$c_0 = -1.129\,29, \qquad c_1 = -0.499\,63, \qquad c_2 = 1.147\,18,$$
$$c_3 = 0.993\,22, \qquad c_4 = -0.205\,59$$

The corresponding value of $s_2(3.141\,59)$ is $-1.002\,27$.   □

EXERCISE 5.5

1. Verify that the function defined by

$$s(x) = \begin{cases} -x^3 + \frac{17}{2}x^2 - 9x + \frac{3}{2} & -1 \leqslant x \leqslant 1 \\ 2x^3 - \frac{1}{2}x^2 - \frac{3}{2} & 1 \leqslant x \leqslant 2 \\ x^3 + \frac{11}{2}x^2 - 12x + \frac{13}{2} & 2 \leqslant x \leqslant 4 \end{cases}$$

is a cubic spline.

2. Show that the coefficient matrices given in equations (5.27) and (5.35) are symmetric and positive definite.

3. Construct the interpolating cubic spline with (a) free (b) clamped boundary conditions to the function $f(x) = x^4$ using the following data:

x_i	-1	-0.9	-0.7	-0.1	0.2	0.6	0.8	1
$f(x_i)$	1.0000	0.6561	0.240	0.0001	0.0016	0.1296	0.4096	1.0000

Which of these two approximations produces the most accurate estimate of $f(0.125)$ and $f(0.95)$? Explain why this might have been expected. $((0.125)^4 = 0.0024$ and $(0.95)^4 = 0.8145$ (4D))

4. Construct the

   (a) Lagrange interpolating polynomial
   (b) Interpolating cubic spline with free boundary conditions
   (c) Interpolating cubic spline with clamped boundary conditions
      $y'(-6) = 74, \qquad y'(7) = 191$

   to the following data:

x_i	-6	-4	-2	-1	0	6	7
$f(x_i)$	-121	-25	-1	-1	-1	335	503

   Compare the corresponding estimates of $f(-3)$, $f(1)$ and $f(5)$. Explain your results.

5. Sketch a graph of the interpolating cubic spline with clamped boundary

conditions $f'(0) = f'(12) = 0$ to the following data:

x_i	0	1	2	3	4	5	6	7	8	9	10	11	12
$f(x_i)$	0	0	0	0	0	0	1	0	0	0	0	0	0

Compare this graph with those sketched in Exercise 5.1, Question 4 and Exercise 5.4, Question 5.

6. Let

$$s_i(x) = a_i + b_i(x - x_i) + c_i(x - x_i)^2 + d_i(x - x_i)^3$$

for $x \in [x_i, x_{i+1}]$ and $i = 0, 1, \ldots, n - 1$, where $x_{i+1} - x_i = h$. Use property (ii) of Definition 5.1 to show that

$$a_i = y_i$$
$$c_i = z_i/2$$
$$d_i = (z_{i+1} - z_i)/6h$$
$$b_i = [(y_{i+1} - y_i)/h] - [h(2z_i + z_{i+1})/6]$$

where $z_i = s_i''(x_i)$ and $z_{i+1} = s_i''(x_{i+1})$.
Use property (iii) to derive the relationship

$$z_{i+1} + 4z_i + z_{i-1} = \frac{6}{h^2}(y_{i+1} - 2y_i + y_{i-1})$$

Hence find the natural cubic spline which interpolates the following data:

x_i	17	20	23	26
y_i	4.50	7.00	6.10	5.50

5.6 LEAST SQUARES APPROXIMATION TO DISCRETE DATA

If it is known that a given set of n data points (x_i, y_i) $(i = 1, 2, \ldots, n)$ contains experimental errors, then the approximation techniques considered in the previous sections of this chapter may be inappropriate. Consider, for example, the experimental data shown in Fig. 5.7(a). A Lagrangian interpolation polynomial of degree six or a cubic spline approximation could easily be constructed for this data. However, there is no justification for insisting that the data points be reproduced exactly, and such an approximation may well be very misleading since unwanted oscillations are likely. A more satisfactory approach would be to find a straight line which passes close to all seven points. One such possibility is shown in Fig. 5.7(b). Clearly we have to decide what criterion is to be adopted for constructing such an approximation.

Let d_i denote the (signed) vertical distance from the point (x_i, y_i) to the

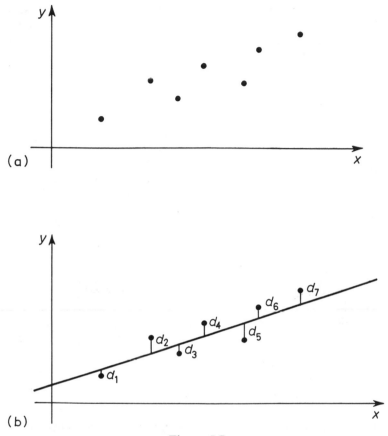

Figure 5.7

line, as indicated in Fig. 5.7(b). One sensible criterion is to find the line which minimizes the quantity $\max_i |d_i|$. This is known as minimax approximation and, although of considerable importance in numerical analysis, it will not be considered here. An extensive treatment of this topic can be found in the books by Watson (1980) and Rice (1964). An alternative criterion is to minimize

$$\sum_{i=1}^{n} w_i d_i^2$$

where the weights w_i $(i = 1, 2, \dots, n)$ are given positive numbers. This approach is known as least squares approximation and is mathematically more tractable. The inclusion of weights allows us to attach varying degrees of importance to different data points. If equal importance is to be assigned to all points, we take $w_i = 1$ $(i = 1, 2, \dots, n)$.

Least squares approximation is not restricted to straight lines. However, in order to motivate the general case we consider this first. Let $p_1(x) = a_0 + a_1 x$, where the coefficients a_0 and a_1 are chosen to minimize the weighted sum of the squares of the errors taken over all data points. If this sum is denoted by $E(a_0, a_1)$, then

$$E(a_0, a_1) = \sum_{i=1}^{n} w_i [y_i - (a_0 + a_1 x_i)]^2$$

From calculus, necessary conditions for a minimum to occur at (a_0, a_1) are

$$\frac{\partial E}{\partial a_0} = \frac{\partial E}{\partial a_1} = 0$$

which give

$$\frac{\partial E}{\partial a_0} = -2 \sum_{i=1}^{n} w_i [y_i - (a_0 + a_1 x_i)] = 0$$

$$\frac{\partial E}{\partial a_1} = -2 \sum_{i=1}^{n} w_i [y_i - (a_0 + a_1 x_i)] x_i = 0$$

These equations can be rearranged to give

$$a_0 \sum_{i=1}^{n} w_i + a_1 \sum_{i=1}^{n} w_i x_i = \sum_{i=1}^{n} w_i y_i$$

$$a_0 \sum_{i=1}^{n} w_i x_i + a_1 \sum_{i=1}^{n} w_i x_i^2 = \sum_{i=1}^{n} w_i x_i y_i$$

(5.38)

and, since the numbers x_i, y_i and w_i are all known, this represents a system of two simultaneous linear equations for a_0 and a_1.

■ Example 5.13

Consider the calculation of a linear least squares approximation to the following data:

x_i	−5	−3	1	3	4	6	8
y_i	18	7	0	7	16	50	67
w_i	1	1	1	1	20	1	1

In this example the point $(4, 16)$ is assumed to be more reliable than the other points. A weight of 20 is therefore used to attach significantly more importance to it.

Evaluation of the summations in (5.38) produces

$$26 a_0 + 90 a_1 = 469$$
$$90 a_0 + 464 a_1 = 2026$$

which may be solved to give $a_0 = 8.8991$ and $a_1 = 2.6403$. The approximating polynomial is therefore

$$p_1(x) = 8.8991 + 2.6403x$$

which is sketched in Fig. 5.8(a). It is instructive to compare this graph with the one shown in Fig. 5.8(b), which represents the linear least squares approximation to the same data but with $w_i = 1$ for all i. In the former case, the line passes much closer to the point $(4, 16)$.   □

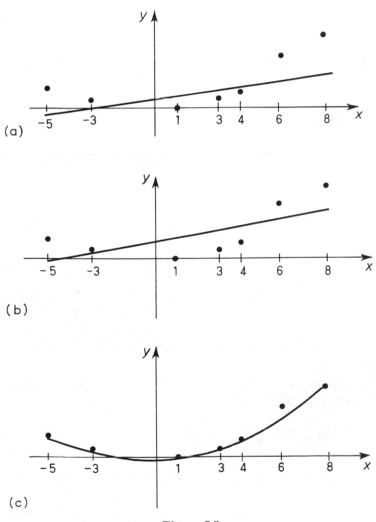

(a)

(b)

(c)

Figure 5.8

We now consider the least squares approximation of a set of n data points (x_i, y_i) $(i = 1, 2, \ldots, n)$ by a polynomial of degree $m < n$. If $p_m(x)$ is written as

$$a_0 + a_1 x + \cdots + a_m x^m$$

then the weighted least squares error is

$$E(a_0, a_1, \ldots, a_m) = \sum_{i=1}^{n} w_i [y_i - (a_0 + a_1 x_i + \cdots + a_m x_i^m)]^2$$

Necessary conditions for a minimum to occur at $(a_0, a_1, \ldots, a_m)$ are

$$\frac{\partial E}{\partial a_0} = \frac{\partial E}{\partial a_1} = \cdots = \frac{\partial E}{\partial a_m} = 0$$

Carrying out the differentiation with respect to $a_0, a_1, \ldots, a_m$ in turn and rearranging gives

$$a_0 \sum_{i=1}^{n} w_i + a_1 \sum_{i=1}^{n} w_i x_i + \cdots + a_m \sum_{i=1}^{n} w_i x_i^m = \sum_{i=1}^{n} w_i y_i$$

$$a_0 \sum_{i=1}^{n} w_i x_i + a_1 \sum_{i=1}^{n} w_i x_i^2 + \cdots + a_m \sum_{i=1}^{n} w_i x_i^{m+1} = \sum_{i=1}^{n} w_i x_i y_i$$

$$\vdots \qquad \vdots \qquad \qquad \vdots \qquad \qquad \vdots$$

$$a_0 \sum_{i=1}^{n} w_i x_i^m + a_1 \sum_{i=1}^{n} w_i x_i^{m+1} + \cdots + a_m \sum_{i=1}^{n} w_i x_i^{2m} = \sum_{i=1}^{n} w_i x_i^m y_i$$

These are referred to as the *normal equations*, and they represent a system of $m + 1$ simultaneous linear equations for the $m + 1$ unknowns $a_0, a_1, \ldots, a_m$. It can be proved (see Powell, 1981) that, provided the x_i are distinct with $m < n$, the coefficient matrix of this system is symmetric and positive definite. Hence the normal equations possess a unique solution (which could, for example, be computed via the Choleski decomposition method described in Section 2.2).

■ **Example 5.14**

The normal equations for the quadratic least squares approximation to the data given in Example 5.13 are

$$a_0 \sum_{i=1}^{7} w_i + a_1 \sum_{i=1}^{7} w_i x_i + a_2 \sum_{i=1}^{7} w_i x_i^2 = \sum_{i=1}^{7} w_i y_i$$

$$a_0 \sum_{i=1}^{7} w_i x_i + a_1 \sum_{i=1}^{7} w_i x_i^2 + a_2 \sum_{i=1}^{7} w_i x_i^3 = \sum_{i=1}^{7} w_i x_i y_i$$

$$a_0 \sum_{i=1}^{7} w_i x_i^2 + a_1 \sum_{i=1}^{7} w_i x_i^3 + a_2 \sum_{i=1}^{7} w_i x_i^4 = \sum_{i=1}^{7} w_i x_i^2 y_i$$

Evaluation of the summations gives

$$\begin{pmatrix} 26 & 90 & 464 \\ 90 & 464 & 1\,884 \\ 464 & 1\,884 & 11\,300 \end{pmatrix} \begin{pmatrix} a_0 \\ a_1 \\ a_2 \end{pmatrix} = \begin{pmatrix} 469 \\ 2\,026 \\ 11\,784 \end{pmatrix}$$

which has solution

$$a_0 = -3.4079, \qquad a_1 = 0.6964, \qquad a_2 = 1.0667$$

A graph of

$$p_2(x) = -3.4079 + 0.6964x + 1.0667x^2$$

is sketched in Fig. 5.8(c). ☐

Although polynomials are frequently used as the approximating function they are by no means the only possibility. By way of an introduction to other types of function, we shall consider the least squares approximation of the form

$$y = ae^{bx}$$

The least squares error is given by

$$E(a, b) = \sum_{i=1}^{n} w_i(y_i - ae^{bx_i})^2$$

with associated normal equations

$$\frac{\partial E}{\partial a} = -2 \sum_{i=1}^{n} w_i(y_i - ae^{bx_i})e^{bx_i} = 0$$

$$\frac{\partial E}{\partial b} = -2 \sum_{i=1}^{n} w_i(y_i - ae^{bx_i})ax_ie^{bx_i} = 0$$

(5.39)

This represents a system of two equations in the two unknowns a and b. Unfortunately, this system is non-linear and must be solved using one of the methods discussed in Section 3.4. However, it is possible to avoid this difficulty by linearizing the original problem. Taking logarithms of both sides of the equation

$$y = ae^{bx}$$

gives

$$\ln y = \ln a + bx$$

which may be written as

$$Y = A + BX$$

with $A = \ln a$, $B = b$, $X = x$ and $Y = \ln y$. The values of A and B can be chosen to minimize

$$\sum_{i=1}^{n} w_i [Y_i - (A + BX_i)]^2$$

where $X_i = x_i$ and $Y_i = \ln y_i$. This time the normal equations are linear. They can be trivially solved for A and B, and the values of a and b deduced from the relations $a = e^A$ and $b = B$.

■ **Example 5.15**
Consider the least squares approximation of the form $y = ae^{bx}$ to the following data:

x_i	0	0.25	0.4	0.5
y_i	9.532	7.983	4.826	5.503
w_i	1	1	1	1

The data corresponding to the linearized problem is

X_i	0	0.25	0.4	0.5
Y_i	2.255	2.077	1.574	1.705
w_i	1	1	1	1

The normal equations (5.38) are then

$$4.000A + 1.150B = 7.611$$
$$1.150A + 0.473B = 2.001$$

which have solution $A = 2.281$, $B = -1.315$. Hence

$$a = 9.786, \qquad b = -1.315$$

Of course, the values of a and b calculated for the linearized problem will not necessarily be the same as the values obtained for the original least squares problem. In this example the non-linear system (5.39) becomes

$$9.532 - a + 7.983e^{0.25b} - ae^{0.5b} + 4.826e^{0.4b}$$
$$- ae^{0.8b} + 5.503e^{0.5b} - ae^b = 0$$
$$1.996ae^{0.25b} - 0.25a^2e^{0.5b} + 1.930ae^{0.4b}$$
$$- 0.4a^2e^{0.8b} + 2.752ae^{0.5b} - 0.5a^2e^b = 0$$

Newton's method can now be applied to these equations to get

$$a = 9.731, \qquad b = -1.265 \qquad \qquad \square$$

EXERCISE 5.6

1. Find the least squares polynomial approximations of degree 1, 2 and 3 to the following data:

x_i	0	1	3	4	6
y_i	1	2	7	14	20
w_i	1	3	2	5	1

Compare the accuracy of these approximations at each of the data points.

2. Find the linear least squares approximation to the following data:

x_i	-1	2	5	6
y_i	-3	5	12	21
w_i	1	1	α	1

in the case when (a) $\alpha = 1$ (b) $\alpha = 10$ (c) $\alpha = 100$.
Compare the accuracy of these approximations at $x = 5$.

3. The least squares approximation of the form

$$y = ae^x + be^{-x}$$

to a set of n data points (x_i, y_i) with positive weights w_i is obtained by minimizing

$$\sum_{i=1}^{n} w_i [y_i - (ae^{x_i} + be^{-x_i})]^2$$

Derive the corresponding normal equations and hence find such an approximation to the following data:

x_i	0	0.5	1.0	1.5	2.0	2.5
y_i	5.02	5.21	6.49	9.54	16.02	24.53
w_i	1	1	1	1	1	1

4. Derive the normal equations for the least squares approximation of the form

$$y = ax^b$$

to a set of data points with unit weights.
Explain why an exact solution of these equations cannot always be found.

One way of avoiding this difficulty is to take logarithms to obtain

$$\ln y = \ln a + b \ln x$$

i.e.

$$Y = A + BX$$

where $Y = \ln y$, $X = \ln x$, $A = \ln a$ and $B = b$. Find the values of A and B which minimize

$$\sum_{i=1}^{4} [Y_i - (A + BX_i)]^2$$

for the following data, and deduce an approximation of the form $y = ax^b$:

x_i	1	2	4	10
y_i	2.87	4.51	6.11	9.43

5. The following data is based on a polynomial of degree m. It is known that nine of these points are at most 20% in error and that one of them has an error much greater than this. Find the value of m and identify the point which is grossly in error.

x_i	0	3	6	-3.5	2.5	-1
y_i	1.15	19.55	214.26	-29.98	103.5	1.87

x_i	-4.75	5.25	4.7	6.85
y_i	-85.15	155.01	77.14	261.33

6. Using the method of Lagrange multipliers, or otherwise, find the least squares fit of a quadratic which passes through the point $(-1, 1)$ to the following data:

x_i	1	2	3	4
y_i	5	7	13	17
w_i	1	1	1	1

5.7 LEAST SQUARES APPROXIMATION TO CONTINUOUS FUNCTIONS

The previous section was concerned with the least squares approximation of discrete data. We now extend this idea to the polynomial approximation of a given continuous function f defined on an interval $[a, b]$. The summations of Section 5.6 are replaced by their continuous counterparts, namely definite integrals, and a system of normal equations is obtained as before. Weight functions are again introduced to enable us to attach more importance to some parts of the interval than others. Any non-negative, integrable function w may be used as a weight function, provided that it is not identically zero on any subinterval of $[a, b]$. For the least squares polynomial approximation of degree m, the coefficients in

$$p_m(x) = a_0 + a_1 x + \cdots + a_m x^m$$

are chosen to minimize

$$E(a_0, a_1, \ldots, a_m) = \int_a^b w(x)\{f(x) - (a_0 + a_1 x + \cdots + a_m x^m)\}^2 \, dx$$

From calculus, necessary conditions for $(a_0, a_1, \ldots, a_m)$ to be a minimum are

$$\frac{\partial E}{\partial \bar{a}_0} = \frac{\partial E}{\partial a_1} = \cdots = \frac{\partial E}{\partial a_m} = 0$$

i.e.

$$-2 \int_a^b w(x)\{f(x) - (a_0 + a_1 x + \cdots + a_m x^m)\} \, dx \quad = 0$$

$$-2 \int_a^b w(x)\{f(x) - (a_0 + a_1 x + \cdots + a_m x^m)\} x \, dx \quad = 0$$

$$\vdots$$

$$-2 \int_a^b w(x)\{f(x) - (a_0 + a_1 x + \cdots + a_m x^m)\} x^m \, dx = 0$$

which may be rearranged to give

$$a_0 \int_a^b w(x)\,dx \; + a_1 \int_a^b w(x)x\,dx \; + \cdots + a_m \int_a^b w(x)x^m\,dx \; = \int_a^b w(x)f(x)\,dx$$

$$a_0 \int_a^b w(x)x\,dx \; + a_1 \int_a^b w(x)x^2\,dx \; + \cdots + a_m \int_a^b w(x)x^{m+1}\,dx = \int_a^b w(x)f(x)x\,dx$$

$$\vdots \qquad\qquad\qquad\qquad\qquad\qquad\qquad\qquad \vdots$$

$$a_0 \int_a^b w(x)x^m\,dx + a_1 \int_a^b w(x)x^{m+1}\,dx + \cdots + a_m \int_a^b w(x)x^{2m}\,dx = \int_a^b w(x)f(x)x^m\,dx$$

These are called the *normal equations*, and it can be shown (see Powell, 1981) that they always possess a unique solution.

■ Example 5.16
The normal equations corresponding to the cubic polynomial least squares

approximation of e^x on $[0,1]$ with a uniform weight $w(x) = 1$ are

$$a_0 \int_0^1 dx + a_1 \int_0^1 x \, dx + a_2 \int_0^1 x^2 \, dx + a_3 \int_0^1 x^3 \, dx = \int_0^1 e^x \, dx$$

$$a_0 \int_0^1 x \, dx + a_1 \int_0^1 x^2 \, dx + a_2 \int_0^1 x^3 \, dx + a_3 \int_0^1 x^4 \, dx = \int_0^1 x e^x \, dx$$

$$a_0 \int_0^1 x^2 \, dx + a_1 \int_0^1 x^3 \, dx + a_2 \int_0^1 x^4 \, dx + a_3 \int_0^1 x^5 \, dx = \int_0^1 x^2 e^x \, dx$$

$$a_0 \int_0^1 x^3 \, dx + a_1 \int_0^1 x^4 \, dx + a_2 \int_0^1 x^5 \, dx + a_3 \int_0^1 x^6 \, dx = \int_0^1 x^3 e^x \, dx$$

Evaluation of these integrals gives

$$\begin{pmatrix} 1 & 1/2 & 1/3 & 1/4 \\ 1/2 & 1/3 & 1/4 & 1/5 \\ 1/3 & 1/4 & 1/5 & 1/6 \\ 1/4 & 1/5 & 1/6 & 1/7 \end{pmatrix} \begin{pmatrix} a_0 \\ a_1 \\ a_2 \\ a_3 \end{pmatrix} = \begin{pmatrix} e-1 \\ 1 \\ e-2 \\ 6-2e \end{pmatrix}$$

which has solution

$$a_0 = -1456 + 536e, \quad a_1 = 16\,800 - 6180e, \quad a_2 = -41\,100 + 15\,120e,$$
$$a_3 = 27\,020 - 9940e$$

i.e.

$$p_3(x) = 0.278\,63x^3 + 0.421\,25x^2 + 1.018\,30x + 0.999\,06$$

Values of e^x, $p_3(x)$ and $e^x - p_3(x)$ at selected points are listed in Table 5.6.

Table 5.6

x	e^x	$p_3(x)$	$e^x - p_3(x)$
0.0	1.000 00	0.999 06	0.000 94
0.1	1.105 17	1.105 38	-0.000 21
0.2	1.221 40	1.221 80	-0.000 40
0.3	1.349 86	1.349 99	-0.000 13
0.4	1.491 82	1.491 61	0.000 21
0.5	1.648 72	1.648 35	0.000 37
0.6	1.822 12	1.821 87	0.000 25
0.7	2.013 75	2.013 85	-0.000 10
0.8	2.225 54	2.225 96	-0.000 42
0.9	2.459 60	2.459 86	-0.000 26
1.0	2.718 28	2.717 24	0.001 04

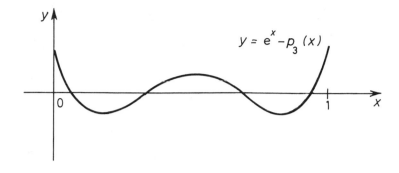

Figure 5.9

A graph of the error function $e^x - p_3(x)$ on $[0, 1]$ is shown in Fig. 5.9 to be oscillatory with five extrema of alternating sign. In fact, it can be proved (see Cheney, 1966) that the error function associated with the polynomial least squares approximation of degree n oscillates with at least $n + 2$ alternating extrema on $[a, b]$. □

In Example 5.16, with $a = 0$, $b = 1$ and $w(x) = 1$, the (i, j) entry of the coefficient matrix is

$$\int_0^1 x^{i+j-2} \, dx = \frac{1}{i+j-1}$$

Matrices of this form are known as *Hilbert matrices* and are notoriously ill-conditioned (see Exercise 2.1, Question 5). Numerical results obtained using these matrices are sensitive to rounding errors and must be considered suspect. Unfortunately this property is shared by the normal equations associated with other values of a and b and weight functions w, the degree of ill-conditioning generally increasing with the order of the matrix. This difficulty can, however, be avoided by using an alternative representation of the polynomial p_m. We write

$$p_m(x) = a_0 \varphi_0(x) + a_1 \varphi_1(x) + \cdots + a_m \varphi_m(x)$$

where φ_i $(i = 0, 1, \ldots, m)$ is a polynomial of degree i. The foregoing is a special case of this with $\varphi_i(x) = x^i$. We show that, by judicious choice of the φ_i, the least squares approximation can be determined without having to solve a system of linear equations.

The values of a_i $(i = 0, 1, \ldots, m)$ are chosen to minimize

$$E(a_0, a_1, \ldots, a_m) = \int_a^b w(x)\{f(x) - [a_0 \varphi_0(x) + a_1 \varphi_1(x) + \cdots + a_m \varphi_m(x)]^2\} \, dx$$

Proceeding as before gives

$$\frac{\partial E}{\partial a_i} = -2 \int_a^b w(x)\{f(x) - [a_0\varphi_0(x) + a_1\varphi_1(x) + \cdots + a_m\varphi_m(x)]\}\varphi_i(x)\,dx = 0$$

The normal equations are therefore given by

$$a_0 \int_a^b w(x)\varphi_0(x)\varphi_i(x)\,dx + a_1 \int_a^b w(x)\varphi_1(x)\varphi_i(x)\,dx + \cdots$$

$$+ a_m \int_a^b w(x)\varphi_m(x)\varphi_i(x)\,dx = \int_a^b w(x)f(x)\varphi_i(x)\,dx \qquad i = 0, 1, \ldots, m \quad (5.40)$$

This system will be particularly easy to solve if the polynomials φ_i are chosen to satisfy

$$\int_a^b w(x)\varphi_j(x)\varphi_i(x)\,dx \quad \begin{cases} = 0 & i \neq j \\ \neq 0 & i = j \end{cases} \qquad (5.41)$$

In this case, all but one of the integrals on the left hand side of equation (5.40) are zero, and the solution can be obtained immediately as

$$a_i = \frac{\int_a^b w(x)f(x)\varphi_i(x)\,dx}{\int_a^b w(x)\varphi_i^2(x)\,dx} \qquad i = 0, 1, \ldots, m \qquad (5.42)$$

Polynomials satisfying (5.41) are said to be *orthogonal* on the interval $[a, b]$ with respect to the weight function w. The calculation of the a_i from (5.42) is trivial once a suitable set of orthogonal polynomials has been determined. Although the problem of constructing orthogonal polynomials is not an insignificant one, notice that they only depend on a, b and w and so can be reused for different functions f.

A procedure exists for the construction of orthogonal polynomials φ_i with leading term x^i corresponding to any interval $[a, b]$ and any weight function w. This is known as the *Gram-Schmidt orthogonalization process* and can be described as follows:

STEP 0   Take $\varphi_0(x) = 1$.

STEP 1   Any linear polynomial φ_1 with leading term x can be written as

$$\varphi_1(x) = x + K_{1,0}\varphi_0(x)$$

for some constant $K_{1,0}$. The value of $K_{1,0}$ is chosen to ensure that φ_0 and φ_1 are orthogonal. This requires

$$\int_a^b w(x)\varphi_0(x)\{x + K_{1,0}\varphi_0(x)\}\,dx = 0$$

i.e.

$$K_{1,0} = -\frac{\int_a^b w(x)\varphi_0(x)x\,dx}{\int_a^b w(x)\varphi_0^2(x)\,dx}$$

STEP 2 Any quadratic polynomial φ_2 with leading term x^2 can be written as

$$\varphi_2(x) = x^2 + K_{2,0}\varphi_0(x) + K_{2,1}\varphi_1(x)$$

for some constants $K_{2,0}$ and $K_{2,1}$. The values of $K_{2,0}$ and $K_{2,1}$ are chosen to ensure that φ_2 is orthogonal to both φ_0 and φ_1. If φ_0 and φ_2 are orthogonal then

$$\int_a^b w(x)\varphi_0(x)\{x^2 + K_{2,0}\varphi_0(x) + K_{2,1}\varphi_1(x)\}\,dx = 0$$

However, from step 1,

$$\int_a^b w(x)\varphi_0(x)\varphi_1(x)\,dx = 0$$

since φ_0 and φ_1 are orthogonal. Hence we require

$$\int_a^b w(x)\varphi_0(x)\{x^2 + K_{2,0}\varphi_0(x)\}\,dx = 0$$

i.e.

$$K_{2,0} = -\frac{\int_a^b w(x)\varphi_0(x)x^2\,dx}{\int_a^b w(x)\varphi_0^2(x)\,dx}$$

Similarly, the requirement that φ_1 and φ_2 are orthogonal leads to

$$K_{2,1} = -\frac{\int_a^b w(x)\varphi_1(x)x^2\,dx}{\int_a^b w(x)\varphi_1^2(x)\,dx}$$

The general step of the process is given by

STEP i We write

$$\varphi_i(x) = x^i + K_{i,0}\varphi_0(x) + K_{i,1}\varphi_1(x) + \cdots + K_{i,i-1}\varphi_{i-1}(x)$$

where the constants $K_{i,j}$ are chosen to ensure that φ_i is orthogonal to φ_j $(j = 0, 1, \ldots, i - 1)$. These conditions lead to

$$K_{i,j} = -\frac{\displaystyle\int_a^b w(x)\varphi_j(x)x^i \, dx}{\displaystyle\int_a^b w(x)\varphi_j^2(x) \, dx}$$

■ **Example 5.17**

As an illustration of the Gram-Schmidt process, consider the construction of orthogonal polynomials $\varphi_0, \varphi_1, \varphi_2$ and φ_3 on $[0, 1]$ with respect to the weight function $w(x) = 1$.

STEP 0

$$\varphi_0(x) = 1$$

STEP 1

$$\varphi_1(x) = x + K_{1,0}1$$

with

$$K_{1,0} = -\frac{\displaystyle\int_0^1 1x \, dx}{\displaystyle\int_0^1 1^2 \, dx} = -\frac{1/2}{1} = -1/2$$

Hence

$$\varphi_1(x) = x - 1/2$$

STEP 2

$$\varphi_2(x) = x^2 + K_{2,0}1 + K_{2,1}(x - 1/2)$$

with

$$K_{2,0} = -\frac{\displaystyle\int_0^1 1x^2 \, dx}{\displaystyle\int_0^1 1^2 \, dx} = -\frac{1/3}{1} = -1/3$$

and

$$K_{2,1} = -\frac{\displaystyle\int_0^1 (x - 1/2)x^2 \, dx}{\displaystyle\int_0^1 (x - 1/2)^2 \, dx} = -\frac{1/12}{1/12} = -1$$

Hence

$$\varphi_2(x) = x^2 - x + 1/6$$

STEP 3

$$\varphi_3(x) = x^3 + K_{3,0}1 + K_{3,1}(x - 1/2) + K_{3,2}(x^2 - x + 1/6)$$

with

$$K_{3,0} = -\frac{\displaystyle\int_0^1 1x^3\,dx}{\displaystyle\int_0^1 1^2\,dx} = -\frac{1/4}{1} = -1/4$$

$$K_{3,1} = -\frac{\displaystyle\int_0^1 (x-1/2)x^3\,dx}{\displaystyle\int_0^1 (x-1/2)^2\,dx} = -\frac{3/40}{1/12} = -9/10$$

and

$$K_{3,2} = -\frac{\displaystyle\int_0^1 (x^2 - x + 1/6)x^3\,dx}{\displaystyle\int_0^1 (x^2 - x + 1/6)^2\,dx} = -\frac{1/120}{1/180} = -3/2$$

Hence

$$\varphi_3(x) = x^3 - 3/2x^2 + 3/5x - 1/20 \qquad \square$$

We now return to the problem of least squares approximation by a polynomial of the form

$$p_m(x) = a_0\varphi_0(x) + a_1\varphi_1(x) + \cdots + a_m\varphi_m(x)$$

where the φ_i are orthogonal. The coefficients a_i can be calculated from (5.42), and we illustrate this formula by reconsidering the problem posed in Example 5.16.

■ **Example 5.18**

The orthogonal polynomials φ_i $(i = 0, 1, 2, 3)$ with leading term x^i for the interval $[0, 1]$ and weight function $w(x) = 1$ were found in Example 5.17. Hence, the cubic polynomial least squares approximation to e^x is

$$p_3(x) = a_0\varphi_0(x) + a_1\varphi_1(x) + a_2\varphi_2(x) + a_3\varphi_3(x)$$

where

$$a_0 = \frac{\displaystyle\int_0^1 e^x 1\,dx}{\displaystyle\int_0^1 1^2\,dx} = \frac{e-1}{1} = e - 1$$

$$a_1 = \frac{\displaystyle\int_0^1 e^x(x-1/2)\,dx}{\displaystyle\int_0^1 (x-1/2)^2\,dx} = \frac{(1/2)(3-e)}{1/12} = -6(e-3)$$

$$a_2 = \frac{\displaystyle\int_0^1 e^x(x^2 - x + 1/6)\,dx}{\displaystyle\int_0^1 (x^2 - x + 1/6)^2\,dx} = \frac{(1/6)(7e - 19)}{1/180} = 30(7e - 19)$$

$$a_3 = \frac{\displaystyle\int_0^1 e^x[x^3 - (3/2)x^2 + (3/5)x - 1/20]\,dx}{\displaystyle\int_0^1 [x^3 - (3/2)x^2 + (3/5)x - 1/20]^2\,dx} = \frac{(1/20)(193 - 71e)}{1/2800}$$

$$= - 140(71e - 193)$$

(For this simple example, the exact values of the integrals are easily found. In practice, however, they are usually calculated using the numerical techniques described in Chapter 6.)

Therefore

$$p_3(x) = (e - 1)\varphi_0(x) - 6(e - 3)\varphi_1(x) + 30(7e - 19)\varphi_2(x) - 140(71e - 193)\varphi_3(x)$$
$$= 0.278\,63x^3 + 0.421\,25x^2 + 1.018\,30x + 0.999\,06 \qquad \square$$

EXERCISE 5.7

1. Use the normal equations to find the linear least squares approximation of $\sin x$ on $[0, \pi/2]$ with respect to the weight function $w(x) = x$. Sketch a graph of the corresponding error function.

2. Use orthogonal polynomials to find the least squares approximation of Question 1.

3. One of the most commonly used sets of orthogonal polynomials are the *Legendre polynomials* which are orthogonal on $[-1, 1]$ with respect to the weight function $w(x) = 1$. Find the first four Legendre polynomials with leading term x^i using the Gram-Schmidt process.

4. Use the result of Question 3 to show that the quadratic least squares approximation of $f(x)$ on $[-1, 1]$ with respect to the weight function $w(x) = 1$ is given by

$$p_2(x) = \frac{3}{8}\int_{-1}^1 (3 + 4xt + 15x^2t^2 - 5x^2 - 5t^2)f(t)\,dt$$

5. If a set of orthogonal polynomials has the additional property that

$$\int_a^b w(x)\varphi_i^2(x)\,dx = 1 \qquad \text{for all } i$$

then it is said to be *orthonormal*. Construct such a set from the polynomials

$$1, \qquad x - \tfrac{1}{2}, \qquad x^2 - x + \tfrac{1}{6}, \qquad x^3 - \tfrac{3}{2}x^2 + \tfrac{3}{5}x - \tfrac{1}{20}$$

which are known to be orthogonal on $[0, 1]$ with respect to the weight function $w(x) = 1$.

6. Defining the *Laguerre polynomials* $P_n(x)$ by

$$P_n(x) = (-1)^n e^x \frac{d^n}{dx^n} [x^n e^{-x}]$$

show that $P_n(x)$ is a polynomial of degree n. By repeated integration by parts, show that $P_n(x)$ $(n = 0, 1, 2, \ldots)$ is orthogonal on $[0, \infty)$ with respect to the weight function $w(x) = e^{-x}$.

Find the Laguerre polynomials of degree 1, 2, 3 and 4.

6
Numerical
differentiation
and integration

In this chapter we consider numerical methods for differentiation and integration. Both of these problems may be approached in the same way. A function f, known either explicitly or as a set of data points, is replaced by a simpler function. A polynomial p is the obvious choice of approximating function, since the operations of differentiation and integration are then easily performed.

The polynomial is differentiated to obtain $p'(x)$, which is taken as an approximation to $f'(x)$ for any numerical value of x. Geometrically, this is equivalent to replacing the slope of f, at x, by that of p. Similarly, p is integrated to obtain

$$\int_a^b p(x)\,\mathrm{d}x$$

which is taken as an approximation to the definite integral

$$\int_a^b f(x)\,\mathrm{d}x$$

for any interval $[a, b]$. Geometrically, this is equivalent to replacing the area under the graph of f between the ordinates $x = a$ and $x = b$ by that of p and, as such, is known as *numerical quadrature*.

Methods derived in this way for numerical quadrature are generally more accurate than their counterparts for differentiation. To see this, consider Fig. 6.1. Here, a function f is approximated by the quadratic interpolation polynomial p_2 based on the points $x = x_0$, x_1 and x_2. It is clear from this diagram that the slopes of the tangents to the two curves at x_1, for example, are very different. On the other hand, the areas under the two curves over the range $[x_0, x_2]$ are nearly the same because local errors average out. Integration is a smoothing process which irons out local detail, whereas differentiation highlights it.

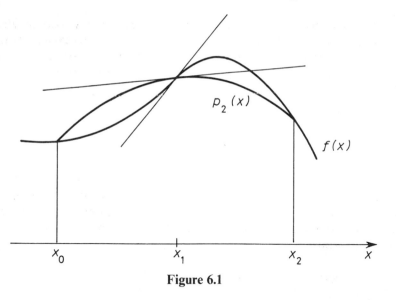

Figure 6.1

6.1 NUMERICAL DIFFERENTIATION

The derivation and analysis of formulas for numerical differentiation is considerably simplified when the data is equally spaced. It will be assumed, therefore, that the points x_i are given by $x_i = x_0 + ih$ for some fixed tabular interval h. The more general situation is considered briefly in the exercises.

Newton's forward difference interpolating polynomial p_k, of degree at most k, to data (x_i, f_i) $(i = 0, 1, \ldots, k)$ is given from (5.10) as

$$p_k(x) = f_0 + \Delta f_0 \binom{s}{1} + \Delta^2 f_0 \binom{s}{2} + \cdots + \Delta^k f_0 \binom{s}{k} \qquad (6.1)$$

where $x = x_0 + sh$. The corresponding error term $e_k = f - p_k$ is given from (5.11) as

$$e_k(x) = s(s-1)\cdots(s-k)\frac{h^{k+1}}{(k+1)!} f^{(k+1)}(x_0 + \theta h) \qquad (6.2)$$

for some number $\theta \in (0, k)$ which depends on s. Now, by the chain rule,

$$\frac{dp_k}{dx} = \frac{dp_k}{ds}\frac{ds}{dx} = \frac{dp_k}{ds}\frac{1}{h}$$

because $s = (x - x_0)/h$. We may, therefore, differentiate (6.1) with respect to x to obtain

$$p'_k(x) = \frac{1}{h}\left\{\Delta f_0 \frac{d}{ds}\binom{s}{1} + \Delta^2 f_0 \frac{d}{ds}\binom{s}{2} + \cdots + \Delta^k f_0 \frac{d}{ds}\binom{s}{k}\right\} \qquad (6.3)$$

The derivatives in this expression are easily evaluated and, by taking different values of k, various formulas for the approximation of $f'(x)$ may be obtained. The error in this approximation, known as the *truncation* or *discretization error*, is obtained by differentiating (6.2) with respect to x to get

$$\frac{de_k}{dx} = \frac{1}{h}\frac{de_k}{ds} = \frac{d}{ds}\{s(s-1)\cdots(s-k)\}\frac{h^k}{(k+1)!}f^{(k+1)}(x_0+\theta h)$$

$$+ s(s-1)\cdots(s-k)\frac{h^k}{(k+1)!}\frac{d}{ds}\{f^{(k+1)}(x_0+\theta h)\}$$

Unfortunately, this expression is of little use as an estimate of the error because θ is an unknown function of s and it is impossible to evaluate the second term on the right hand side. However, by imposing the restriction that s takes an integer value in the range $0, 1, \ldots, k$, the offending term can be made to vanish, leaving

$$e_k'(x) = \frac{d}{ds}\{s(s-1)\cdots(s-k)\}\frac{h^k}{(k+1)!}f^{(k+1)}(x_0+\theta h) \qquad (6.4)$$

for $x = x_i$ $(i = 0, 1, \ldots, k)$.

To illustrate how the above may be used in practice, consider the interpolating polynomial (6.1), of degree $k = 1$, to the two points (x_0, f_0) and (x_1, f_1). From (6.3) and (6.4),

$$f'(x) = \frac{\Delta f_0}{h}\frac{ds}{ds} + \frac{d}{ds}\{s(s-1)\}\frac{h}{2!}f''(x_0+\theta h)$$

$$= \frac{1}{h}(f_1 - f_0) + \frac{(2s-1)}{2}hf''(x_0+\theta h)$$

for some $\theta \in (0, 1)$ and $s = 0, 1$. For example, if $s = 0$ then

$$f'(x_0) = \frac{1}{h}(f_1 - f_0) - \frac{h}{2}f''(x_0+\theta h) \qquad (6.5)$$

The quantity $(f_1 - f_0)/h$ provides an approximation to $f'(x_0)$ and $-(h/2)f''(x_0+\theta h)$ is the corresponding truncation error. This forward difference approximation has a simple geometric interpretation, illustrated in Fig. 6.2; $(f_1 - f_0)/h$ is the slope of the chord joining (x_0, f_0) and (x_1, f_1).

It is interesting to observe that (6.5) may also be derived from Taylor's theorem. Expansion of $f(x_1)$ about x_0 as far as terms involving h^2 gives

$$f(x_1) = f(x_0) + hf'(x_0) + \frac{h^2}{2!}f''(x_0+\theta h)$$

The result follows by subtracting $f(x_0)$ from both sides and dividing both sides by h.

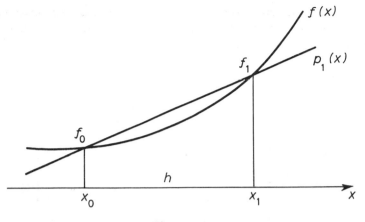

Figure 6.2

■ Example 6.1

As an illustration of (6.5), consider the calculation of $f'(1)$ for $f(x) = e^{-x}$. In this case $x_0 = 1$ and $x_1 = 1 + h$. Moreover, since $f''(x) = e^{-x} \leqslant e^{-1}$ on $(1, 1 + h)$, the magnitude of the truncation error does not exceed $(he^{-1})/2$. Numerical results for various values of h are listed in Table 6.1. The values of f are rounded to 6D. The fifth column represents the truncation error and the sixth column represents the difference between the exact value $f'(1) = -e^{-1}$ and the approximate value $(f_1 - f_0)/h$.

There are two important points to notice about these results. The actual error decreases at first with decreasing h, as we might have expected, but then starts to increase for values of h below 0.002. Also, the actual error is greater than the truncation error if $h < 0.01$. An explanation of this behaviour is given at the end of this section. □

Table 6.1

h	f_0	f_1	$(f_1 - f_0)/h$	$(he^{-1})/2$	|Actual error|
1	0.367 879	0.135 335	− 0.232 544	0.183 940	0.135 335
0.2	0.367 879	0.301 194	− 0.333 425	0.036 788	0.034 454
0.1	0.367 879	0.332 871	− 0.350 080	0.018 394	0.017 799
0.02	0.367 879	0.360 595	− 0.364 200	0.003 679	0.003 679
0.01	0.367 879	0.364 219	− 0.366 000	0.001 839	0.001 879
0.002	0.367 879	0.367 144	− 0.367 500	0.000 368	0.000 379
0.001	0.367 879	0.367 512	− 0.367 000	0.000 184	0.000 879
0.0002	0.367 879	0.367 806	− 0.365 000	0.000 037	0.002 879

To derive more accurate formulas, consider the quadratic interpolating polynomial p_2 to the three points (x_0, f_0), (x_1, f_1) and (x_2, f_2). From (6.3) and (6.4),

$$f'(x) = \frac{1}{h}(f_1 - f_0) + \frac{1}{2h}(f_2 - 2f_1 + f_0)(2s - 1)$$

$$+ \frac{h^2}{6}\{s(s-1) + s(s-2) + (s-1)(s-2)\}f'''(x_0 + \theta h)$$

for $\theta \in (0, 2)$ and $s = 0, 1, 2$. If $s = 0$ then

$$f'(x_0) = \frac{1}{h}(f_1 - f_0) - \frac{1}{2h}(f_2 - 2f_1 + f_0) + \frac{h^2}{3}f'''(x_0 + \theta_0 h)$$

$$= \frac{1}{2h}(-f_2 + 4f_1 - 3f_0) + \frac{h^2}{3}f'''(x_0 + \theta_0 h) \qquad (6.6)$$

Similarly, taking $s = 1$ and $s = 2$ gives

$$f'(x_1) = \frac{1}{2h}(f_2 - f_0) - \frac{h^2}{6}f'''(x_0 + \theta_1 h) \qquad (6.7)$$

and

$$f'(x_2) = \frac{1}{2h}(3f_2 - 4f_1 + f_0) + \frac{h^2}{3}f'''(x_0 + \theta_2 h) \qquad (6.8)$$

The notation θ_i $(i = 0, 1, 2)$ emphasizes the fact that the value of θ may be different in each case.

These formulas may also be derived using Taylor's theorem, and the details are left as an exercise. A comparison of the truncation errors for (6.6), (6.7)

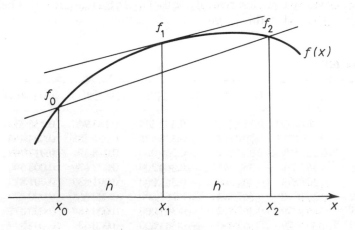

Figure 6.3

and (6.8) shows that (6.7) is likely to be the most accurate. This central difference approximation has a simple geometric interpretation, which is illustrated in Fig. 6.3; $(f_2 - f_0)/2h$ is the slope of the chord joining (x_0, f_0) and (x_2, f_2). Notice that, for the particular function sketched in Fig. 6.3, the slope of the chord is an extremely accurate approximation to the slope of the tangent passing through (x_1, f_1).

■ **Example 6.2**

As an illustration of (6.7), consider the calculation of $f'(1)$ for $f(x) = e^{-x}$. In this case $x_0 = 1 - h$, $x_1 = 1$ and $x_2 = 1 + h$. Moreover, since $|f'''(x)| = e^{-x} \leqslant e^{h-1}$ on $(1 - h, 1 + h)$, the magnitude of the truncation error does not exceed $h^2 e^{h-1}/6$.

Numerical results are listed in Table 6.2. These results are more accurate than those given in Table 6.1. This is to be expected since the truncation error in (6.5) is $O(h)$ compared with $O(h^2)$ in (6.7). Notice that, as in Example 6.1, it is not wise to make h too small. In this example, the actual error increases for $h < 0.01$ and it exceeds the truncation error for $h < 0.01$.   □

The second derivative of f can be approximated in a similar way. If (6.1) is differentiated twice with respect to x, then

$$p_k''(x) = \frac{1}{h^2}\left\{\Delta^2 f_0 + \Delta^3 f_0 \frac{d^2}{ds^2}\binom{s}{3} + \cdots + \Delta^k f_0 \frac{d^2}{ds^2}\binom{s}{k}\right\}$$

Taking $k = 2$ and $s = 1$ gives

$$p_2''(x_1) = \frac{1}{h^2}(f_2 - 2f_1 + f_0)$$

which may be used as an approximation to $f''(x_1)$. The truncation error

Table 6.2

h	f_0	f_2	$(f_2 - f_0)/2h$	$h^2 e^{h-1}/6$	|Actual error|
1	1.000 000	0.135 335	−0.432 333	0.166 667	0.064 454
0.2	0.449 329	0.301 194	−0.370 338	0.002 996	0.002 459
0.1	0.406 570	0.332 871	−0.368 495	0.000 678	0.000 616
0.02	0.375 311	0.360 595	−0.367 900	0.000 025	0.000 021
0.01	0.371 577	0.364 219	−0.367 900	0.000 006	0.000 021
0.002	0.368 616	0.367 144	−0.368 000	0.000 000	0.000 121
0.001	0.368 248	0.367 512	−0.368 000	0.000 000	0.000 121
0.0002	0.367 953	0.367 806	−0.367 500	0.000 000	0.000 379

corresponding to this formula is obtained by differentiating e_2 twice to get

$$e_2''(x) = \frac{1}{h^2}\frac{d^2}{ds^2}\left\{s(s-1)(s-2)\frac{h^3}{6}f'''(x_0+\theta h)\right\}$$

$$= (s-1)hf'''(x_0+\theta h) + (3s^2-6s+2)\frac{h}{3}\frac{d}{ds}\{f'''(x_0+\theta h)\}$$

$$+ s(s-1)(s-2)\frac{h}{6}\frac{d^2}{ds^2}\{f'''(x_0+\theta h)\}$$

Unfortunately the second term does not vanish when $s = 1$, so this expression cannot be used to estimate the truncation error. A more fruitful approach is to use Taylor's theorem. Expansion of $f(x_0)$ and $f(x_2)$ about $x = x_1$ gives

$$f(x_0) = f(x_1-h) = f_1 - hf_1' + \frac{h^2}{2!}f_1'' - \frac{h^3}{3!}f_1''' + \frac{h^4}{4!}f^{(4)}(x_1+\theta h)$$

$$f(x_2) = f(x_1+h) = f_1 + hf_1' + \frac{h^2}{2!}f_1'' + \frac{h^3}{3!}f_1''' + \frac{h^4}{4!}f^{(4)}(x_1+\varphi h)$$

for $\theta\in(-1, 0)$ and $\varphi\in(0, 1)$. Terms involving odd powers of h can be eliminated by adding these two equations to get

$$f_0 + f_2 = 2f_1 + h^2 f_1'' + \frac{h^4}{4!}\{f^{(4)}(x_1+\theta h) + f^{(4)}(x_1+\varphi h)\}$$

Now, the average of a set of numbers is known to lie between its extremes, so we can apply the intermediate value theorem to deduce that

$$\tfrac{1}{2}\{f^{(4)}(x_1+\theta h) + f^{(4)}(x_1+\varphi h)\} = f^{(4)}(x_1+\psi h)$$

for some $\psi\in(-1, 1)$. Hence

$$f_0 + f_2 = 2f_1 + h^2 f_1'' + \frac{h^4}{12}f^{(4)}(x_1+\psi h)$$

which may be rearranged as

$$f_1'' = \frac{1}{h^2}(f_2 - 2f_1 + f_0) - \frac{h^2}{12}f^{(4)}(x_1+\psi h) \qquad (6.9)$$

The truncation error in the central difference approximation

$$f_1'' \approx \frac{1}{h^2}(f_2 - 2f_1 + f_0)$$

is therefore of order h^2. A numerical example illustrating the use of this formula is given in Exercise 6.1.

It was found in Examples 6.1 and 6.2 that, as the interval size h is progressively decreased, a point is reached after which the errors begin to increase. Now, every formula for numerical differentiation has an associated truncation error which becomes smaller as h decreases. In addition, however, there are usually rounding errors present in the tabulated values of f_i, which can have a disastrous effect on the computed value of $f'(x)$ when h is small. To see this, consider the approximation

$$f'(x_0) = \frac{1}{h}(f_1 - f_0)$$

which has truncation error $-(h/2)f''(x_0 + \theta h)$ with $\theta \in (0, 1)$, and suppose that the magnitudes of the rounding errors in f_i do not exceed ε. If we are unlucky and the errors in f_0 and f_1 are of opposite signs, then the rounding error in $(f_1 - f_0)/h$ could be as large as $2\varepsilon/h$. Therefore there are two contributions to the error of this approximation – one due to discretization, the other due to rounding. Whilst the truncation error converges to zero as $h \to 0$, the rounding error grows without bound. If $|f''(x)| \leqslant M$ on $[x_0, x_1]$, then an upper bound on the total error is

$$E(h) = \frac{Mh}{2} + \frac{2\varepsilon}{h}$$

An obvious choice of h is that which minimizes $E(h)$. From calculus,

$$\frac{dE}{dh} = \frac{M}{2} - \frac{2\varepsilon}{h^2} = 0$$

which gives

$$h = \sqrt{(4\varepsilon/M)}$$

In Example 6.1 $M = e^{-1}$ and, since the values of f_i are rounded to 6D, $\varepsilon = (1/2) \times 10^{-6}$. The optimal value of h is therefore

$$\sqrt{[4 \times (1/2) \times 10^{-6}/e^{-1}]} = 0.0023$$

which is in agreement with the results in Table 6.1.

EXERCISE 6.1

1. The values of a function f are tabulated at equal intervals h, so that $f_i = f(x_i)$ with $x_i = x_0 + ih$. By considering an appropriate interpolating polynomial, derive the approximation

$$f'(x_2) \approx \frac{1}{12h}(f_0 - 8f_1 + 8f_3 - f_4)$$

If $|f^{(5)}(x)| \leqslant M$ and the tabulated values have rounding errors not exceeding ε, show that the error in the computed value of $f'(x_2)$ does not exceed

$$\frac{h^4 M}{30} + \frac{3\varepsilon}{2h}$$

2. Use Taylor's theorem to show that

$$f'(x_0) = \frac{1}{2h}(-f_2 + 4f_1 - 3f_0) + O(h^2)$$

$$f'(x_1) = \frac{1}{2h}(f_2 - f_0) + O(h^2)$$

$$f'(x_2) = \frac{1}{2h}(3f_2 - 4f_1 + f_0) + O(h^2)$$

where $f_i = f(x_i)$ with $x_i = x_0 + ih$.

Use these formulas to estimate $f'(0.01)$, $f'(0.05)$ and $f'(0.1)$ from the following table of values of $f(x) = x^{1/2} e^{-2x}$:

x_i	0.01	0.02	0.03	0.04	0.05
f_i	0.098 020	0.135 876	0.163 118	0.184 623	0.202 328
x_i	0.06	0.07	0.08	0.09	0.10
f_i	0.217 250	0.230 011	0.241 023	0.250 581	0.258 905

3. Estimate $f''(1)$ for the function $f(x) = \sin 4x$ using

$$f_1'' \approx \frac{1}{h^2}(f_2 - 2f_1 + f_0)$$

with $h = 0.05$, 0.04, 0.03, 0.02 and 0.01. Work to 5D.

4. Consider using the formula $(f_2 - 2f_1 + f_0)/h^2$ to estimate $f''(x_1)$. If $|f^{(4)}(x)| \leqslant M$ on some interval containing the points x_i ($i = 0, 1, 2$) and if the function values f_i have rounding errors not exceeding ε, show that the error in the computed value of $f''(x_1)$ does not exceed

$$\frac{h^2 M}{12} + \frac{4\varepsilon}{h^2}$$

Find the value of h which minimizes this bound. Hence, calculate the optimal value of h in the case when $f(x) = \sin 4x$ ($x_i \in [0.95, 1.05]$) and the function values are rounded to 5D. Verify that this is in agreement with the numerical results obtained in Question 3.

5. By considering the Lagrange form of the interpolating polynomial p_2 based on the points x_0, x_1 and x_2 (not necessarily equally spaced), derive

an expression for $p'_2(x_1)$. Values of the function $f(x) = \sqrt{x}$ are given in the following table. Use this data to estimate $f'(0.6)$.

x_i	0.5	0.6	1.0
f_i	0.70711	0.77460	1.00000

Obtain a theoretical bound on the error in this estimate and compare this with the actual error. ($f'(0.6) = 0.64550$ (5D))

6.2 NUMERICAL INTEGRATION: NEWTON–COTES FORMULAS

One problem which frequently arises in mathematics is that of evaluating

$$\int_a^b f(x)\,dx \tag{6.10}$$

over a finite interval $[a, b]$. If a function F exists such that $F' = f$, then the value of this integral is $F(b) - F(a)$. Unfortunately in practical problems it is extremely difficult, if not impossible, to obtain an explicit expression for F. Indeed, the values of f may only be known at a discrete set of points and in this situation a numerical approach is essential.

Suppose that $a = x_0$ and $h = b - a$. The simplest way of approximating (6.10) is to replace $f(x)$ by its 0th degree interpolating polynomial with error term. From (6.1) and (6.2),

$$f(x_0 + sh) = f_0 + hs f'(x_0 + \theta h)$$

for some number $\theta \in (0, 1)$ which depends on s. The change of variable $x = x_0 + sh$ produces

$$\int_a^b f(x)\,dx = \int_0^1 f(x_0 + sh)h\,ds$$

$$= h\int_0^1 f_0\,ds + h^2\int_0^1 sf'(x_0 + \theta h)\,ds$$

$$= hf_0 + h^2\int_0^1 sf'(x_0 + \theta h)\,ds$$

Now s does not change sign on $[0, 1]$ so, provided f' is continuous on $[a, b]$, we can apply the integral mean value theorem to deduce the existence of a number $\xi \in (0, 1)$ such that

$$\int_0^1 sf'(x_0 + \theta h)\,ds = f'(x_0 + \xi h)\int_0^1 s\,ds = \tfrac{1}{2}f'(x_0 + \xi h)$$

Hence

$$\int_a^b f(x)\,dx = hf_0 + \frac{h^2}{2} f'(x_0 + \xi h) \tag{6.11}$$

The first term on the right hand side of (6.11) gives the quadrature rule and the second term gives the truncation error. This rule has a simple geometric interpretation, illustrated in Fig. 6.4; hf_0 is the area of the rectangle of width $h = b - a$ and height f_0. For this reason, (6.11) is known as the *rectangular rule*.

To derive more accurate formulas, consider the points $x_i = x_0 + ih$ ($i = 0, 1, \ldots, k$) with $a = x_0$, $b = x_k$ and $h = (b - a)/k$. The integral (6.10) is approximated by

$$\int_a^b p_k(x)\,dx$$

where p_k is the interpolating polynomial of degree at most k passing through the points (x_i, f_i) ($i = 0, 1, \ldots, k$). The change of variable $x = x_0 + sh$ produces

$$\int_a^b p_k(x)\,dx = \int_0^k p_k(x_0 + sh)h\,ds$$

$$= h \int_0^k f_0 + \Delta f_0 \binom{s}{1} + \Delta^2 f_0 \binom{s}{2} + \cdots + \Delta^k f_0 \binom{s}{k}\,ds \tag{6.12}$$

using the forward difference form of the interpolating polynomial (6.1). The integrals in this expression are easily evaluated and, by taking different values of k, various integration rules for the approximation of (6.10) are obtained. The truncation error in this approximation is obtained by integrating the

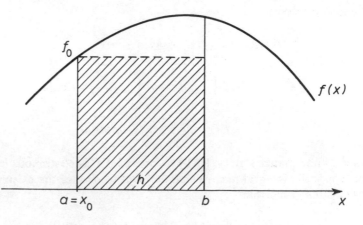

Figure 6.4

interpolation error term (6.2) to get

$$\frac{h^{k+2}}{(k+1)!}\int_0^k s(s-1)\cdots(s-k)f^{(k+1)}(x_0+\theta h)\,ds \tag{6.13}$$

To illustrate how (6.12) and (6.13) may be used to derive quadrature rules and associated truncation errors, we consider the cases $k=1$, 2 and 3. If $k=1$ then, from (6.12) and (6.13),

$$\int_a^b f(x)\,dx = h\int_0^1 f_0\,ds + h\int_0^1 s(f_1-f_0)\,ds$$

$$+ \frac{h^3}{2}\int_0^1 s(s-1)f''(x_0+\theta h)\,ds$$

$$= hf_0 + \frac{h}{2}(f_1-f_0) + \frac{h^3}{2}\int_0^1 s(s-1)f''(x_0+\theta h)\,ds$$

Now $s(s-1)$ does not change sign on $[0,1]$ so, provided f'' is continuous on $[a,b]$, we can apply the integral mean value theorem to deduce the existence of a number $\xi \in (0,1)$ such that

$$\int_0^1 s(s-1)f''(x_0+\theta h)\,ds = f''(x_0+\xi h)\int_0^1 s(s-1)\,ds = -\tfrac{1}{6}f''(x_0+\xi h)$$

Hence

$$\int_a^b f(x)\,dx = \frac{h}{2}(f_0+f_1) - \frac{h^3}{12}f''(x_0+\xi h) \tag{6.14}$$

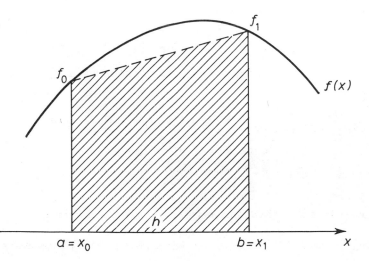

Figure 6.5

This rule also has a simple geometric interpretation, shown in Fig. 6.5; $h(f_0 + f_1)/2$ is the area of the trapezium of width $h = b - a$ with vertical sides of height f_0 and f_1. For this reason, (6.14) is known as the *trapezium rule*.

If $k = 2$ then, from (6.12) and (6.13),

$$\int_a^b f(x)\,dx = h\int_0^2 f_0\,ds + h\int_0^2 s(f_1 - f_0)\,ds$$

$$+ \frac{h}{2}\int_0^2 s(s-1)(f_2 - 2f_1 + f_0)\,ds$$

$$+ \frac{h^4}{6}\int_0^2 s(s-1)(s-2)f'''(x_0 + \theta h)\,ds$$

$$= 2hf_0 + 2h(f_1 - f_0) + \frac{h}{3}(f_2 - 2f_1 + f_0)$$

$$+ \frac{h^4}{6}\int_0^2 s(s-1)(s-2)f'''(x_0 + \theta h)\,ds$$

$$= \frac{h}{3}(f_0 + 4f_1 + f_2) + \frac{h^4}{6}\int_0^2 s(s-1)(s-2)f'''(x_0 + \theta h)\,ds$$

Now $s(s-1)(s-2)$ does change sign on $[0, 2]$, so it is not possible to use the integral mean value theorem in this case. However, a more detailed mathematical analysis of the truncation error can be performed to obtain the following general result.

■ Theorem 6.1
The truncation error of (6.12) is

$$\frac{h^{k+3}}{(k+2)!}f^{(k+2)}(x_0 + \xi h)\int_0^k s^2(s-1)\cdots(s-k)\,ds$$

$$k \text{ even}, \quad f^{(k+2)} \text{ continuous on } [a, b]$$

$$(6.15)$$

and

$$\frac{h^{k+2}}{(k+1)!}f^{(k+1)}(x_0 + \xi h)\int_0^k s(s-1)\cdots(s-k)\,ds$$

$$k \text{ odd}, \quad f^{(k+1)} \text{ continuous on } [a, b]$$

$$(6.16)$$

for some $\xi \in (0, k)$.

■ Proof
A proof of this can be found in Isaacson and Keller (1966).   □

From Theorem 6.1, the truncation error associated with the case $k = 2$ is

$$\frac{h^5}{4!}f^{(4)}(x_0 + \xi h)\int_0^2 s^2(s-1)(s-2)\,ds = -\frac{h^5}{90}f^{(4)}(x_0 + \xi h)$$

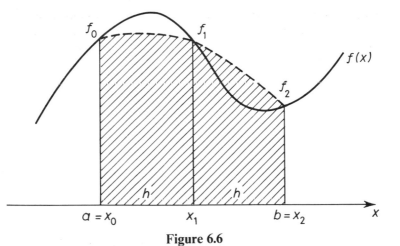

Figure 6.6

for $\xi \in (0, 2)$. Hence

$$\int_a^b f(x)\,dx = \frac{h}{3}(f_0 + 4f_1 + f_2) - \frac{h^5}{90} f^{(4)}(x_0 + \xi h) \qquad (6.17)$$

This rule, known as *Simpson's rule*, has a simple geometric interpretation, illustrated in Fig. 6.6; $h(f_0 + 4f_1 + f_2)/3$ is the area under the parabola passing through (x_0, f_0), (x_1, f_1) and (x_2, f_2).

The quadrature rule based on $k = 3$, known as *Simpson's 3/8 rule*, is

$$\int_a^b f(x)\,dx = \frac{3h}{8}(f_0 + 3f_1 + 3f_2 + f_3) - \frac{3h^5}{80} f^{(4)}(x_0 + \xi h)$$

for $\xi \in (0, 3)$. The derivation of this rule via (6.12) and (6.16) is left as an exercise. It is interesting to note that an increase in the degree of the interpolating polynomial from two to three does not produce an improvement in the order of the truncation error; both are $O(h^5)$. Consideration of (6.15) and (6.16) shows that, in general, to improve upon a quadrature rule for which k is even it is necessary to add data points in multiples of two; little will be gained by adding a single point. In confirmation of this, the rule for $k = 4$ is

$$\int_a^b f(x)\,dx = \frac{2h}{45}(7f_0 + 32f_1 + 12f_2 + 32f_3 + 7f_4) - \frac{8h^7}{945} f^{(6)}(x_0 + \xi h)$$

for $\xi \in (0, 4)$.

■ **Example 6.3**
To illustrate these formulas, consider the estimation of

$$\int_0^{\pi/4} x \cos x\,dx$$

Rectangular rule:

$$\frac{\pi}{4} f(0) = 0.000\,00$$

Trapezium rule:

$$\frac{1}{2}\frac{\pi}{4}\left[f(0) + f\left(\frac{\pi}{4}\right) \right] = 0.218\,090$$

Simpson's rule:

$$\frac{1}{3}\frac{\pi}{8}\left[f(0) + 4f\left(\frac{\pi}{8}\right) + f\left(\frac{\pi}{4}\right) \right] = 0.262\,662$$

Simpson's 3/8 rule:

$$\frac{3}{8}\frac{\pi}{12}\left[f(0) + 3f\left(\frac{\pi}{12}\right) + 3f\left(\frac{\pi}{6}\right) + f\left(\frac{\pi}{4}\right) \right] = 0.262\,553$$

In fact the exact value of this integral is 0.262 467 (6D), and so the corresponding errors are 0.262 467, 0.044 377, $-0.000\,195$ and $-0.000\,086$, respectively. ☐

Quadrature rules derived from (6.12) are known as *Newton–Cotes formulas* and are said to be *closed* because the approximating polynomial interpolates the end points a and b as well as the points between a and b. An alternative approach is to approximate f by a polynomial which only interpolates the internal points; this is illustrated in Fig. 6.7. The internal points are denoted by $x_i = x_0 + ih$ $(i = 0, 1, \ldots, k)$ with $x_0 = a + h$, $x_k = b - h$ and $h = (b-a)/(k+2)$. To be consistent, we shall write $a = x_{-1}$ and $b = x_{k+1}$. The integral

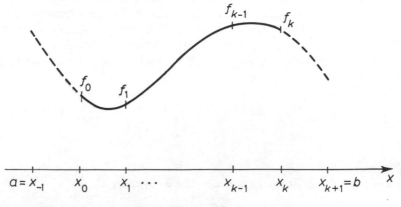

Figure 6.7

(6.10) is approximated by

$$\int_a^b p_k(x)\,dx$$

where p_k is the interpolating polynomial, of degree at most k, passing through the points (x_i, f_i) $(i = 0, 1, \ldots, k)$. The change of variable $x = x_0 + sh$ produces

$$\int_a^b p_k(x)\,dx = h\int_{-1}^{k+1} f_0 + \Delta f_0\binom{s}{1} + \Delta^2 f_0\binom{s}{2} + \cdots + \Delta^k f_0\binom{s}{k}\,ds$$

(6.18)

Quadrature rules derived in this way are known as *open* Newton–Cotes formulas. The truncation errors of such formulas are given by the following theorem, which is proved in Isaacson and Keller (1966).

■ **Theorem 6.2**
The truncation error of (6.18) is

$$\frac{h^{k+3}}{(k+2)!} f^{(k+2)}(x_0 + \xi h)\int_{-1}^{k+1} s^2(s-1)\cdots(s-k)\,ds$$
$$k \text{ even,} \quad f^{k+2} \text{ continuous on } [a, b] \tag{6.19}$$

and

$$\frac{h^{k+2}}{(k+1)!} f^{(k+1)}(x_0 + \xi h)\int_{-1}^{k+1} s(s-1)\cdots(s-k)\,ds$$
$$k \text{ odd,} \quad f^{(k+1)} \text{ continuous on } [a, b] \tag{6.20}$$

for some $\xi \in (-1, k+1)$.   □

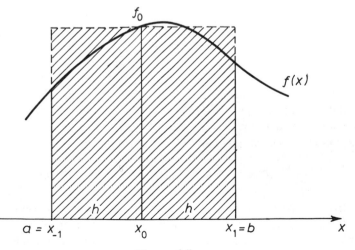

Figure 6.8

To illustrate the use of (6.18), (6.19) and (6.20) we consider the cases $k = 0, 1, 2$ and 3.

If $k = 0$ then, from (6.18) and (6.19),

$$\int_a^b f(x)\,dx = h \int_{-1}^1 f_0\,ds + \frac{h^3}{2!} f''(x_0 + \xi h) \int_{-1}^1 s^2\,ds$$

$$= 2hf_0 + \frac{h^3}{3} f''(x_0 + \xi h) \qquad \xi \in (-1, 1) \tag{6.21}$$

Figure 6.8 shows the geometric interpretation of this rule. In this case, $2hf_0$ is the area of the rectangle of width $2h$ and height f_0, the value of f at the mid-point of the interval. For this reason, (6.21) is known as the *mid-point rule*.

If $k = 1$ then, from (6.18) and (6.20),

$$\int_a^b f(x)\,dx = h \int_{-1}^2 f_0\,ds + h \int_{-1}^2 s(f_1 - f_0)\,ds$$

$$+ \frac{h^3}{2!} f''(x_0 + \xi h) \int_{-1}^2 s(s - 1)\,ds$$

$$= \frac{3h}{2}(f_0 + f_1) + \frac{3h^3}{4} f''(x_0 + \xi h) \qquad \xi \in (-1, 2)$$

Similarly, the formulas corresponding to $k = 2$ and 3 are

$$\int_a^b f(x)\,dx = \frac{4h}{3}(2f_0 - f_1 + 2f_2) + \frac{14h^5}{45} f^{(4)}(x_0 + \xi h) \qquad \xi \in (-1, 3)$$

and

$$\int_a^b f(x)\,dx = \frac{5h}{24}(11f_0 + f_1 + f_2 + 11f_3) + \frac{95h^5}{144} f^{(4)}(x_0 + \xi h) \qquad \xi \in (-1, 4)$$

respectively.

It is interesting to observe that, for each k, the truncation errors of the open and closed formulas are of the same order but with the closed formulas possessing a smaller multiplicative constant. For example, with $k = 1$ the truncation errors of the open and closed formulas are

$$\frac{3h^3}{4} f''(x_0 + \xi h) \qquad \text{and} \qquad \frac{-h^3}{12} f''(x_0 + \xi h)$$

respectively. However, the values of ξ are different for these two terms, and it sometimes happens that these numbers are such that the error in the open formula is actually smaller than that of the corresponding closed formula. As an illustration of this, we return to the problem given in Example 6.3.

■ **Example 6.4**

Results based on open Newton–Cotes formulas for the integral

$$\int_0^{\pi/4} x \cos x \, dx$$

are obtained as follows:

$k = 0$:

$$2\frac{\pi}{8} f\left(\frac{\pi}{8}\right) = 0.284\,948$$

$k = 1$:

$$\frac{3}{2}\frac{\pi}{12}\left[f\left(\frac{\pi}{12}\right) + f\left(\frac{\pi}{6}\right)\right] = 0.277\,375$$

$k = 2$:

$$\frac{4}{3}\frac{\pi}{16}\left[2f\left(\frac{\pi}{16}\right) - f\left(\frac{\pi}{8}\right) + 2f\left(\frac{3\pi}{16}\right)\right] = 0.262\,297$$

$k = 3$:

$$\frac{5}{24}\frac{\pi}{20}\left[11f\left(\frac{\pi}{20}\right) + f\left(\frac{\pi}{10}\right) + f\left(\frac{3\pi}{20}\right) + 11f\left(\frac{\pi}{5}\right)\right] = 0.262\,349$$

The corresponding errors are $-0.022\,481$, $-0.014\,908$, $-0.000\,170$ and $0.000\,118$, respectively.

Comparison of the above results with those of Example 6.3 shows that for this particular problem the estimates obtained from the open formulas with $k = 1$ and 2 are more accurate than the results obtained from the corresponding closed formulas.   □

EXERCISE 6.2

1. Use equations (6.12) and (6.16) to derive Simpson's 3/8 rule

$$\int_a^b f(x)\,dx = \frac{3h}{8}(f_0 + 3f_1 + 3f_2 + f_3) - \frac{3h^5}{80} f^{(4)}(x + \xi h)$$

for some $\xi \in (0, 3)$.

2. Use the open and closed Newton–Cotes formulas for $k = 0, 1, 2$ and 3 to approximate

   (a) $\displaystyle\int_0^{\pi/6} e^{-x} \sin x \, dx$ (b) $\displaystyle\int_0^1 e^{x^2} \, dx$

3. Verify that the Simpson's rule estimate of

$$\int_a^b f(x)\,dx$$

is exact for the functions $f(x) = 1$, x, x^2 and x^3. Deduce that Simpson's rule is exact for any polynomial of degree 3 or less. Give a second proof of this result by considering the error term.

4. Consider the evaluation of

$$I = \int_{0.5}^{1.5} x^{1/2}\,dx$$

by

(a) Using the trapezium rule
(b) Writing

$$I = \int_{0.5}^{1.0} x^{1/2}\,dx + \int_{1.0}^{1.5} x^{1/2}\,dx$$

and applying the trapezium rule separately to both of these integrals.

Verify that (b) produces a more accurate approximation than (a), and explain why this is to be expected.

5. Use Theorem 5.1 to show that provided f is sufficiently smooth there is a number ξ between 0 and 1 such that

$$f(x) = xf(1) - (x-1)f(0) + \tfrac{1}{2}x(x-1)f''(\xi)$$

Hence show that, for each $\alpha > -1$,

$$\int_0^1 x^\alpha f(x)\,dx \approx \frac{1}{2+\alpha}\left[f(1) + \frac{1}{1+\alpha}f(0)\right]$$

with error term $-f''(\eta)/[2(3+\alpha)(2+\alpha)]$ for some $\eta \in (0,1)$.
    The integral

$$\int_0^1 x^{-1/2}e^{x/2}\,dx$$

can be approximated by

(a) Taking $\alpha = -1/2$ and $f(x) = e^{x/2}$
(b) Changing the variable to $t = x^{1/2}$ and taking $\alpha = 0$.

Obtain upper and lower bounds to the errors of both methods and deduce that (a) produces a more accurate approximation than (b).

6. By considering the interpolating polynomial p_2 passing through the points (x_i, f_i) $(i = 0, 1, 2)$, where $x_i = x_0 + ih$, show that

$$\int_{x_0}^{x_1} f(x)\,dx = \frac{h}{12}(5f_0 + 8f_1 - f_2) + \frac{h^4}{24}f'''(\eta)$$

for some $\eta \in (x_0, x_2)$.

Hence estimate

$$\int_0^{0.5} \sin x \, dx$$

and obtain an upper bound on the error of this approximation.

6.3 QUADRATURE RULES IN COMPOSITE FORM

The Newton–Cotes formulas, derived in Section 6.2, produce very inaccurate results when the integration interval is large. Formulas based on low degree interpolating polynomials are clearly unsuitable since it is then necessary to use large values of h. Moreover, those based on high degree polynomials are also inaccurate because, as we found in Chapter 5, they oscillate wildly between data points. This difficulty can be avoided by adopting a piecewise approach; the integration interval is divided into subintervals and low order formulas are applied on each of these. The corresponding quadrature rules are said to be in composite form, and the most successful formulas of this type make use of the trapezium and Simpson rules.

For the composite form of the trapezium rule, the interval $[a, b]$ is divided into n equal subintervals of width h, so that $h = (b - a)/n$. This is illustrated in Fig. 6.9. The ends of the subintervals are denoted by $x_i = x_0 + ih$, $(i = 0, 1, \dots, n)$ with $x_0 = a$ and $x_n = b$. Now

$$\int_a^b f(x) \, dx = \int_{x_0}^{x_1} f(x) \, dx + \int_{x_1}^{x_2} f(x) \, dx + \cdots + \int_{x_{n-1}}^{x_n} f(x) \, dx$$

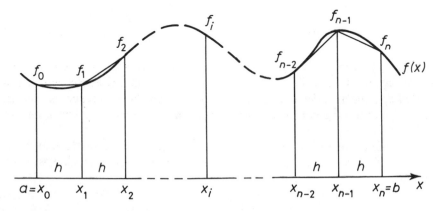

Figure 6.9

Applying the trapezium rule (6.14) to each of these integrals gives

$$\int_a^b f(x)\,dx = \frac{h}{2}(f_0 + f_1) - \frac{h^3}{12} f''(x_0 + \xi_0 h)$$

$$+ \frac{h}{2}(f_1 + f_2) - \frac{h^3}{12} f''(x_1 + \xi_1 h) + \cdots$$

$$+ \frac{h}{2}(f_{n-1} + f_n) - \frac{h^3}{12} f''(x_{n-1} + \xi_{n-1} h)$$

$$= \frac{h}{2}\{f_0 + 2f_1 + 2f_2 + \cdots + 2f_{n-1} + f_n\}$$

$$- \frac{h^3}{12}\{f''(x_0 + \xi_0 h) + f''(x_1 + \xi_1 h) + \cdots + f''(x_{n-1} + \xi_{n-1} h)\}$$

where each $\xi_i \in (0, 1)$. The quantity

$$\frac{1}{n}\sum_{i=0}^{n-1} f''(x_i + \xi_i h)$$

is the average of the n numbers $f''(x_i + \xi_i h)$ $(i = 0, 1, \ldots, n - 1)$ and so must lie between $\min_i f''(x_i + \xi_i h)$ and $\max_i f''(x_i + \xi_i h)$. If f'' is continuous on $[a, b]$ we can use the intermediate value theorem to deduce the existence of a number $\eta \in (0, n)$ such that

$$\frac{1}{n}\sum_{i=0}^{n-1} f''(x_i + \xi_i h) = f''(x_0 + \eta h)$$

Hence

$$\int_a^b f(x)\,dx = \frac{h}{2}\{f_0 + 2f_1 + 2f_2 + \cdots + 2f_{n-1} + f_n\} - \frac{nh^3}{12} f''(x_0 + \eta h)$$

$$= \frac{h}{2}\{\text{'ends'} + (2 \times \text{'middles'})\} - \frac{(b-a)h^2}{12} f''(x_0 + \eta h) \qquad (6.22)$$

Notice that whereas the local trapezium rule has a truncation error of order h^3, the composite rule has an error of order h^2. This means that when h is halved and the number of subintervals is doubled, the error decreases by a factor of approximately four (assuming that f'' remains fairly constant throughout $[a, b]$).

It is interesting to observe that, if the rounding errors in the tabulated values of f_i are less than ε, the rounding error in the trapezium rule estimate is at most

$$\frac{h}{2}\left\{\varepsilon + \varepsilon + 2\sum_{i=1}^{n-1}\varepsilon\right\} = nh\varepsilon = (b-a)\varepsilon$$

which is bounded as $h \to 0$. This shows that the evaluation of an integral via the trapezium rule is a stable process, which contrasts with the unstable behaviour of the differentiation formulas derived in Section 6.1.

■ **Example 6.5**

To illustrate the error reduction of the trapezium rule, consider the estimation of

$$\int_0^2 e^x \, dx$$

using 1, 2, 4 and 8 subintervals.

$n = 1$:

$$\tfrac{1}{2}(2)\{f(0) + f(2)\} = 8.389\,056\,1$$

$n = 2$:

$$\tfrac{1}{2}(1)\{f(0) + f(2) + 2f(1)\} = 6.912\,809\,9$$

$n = 4$:

$$\tfrac{1}{2}(\tfrac{1}{2})\{f(0) + f(2) + 2[f(\tfrac{1}{2}) + f(1) + f(\tfrac{3}{2})]\} = 6.521\,610\,1$$

$n = 8$:

$$\tfrac{1}{2}(\tfrac{1}{4})\{f(0) + f(2)$$
$$+ 2[f(\tfrac{1}{4}) + f(\tfrac{1}{2}) + f(\tfrac{3}{4}) + f(1) + f(\tfrac{5}{4}) + f(\tfrac{3}{2}) + f(\tfrac{7}{4})]\} = 6.422\,297\,8$$

The exact value of this integral is 6.389 056 1, and so the corresponding errors are $-2.000\,000\,0$, $-0.523\,753\,8$, $-0.132\,554\,0$ and $-0.033\,241\,7$, respectively, which decrease by a factor of about four at each stage.   □

For the composite form of Simpson's rule, the interval $[a, b]$ is divided into

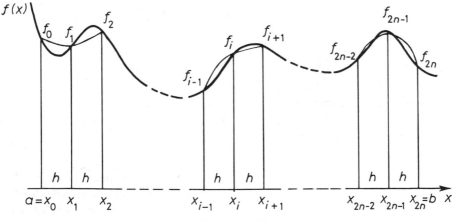

Figure 6.10

an even number $2n$ of equal subintervals of width h, so that $h = (b - a)/2n$. This is illustrated in Fig. 6.10. The ends of the subintervals are denoted by $x_i = x_0 + ih$ $(i = 0, 1, \ldots, 2n)$ with $x_0 = a$ and $x_{2n} = b$. Now

$$\int_a^b f(x)\,dx = \int_{x_0}^{x_2} f(x)\,dx + \int_{x_2}^{x_4} f(x)\,dx + \cdots + \int_{x_{2n-2}}^{x_{2n}} f(x)\,dx$$

Applying Simpson's rule (6.17) to each of these integrals gives

$$\int_a^b f(x)\,dx = \frac{h}{3}(f_0 + 4f_1 + f_2) - \frac{h^5}{90} f^{(4)}(x_0 + \xi_0 h)$$

$$+ \frac{h}{3}(f_2 + 4f_3 + f_4) - \frac{h^5}{90} f^{(4)}(x_2 + \xi_2 h) + \cdots$$

$$+ \frac{h}{3}(f_{2n-2} + 4f_{2n-1} + f_{2n}) - \frac{h^5}{90} f^{(4)}(x_{2n-2} + \xi_{2n-2} h)$$

$$= \frac{h}{3}\{f_0 + 4f_1 + 2f_2 + 4f_3 + \cdots + 2f_{2n-2} + 4f_{2n-1} + f_{2n}\}$$

$$- \frac{h^5}{90}\{f^{(4)}(x_0 + \xi_0 h) + f^{(4)}(x_2 + \xi_2 h) + \cdots + f^{(4)}(x_{2n-2} + \xi_{2n-2} h)\}$$

where each $\xi_{2i} \in (0, 2)$. If $f^{(4)}$ is continuous on $[a, b]$ we can use the intermediate value theorem as before to deduce that

$$\frac{1}{n} \sum_{i=0}^{n-1} f^{(4)}(x_{2i} + \xi_{2i} h) = f^{(4)}(x_0 + \eta h)$$

for some $\eta \in (0, 2n)$. Hence

$$\int_a^b f(x)\,dx = \frac{h}{3}\{f_0 + 4f_1 + 2f_2 + 4f_3 + \cdots + 2f_{2n-2} + 4f_{2n-1} + f_{2n}\}$$

$$- \frac{nh^5}{90} f^{(4)}(x_0 + \eta h)$$

$$= \frac{h}{3}\{\text{'ends'} + (4 \times \text{'odds'}) + (2 \times \text{'evens'})\}$$

$$- \frac{(b - a)h^4}{180} f^{(4)}(x_0 + \eta h) \tag{6.23}$$

The truncation error in the composite Simpson rule is therefore of order h^4. This means that when h is halved and the number of subintervals is doubled, the error decreases by a factor of approximately 16, considerably better than that of the trapezium rule.

■ **Example 6.6**

To illustrate the error reduction of Simpson's rule, consider the estimation of

$$\int_0^2 e^x \, dx$$

using 2, 4 and 8 subintervals, i.e. with $n = 1$, 2 and 4.

$n = 1$:

$$\tfrac{1}{3}(1)\{f(0) + f(2) + 4f(1)\} = 6.420\,727\,8$$

$n = 2$:

$$\tfrac{1}{3}(\tfrac{1}{2})\{f(0) + f(2) + 4[f(\tfrac{1}{2}) + f(\tfrac{3}{2})] + 2f(1)\} = 6.391\,210\,2$$

$n = 4$:

$$\tfrac{1}{3}(\tfrac{1}{4})\{f(0) + f(2)$$
$$+ 4[f(\tfrac{1}{4}) + f(\tfrac{3}{4}) + f(\tfrac{5}{4}) + f(\tfrac{7}{4})] + 2[f(\tfrac{1}{2}) + f(1) + f(\tfrac{3}{2})]\} = 6.389\,193\,7$$

The corresponding errors are $-0.031\,671\,7$, $-0.002\,154\,1$ and $-0.000\,137\,6$, respectively, which decrease by a factor of about 16 at each stage. Notice that these results are more accurate than those obtained in Example 6.5 using the trapezium rule (for the same number of function evaluations). The main disadvantage of Simpson's rule is that it can only be used when $[a, b]$ is divided into an even number of subintervals. ☐

EXERCISE 6.3

1. Use equation (6.11) to derive the composite rectangular rule

$$\int_a^b f(x) \, dx = h \sum_{i=0}^{n-1} f_i + \frac{(b-a)h}{2} f'(x_0 + \eta h)$$

for some $\eta \in (0, n)$, where $x_i = x_0 + ih$, $x_0 = a$ and $x_n = b$.
   Use the composite rectangular rule to estimate the value of

$$\int_1^2 \ln x \, dx$$

with 1, 2, 4 and 8 subintervals, and verify that the error term is indeed $O(h)$.

2. Use the trapezium rule to estimate the value of the following integrals. How many subintervals are needed in each case to obtain results correct to 1, 2 and 5 decimal places?

(a) $\displaystyle\int_0^2 e^{-3x} \sin 3x \, dx$ (b) $\displaystyle\int_0^2 x^2(1 + x^2)^{1/2} \, dx$

(c) $\displaystyle\int_0^{2\pi} x\cos\left(x/2\right)dx$ (d) $\displaystyle\int_2^5 \frac{(x-2)(x-1)}{(x+5)^2}dx$

Repeat using Simpson's rule, and compare your results.

3. If the rounding errors in the tabulated values of f_i do not exceed ε, show that the total rounding error in the Simpson's rule estimate of

$$\int_a^b f(x)\,dx$$

does not exceed $(b-a)\varepsilon$.

4. Find the number of subintervals required to estimate

$$\int_\alpha^1 x^{-1/2}\,dx$$

correct to 4D, using Simpson's rule, when (a) $\alpha = 0.1$ (b) $\alpha = 0.01$ (c) $\alpha = 0.001$ (d) $\alpha = 0$.

5. Use integration by parts to show that

$$\int_0^1 x^2(1-x)^2 f^{(4)}(x)\,dx = -12\{f(1)+f(0)\} + 2\{f'(1)-f'(0)\}$$

$$+24\int_0^1 f(x)\,dx$$

Deduce that

$$\int_0^1 f(x)\,dx = \tfrac{1}{2}\{f(0)+f(1)\} - \tfrac{1}{12}\{f'(1)-f'(0)\} + \tfrac{1}{720}f^{(4)}(\xi)$$

for some $\xi\in(0,1)$.

Hence show that

$$\int_a^b f(x)\,dx = \frac{h}{2}\{f_0 + 2(f_1 + \cdots + f_{n-1}) + f_n\} + \frac{h^2}{12}\{f'(a)-f'(b)\}$$

$$+\frac{(b-a)^5 h^4}{720}f^{(4)}(\eta)$$

where $f_i = f(a+ih)$, $h = (b-a)/n$ and $\eta\in(a,b)$.

Use this rule (known as the corrected trapezium rule) to estimate

$$\int_0^2 e^x\,dx$$

with $1, 2, 4$ and 8 subintervals. Compare these results with those obtained in Example 6.5 using the ordinary trapezium rule.

6. Derive the composite mid-point rule

$$\int_a^b f(x)\,dx = 2h\sum_{i=0}^{n-1} f_{2i+1} + \frac{(b-a)h^2}{6} f''(x_0 + \eta h)$$

for some $\eta \in (0, 2n)$, where $x_i = x_0 + ih$, $x_0 = a$ and $x_{2n} = b$.
  Use this rule to estimate the value of

$$I = \int_1^2 \ln x\,dx$$

using 2, 4 and 8 subintervals. Use the error term to obtain an upper bound on the number of subintervals required to estimate I to within $(1/2) \times 10^{-8}$.

6.4 ROMBERG'S METHOD

The trapezium rule is one of the easiest composite Newton–Cotes formulas to apply, but is not very accurate since its truncation error is only of order h^2. The rate of convergence can be considerably improved by the use of Romberg's method. The trapezium rule is applied in composite form with $1, 2, 4, 8, \ldots, 2^k, \ldots$ subintervals to obtain preliminary estimates of

$$I = \int_a^b f(x)\,dx$$

Romberg's method takes appropriate linear combinations of these estimates to produce approximations of high order accuracy, and is based on the following theorem.

■ **Theorem 6.3**
If all derivatives of f exist and are continuous on $[a, b]$, then the composite trapezium rule may be expressed in the form

$$\int_a^b f(x)\,dx = \frac{h}{2}\left\{ f_0 + f_n + 2\sum_{i=1}^{n-1} f_i \right\} + \sum_{j=1}^{\infty} \alpha_j h^{2j}$$

where $h = (b-a)/n$, $a = x_0$, $b = x_n$, $x_i = x_0 + ih$ $(i = 1, 2, \ldots, n-1)$, and the α_j $(j = 1, 2, \ldots)$ are independent of h for sufficiently small h.

■ **Proof**
The proof of this theorem, which is rather complicated, can be found in Ralston and Rabinowitz (1978).   □

Explicit expressions for the coefficients α_j can be obtained in terms of f and its derivatives. However, for the purposes of the following derivation, it is not necessary to know their precise values.
  Let $T_{k,0}$ $(k = 0, 1, 2, \ldots)$ denote the trapezium rule estimate of I using

2^k subintervals of width $h_k = (b - a)/2^k$. From Theorem 6.3,

$$I = T_{k,0} + \alpha_1 h_k^2 + \alpha_2 h_k^4 + \alpha_3 h_k^6 + \cdots \tag{6.24}$$

If we now halve the value of h and double the number of subintervals, then (6.24) becomes

$$I = T_{k+1,0} + \alpha_1 \left(\frac{h_k}{2}\right)^2 + \alpha_2 \left(\frac{h_k}{2}\right)^4 + \alpha_3 \left(\frac{h_k}{2}\right)^6 + \cdots$$

$$= T_{k+1,0} + \tfrac{1}{4}\alpha_1 h_k^2 + \tfrac{1}{16}\alpha_2 h_k^4 + \tfrac{1}{64}\alpha_3 h_k^6 + \cdots \tag{6.25}$$

Terms involving h_k^2 may be eliminated by multiplying (6.25) by four and subtracting (6.24) to get

$$3I = 4T_{k+1,0} - T_{k,0} - \tfrac{3}{4}\alpha_2 h_k^4 - \tfrac{15}{16}\alpha_3 h_k^6 - \cdots$$

or equivalently

$$I = \tfrac{1}{3}(4T_{k+1,0} - T_{k,0}) - \tfrac{1}{4}\alpha_2 h_k^4 - \tfrac{5}{16}\alpha_3 h_k^6 - \cdots \tag{6.26}$$

The quantity $\tfrac{1}{3}(4T_{k+1,0} - T_{k,0})$, called the first Romberg extrapolate, is denoted by $T_{k+1,1}$. Equation (6.26) shows that this is a fourth order approximation to I and so should be more accurate than either $T_{k,0}$ or $T_{k+1,0}$, which are only second order approximations.

The extrapolation process can now be repeated to produce sixth order estimates of I. From (6.26),

$$I = T_{k,1} + \beta_1 h_k^4 + \beta_2 h_k^6 + \cdots \tag{6.27}$$

for some constants β_j which are independent of h. If the value of h is again halved and the number of subintervals is doubled, then

$$I = T_{k+1,1} + \beta_1 \left(\frac{h_k}{2}\right)^4 + \beta_2 \left(\frac{h_k}{2}\right)^6 + \cdots$$

$$= T_{k+1,1} + \tfrac{1}{16}\beta_1 h_k^4 + \tfrac{1}{64}\beta_2 h_k^6 + \cdots \tag{6.28}$$

Terms involving h_k^4 may be eliminated by multiplying (6.28) by 16 and subtracting (6.27) to get

$$15I = 16T_{k+1,1} - T_{k,1} - \tfrac{3}{4}\beta_2 h_k^6 - \cdots$$

or equivalently

$$I = \tfrac{1}{15}(16T_{k+1,1} - T_{k,1}) - \tfrac{1}{20}\beta_2 h_k^6 - \cdots$$

The quantity $\tfrac{1}{15}(16T_{k+1,1} - T_{k,1})$, called the second Romberg extrapolate, is denoted by $T_{k+1,2}$ and is a sixth order approximation to I.

This process can clearly be continued. If, in general,

$$I = T_{k,j} + \varepsilon_1 h_k^{2j+2} + \varepsilon_2 h_k^{2j+4} + \cdots \tag{6.29}$$

then

$$I = T_{k+1,j} + \varepsilon_1 \left(\frac{h_k}{2}\right)^{2j+2} + \varepsilon_2 \left(\frac{h_k}{2}\right)^{2j+4} + \cdots \tag{6.30}$$

The term involving h_k^{2j+2} can be eliminated by multiplying (6.30) by $2^{2j+2} = 4^{j+1}$ and subtracting (6.29) to get

$$(4^{j+1} - 1)I = 4^{j+1} T_{k+1,j} - T_{k,j} - \tfrac{3}{4}\varepsilon_2 h_k^{2j+4} - \cdots$$

or equivalently

$$I = T_{k+1,j+1} + O(h^{2j+4})$$

where

$$T_{k+1,j+1} = (4^{j+1} T_{k+1,j} - T_{k,j})/(4^{j+1} - 1)$$

The Romberg extrapolates are usually set out in a triangular array of the form

$$
\begin{array}{llll}
T_{0,0} & & & \\
T_{1,0} & T_{1,1} & & \\
T_{2,0} & T_{2,1} & T_{2,2} & \\
T_{3,0} & T_{3,1} & T_{3,2} & T_{3,3} \\
\vdots & & &
\end{array}
$$

The table is constructed row by row. A new trapezium rule estimate $T_{k,0}$ is appended to the first column and the previously computed values in the $(k-1)$th row are used to calculate $T_{k,1}, T_{k,2}, \ldots, T_{k,k}$ in turn.

■ **Example 6.7**

The trapezium rule estimates of

$$\int_0^2 e^x \, dx$$

with $1, 2, 4$ and 8 subintervals were calculated in Example 6.5 as

$$T_{0,0} = 8.389\,056\,1$$
$$T_{1,0} = 6.912\,809\,9$$
$$T_{2,0} = 6.521\,610\,1$$
$$T_{3,0} = 6.422\,297\,8$$

The first Romberg extrapolates are given by

$$T_{1,1} = \tfrac{1}{3}(4T_{1,0} - T_{0,0}) = 6.420\,727\,8$$
$$T_{2,1} = \tfrac{1}{3}(4T_{2,0} - T_{1,0}) = 6.391\,210\,2$$
$$T_{3,1} = \tfrac{1}{3}(4T_{3,0} - T_{2,0}) = 6.389\,193\,7$$

The second Romberg extrapolates are

$$T_{2,2} = \tfrac{1}{15}(16T_{2,1} - T_{1,1}) = 6.389\,242\,4$$
$$T_{3,2} = \tfrac{1}{15}(16T_{3,1} - T_{2,1}) = 6.389\,059\,3$$

The third Romberg extrapolate is

$$T_{3,3} = \tfrac{1}{63}(64T_{3,2} - T_{2,2}) = 6.389\,056\,4$$

which has an error of only $-0.000\,000\,3$.
The corresponding Romberg table is

8.389 056 1

6.912 809 9   6.420 727 8

6.521 610 1   6.391 210 2   6.389 242 4

6.422 297 8   6.389 193 7   6.389 059 3   6.389 056 4

Notice that the first Romberg extrapolates are the same as the Simpson rule estimates computed in Example 6.6. In general, it may be shown (see Question 1 in Exercise 6.4) that $T_{k,1} = S_k$, where S_k denotes the Simpson rule estimate of I using 2^k subintervals.   □

Most of the computational effort associated with Romberg's method goes into the calculation of the original trapezium rule estimates. This is especially so when f is complicated, since in this situation the total computer time is dominated by the time taken to evaluate $f(x_i)$. The number of function evaluations is considerably reduced if $T_{k,0}$ is computed from a known value of $T_{k-1,0}$ using the formula

$$T_{k,0} = \tfrac{1}{2}T_{k-1,0} + \frac{h_{k-1}}{2}\{f(x_0 + \tfrac{1}{2}h_{k-1}) + f(x_0 + \tfrac{3}{2}h_{k-1})$$
$$+ \cdots + f[x_0 + (2^{k-1} - \tfrac{1}{2})h_{k-1}]\} \qquad (6.31)$$

This formula enables us to calculate $T_{k,0}$ using only 2^{k-1} additional function evaluations compared with the $2^k + 1$ evaluations needed if $T_{k,0}$ is calculated directly from the trapezium rule. The case $k = 2$ is illustrated in Fig. 6.11. To prove this formula, consider $T_{k,0} - \tfrac{1}{2}T_{k-1,0}$. From (6.22),

$$T_{k,0} - \tfrac{1}{2}T_{k-1,0}$$

$$= \frac{h_k}{2}\{f(a) + f(b) + 2[f(x_0 + h_k) + f(x_0 + 2h_k) + \cdots + f(x_0 + (2^k - 1)h_k)]\}$$

$$- \frac{h_{k-1}}{4}\{f(a) + f(b) + 2[f(x_0 + h_{k-1}) + f(x_0 + 2h_{k-1}) + \cdots$$
$$+ f(x_0 + (2^{k-1} - 1)h_{k-1})]\}$$

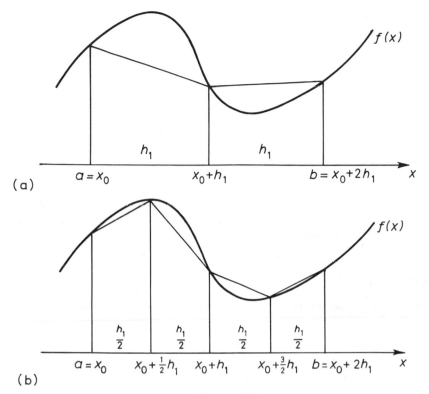

Figure 6.11

$$= \frac{h_{k-1}}{4} \{ f(a) + f(b) + 2[f(x_0 + \tfrac{1}{2}h_{k-1}) + f(x_0 + h_{k-1}) + \cdots$$

$$+ f(x_0 + (2^{k-1} - \tfrac{1}{2})h_{k-1})]\}$$

$$- \frac{h_{k-1}}{4} \{ f(a) + f(b) + 2[f(x_0 + h_{k-1}) + f(x_0 + 2h_{k-1}) + \cdots$$

$$+ f(x_0 + (2^{k-1} - 1)h_{k-1})]\}$$

since $h_k = \tfrac{1}{2}h_{k-1}$. All of the terms corresponding to integer multiples of h_{k-1} cancel, leaving

$$\frac{h_{k-1}}{2} \{ f(x_0 + \tfrac{1}{2}h_{k-1}) + f(x_0 + \tfrac{3}{2}h_{k-1}) + \cdots + f(x_0 + (2^{k-1} - \tfrac{1}{2})h_{k-1})\}$$

as required.

■ **Example 6.8**

To illustrate the use of equation (6.31), consider the calculation of the first four trapezium rule estimates of

$$\int_0^2 e^x \, dx$$

Now

$$T_{0,0} = \tfrac{1}{2}(2)[f(0) + f(2)] = 8.389\,056\,1$$

From (6.31),

$$T_{1,0} = \tfrac{1}{2}T_{0,0} + \tfrac{1}{2}(2)f(1) = 6.912\,809\,9$$
$$T_{2,0} = \tfrac{1}{2}T_{1,0} + \tfrac{1}{2}(1)[f(\tfrac{1}{2}) + f(\tfrac{3}{2})] = 6.521\,610\,1$$
$$T_{3,0} = \tfrac{1}{2}T_{2,0} + \tfrac{1}{2}(\tfrac{1}{2})[f(\tfrac{1}{4}) + f(\tfrac{3}{4}) + f(\tfrac{5}{4}) + f(\tfrac{7}{4})] = 6.422\,297\,8$$

Notice that the calculation of $T_{3,0}$ has involved only four additional function evaluations, whereas the direct application of the trapezium rule involves nine function evaluations. □

EXERCISE 6.4

1. If S_k denotes the Simpson's rule estimate of I using 2^k subintervals, show that

$$T_{k,1} = S_k$$

2. Find the trapezium rule estimates of

$$I = \int_1^2 \ln x \, dx$$

using $1, 2, 4$ and 8 subintervals. Work to 7D.

   Construct the corresponding table of Romberg extrapolates and hence estimate I as accurately as possible.

3. Use Romberg's method to evaluate the integrals given in Exercise 6.3, Question 2. How many Romberg extrapolates are needed in each case to obtain results correct to $1, 2$ and 5 decimal places?

4. How many Romberg extrapolates are required to estimate

$$\int_\alpha^1 x^{-1/2} \, dx$$

correct to 4D, when (a) $\alpha = 0.1$ (b) $\alpha = 0.01$ (c) $\alpha = 0.001$?

5. Let $R_{k,0}$ denote the composite rectangular rule estimate of

$$I = \int_a^b f(x) \, dx$$

using 2^k subintervals of width h_k. Assuming that the corresponding error can be expressed in the form

$$\alpha_1 h_k + \alpha_2 h_k^2 + \alpha_3 h_k^3 + \cdots$$

show that

$$R_{k+1,1} = 2R_{k+1,0} - R_{k,0}$$
$$R_{k+1,2} = \tfrac{1}{3}(4R_{k+1,1} - R_{k,1})$$
$$R_{k+1,3} = \tfrac{1}{7}(8R_{k+1,2} - R_{k,2})$$

are $O(h^2)$, $O(h^3)$ and $O(h^4)$ approximations to I, respectively.
   Calculate $R_{3,3}$ for

$$I = \int_1^2 \ln x \, dx$$

(see Exercise 6.3, Question 1.)

6. If M_k denotes the mid-point rule estimate of I using 2^k subintervals, show that

$$R_{k,1} = M_k$$

6.5 SIMPSON'S ADAPTIVE QUADRATURE

The methods of numerical integration considered so far in this chapter are based on equally spaced points. We now relax this restriction and allow the step size h to vary. This is particularly appropriate for integrands which have regions of both large and small function variation. In regions of large variation smaller steps are taken than in regions of small variation. As an example,

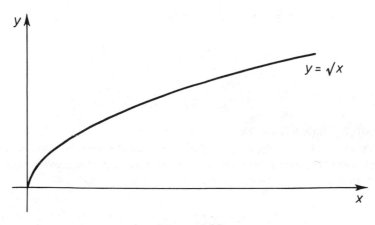

Figure 6.12

consider the function $f(x) = \sqrt{x}$ on $[0, 1]$, illustrated in Fig. 6.12. The graph of f has an infinite slope at $x = 0$, so it makes sense to use a smaller step size near $x = 0$ than is used near $x = 1$.

Adaptive methods automatically vary the step size in order to minimize the number of function evaluations required for the approximation of

$$I = \int_a^b f(x)\,dx$$

to any prescribed accuracy. Adaptive methods may be applied to any of the composite Newton–Cotes formulas. However, we shall only consider Simpson's rule, and we begin by describing a way of estimating the truncation error. Let S_1 denote the Simpson's rule approximation to I using two subintervals of width h. From equation (6.17)

$$S_1 = \frac{h}{3}\left[f(a) + 4f\left(\frac{a+b}{2}\right) + f(b) \right]$$

with truncation error

$$I - S_1 = -\frac{h^5}{90} f^{(4)}(a + \eta_1 h) \tag{6.32}$$

for $\eta_1 \in (0, 2)$, where $h = (b - a)/2$. In addition, let S_2 denote the Simpson's rule approximation to I using four subintervals of width $h/2$. From equation (6.23)

$$S_2 = \frac{1}{3}\frac{h}{2}\left[f(a) + 4f\left(\frac{3a+b}{4}\right) + 2f\left(\frac{a+b}{2}\right) + 4f\left(\frac{a+3b}{4}\right) + f(b) \right]$$

The truncation error in this expression is

$$I - S_2 = -\frac{2h}{180}\left(\frac{h}{2}\right)^4 f^{(4)}\left(a + \eta_2\frac{h}{2}\right)$$

$$= -\frac{h^5}{90 \times 16} f^{(4)}\left(a + \eta_2\frac{h}{2}\right) \tag{6.33}$$

for $\eta_2 \in (0, 4)$, since $b - a = 2h$.

Equations (6.32) and (6.33) are of little practical value because η_1 and η_2 are unknown. However, it is possible to estimate the truncation error if we assume that $f^{(4)}$ is reasonably constant on $[a, b]$, so that

$$f^{(4)}(a + \eta_1 h) \approx f^{(4)}\left(a + \eta_2\frac{h}{2}\right) \approx f^{(4)}(\eta)$$

for some $\eta \in (a, b)$. Subtraction of (6.33) from (6.32) then gives

$$S_2 - S_1 \approx -\frac{h^5}{90} f^{(4)}(\eta) + \frac{1}{16} \frac{h^5}{90} f^{(4)}(\eta)$$

$$= -\frac{15}{16} \frac{h^5}{90} f^{(4)}(\eta)$$

$$\approx 15(I - S_2)$$

Hence

$$|I - S_2| \approx \tfrac{1}{15} |S_2 - S_1| \qquad (6.34)$$

and, since $\tfrac{1}{15}|S_2 - S_1|$ is trivially computed, this provides a simple means of estimating the error in S_2.

■ **Example 6.9**

To illustrate (6.34), consider the evaluation of

$$I = \int_{1/2}^{1} e^x \, dx$$

The corresponding values of S_1 and S_2 are

$$S_1 = \tfrac{1}{3}(\tfrac{1}{4})[f(\tfrac{1}{2}) + 4f(\tfrac{3}{4}) + f(1)] = 1.069\,583\,6$$
$$S_2 = \tfrac{1}{3}(\tfrac{1}{8})[f(\tfrac{1}{2}) + 4f(\tfrac{5}{8}) + 2f(\tfrac{3}{4}) + 4f(\tfrac{7}{8}) + f(1)] = 1.069\,562\,0$$

An estimate of the error in S_2 is obtained from (6.34) as

$$\tfrac{1}{15}|1.069\,562\,0 - 1.069\,583\,6| = 0.000\,001\,4$$

For this particular problem the estimate is extremely good, since the actual error in S_2 is

$$|(e - e^{1/2}) - 1.069\,562\,0| = 0.000\,001\,4 \qquad (7D) \qquad \square$$

It is possible to use (6.34) to calculate I to within a specified tolerance ε. The first step is to compute S_1 and S_2, the Simpson rule approximations to I based on two and four subintervals. If $|S_1 - S_2| < 15\varepsilon$ then we have achieved our desired aim since, from (6.34), $|I - S_2| < \varepsilon$. The value of S_2 is taken as the approximation to I with an error at most ε. If, on the other hand, $|S_1 - S_2| \nless 15\varepsilon$, then the process is applied separately to the two integrals

$$\int_{a}^{(a+b)/2} f(x) \, dx, \qquad \int_{(a+b)/2}^{b} f(x) \, dx$$

using a tolerance $\varepsilon/2$. If necessary the algorithm is repeated, subdividing at each stage with tolerances of $\varepsilon/4, \varepsilon/8, \ldots$ at successive levels. This is to ensure that when the subdivision ceases the total error over the entire interval $[a, b]$

will be less than ε. An estimate of I to within ε is then given by the sum of the individual approximations.

■ **Example 6.10**

As an illustration of the method, consider the evaluation of

$$\int_0^\pi e^{-3x} \sin 3x \, dx$$

using a tolerance $\varepsilon = 0.0005$.

For the first step,

$$S_1 = \frac{1}{3}\frac{\pi}{2}\left[f(0) + 4f\left(\frac{\pi}{2}\right) + f(\pi) \right] = -0.0188$$

$$S_2 = \frac{1}{3}\frac{\pi}{4}\left[f(0) + 4f\left(\frac{\pi}{4}\right) + 2f\left(\frac{\pi}{2}\right) + 4f\left(\frac{3\pi}{4}\right) + f(\pi) \right] = 0.0661$$

Now $|S_1 - S_2| \nleq 15\varepsilon$, so $[0, \pi]$ must be subdivided.

On the interval $[\pi/2, \pi]$,

$$S_1 = \frac{1}{3}\frac{\pi}{4}\left[f\left(\frac{\pi}{2}\right) + 4f\left(\frac{3\pi}{4}\right) + f(\pi) \right] = -0.0017$$

$$S_2 = \frac{1}{3}\frac{\pi}{8}\left[f\left(\frac{\pi}{2}\right) + 4f\left(\frac{5\pi}{8}\right) + 2f\left(\frac{3\pi}{4}\right) + 4f\left(\frac{7\pi}{8}\right) + f(\pi) \right] = -0.0014$$

In this case $|S_1 - S_2| < 15(\varepsilon/2)$, so the value of $\int_{\pi/2}^\pi f(x) \, dx$ is taken as -0.0014, which has an error of at most $\varepsilon/2$.

On the interval $[0, \pi/2]$, $S_1 = 0.0678$, $S_2 = 0.1595$ with $|S_1 - S_2| \nleq 15(\varepsilon/2)$, and so $[0, \pi/2]$ must be subdivided.

On the interval $[\pi/4, \pi/2]$, $S_1 = 0.0018$, $S_2 = 0.0015$ with $|S_1 - S_2| < 15(\varepsilon/4)$, and so the value of $\int_{\pi/4}^{\pi/2} f(x) \, dx$ is taken as 0.0015, which has an error of at most $\varepsilon/4$.

On the interval $[0, \pi/4]$, $S_1 = 0.1577$, $S_2 = 0.1662$ with $|S_1 - S_2| \nleq 15(\varepsilon/4)$, and so $[0, \pi/4]$ must be further subdivided.

On the interval $[\pi/8, \pi/4]$, $S_1 = 0.0669$, $S_2 = 0.0670$ with $|S_1 - S_2| < 15(\varepsilon/8)$, and so the value of $\int_{\pi/8}^{\pi/4} f(x) \, dx$ is taken as 0.0670, which has an error of at most $\varepsilon/8$.

On the interval $[0, \pi/8]$, $S_1 = 0.0993$, $S_2 = 0.0996$ with $|S_1 - S_2| < 15(\varepsilon/8)$, and so the value of $\int_0^{\pi/8} f(x) \, dx$ is taken as 0.0996, which has an error of at most $\varepsilon/8$.

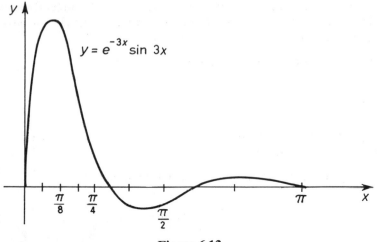

Figure 6.13

Finally, summation of the individual approximations gives

$$\int_0^\pi f(x)\,dx = \int_0^{\pi/8} f(x)\,dx + \int_{\pi/8}^{\pi/4} f(x)\,dx + \int_{\pi/4}^{\pi/2} f(x)\,dx + \int_{\pi/2}^\pi f(x)\,dx$$
$$\approx 0.0996 + 0.0670 + 0.0015 - 0.0014$$
$$= 0.1667$$

The error in this approximation does not exceed

$$\frac{\varepsilon}{8} + \frac{\varepsilon}{8} + \frac{\varepsilon}{4} + \frac{\varepsilon}{2} = \varepsilon = 0.0005$$

Hence

$$\int_0^\pi e^{-3x} \sin 3x \, dx = 0.1667 \pm 0.0005$$

The subintervals used are shown in Fig. 6.13.   □

EXERCISE 6.5

1. Use Simpson's adaptive quadrature to estimate

$$\int_0^2 \sqrt{x}\,dx$$

using (a) $\varepsilon = 0.005$ (b) $\varepsilon = 0.001$.
    Find the exact value of this integral and comment on what you find.

2. Use Simpson's adaptive quadrature to evaluate the integrals given in Exercise 6.3, Question 2. How many subintervals are needed in each case to obtain results using $\varepsilon = 0.05$, 0.005 and $0.000\,005$?

3. How many subintervals are required to estimate

$$\int_{\alpha}^{1} x^{-1/2}\,dx$$

using Simpson's adaptive quadrature with $\varepsilon = (1/2) \times 10^{-4}$ when (a) $\alpha = 0.1$ (b) $\alpha = 0.01$ (c) $\alpha = 0.001$?

4. If T_1, T_2 denote the trapezium rule approximations to

$$I = \int_{a}^{b} f(x)\,dx$$

with 1, 2 subintervals respectively, show that

$$|I - T_2| \approx \tfrac{1}{3}|T_2 - T_1|$$

Hence estimate the error incurred when $\int_{1/2}^{1} e^x\,dx$ is approximated by $\tfrac{1}{8}(e^{1/2} + 2e^{3/4} + e)$.

5. Evaluate

$$\int_{0}^{1} x^n e^{x-1}\,dx$$

using Simpson's adaptive quadrature with $\varepsilon = 10^{-6}$ for (a) $n = 1$ (b) $n = 7$ (c) $n = 20$ (d) $n = 60$, and compare the number of subintervals required.

6. One way of evaluating integrals of the form

$$I = \int_{0}^{\infty} f(x)\,dx$$

is to approximate I by $\int_{0}^{k} f(x)\,dx$, where k is chosen so that $|f(x)| < \delta$ for all $x > k$ and some prescribed number δ.

Illustrate this approach by calculating $\Gamma(1.5)$, $\Gamma(3)$ and $\Gamma(6)$ using Simpson's adaptive quadrature, where

$$\Gamma(m) = \int_{0}^{\infty} x^{m-1} e^{-x}\,dx$$

Take $\varepsilon = \delta = 10^{-4}$.

6.6 GAUSSIAN QUADRATURE

All of the Newton–Cotes formulas derived in Section 6.2 are of the form

$$\int_a^b f(x)\,dx \approx \sum_{i=0}^k c_i f(x_i) \tag{6.35}$$

The points x_i are equally spaced in the interval $[a, b]$, and the coefficients c_i are obtained via the integration of an appropriate interpolating polynomial. Quadrature rules derived in this way are exact for polynomials of a certain degree. The trapezium and Simpson rules, for example, are exact for polynomials of degree one and three, respectively.

Gauss showed that, by a judicious choice of points x_i, it is possible to construct formulas which are significantly more accurate than the corresponding Newton–Cotes formulas. In Gaussian quadrature, the x_i and c_i are treated as unknowns. These numbers are chosen so that (6.35) is exact for polynomials of the highest possible degree. Now this formula is exact for polynomials of degree m if and only if it is exact for the monomials $1, x, x^2, \ldots, x^m$. This follows from the linearity of both sides of equation (6.35). Hence we require

$$\sum_{i=0}^k c_i x_i^r = \int_a^b x^r\,dx = \frac{b^{r+1} - a^{r+1}}{r+1} \qquad r = 0, 1, \ldots, m \tag{6.36}$$

The quantities x_i and c_i ($i = 0, 1, \ldots, k$) comprise a set of $2k + 2$ unknowns, so if $m = 2k + 1$ then (6.36) represents a system of $2k + 2$ equations in the same number of unknowns. However, since this system is non-linear in the x_i we have no guarantee that it possesses a unique solution. In fact, it is possible to solve this system directly. Details of this can be found in Davis and Rabinowitz (1984). The case $k = 1$ is considered in Question 1 of Exercise 6.6.

We adopt an alternative approach to the construction of Gaussian quadrature formulas and make use of the orthogonal polynomials discussed in Section 5.7. Let us assume for the moment that (6.36) does have a unique solution, and consider the polynomial of degree $k + 1$, with leading term x^{k+1}, defined by

$$\varphi_{k+1}(x) = (x - x_0)(x - x_1)\cdots(x - x_k)$$

Let q_k denote any polynomial of degree k or less. It follows that $\varphi_{k+1}q_k$ is a polynomial of degree at most $2k + 1$, and so may be integrated exactly using (6.35). Hence

$$\int_a^b \varphi_{k+1}(x)q_k(x)\,dx = \sum_{i=0}^k c_i \varphi_{k+1}(x_i)q_k(x_i) = 0$$

since φ_{k+1} vanishes at the points x_i ($i = 0, 1, \ldots, k$). This means that φ_{k+1} is orthogonal to every polynomial of lower degree. Therefore φ_{k+1} is a member of the unique sequence of orthogonal polynomials corresponding to the weight function $w(x) = 1$ and interval $[a, b]$. Moreover, it can be proved (see Ralston, 1965) that such polynomials have real, distinct, zeros lying in the interval (a, b). We have therefore shown that if a suitable quadrature formula exists, then the numbers x_i must be uniquely determined as the $k + 1$ zeros of the orthogonal polynomial φ_{k+1}.

Once the x_i have been calculated, the c_i can be found by solving the first $k + 1$ equations of (6.36), which take the form

$$
\begin{pmatrix}
1 & 1 & 1 & \cdots & 1 \\
x_0 & x_1 & x_2 & \cdots & x_k \\
x_0^2 & x_1^2 & x_2^2 & \cdots & x_k^2 \\
\vdots & & & & \vdots \\
x_0^k & x_1^k & x_2^k & \cdots & x_k^k
\end{pmatrix}
\begin{pmatrix}
c_0 \\ c_1 \\ c_2 \\ \vdots \\ c_k
\end{pmatrix}
=
\begin{pmatrix}
(b - a) \\
(b^2 - a^2)/2 \\
(b^3 - a^3)/3 \\
\vdots \\
(b^{k+1} - a^{k+1})/(k + 1)
\end{pmatrix}
\tag{6.37}
$$

The coefficient matrix of (6.37) is known as the *Vandermonde matrix*. It can be proved (see Davis, 1975) that this has determinant

$$
\prod_{i>j} (x_i - x_j)
$$

which is non-zero because the points x_i are distinct. Hence the linear equations (6.37) possess a unique solution. We have therefore shown that if a suitable quadrature formula exists, then the numbers c_i must be uniquely determined as the solution of (6.37).

Finally, it remains to check that if the x_i and c_i are obtained in the way described above, then (6.35) is indeed exact for polynomials of degree $2k + 1$. Now any polynomial p_{2k+1} of degree $2k + 1$ can be written as

$$
p_{2k+1}(x) = \varphi_{k+1}(x)q_k(x) + r_k(x)
$$

where q_k and r_k are the quotient and remainder when p_{2k+1} is divided by φ_{k+1} and as such are polynomials of degree at most k. Equation (6.35) is exact for $\varphi_{k+1}q_k$ since both sides of the equation

$$
\int_a^b \varphi_{k+1}(x)q_k(x)\,dx = \sum_{i=0}^{k} c_i\varphi_{k+1}(x_i)q_k(x_i)
\tag{6.38}
$$

are zero. The left hand side is zero because, by choice of x_i, φ_{k+1} is orthogonal to any polynomial of degree k or less; the right hand side is zero because the x_i are the zeros of φ_{k+1}.

The c_i are chosen in (6.37) to ensure that (6.35) is exact for any polynomial of degree k or less, so in particular

$$
\int_a^b r_k(x)\,dx = \sum_{i=0}^{k} c_i r_k(x_i)
\tag{6.39}
$$

Addition of (6.38) and (6.39) gives

$$\int_a^b \{\varphi_{k+1}(x)q_k(x) + r_k(x)\}\,dx = \sum_{i=0}^k c_i\{\varphi_{k+1}(x_i)q_k(x_i) + r_k(x_i)\}$$

i.e.

$$\int_a^b p_{2k+1}(x)\,dx = \sum_{i=0}^k c_i p_{2k+1}(x_i)$$

as required.

■ **Example 6.11**

Consider the derivation of the numbers x_0, x_1, c_0 and c_1 such that

$$\int_0^1 f(x)\,dx \approx c_0 f(x_0) + c_1 f(x_1)$$

is exact for polynomials of degree three or less.

From Example 5.17, the polynomial φ_2 is given by

$$\varphi_2(x) = x^2 - x + 1/6$$

which has zeros

$$x_0 = \frac{1}{2}\left(1 + \frac{1}{\sqrt{3}}\right), \qquad x_1 = \frac{1}{2}\left(1 - \frac{1}{\sqrt{3}}\right)$$

The system of equations (6.37) becomes

$$\begin{pmatrix} 1 & 1 \\ \frac{1}{2}\left(1 + \frac{1}{\sqrt{3}}\right) & \frac{1}{2}\left(1 - \frac{1}{\sqrt{3}}\right) \end{pmatrix} \begin{pmatrix} c_0 \\ c_1 \end{pmatrix} = \begin{pmatrix} 1 \\ \frac{1}{2} \end{pmatrix}$$

which has solution $c_0 = 1/2$ and $c_1 = 1/2$.

Hence the corresponding quadrature rule is

$$\int_0^1 f(x)\,dx \approx \frac{1}{2}\left\{f\left[\frac{1}{2}\left(1 + \frac{1}{\sqrt{3}}\right)\right] + f\left[\frac{1}{2}\left(1 - \frac{1}{\sqrt{3}}\right)\right]\right\} \qquad \square$$

One of the unfortunate consequences of Gaussian integration is the dependence of the x_i and c_i on a and b, so that each time a new integration interval is encountered they have to be recalculated. This problem is greatly simplified if the integrals are transformed using the change of variable $t = (2x - a - b)/(b - a)$ to obtain

$$\int_a^b f(x)\,dx = \int_{-1}^1 f\left(\frac{(b - a)t + a + b}{2}\right)\frac{(b - a)}{2}\,dt$$

Polynomials orthogonal on $[-1, 1]$ with respect to the weight function $w(x) = 1$ are known as *Legendre polynomials*. They have many important theoretical properties, most of which can be found in Bell (1968). The first

Table 6.3

k	x_i	c_i
0	0	2
1	$-1/\sqrt{3}$	1
	$1/\sqrt{3}$	1
2	$-\sqrt{(3/5)}$	5/9
	0	8/9
	$\sqrt{(3/5)}$	5/9
3	$-0.861\,136\,311\,6$	$0.347\,854\,845\,1$
	$-0.339\,981\,043\,6$	$0.652\,145\,154\,9$
	$0.339\,981\,043\,6$	$0.652\,145\,154\,9$
	$0.861\,136\,311\,6$	$0.347\,854\,845\,1$

few Legendre polynomials are

$$\varphi_0(x) = 1$$
$$\varphi_1(x) = x$$
$$\varphi_2(x) = x^2 - \tfrac{1}{3}$$
$$\varphi_3(x) = x^3 - \tfrac{3}{5}x$$
$$\varphi_4(x) = x^4 - \tfrac{6}{7}x^2 + \tfrac{3}{35}$$

The corresponding zeros x_i and coefficients c_i are listed in Table 6.3. The numbers corresponding to φ_4 are rounded to 10D.

■ **Example 6.12**
Consider the estimation of

$$\int_0^2 e^x\,dx$$

using the Gaussian quadrature formula with $k = 3$.
The change of variable $t = x - 1$ gives

$$\int_0^2 e^x\,dx = \int_{-1}^1 e^{t+1}\,dt$$

From Table 6.3,

$$\int_{-1}^1 f(t)\,dt \approx 0.347\,854\,8\,f(-0.861\,136\,3) + 0.652\,145\,2\,f(-0.339\,981\,0)$$
$$+ 0.652\,145\,2\,f(0.339\,981\,0) + 0.347\,854\,8\,f(0.861\,136\,3)$$
$$= 6.389\,055\,1 \qquad \text{(7D)}$$

This has an error, $0.000\,000\,9$, which is much smaller than the errors obtained in Examples 6.5 and 6.6 using the composite trapezium and Simpson rules.

Indeed, this estimate is almost as accurate as that obtained in Example 6.7 using Romberg's method. The main disadvantage of Gaussian quadrature is that it cannot be used when the integrand is given as a set of discrete experimental readings, since it is most unlikely that the values of f are available at the points x_i. □

EXERCISE 6.6

1. Show that the quadrature rule

$$\int_{-1}^{1} f(x)\,dx \approx c_0 f(x_0) + c_1 f(x_1)$$

   is exact for polynomials of degree three if and only if

$$
\begin{aligned}
c_0 \quad + c_1 \quad &= 2 \\
c_0 x_0 + c_1 x_1 &= 0 \\
c_0 x_0^2 + c_1 x_1^2 &= 2/3 \\
c_0 x_0^3 + c_1 x_1^3 &= 0
\end{aligned}
$$

   Suppose that x_0 and x_1 are the roots of

$$x^2 + ux + v = 0$$

   By multiplying the four equations by v, u, 1 and 0, respectively, and adding, show that $v = -1/3$.
   Similarly, by multiplying them by 0, v, u and 1, show that $u = 0$. Hence find x_0, x_1, c_0 and c_1.

2. Use the Gaussian quadrature rules whose coefficients are listed in Table 6.3 to evaluate the integrals given in Exercise 6.3, Question 2.

3. Find the numbers x_i, c_i ($i = 0, 1, 2$) correct to 5D, such that

$$\int_{0}^{1} f(x)\,dx = c_0 f(x_0) + c_1 f(x_1) + c_2 f(x_2)$$

   for polynomials of degree five. (From Example 5.17,

$$\varphi_3(x) = x^3 - \tfrac{3}{2}x^2 + \tfrac{3}{5}x - \tfrac{1}{20})$$

   Hence estimate

$$\int_{0}^{1} c^x\,dx$$

4. The quadrature rule

$$\int_{-1}^{1} f(x)\,dx \approx f\left(\frac{-1}{\sqrt{3}}\right) + f\left(\frac{1}{\sqrt{3}}\right)$$

is known to be exact for polynomials of degree three. By replacing $f(x)$ by its cubic Taylor polynomial approximation with error term, namely

$$p_3(x) + e_3(x)$$

where

$$p_3(x) = f(0) + xf'(0) + \frac{x^2}{2!}f''(0) + \frac{x^3}{3!}f'''(0)$$

$$e_3(x) = \frac{x^4}{4!}f^{(4)}(\xi)$$

show that the error in this rule is given by

$$E = \int_{-1}^{1} e_3(x)\,dx - e_3\left(\frac{-1}{\sqrt{3}}\right) - e_3\left(\frac{1}{\sqrt{3}}\right)$$

Deduce that $|E| \leqslant 7M/270$, where

$$M = \max_{-1 \leqslant x \leqslant 1} |f^{(4)}(x)|$$

5. Verify that the polynomials

$$\varphi_0(x) = 1$$

$$\varphi_1(x) = x - \frac{1}{3}$$

$$\varphi_2(x) = x^2 - \frac{6x}{7} + \frac{3}{35}$$

are orthogonal on $[0, 1]$ with respect to the weight function $w(x) = x^{-1/2}$.
   Show that if x_0, x_1 are the roots of φ_2 then there exist numbers c_0, c_1 such that

$$\int_0^1 x^{-1/2} f(x)\,dx = c_0 f(x_0) + c_1 f(x_1)$$

for any polynomial f of degree three.
   Use this quadrature rule to estimate

$$\int_0^1 x^{-1/2} e^{x/2}\,dx$$

6. Obtain the values of x_1, c_0 and c_1 such that the quadrature rule

$$\int_0^1 f(x)\,dx \approx c_0 f(0) + c_1 f(x_1)$$

is exact for polynomials of the highest possible degree. What is this degree?

7

Ordinary differential equations: initial value problems

Many mathematical models of physical problems result in the formulation of a first order ordinary differential equation of the form

$$y' = f(x, y) \tag{7.1}$$

Here f is a given function of two real variables and y is an unknown function of the independent variable x. The general solution of (7.1) contains an arbitrary constant. In order to determine the solution uniquely, it is necessary to impose an additional condition on y. This usually takes the form

$$y(x_0) = y_0 \tag{7.2}$$

for given numbers x_0 and y_0 and is known as an *initial condition*. Problems specified by (7.1) and (7.2) are called initial value problems. Sufficient conditions for the existence and uniqueness of such problems may be found in Rao (1981).

Although analytical techniques can sometimes be used to solve differential equations, their applicability is restricted to certain special types which rarely include those arising in practice. The aim of this chapter is to derive and analyse numerical techniques for the solution of (7.1) and (7.2) at a sequence of points $x_i = x_0 + ih$. We shall let y_i denote the numerical value obtained as an approximation to the exact solution $y(x_i)$. This is illustrated in Fig. 7.1. The size of the interval h is usually called the *step length*, and the points x_i are referred to as *mesh points*. A continuous approximation to y may be determined, if desired, by interpolating the data points (x_i, y_i) as described in Chapter 5.

7.1 DERIVATION OF LINEAR MULTISTEP METHODS

We now show how the differential equation and given initial condition can be used to calculate values of y_1, y_2,... in a step-by-step manner. At

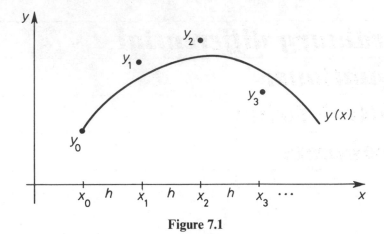

Figure 7.1

$x = x_n$ the unknown function y satisfies

$$y'(x_n) = f(x_n, y(x_n))$$

The simplest way of approximating this equation is to replace the derivative $y'(x_n)$ by its forward difference approximation $(y_{n+1} - y_n)/h$, derived in Section 6.1. This gives

$$(y_{n+1} - y_n)/h = f(x_n, y_n)$$

which can be rearranged as

$$y_{n+1} = y_n + hf(x_n, y_n) \tag{7.3}$$

Equation (7.3), known as *Euler's formula*, provides a means of computing y_{n+1} from a known value of y_n and so can be used recursively, as the following example demonstrates.

■ **Example 7.1**
Consider the use of Euler's method, with a step length $h = 0.05$, to estimate $y(0.2)$, where y satisfies

$$y' = \frac{-y^2}{1+x}; \qquad y(0) = 1$$

The above notation is illustrated diagrammatically as follows:

$$
\begin{array}{cccccc}
y_0 = 1 & y_1 = ? & y_2 = ? & y_3 = ? & y_4 = ? & y \\
\hline
x_0 = 0 & x_1 = 0.05 & x_2 = 0.1 & x_3 = 0.15 & x_4 = 0.2 & x
\end{array}
$$

For this example, equation (7.3) becomes

$$y_{n+1} = y_n - \frac{0.05 y_n^2}{1 + x_n}$$

Hence

$$y_1 = y_0 - \frac{0.05y_0^2}{1 + x_0} = 1 - \frac{0.05(1)^2}{1 + 0} = 0.95$$

$$y_2 = y_1 - \frac{0.05y_1^2}{1 + x_1} = 0.95 - \frac{0.05(0.95)^2}{1 + 0.05} = 0.907\,02$$

$$y_3 = y_2 - \frac{0.05y_2^2}{1 + x_2} = 0.907\,02 - \frac{0.05(0.907\,02)^2}{1 + 0.1} = 0.869\,63$$

$$y_4 = y_3 - \frac{0.05y_3^2}{1 + x_3} = 0.869\,63 - \frac{0.05(0.869\,63)^2}{1 + 0.15} = 0.836\,75$$

It can be verified that the analytical solution is

$$\frac{1}{1 + \ln(1 + x)}$$

and so the error in the estimated value of $y(0.2)$ is

$$0.845\,79 - 0.836\,75 = 0.009\,04. \qquad \square$$

An alternative approach to the construction of step-by-step formulas is to make use of the integration rules derived in Chapter 6. If both sides of (7.1) are integrated with respect to x between the limits x_n and x_{n+1}, then

$$y(x_{n+1}) - y(x_n) = \int_{x_n}^{x_{n+1}} f(x, y(x))\,dx \qquad (7.4)$$

because

$$\int_{x_n}^{x_{n+1}} y'(x)\,dx = [y(x)]_{x_n}^{x_{n+1}}$$

It is interesting to observe that when the integral on the right hand side of equation (7.4) is approximated by the rectangular rule (6.11), Euler's formula is obtained. A more accurate integration technique is the trapezium rule (6.14), which produces

$$y_{n+1} - y_n = \frac{h}{2}[f(x_n, y_n) + f(x_{n+1}, y_{n+1})] \qquad (7.5)$$

This is called the *trapezoidal formula*, and it enables us to calculate y_{n+1} from y_n as before. There is, however, an important difference between equations (7.3) and (7.5). In the latter case the unknown y_{n+1} is an argument of the function f and so, in general, (7.5) represents a non-linear algebraic equation which needs to be solved at each step. The most convenient way of doing this is to use the fixed point iteration of Section 3.2, namely

$$y_{n+1}^{[m+1]} = y_n + \frac{h}{2}[f(x_n, y_n) + f(x_{n+1}, y_{n+1}^{[m]})] \tag{7.6}$$

A starting value $y_{n+1}^{[0]}$ has to be supplied, which may be calculated from Euler's formula as

$$y_{n+1}^{[0]} = y_n + hf(x_n, y_n) \tag{7.7}$$

To compute y_{n+1} from a known value of y_n we first calculate $y_{n+1}^{[0]}$ using (7.7) and then iterate with (7.6) using this as the starting value. The iteration may be stopped either after a fixed number of iterations or when the quantity $|y_{n+1}^{[m+1]} - y_{n+1}^{[m]}|$ is less than a prescribed tolerance. Used in this way, Euler's formula is called a *predictor* and the trapezoidal formula a *corrector*. Although Euler's method is generally less accurate than the trapezoidal method, it does provide a reasonable approximation and so only a few iterations should be needed for the convergence of (7.6).

■ **Example 7.2**

Consider the solution of the problem given in Example 7.1 using the trapezoidal method with the same step length as before. The Euler-trapezoidal predictor-corrector pair is given by

$$y_{n+1}^{[0]} = y_n - \frac{0.05 y_n^2}{1 + x_n}$$

and

$$y_{n+1}^{[m+1]} = y_n - 0.025 \left[\frac{y_n^2}{1 + x_n} + \frac{(y_{n+1}^{[m]})^2}{1 + x_{n+1}} \right]$$

The predicted value of y_1 is

$$y_1^{[0]} = y_0 - \frac{0.05 y_0^2}{1 + x_0} = 1 - \frac{0.05(1)^2}{1 + 0} = 0.95$$

If two corrections are applied per step, then

$$y_1^{[1]} = y_0 - 0.025 \left[\frac{y_0^2}{1 + x_0} + \frac{(y_1^{[0]})^2}{1 + x_1} \right]$$

$$= 1 - 0.025 \left[\frac{(1)^2}{1 + 0} + \frac{(0.95)^2}{1 + 0.05} \right] = 0.953\,51$$

and

$$y_1^{[2]} = y_0 - 0.025 \left[\frac{y_0^2}{1 + x_0} + \frac{(y_1^{[1]})^2}{1 + x_1} \right]$$

$$= 1 - 0.025 \left[\frac{(1)^2}{1 + 0} + \frac{(0.953\,51)^2}{1 + 0.05} \right] = 0.953\,35$$

i.e. $y_1 = 0.953\,35$.

Similarly,

$$y_2^{[0]} = 0.910\,07, \qquad y_2^{[1]} = 0.912\,89, \qquad y_2^{[2]} = 0.912\,77$$

i.e. $y_2 = 0.912\,77$.

$$y_3^{[0]} = 0.874\,90, \qquad y_3^{[1]} = 0.877\,19, \qquad y_3^{[2]} = 0.877\,11,$$

i.e. $y_3 = 0.877\,11$.

$$y_4^{[0]} = 0.843\,66, \qquad y_4^{[1]} = 0.845\,56, \qquad y_4^{[2]} = 0.845\,49,$$

i.e. $y_4 = 0.845\,49$.
The error in the estimated value y_4 is 0.000 30. □

An integration technique with the same order of convergence as the trapezium rule is the mid-point rule (6.21). In order to apply this to the differential equation (7.1), we integrate over the interval $[x_{n-1}, x_{n+1}]$ of width $2h$ to obtain

$$y(x_{n+1}) - y(x_{n-1}) = \int_{x_{n-1}}^{x_{n+1}} f(x, y(x))\,dx \qquad (7.8)$$

Approximation of this integral by the mid-point rule produces the recurrence relation

$$y_{n+1} = y_{n-1} + 2hf(x_n, y_n)$$

This is called the mid-point formula and it enables us to calculate y_{n+1} in terms of two back values y_n and y_{n-1}. No special iteration is needed for the calculation of y_{n+1}, but there is a slight difficulty in getting the method started. The smallest value of n that can be taken so as to avoid negative subscripts is one. The first step is therefore given by

$$y_2 = y_0 + 2hf(x_1, y_1)$$

and, although y_0 is known, y_1 is undetermined. Consequently a different technique, such as the trapezoidal method, has to be used to estimate $y(x_1)$. Once this has been found, the mid-point method can be applied to compute $y_2, y_3, \ldots$ in turn.

■ Example 7.3
Consider the solution of the problem given in Example 7.1, using the mid-point method, with the same step length as before. The mid-point formula is

$$y_{n+1} = y_{n-1} - \frac{0.1y_n^2}{1 + x_n}$$

The trapezoidal estimate of $y(x_1)$, obtained in Example 7.2, is 0.953 35.

Hence

$$y_2 = y_0 - \frac{0.1y_1^2}{1 + x_1} = 1 - \frac{0.1(0.953\,35)^2}{1 + 0.05} = 0.913\,44$$

Similarly,

$$y_3 = 0.877\,50, \qquad y_4 = 0.846\,48$$

The error in the estimate y_4 is $-0.000\,69$.   □

The three formulas derived so far in this section for the step-by-step solution of initial value problems are all special cases of

$$y_{n+1} = \sum_{i=1}^{j} \alpha_i y_{n+1-i} + h \sum_{i=0}^{j} \beta_i f_{n+1-i} \qquad (7.9)$$

where $f_k = f(x_k, y_k)$ and α_k, β_k are constants specifying the method. Algorithms defined by (7.9) are called *j-step linear multistep methods*, since they express y_{n+1} as a linear combination of y_{n+1-i} and f_{n+1-i}. These methods are said to be *explicit* if $\beta_0 = 0$ and *implicit* if $\beta_0 \neq 0$, where β_0 is the coefficient of f_{n+1}. In this terminology, Euler's method is 1-step explicit, the trapezoidal method is 1-step implicit and the mid-point method is 2-step explicit.

Equation (7.9) provides a formula for the calculation of y_{n+1} in terms of j back values $y_n, y_{n-1}, \ldots, y_{n+1-j}$. When $j > 1$ a different technique has to be used to determine the starting values $y_1, y_2, \ldots, y_{j-1}$. Clearly any 1-step method, such as the trapezoidal method, can be used for this purpose. However, it is essential that the accuracy of the starting procedure is at least as high as that of the main method. It will be shown in the next section that 1-step methods tend to be less accurate than j-step ($j > 1$) methods. It may, therefore, be necessary to use a considerably smaller step length in the starting procedure than in the main method to attain a comparable accuracy. An alternative approach is to construct the Taylor series expansion of $y(x)$ about $x = x_0$ and hence calculate $y(x_r)$ ($r = 1, 2, \ldots, j-1$) from

$$y(x_r) = y(x_0 + rh) = y(x_0) + rhy'(x_0) + \frac{(rh)^2}{2!} y''(x_0) + \frac{(rh)^3}{3!} y'''(x_0) + \cdots$$

The value of $y(x_0)$ is the given initial condition and the values of $y'(x_0)$, $y''(x_0), \ldots$ can be obtained from the differential equation and its derivatives.

■ **Example 7.4**
As an illustration of the use of Taylor series, consider the calculation of $y(x_1)$ and $y(x_2)$, correct to 5D, for the initial value problem

$$y' = \frac{-y^2}{1 + x}; \qquad y(0) = 1$$

using a step length $h = 0.05$. Here

$$x_0 = 0, \qquad y(0) = 1, \qquad y'(0) = \frac{-[y(0)]^2}{1 + 0} = -1$$

To find y'' we differentiate both sides of the differential equation with respect to x, making use of the quotient and chain rules, to obtain

$$y'' = \frac{-2yy'(1+x) - (-y^2)}{(1+x)^2}$$

This expression simplifies to $(2y^3 + y^2)/(1+x)^2$ because $y' = -y^2/(1+x)$. Hence

$$y''(0) = \frac{2[y(0)]^3 + [y(0)]^2}{(1+0)^2} = 3$$

Differentiating again gives

$$y''' = \frac{(6y^2y' + 2yy')(1+x)^2 - 2(1+x)(2y^3 + y^2)}{(1+x)^4} = -\frac{6y^4 + 6y^3 + 2y^2}{(1+x)^3}$$

and so $y'''(0) = -14$.

This process can be repeated to obtain

$$y^{(4)}(0) = 88, \qquad y^{(5)}(0) = -694, \qquad y^{(6)}(0) = 6578, \qquad y^{(7)}(0) = -72\,792$$

Hence

$$y(x_r) = 1 - rh + \tfrac{3}{2}(rh)^2 - \tfrac{14}{6}(rh)^3 + \tfrac{88}{24}(rh)^4$$
$$- \tfrac{694}{120}(rh)^5 + \tfrac{6578}{720}(rh)^6 - \tfrac{72\,792}{5040}(rh)^7 + \cdots$$

Taking $h = 0.05$ and $r = 1$ gives

$$y(x_1) = 1 - 0.05 + 0.003\,75 - 0.000\,291\,7 + 0.000\,022\,9$$
$$- 0.000\,001\,8 + 0.000\,000\,1 - 0.000\,000\,0 + \cdots$$

Notice that successive terms decrease in magnitude with alternating sign. Hence the error incurred in truncating this series after a finite number of terms does not exceed the first neglected term. In this case, we can sum the first five terms to get

$$y(x_1) = 0.953\,48 \qquad (5\mathrm{D})$$

Taking $h = 0.05$ and $r = 2$ gives

$$y(x_2) = 1 - 0.1 + 0.015 - 0.002\,333\,3 + 0.000\,366\,7$$
$$- 0.000\,057\,8 + 0.000\,009\,1 - 0.000\,001\,4 + \cdots$$

This time we need to sum the first seven terms to obtain 5D accuracy, which gives

$$y(x_2) = 0.912\,98 \qquad (5\mathrm{D}) \qquad\qquad \square$$

There are two disadvantages of the Taylor series method. Firstly, it is extremely tedious to perform the differentiation, particularly if f is complicated. Secondly, Taylor series only provide accurate approximations close to the points about which they are expanded, and so when a larger number of starting values are required this approach may be inappropriate. To a certain extent this difficulty can be overcome by expanding about the points $x_0, x_1, \ldots, x_{j-2}$ in turn; we expand about x_0 to calculate $y(x_1)$, then expand about x_1 to calculate $y(x_2)$, and so on until $y(x_{j-1})$ is determined. Further information about the Taylor series method is given in Lapidus and Seinfeld (1971).

When a linear multistep formula is implicit, equation (7.9) represents a non-linear algebraic equation for y_{n+1}, and is solved using the iteration

$$y_{n+1}^{[m+1]} = h\beta_0 f(x_{n+1}, y_{n+1}^{[m]}) + \sum_{i=1}^{j} (\alpha_i y_{n+1-i} + h\beta_i f_{n+1-i}) \qquad (7.10)$$

called the corrector. An initial estimate $y_{n+1}^{[0]}$ may be calculated using any explicit formula, which in this context is called the predictor.

A family of *predictor-corrector pairs* can be derived using the backward difference interpolating polynomials considered in Section 5.3. From equations (7.1) and (7.4),

$$y(x_{n+1}) - y(x_n) = \int_{x_n}^{x_{n+1}} y'(x)\,dx \qquad (7.11)$$

which may be written as

$$y(x_{n+1}) - y(x_n) = h \int_0^1 y'(x_n + sh)\,ds$$

using the substitution $x = x_n + sh$. If $y'(x_n + sh)$ is replaced by its backward difference expansion (5.12), then

$$y_{n+1} - y_n = h \int_0^1 \left\{ y_n' + s\nabla y_n' + \frac{s(s+1)}{2!}\nabla^2 y_n' + \frac{s(s+1)(s+2)}{3!}\nabla^3 y_n' + \cdots \right\} ds$$

$$= h\{y_n' + \tfrac{1}{2}\nabla y_n' + \tfrac{5}{12}\nabla^2 y_n' + \tfrac{3}{8}\nabla^3 y_n' + \cdots\}$$

Terminating this expansion after the mth difference and replacing y_n' by f_n gives, for $m = 0, 1, 2$ and 3,

$$y_{n+1} = y_n + hf_n$$

$$y_{n+1} = y_n + \frac{h}{2}\{3f_n - f_{n-1}\}$$

$$y_{n+1} = y_n + \frac{h}{12}\{23f_n - 16f_{n-1} + 5f_{n-2}\}$$

$$y_{n+1} = y_n + \frac{h}{24}\{55f_n - 59f_{n-1} + 37f_{n-2} - 9f_{n-3}\}$$

These formulas define explicit methods and are known as *Adams–Bashforth techniques*.

Implicit methods can be derived in an analogous way. Using the substitution $x = x_{n+1} + sh$ in (7.11), we obtain

$$y(x_{n+1}) - y(x_n) = h \int_{-1}^{0} y(x_{n+1} + sh)\, ds$$

Proceeding as before gives

$$y_{n+1} - y_n = h \int_{-1}^{0} \left\{ y'_{n+1} + s\nabla y'_{n+1} + \frac{s(s+1)}{2!} \nabla^2 y'_{n+1} \right.$$

$$\left. + \frac{s(s+1)(s+2)}{3!} \nabla^3 y'_{n+1} + \cdots \right\} ds$$

$$= h\{ y'_{n+1} - \tfrac{1}{2}\nabla y'_{n+1} - \tfrac{1}{12}\nabla^2 y'_{n+1} - \tfrac{1}{24}\nabla^3 y'_{n+1} - \cdots \}$$

which produces

$$y_{n+1} = y_n + h f_{n+1}$$

$$y_{n+1} = y_n + \frac{h}{2}\{ f_{n+1} + f_n \}$$

$$y_{n+1} = y_n + \frac{h}{12}\{ 5 f_{n+1} + 8 f_n - f_{n-1} \}$$

$$y_{n+1} = y_n + \frac{h}{24}\{ 9 f_{n+1} + 19 f_n - 5 f_{n-1} + f_{n-2} \}$$

These formulas define implicit methods and are known as *Adams–Moulton techniques*. It will be shown in the next section that the errors associated with the *j*-step Adams–Bashforth and $(j-1)$-step Adams–Moulton methods are roughly proportional to h^j, with the latter method having the smaller constant of proportionality. It therefore makes sense to couple these formulas as predictor-corrector pairs. Two of the most popular methods of this type are the cases $j = 3$ and $j = 4$, i.e.

$$y_{n+1}^{[0]} = y_n + \frac{h}{12}\{ 23 f_n - 16 f_{n-1} + 5 f_{n-2} \}$$

$$y_{n+1}^{[m+1]} = y_n + \frac{h}{12}\{ 5 f(x_{n+1}, y_{n+1}^{[m]}) + 8 f_n - f_{n-1} \}$$

and

$$y_{n+1}^{[0]} = y_n + \frac{h}{24}\{ 55 f_n - 59 f_{n-1} + 37 f_{n-2} - 9 f_{n-3} \}$$

$$y_{n+1}^{[m+1]} = y_n + \frac{h}{24}\{ 9 f(x_{n+1}, y_{n+1}^{[m]}) + 19 f_n - 5 f_{n-1} + f_{n-2} \}$$

■ **Example 7.5**

Consider the solution of the problem specified in Example 7.1, using the 3-step Adams–Bashforth and 2-step Adams–Moulton formulas with the same step length as before. The predictor-corrector pair is given by

$$y_{n+1}^{[0]} = y_n - \frac{0.05}{12}\left\{\frac{23y_n^2}{1+x_n} - \frac{16y_{n-1}^2}{1+x_{n-1}} + \frac{5y_{n-2}^2}{1+x_{n-2}}\right\}$$

$$y_{n+1}^{[m+1]} = y_n - \frac{0.05}{12}\left\{\frac{5(y_{n+1}^{[m]})^2}{1+x_{n+1}} + \frac{8y_n^2}{1+x_n} - \frac{y_{n-1}^2}{1+x_{n-1}}\right\}$$

From Example 7.4, $y(x_1) = 0.953\,48$ and $y(x_2) = 0.912\,98$ (5D). Hence

$$y_3^{[0]} = y_2 - \frac{0.05}{12}\left\{\frac{23y_2^2}{1+x_2} - \frac{16y_1^2}{1+x_1} + \frac{5y_0^2}{1+x_0}\right\}$$

$$= 0.912\,98 - \frac{0.05}{12}\left\{\frac{23(0.912\,98)^2}{1+0.1} - \frac{16(0.953\,48)^2}{1+0.05} + \frac{5(1)^2}{1+0}\right\}$$

$$= 0.877\,25$$

Now

$$y_3^{[m+1]} = y_2 - \frac{0.05}{12}\left\{\frac{5(y_3^{[m]})^2}{1+x_3} + \frac{8y_2^2}{1+x_2} - \frac{y_1^2}{1+x_1}\right\}$$

$$= 0.912\,98 - \frac{0.05}{12}\left\{\frac{5(y_3^{[m]})^2}{1+0.15} + \frac{8(0.912\,98)^2}{1+0.1} - \frac{(0.953\,48)^2}{1+0.05}\right\}$$

$$= 0.891\,33 - 0.018\,12(y_3^{[m]})^2$$

If two corrections are applied per step, then

$$y_3^{[1]} = 0.891\,33 - 0.018\,12(0.877\,25)^2 = 0.877\,39$$
$$y_3^{[2]} = 0.891\,33 - 0.018\,12(0.877\,39)^2 = 0.877\,38$$

i.e. $y_3 = 0.877\,38$.

Similarly,

$$y_4^{[0]} = 0.845\,71, \qquad y_4^{[1]} = 0.845\,80, \qquad y_4^{[2]} = 0.845\,80$$

i.e. $y_4 = 0.845\,80$.

The error in the estimate y_4 is $-0.000\,01$.   □

EXERCISE 7.1

1. Use Euler's method to estimate the solution of

$$y' = (1-x)y^2 - y; \qquad y(0) = 1$$

at $x = 1$, using (a) $h = 0.1$ (b) $h = 0.01$ (c) $h = 0.001$.

Verify, by direct substitution into the differential equation, that the analytical solution is

$$y(x) = \frac{1}{e^x - x}$$

Hence write down the errors of the estimates obtained in parts (a), (b) and (c).

It is known that the error associated with Euler's method is roughly proportional to h^p, where p is an integer. Use your results to deduce the value of p, and hence find the largest value of h which could be used to estimate $y(1)$ correct to 8D.

2. Use the Euler-trapezoidal predictor-corrector pair to estimate the solution of

$$y' = \frac{-y^2}{1+x}; \qquad y(0) = 1$$

at $x = 1$, with $h = 0.1$, applying (a) 0 (b) 1 (c) 2 (d) 3 corrections per step. Find the corresponding errors in these estimates, and hence write down the optimal number of corrections in this case.

3. Consider the solution of the non-linear algebraic equation

$$y_{n+1} = h\beta_0 f(x_{n+1}, y_{n+1}) + \sum_{i=1}^{j} (\alpha_i y_{n+1-i} + h\beta_i f_{n+1-i}) \qquad (1)$$

using the iteration

$$y_{n+1}^{[m+1]} = h\beta_0 f(x_{n+1}, y_{n+1}^{[m]}) + \sum_{i=1}^{j} (\alpha_i y_{n+1-i} + h\beta_i f_{n+1-i}) \qquad (2)$$

If $|\partial f/\partial y| \leq K$ for some constant K, prove that $y_{n+1}^{[m]} \to y_{n+1}$ as $m \to \infty$, provided $h < 1/K|\beta_0|$.

4. Consider the application of a single step of the Euler-trapezoidal predictor-corrector pair to the initial value problem

$$y' = (1-x)y^2 - y; \qquad y(0) = 1$$

Find the number of iterations required for the convergence of the corrector in the case when (a) $h = 0.01$ (b) $h = 0.1$ (c) $h = 1$ using the stopping criterion

$$|y_1^{[m+1]} - y_1^{[m]}| \leq \tfrac{1}{2} \times 10^{-4}$$

Explain the general behaviour exhibited by these results.

5. Use an appropriate integration rule to derive the recurrence relation

$$y_{n+1} = y_{n-1} + \frac{h}{3}(f_{n+1} + 4f_n + f_{n-1})$$

for the step-by-step solution of initial value problems.

6. Construct the Taylor polynomial approximation of degree 7 about $x = 0$ for the solution of the initial value problem

$$y' = -y^3 - y + e^{-3x}; \qquad y(0) = 1$$

Hence find the values of $y(0.1)$, $y(0.2)$ and $y(0.3)$, correct to 6D. Use the 4-step Adams–Bashforth and 3-step Adams–Moulton methods as a predictor-corrector pair to extend the numerical solution to $x = 0.4$, performing exactly two corrections.

7.2 ANALYSIS OF LINEAR MULTISTEP METHODS

We have derived several linear multistep methods for the numerical solution of initial value problems. In this section we investigate the stability and accuracy of these techniques. However, it is not possible in a text of this length to produce a comprehensive treatment of the theory of linear multistep methods, and the interested reader is advised to consult the books by Henrici (1962) and Lambert (1973).

In order to motivate the definition of stability, consider the solution of

$$y' = -\lambda y; \qquad y(0) = \alpha \tag{7.12}$$

for some positive constant λ, using Euler's formula on a computer. In theory the estimates y_i satisfy

$$y_{n+1} = y_n - \lambda h y_n; \qquad y_0 = \alpha \tag{7.13}$$

In practice, however, the computer will introduce rounding errors into the calculations. For simplicity, suppose that there is a rounding error ε in the stored value of y_0 and that all subsequent calculations are performed exactly. The computed estimates, $\bar{y}_i$ say, satisfy

$$\bar{y}_{n+1} = \bar{y}_n - \lambda h \bar{y}_n; \qquad \bar{y}_0 = \alpha + \varepsilon \tag{7.14}$$

Hence if we let $\varepsilon_i = y_i - \bar{y}_i$, we can subtract (7.14) from (7.13) to deduce that

$$\varepsilon_{n+1} = \varepsilon_n - \lambda h \varepsilon_n; \qquad \varepsilon_0 = \varepsilon \tag{7.15}$$

Comparison of (7.13) with (7.15) shows that rounding errors are propagated in exactly the same way as the exact solution of Euler's formula and, since it is essential that the growth in rounding errors is bounded as $n \to \infty$, we are led to the following definition of stability.

■ **Definition 7.1**
A linear multistep method is *stable* if the solution y_n of the linear multistep formula is bounded as $n \to \infty$. □

For example, note from (7.13) that y_{n+1} is obtained by multiplying y_n by

$(1 - \lambda h)$. After n steps we have $y_n = (1 - \lambda h)^n \alpha$, since $y_0 = \alpha$. Hence y_n is bounded if and only if $|1 - \lambda h| \leqslant 1$, or equivalently

$$-1 \leqslant 1 - \lambda h \leqslant 1$$

The right hand inequality is true for all h since $\lambda > 0$, and the left hand inequality is true provided $h \leqslant 2/\lambda$. Therefore Euler's method is stable if and only if $h \leqslant 2/\lambda$. Methods which are stable for a restricted range of step lengths are called partially stable. More formally:

■ **Definition 7.2**
A linear multistep method which is stable if $h \leqslant \bar{h}$ and unstable if $h > \bar{h}$ is said to be *partially stable*. □

Euler's method is partially stable with $\bar{h} = 2/\lambda$.
    As a second example, consider the solution of (7.12) using the trapezoidal method, namely

$$y_{n+1} = y_n + \tfrac{1}{2}h(- \lambda y_n - \lambda y_{n+1})$$

which can be rearranged as

$$y_{n+1} = \left(\frac{1 - \lambda h/2}{1 + \lambda h/2} \right) y_n$$

This has solution

$$y_n = \left(\frac{1 - \lambda h/2}{1 + \lambda h/2} \right)^n \alpha$$

and is bounded if and only if

$$\left| \frac{1 - \lambda h/2}{1 + \lambda h/2} \right| \leqslant 1$$

This inequality is true for all values of h because $\lambda > 0$. Methods of this type are called A-stable. More formally:

■ **Definition 7.3**
A linear multistep method which is stable for all $h \geqslant 0$ is said to be *A-stable*. □

The stability properties of Euler's method and the trapezoidal method are fairly typical. Implicit methods generally have larger intervals of stability than explicit methods. This means that although implicit methods require more computational effort per step (owing to the corrector iteration), it is possible to use a larger step length and so reach any particular value of x in fewer steps.

In a sense, A-stability is an extreme case of partial stability with $\bar{h} = \infty$. At the other end of the spectrum we have:

■ **Definition 7.4**
A linear multistep method which is stable for $h = 0$ is said to be *zero stable*.   □

Clearly zero stability is not a great deal to ask of a linear multistep method, and methods which fail to satisfy this property are useless.

So far we have only considered the stability of 1-step methods. We now extend our investigation to 2-step methods and, since zero stability is the easiest type of stability to check, we consider this first. From (7.9), the general 2-step formula is

$$y_{n+1} - \alpha_1 y_n - \alpha_2 y_{n-1} = h\{\beta_0 f_{n+1} + \beta_1 f_n + \beta_2 f_{n-1}\} \tag{7.16}$$

and putting $h = 0$ gives

$$y_{n+1} - \alpha_1 y_n - \alpha_2 y_{n-1} = 0 \tag{7.17}$$

This is a linear difference equation with constant coefficients, and has an auxiliary equation

$$m^2 - \alpha_1 m - \alpha_2 = 0 \tag{7.18}$$

If the roots of this quadratic are m_1 and m_2, then the general solution of (7.17) is

$$y_n = \begin{cases} A(m_1)^n + B(m_2)^n & m_1 \neq m_2 \\ (A + Bn)(m_1)^n & m_1 = m_2 \end{cases}$$

for arbitrary constants A and B. In the case of distinct roots, y_n is bounded if and only if $|m_i| \leqslant 1$ $(i = 1, 2)$, whereas in the case of equal roots we require $|m_i| < 1$.

■ **Example 7.6**
Consider the linear multistep formula

$$y_{n+1} - (1 - \alpha_2)y_n - \alpha_2 y_{n-1} = h\{\beta_0 f_{n+1} + \beta_1 f_n + \beta_2 f_{n-1}\}$$

To investigate zero stability, h is set to zero, i.e.

$$y_{n+1} - (1 - \alpha_2)y_n - \alpha_2 y_{n-1} = 0$$

The auxiliary equation is

$$m^2 - (1 - \alpha_2)m - \alpha_2 = 0$$

with roots $m = 1$ and $-\alpha_2$. This formula is therefore zero stable, provided the parameter α_2 lies in the interval $(-1, 1]$. The value $\alpha_2 = -1$ is excluded, since the auxiliary equation would then have two equal roots with unit modulus.   □

We now consider the partial stability of 2-step methods. Unfortunately, for a general function f, (7.16) represents a non-linear difference equation, and it is impossible to solve such equations in the sense of providing a formula for y_n in terms of n. Consequently we restrict our attention to the linear test equation $y' = -\lambda y \, (\lambda > 0)$, with analytical solution $y = Ae^{-\lambda x}$ for any arbitrary constant A. (We are not particularly interested in the case $\lambda < 0$, since then $y \to \infty$ as $x \to \infty$ and an unbounded growth in rounding errors would probably be regarded as acceptable.)

Applying (7.16) to $y' = -\lambda y$ gives

$$y_{n+1} - \alpha_1 y_n - \alpha_2 y_{n-1} = h\{-\beta_0 \lambda y_{n+1} - \beta_1 \lambda y_n - \beta_2 \lambda y_{n-1}\}$$

This can be rearranged as

$$(\beta_0 \lambda h + 1)y_{n+1} + (\beta_1 h\lambda - \alpha_1)y_n + (\beta_2 h\lambda - \alpha_2)y_{n-1} = 0$$

and has auxiliary equation

$$(\beta_0 \lambda h + 1)m^2 + (\beta_1 h\lambda - \alpha_1)m + (\beta_2 h\lambda - \alpha_2) = 0$$

Now, although it is possible to invoke the formula for solving a quadratic equation at this stage, the actual expressions for the roots are rather complicated. Indeed, we are not particularly interested in finding the roots themselves, but only in investigating conditions on h which ensure that they are bounded by one in modulus. A very simple and elegant way of doing this is provided by the following theorem.

■ **Theorem 7.1**
The roots of the quadratic equation

$$Q(x) = ax^2 + bx + c = 0 \qquad a > 0$$

are bounded by one in modulus if and only if

$$Q(1) \geqslant 0, \qquad Q(-1) \geqslant 0, \qquad |c/a| \leqslant 1$$

■ **Proof**
To show the necessity of these conditions, note that since $a > 0$, $Q(x) > 0$ for sufficiently large positive and negative values of x. It follows from the intermediate value theorem that if $Q(1) < 0$ or $Q(-1) < 0$ then Q has a root outside the interval $[-1, 1]$. Moreover, since c/a equals the product of the roots, the condition $|c/a| > 1$ means that at least one of the roots has a modulus exceeding unity.

To show the sufficiency of these conditions, we consider the real and complex cases separately.

Case 1　If the roots m_i $(i = 1, 2)$ are complex then $m_1 = \bar{m}_2$. Hence

$$|c/a| = m_1 m_2 = m_1 \bar{m}_1 = |m_1|^2$$

and the condition $|c/a| \leqslant 1$ ensures that $|m_i| \leqslant 1$.

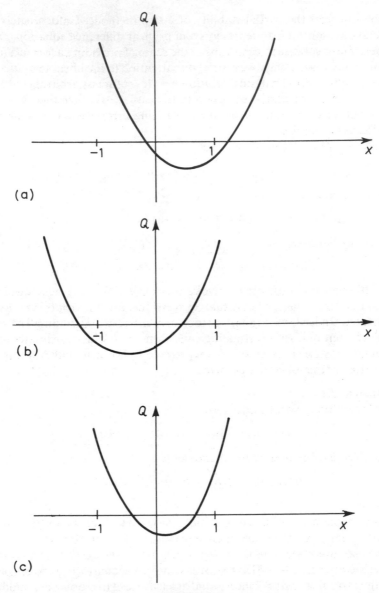

Figure 7.2

Case 2   If the roots are real, the condition $|c/a| \leqslant 1$ means that at least one of the roots has a modulus less than unity. Hence the graph of Q must take one of the three forms indicated in Fig. 7.2(a)–(c). However, the other two conditions, $Q(1) \geqslant 0$ and $Q(-1) \geqslant 0$, rule out the possibility of graphs (a) and (b). Therefore Q takes the form of graph (c), with $|m_i| \leqslant 1$ for $i = 1$ and 2.   □

■ **Example 7.7**

As an illustration of the use of Theorem 7.1, consider the 2-step Adams–Bashforth method

$$y_{n+1} = y_n + \frac{h}{2}\{3f_n - f_{n-1}\}$$

which, when applied to the equation $y' = -\lambda y$, is given by

$$y_{n+1} + \left(\frac{3\lambda h}{2} - 1\right) y_n - \frac{\lambda h}{2} y_{n-1} = 0$$

The corresponding auxiliary equation is

$$Q(m) = m^2 + \left(\frac{3\lambda h}{2} - 1\right) m - \frac{\lambda h}{2} = 0$$

and so

$$Q(1) = \lambda h, \qquad Q(-1) = 2 - 2\lambda h$$

Clearly $Q(1) \geqslant 0$ for all h, and $Q(-1) \geqslant 0$ if and only if $h \leqslant 1/\lambda$. Furthermore, it is easy to see that the condition $|c/a| \leqslant 1$ is automatically satisfied when $h \leqslant 1/\lambda$ because $|c/a| = \lambda h/2 \leqslant 1/2$. Hence we can deduce that the 2-step Adams–Bashforth method is partially stable with $\bar{h} = 1/\lambda$.   □

So far no mention has been made of the accuracy of a linear multistep method. We would clearly like the error $y(x_n) - y_n$ to converge to zero as $h \to 0$ and, moreover, would like this to occur at a reasonable rate. The local truncation error gives us a simple way of quantifying this behaviour for any particular method, and is defined as follows.

■ **Definition 7.5**

If $y(x_i)$ denotes the exact solution of the initial value problem at $x = x_i$, then the *local truncation error* of (7.9) is given by

$$T = y(x_{n+1}) - \sum_{i=1}^{j} \alpha_i y(x_{n+1-i}) - h \sum_{i=0}^{j} \beta_i y'(x_{n+1-i}) \qquad \square$$

The local truncation error is the difference in the two sides of a linear multistep formula when the numerical solution y_i is replaced by the analytical solution $y(x_i)$. Consequently, it measures the extent to which the solution of the differential equation fails to satisfy the approximating difference equation. Each term in the expression for T can be expanded as a power series in h about $x = x_n$ (or some other convenient point). The first few terms in the expansion of T usually cancel to give an expression of the form

$$T = A y^{(p+1)}(x_n) h^{p+1} + O(h^{p+2})$$

for some integer p, where A is a constant independent of h.

■ **Definition 7.6**

If the leading term of the local truncation error is $Ay^{(p+1)}(x_n)h^{p+1}$, then the method is said to be *consistent of order p* and $|A|$ is referred to as the *error constant.*   □

■ **Example 7.8**

As an example of the foregoing, consider the 3-step Adams–Moulton formula

$$y_{n+1} = y_n + \frac{h}{24}\{9f_{n+1} + 19f_n - 5f_{n-1} + f_{n-2}\}$$

The local truncation error is given by

$$T = y(x_{n+1}) - y(x_n) - \frac{h}{24}\{9y'(x_{n+1}) + 19y'(x_n) - 5y'(x_{n-1}) + y'(x_{n-2})\}$$

From Taylor's theorem,

$$y(x_{n+1}) = y(x_n + h)$$
$$= y(x_n) + hy'(x_n) + \frac{h^2}{2!}y''(x_n) + \frac{h^3}{3!}y'''(x_n) + \frac{h^4}{4!}y^{(4)}(x_n) + \frac{h^5}{5!}y^{(5)}(x_n) + \cdots$$

$$y'(x_{n+1}) = y'(x_n + h)$$
$$= y'(x_n) + hy''(x_n) + \frac{h^2}{2!}y'''(x_n) + \frac{h^3}{3!}y^{(4)}(x_n) + \frac{h^4}{4!}y^{(5)}(x_n) + \frac{h^5}{5!}y^{(6)}(x_n) + \cdots$$

$$y'(x_{n-1}) = y'(x_n - h)$$
$$= y'(x_n) - hy''(x_n) + \frac{h^2}{2!}y'''(x_n) - \frac{h^3}{3!}y^{(4)}(x_n) + \frac{h^4}{4!}y^{(5)}(x_n) - \frac{h^5}{5!}y^{(6)}(x_n) + \cdots$$

$$y'(x_{n-2}) = y'(x_n - 2h)$$
$$= y'(x_n) - 2hy''(x_n) + \frac{4h^2}{2!}y'''(x_n) - \frac{8h^3}{3!}y^{(4)}(x_n) + \frac{16h^4}{4!}y^{(5)}(x_n)$$
$$- \frac{32h^5}{5!}y^{(6)}(x_n) + \cdots$$

If these expansions are substituted into the expression for T and $y^{(r)}(x_n)$ is replaced by $y^{(r)}$ for clarity, then

$$T = y + hy' + \frac{h^2}{2!}y'' + \frac{h^3}{3!}y''' + \frac{h^4}{4!}y^{(4)} + \frac{h^5}{5!}y^{(5)} + \cdots$$
$$- y$$
$$- \frac{h}{24}\left\{9\left(y' + hy'' + \frac{h^2}{2!}y''' + \frac{h^3}{3!}y^{(4)} + \frac{h^4}{4!}y^{(5)} + \frac{h^5}{5!}y^{(6)} + \cdots\right)\right.$$
$$+ 19y'$$

$$-5\left(y'-hy''+\frac{h^2}{2!}y'''-\frac{h^3}{3!}y^{(4)}+\frac{h^4}{4!}y^{(5)}-\frac{h^5}{5!}y^{(6)}+\cdots\right)$$
$$+\left(y'-2hy''+2h^2y'''-\frac{4h^3}{3}y^{(4)}+\frac{2h^4}{3}y^{(5)}-\frac{4h^5}{15}y^{(6)}+\cdots\right)\Bigg\}$$

It is easy to check that all terms up to and including those involving h^4 cancel, leaving

$$T=-\frac{19h^5}{720}y^{(5)}+O(h^6)$$

The 3-step Adams–Moulton method is therefore consistent of order four with error constant 19/720. By the same reasoning it is easy to verify that the 4-step Adams–Bashforth method is also consistent of order four, but with the larger error constant 251/720. This confirms the remark made at the end of the previous section concerning the relative accuracy of the Adams–Bashforth and Adams–Moulton methods.   □

Intuitively, one would expect the local truncation error to represent the error incurred by applying a single step of a linear multistep method. In calculating y_n, (approximately) n steps are used. Hence if $T=O(h^{p+1})$, the error in y_n is

$$nO(h^{p+1})=nhO(h^p)=(x_n-x_0)O(h^p)$$

Therefore, if $h\to0$ with x_n fixed, the global error $y(x_n)-y_n$ is $O(h^p)$.

■ **Definition 7.7**
A linear multistep method is said to be *convergent* of order p if the error

$$y(x_n)-y_n=O(h^p)$$

as $h\to0$ with x_n fixed.   □

We have given an intuitive proof that if a method is consistent of order p then it is convergent of order p. In fact, the following theorem can be proved rigorously (see Henrici, 1962).

■ **Theorem 7.2**
A linear multistep method is convergent of order p if and only if it is consistent of order p and zero stable.   □

It is assumed in Theorem 7.2 that the starting values are either exact or are obtained using a method which has an order of convergence of at least p. Note the zero stability condition in Theorem 7.2. An apparently attractive approach for the derivation of high order formulas would be to deliberately choose the coefficients in (7.9) so as to annihilate as many terms as possible in the local truncation error. Unfortunately, methods obtained in this way are rarely zero stable, as the following example demonstrates.

■ **Example 7.9**

Consider the construction of an explicit 2-step method defined by

$$y_{n+1} - \alpha_1 y_n - \alpha_2 y_{n-1} = h\{\beta_1 f_n + \beta_2 f_{n-1}\}$$

This formula has four free parameters, so one might hope to be able to choose these to annihilate the terms of order h^0, h^1, h^2 and h^3 in the local truncation error and so derive a method which is convergent of order three. The local truncation error is

$$T = y(x_{n+1}) - \alpha_1 y(x_n) - \alpha_2 y(x_{n-1}) - h\{\beta_1 y'(x_n) + \beta_2 y'(x_{n-1})\}$$

$$= y(x_n) + hy'(x_n) + \frac{h^2}{2!} y''(x_n) + \frac{h^3}{3!} y'''(x_n) + \cdots$$

$$- \alpha_1 y(x_n)$$

$$- \alpha_2 \left\{ y(x_n) - hy'(x_n) + \frac{h^2}{2!} y''(x_n) - \frac{h^3}{3!} y'''(x_n) + \cdots \right\}$$

$$- h\beta_1 y'(x_n)$$

$$- h\beta_2 \left\{ y'(x_n) - hy''(x_n) + \frac{h^2}{2!} y'''(x_n) - \cdots \right\}$$

The first term in this expression is $(1 - \alpha_1 - \alpha_2)y(x_n)$, which can be set to zero by choosing $\alpha_1 = 1 - \alpha_2$. The linear multistep formula then becomes

$$y_{n+1} - (1 - \alpha_2)y_n - \alpha_2 y_{n-1} = h\{\beta_1 f_n + \beta_2 f_{n-1}\}$$

From Example 7.6, this method is zero stable if and only if $\alpha_2 \in (-1, 1]$.
    The first few terms in the local truncation error are

$$(1 + \alpha_2 - \beta_1 - \beta_2)hy'(x_n) + \left(\frac{1}{2} - \frac{\alpha_2}{2} + \beta_2 \right)h^2 y''(x_n) + \left(\frac{1}{6} + \frac{\alpha_2}{6} - \frac{\beta_2}{2} \right)h^3 y'''(x_n)$$

which are zero provided

$$\alpha_2 - \beta_1 - \beta_2 = -1$$

$$-\frac{\alpha_2}{2} + \beta_2 = -\frac{1}{2}$$

$$\frac{\alpha_2}{6} - \frac{\beta_2}{2} = -\frac{1}{6}$$

Eliminating β_2 from the last two equations gives $\alpha_2 = 5 \notin (-1, 1]$ and so the method is not zero stable. The best we can do is to solve the first two equations and so construct a one parameter family of second order convergent methods,

$$y_{n+1} - (1 - \alpha_2)y_n - \alpha_2 y_{n-1} = \frac{h}{2}\{(1 + \alpha_2)f_n - (1 - \alpha_2)f_{n-1}\}$$

with $\alpha_2 \in (-1, 1]$.   □

EXERCISE 7.2

1. Show that the Adams–Bashforth and Adams–Moulton methods are all zero stable.

2. Investigate the partial stability of the following linear multistep formulas when used to solve $y' = -\lambda y \; (\lambda > 0)$:

   (a) $y_{n+1} = y_n + h f_{n+1}$

   (b) $y_{n+1} = y_n + \dfrac{h}{12}\{5 f_{n+1} + 8 f_n - f_{n-1}\}$

   (c) $y_{n+1} = y_{n-1} + 2 h f_n$

3. Find the leading term in the local truncation error of each of the following linear multistep formulas:

   (a) $y_{n+1} = y_n + h f_n$

   (b) $y_{n+1} = y_n + \dfrac{h}{2}\{f_{n+1} + f_n\}$

   (c) $y_{n+1} = y_n + \dfrac{h}{12}\{5 f_{n+1} + 8 f_n - f_{n-1}\}$

   (d) $y_{n+1} = y_n + \dfrac{h}{12}\{23 f_n - 16 f_{n-1} + 5 f_{n-2}\}$

4. Investigate the partial stability of the Euler-trapezoidal predictor-corrector pair when used to solve $y' = -\lambda y \; (\lambda > 0)$ in the case when exactly one correction is applied per step.

5. Show that the linear multistep formula

$$y_{n+1} - \tfrac{1}{2} y_n - \tfrac{1}{2} y_{n-1} = h\{\beta_0 f_{n+1} + \beta_1 f_n + \beta_2 f_{n-1}\}$$

is zero stable.

   Determine the coefficients β_0, β_1 and β_2 such that this formula has the highest possible order of consistency in the case when it is (a) explicit (b) implicit. Investigate the partial stability of these two formulas when used to solve $y' = -\lambda y \; (\lambda > 0)$.

6. The initial value problem

$$y' = \lambda y \; (\lambda > 0); \qquad y(0) = 1$$

has an exact solution $y(x) = e^{-\lambda x}$. When applied to this problem the following formulas are stable if and only if the step length h satisfies

$$h \leqslant A/\lambda$$

for some number A. In each case determine the value of A, correct to 1D, by taking $\lambda = 1$ and experimenting with various values of h. Use the exact solution to obtain the starting values and perform about 30 steps.

(a) $y_{n+1} = y_n + \dfrac{h}{12}\{23f_n - 16f_{n-1} + 5f_{n-2}\}$

(b) $y_{n+1} = y_n + \dfrac{h}{24}\{55f_n - 59f_{n-1} + 37f_{n-2} - 9f_{n-3}\}$

(c) $y_{n+1} = y_n + \dfrac{h}{24}\{9f_{n+1} + 19f_n - 5f_{n-1} + f_{n-2}\}$

7.3 RUNGE–KUTTA METHODS

One of the computational difficulties associated with linear multistep formulas is the need for an alternative starting procedure. The exceptions to this are 1-step methods, but unfortunately they are of low order accuracy. Indeed, the best 1-step method is the trapezoidal method and even this is only convergent of order two. In this section we describe a class of methods, known as Runge–Kutta methods, which combine the advantage of high order accuracy with the property of being 1-step.

Consider the Euler-trapezoidal predictor-corrector pair (7.6) and (7.7), and suppose that just one correction is applied per step so that

$$y_{n+1}^{[0]} = y_n + hf(x_n, y_n)$$

$$y_{n+1} = y_{n+1}^{[1]} = y_n + \frac{h}{2}[f(x_n, y_n) + f(x_{n+1}, y_{n+1}^{[0]})]$$

Substitution of the expression for $y_{n+1}^{[0]}$ into that of y_{n+1} gives

$$y_{n+1} = y_n + \frac{h}{2}[f(x_n, y_n) + f(x_n + h, y_n + hf(x_n, y_n))]$$

Hence, if we define

$$k_1 = f(x_n, y_n) \tag{7.19}$$
$$k_2 = f(x_n + h, y_n + hk_1) \tag{7.20}$$

then

$$y_{n+1} = y_n + \frac{h}{2}[k_1 + k_2] \tag{7.21}$$

Written in this form, the method is called a 2-stage Runge–Kutta method. The numerical value of k_1, calculated from (7.19), is substituted into (7.20) to determine k_2. These numbers are then used to calculate y_{n+1} from (7.21).

The general R-stage Runge–Kutta method is defined by

$$k_1 = f(x_n, y_n)$$

$$k_s = f\left(x_n + h\sum_{t=1}^{s-1} b_t, y_n + h\sum_{i=1}^{s-1} b_t k_t\right) \qquad s = 2, 3, \ldots, R$$

$$y_{n+1} = y_n + h\sum_{s=1}^{R} c_s k_s$$

for appropriate constants b_t and c_s.

The most popular method of this type is the classical 4-stage method

$$k_1 = f(x_n, y_n)$$
$$k_2 = f(x_n + \tfrac{1}{2}h, y_n + \tfrac{1}{2}hk_1)$$
$$k_3 = f(x_n + \tfrac{1}{2}h, y_n + \tfrac{1}{2}hk_2)$$
$$k_4 = f(x_n + h, y_n + hk_3)$$
$$y_{n+1} = y_n + \frac{h}{6}[k_1 + 2k_2 + 2k_3 + k_4]$$

■ **Example 7.10**

To illustrate the 4-stage method, consider the solution of the problem given in Example 7.1, with the same step length as before. Taking $n = 0$ gives

$$k_1 = f(x_0, y_0) \qquad\qquad = f(0, 1) \qquad = \frac{-(1)^2}{1+0} \qquad = -1$$

$$k_2 = f(x_0 + 0.025, y_0 + 0.025k_1) = f(0.025, 0.975) \quad = \frac{-(0.975)^2}{1 + 0.025} = -0.927\,44$$

$$k_3 = f(x_0 + 0.025, y_0 + 0.025k_2) = f(0.025, 0.976\,81) = \frac{-(0.976\,81)^2}{1 + 0.025} = -0.930\,89$$

$$k_4 = f(x_0 + 0.05, y_0 + 0.05k_3) \quad = f(0.05, 0.953\,46) \quad = \frac{-(0.953\,46)^2}{1 + 0.05} = -0.865\,80$$

$$y_1 = y_0 + \frac{0.05}{6}[k_1 + 2k_2 + 2k_3 + k_4] = 0.953\,48$$

The calculations involved in the next three steps are summarized in Table 7.1. The estimate y_4 agrees with the exact solution to 5D. □

There are no major computational difficulties associated with methods of this form, since they are all explicit and 1-step. Further, if $h = 0$ then $y_{n+1} = y_n$, showing that Runge–Kutta methods are zero stable. The only drawback is that they require comparatively more function evaluations per step; an R-stage method uses R evaluations, whereas an explicit linear multistep method uses only one evaluation.

The theory developed in the previous section extends in an obvious way to

Table 7.1

n	1	2	3
k_1	$-0.865\,83$	$-0.757\,76$	$-0.669\,37$
k_2	$-0.807\,73$	$-0.710\,49$	$-0.630\,38$
k_3	$-0.810\,26$	$-0.712\,37$	$-0.631\,81$
k_4	$-0.757\,74$	$-0.669\,36$	$-0.596\,12$
y_{n+1}	$0.912\,98$	$0.877\,37$	$0.845\,79$

Runge–Kutta methods. In particular, it is possible to investigate partial stability in much the same way as before. For instance, if the 2-stage method (7.19)–(7.21) is applied to the test equation $y' = -\lambda y$ $(\lambda > 0)$, then

$$k_1 = -\lambda y_n$$
$$k_2 = -\lambda(y_n + hk_1) = -\lambda(y_n - \lambda h y_n)$$
$$y_{n+1} = y_n + \frac{h}{2}[k_1 + k_2] = \left(1 - \lambda h + \frac{\lambda^2 h^2}{2}\right)y_n$$

This method is stable if and only if

$$-1 \leqslant 1 - \lambda h + \frac{\lambda^2 h^2}{2} \leqslant 1$$

The left hand inequality can be rearranged as $\frac{1}{2}(\lambda h - 1)^2 + \frac{3}{2} \geqslant 0$ and so is always true, whereas the right hand inequality can be rearranged as $\lambda h \leqslant 2$. Hence, we have the same stability restriction as Euler's method, namely $h \leqslant 2/\lambda$. In fact, it can be shown quite generally that Runge–Kutta methods have similar stability characteristics to explicit linear multistep methods.

Local truncation error can also be determined, but the algebra involved is very messy since we need to use Taylor series for functions of two variables. For the 2-stage method (7.19)–(7.21),

$$T = y(x_{n+1}) - y(x_n) - \frac{1}{2}h[k_1 + k_2]$$

with

$$y(x_{n+1}) = y + hy' + \frac{h^2}{2!}y'' + \frac{h^3}{3!}y''' + \cdots$$

$$y(x_n) = y$$

$$k_1 = f$$

$$k_2 = f + (hf_x + hk_1 f_y) + \frac{1}{2!}(h^2 f_{xx} + 2h^2 k_1 f_{xy} + h^2 k_1^2 f_{yy}) + \cdots$$

The derivatives of y are evaluated at x_n and the partial derivatives of f are evaluated at $(x_n, y(x_n))$. In order to be able to cancel the first few terms in the local truncation error, it is necessary to express the derivatives of y in terms of the partial derivatives of f. By virtue of the differential equation, $y' = f$. Hence

$$y'' = \frac{dy'}{dx} = \frac{df}{dx} = \frac{\partial f}{\partial x}\frac{dx}{dx} + \frac{\partial f}{\partial y}\frac{dy}{dx} = f_x + f_y f$$

using the chain rule for partial differentiation.
  Similarly,

$$y''' = \frac{dy''}{dx} = \frac{d}{dx}(f_x + f_y f) = \frac{df_x}{dx} + f\frac{df_y}{dx} + f_y\frac{df}{dx}$$

$$= (f_{xx} + f_{xy}f) + f(f_{yx} + f_{yy}f) + f_y(f_x + f_y f)$$
$$= f_{xx} + 2f_{xy}f + f^2 f_{yy} + f_y f_x + f_y^2 f$$

Substituting these expansions into T gives

$$T = y + hf + \frac{h^2}{2!}(f_x + f_y f) + \frac{h^3}{3!}(f_{xx} + 2f_{xy}f + f^2 f_{yy} + f_y f_x + f_y^2 f) + \cdots$$

$$- y$$

$$- \frac{h}{2}f$$

$$- \frac{h}{2}[f + (hf_x + hff_y) + \frac{1}{2!}(h^2 f_{xx} + 2h^2 ff_{xy} + h^2 f^2 f_{yy}) + \cdots]$$

$$= \frac{h^3}{12}(-f_{xx} + 2f_x f_y - 2ff_{xy} - f^2 f_{yy} + 2ff_y^2) + O(h^4)$$

The 2-stage method is therefore convergent of order 2.
  Runge–Kutta techniques are recommended as starting procedures for linear multistep methods. The particular Runge–Kutta method to use is chosen so that it has the same order as the corresponding linear multistep formula. For example, the classical 4-stage method is fourth order and so can be used in conjunction with the fourth order Adams–Bashforth, Adams–Moulton predictor-corrector pair.

EXERCISE 7.3

1. Use the classical 4-stage Runge–Kutta method to estimate the solution of

$$y' = (1 - x)y^2 - y; \qquad y(0) = 1$$

at $x = 1$ using (a) $h = 0.2$ (b) $h = 0.1$.
  The analytical solution is $y(x) = (e^x - x)^{-1}$. Write down the errors of

these two estimates and deduce that this method is fourth order convergent. Hence find the largest value of h needed to estimate $y(1)$ correct to 8D.

2. Consider the 2-stage method

$$k_1 = f(x_n, y_n)$$
$$k_2 = f(x_n + \tfrac{2}{3}h, y_n + \tfrac{2}{3}hk_1)$$
$$y_{n+1} = y_n + \frac{h}{4}[k_1 + 3k_2]$$

(a) Investigate the partial stability of this method when used to solve $y' = -\lambda y$ ($\lambda > 0$).
(b) Find the leading term in the local truncation error.

3. A Runge–Kutta method for the solution of $y' = f(x, y)$ is expressed as

$$y_{n+1} = y_n + h[c_1 k_1 + c_2 k_2 + c_3 k_3]$$

where

$$k_1 = f(x_n, y_n)$$

$$k_2 = f(x_n + \tfrac{1}{2}h, y_n + \tfrac{1}{2}hk_1)$$

$$k_3 = f(x_n + h, y_n + h(2k_2 - k_1))$$

Find the values of c_1, c_2 and c_3 if this method is to be consistent of order (at least) three.

4. Show that, when applied to the equation $y' = -\lambda y$ ($\lambda > 0$), the classical 4-stage method is given by

$$y_{n+1} = p(\lambda h) y_n$$

where

$$p(x) = 1 - x + \frac{x^2}{2} - \frac{x^3}{6} + \frac{x^4}{24}$$

By sketching a graph of p, or otherwise, show that this method is stable provided

$$h \leqslant A/\lambda$$

for some number A, which should be calculated to 1D.

7.4 SYSTEMS AND HIGHER ORDER EQUATIONS

So far we have only considered the numerical solution of a single first order ordinary differential equation. The majority of real problems, however, involve systems of equations

$$y'_1 = f_1(x, y_1, y_2, \ldots, y_p)$$
$$y'_2 = f_2(x, y_1, y_2, \ldots, y_p)$$
$$\vdots$$
$$y'_p = f_p(x, y_1, y_2, \ldots, y_p)$$

This may be written more concisely as

$$y' = f(x, y)$$

where y, y' and f are vectors with components y_i, y'_i and f_i ($i = 1, 2, \ldots, p$), respectively. If this system is to possess a unique solution, it is necessary to impose an additional condition on y. This usually takes the form

$$y(x_0) = y_0$$

for a given number x_0 and vector y_0. Sufficient conditions for the existence and uniqueness of such problems may be found in Rao (1981).

We now describe how the methods derived in Sections 7.1 and 7.3 for the solution of a single equation may be extended to systems of equations. We concentrate on the case $p = 2$ and for clarity use the alternative notation

$$u' = f(x, u, v); \qquad u(x_0) = u_0$$
$$v' = g(x, u, v); \qquad v(x_0) = v_0$$

This displays all of the features of the general system, whilst avoiding the use of double subscript notation which is rather cumbersome.

In vector form, Euler's formula is

$$y_{n+1} = y_n + hf(x_n, y_n)$$

where $y_n = (u_n, v_n)^T$ denotes a numerical approximation to $y(x) = (u(x), v(x))^T$ at $x = x_n$. This gives

$$u_{n+1} = u_n + hf(x_n, u_n, v_n)$$
$$v_{n+1} = v_n + hg(x_n, u_n, v_n)$$

which can be applied in a step-by-step manner, as the following example demonstrates.

■ **Example 7.11**

Consider the use of Euler's method, with a step length $h = 0.05$, to estimate the solution of the differential system

$$u' = u^2 - 2uv; \qquad u(0) = 1$$
$$v' = xu + u^2 \sin v; \quad v(0) = -1$$

at $x = 0.1$. For this system, Euler's formula is

$$u_{n+1} = u_n + 0.05(u_n^2 - 2u_n v_n)$$
$$v_{n+1} = v_n + 0.05(x_n u_n + u_n^2 \sin v_n)$$

Hence

$$u_1 = u_0 + 0.05(u_0^2 - 2u_0 v_0) \qquad = 1 + 0.05[1^2 - (2)(1)(-1)] = 1.15$$
$$v_1 = v_0 + 0.05(x_0 u_0 + u_0^2 \sin v_0) = -1 + 0.05[(0)(1) + 1^2 \sin(-1)]$$
$$= -1.042\,07$$

Similarly,

$$u_2 = 1.335\,96, \qquad v_2 = -1.096\,29 \qquad\qquad \square$$

In vector form, the trapezoidal formula is

$$y_{n+1} = y_n + \frac{h}{2}[f(x_n, y_n) + f(x_{n+1}, y_{n+1})]$$

which represents a system of non-linear algebraic equations for the unknown components of y_{n+1}. The corresponding corrector iteration is

$$y_{n+1}^{[m+1]} = y_n + \frac{h}{2}[f(x_n, y_n) + f(x_{n+1}, y_{n+1}^{[m]})]$$

so that

$$u_{n+1}^{[m+1]} = u_n + \frac{h}{2}[f(x_n, u_n, v_n) + f(x_{n+1}, u_{n+1}^{[m]}, v_{n+1}^{[m]})]$$

$$v_{n+1}^{[m+1]} = v_n + \frac{h}{2}[g(x_n, u_n, v_n) + g(x_{n+1}, u_{n+1}^{[m]}, v_{n+1}^{[m]})]$$

The predicted values of u_{n+1} and v_{n+1} may be calculated using Euler's formula.

■ Example 7.12
Consider the solution of the problem given in Example 7.11 using the trapezoidal method with the same step length as before. The Euler-trapezoidal predictor-corrector pair is

$$u_{n+1}^{[0]} = u_n + 0.05(u_n^2 - 2u_n v_n)$$
$$v_{n+1}^{[0]} = v_n + 0.05(x_n u_n + u_n^2 \sin v_n)$$
$$u_{n+1}^{[m+1]} = u_n + 0.025[u_n^2 - 2u_n v_n + (u_{n+1}^{[m]})^2 - 2u_{n+1}^{[m]} v_{n+1}^{[m]}]$$
$$v_{n+1}^{[m+1]} = v_n + 0.025\,[x_n u_n + u_n^2 \sin v_n + x_{n+1} u_{n+1}^{[m]} + (u_{n+1}^{[m]})^2 \sin v_{n+1}^{[m]}]$$

From Example 7.11, $u_1^{[0]} = 1.15$, $v_1^{[0]} = -1.042\,07$.

Now

$$u_1^{[m+1]} = u_0 + 0.025[u_0^2 - 2u_0v_0 + (u_1^{[m]})^2 - 2u_1^{[m]}v_1^{[m]}]$$
$$= 1.075 + 0.025[(u_1^{[m]})^2 - 2u_1^{[m]}v_1^{[m]}]$$
$$v_1^{[m+1]} = v_0 + 0.025[x_0u_0 + u_0^2 \sin v_0 + x_1u_1^{[m]} + (u_1^{[m]})^2 \sin v_1^{[m]}]$$
$$= -1.021\,04 + 0.025[0.05u_1^{[m]} + (u_1^{[m]})^2 \sin v_1^{[m]}]$$

If two corrections are applied per step then

$$u_1^{[1]} = 1.075 + 0.025[(1.15)^2 - (2)(1.15)(-1.042\,07)] = 1.167\,98$$
$$v_1^{[1]} = -1.021\,04 + 0.025[(0.05)(1.15) + (1.15)^2 \sin(-1.042\,07)]$$
$$= -1.048\,15$$

and

$$u_1^{[2]} = 1.075 + 0.025[(1.167\,98)^2 - (2)(1.167\,98)(-1.048\,15)]$$
$$= 1.170\,32$$
$$v_1^{[2]} = -1.021\,04 + 0.025[(0.05)(1.167\,98) + (1.167\,98)^2 \sin(-1.048\,15)]$$
$$= -1.049\,13$$

i.e. $u_1 = 1.170\,32$, $v_2 = -1.049\,13$.
    Similarly,

$$u_2^{[0]} = 1.361\,58, \qquad v_2^{[0]} = -1.105\,58$$
$$u_2^{[1]} = 1.387\,56, \qquad v_2^{[1]} = -1.115\,37$$
$$u_2^{[2]} = 1.391\,47, \qquad v_2^{[2]} = -1.117\,11$$

i.e. $u_2 = 1.391\,47$, $v_2 = -1.117\,11$.   □

In vector form, the classical 4-stage Runge–Kutta method is

$$k_1 = f(x_n, y_n)$$
$$k_2 = f(x_n + \tfrac{1}{2}h, y_n + \tfrac{1}{2}hk_1)$$
$$k_3 = f(x_n + \tfrac{1}{2}h, y_n + \tfrac{1}{2}hk_2)$$
$$k_4 = f(x_n + h, y_n + hk_3)$$

$$y_{n+1} = y_n + \frac{h}{6}[k_1 + 2k_2 + 2k_3 + k_4]$$

If $k_i = (k_i, l_i)^T$, then in component form this is given by

$$k_1 = f(x_n, u_n, v_n)$$
$$l_1 = g(x_n, u_n, v_n)$$
$$k_2 = f(x_n + \tfrac{1}{2}h, u_n + \tfrac{1}{2}hk_1, v_n + \tfrac{1}{2}hl_1)$$
$$l_2 = g(x_n + \tfrac{1}{2}h, u_n + \tfrac{1}{2}hk_1, v_n + \tfrac{1}{2}hl_1)$$
$$k_3 = f(x_n + \tfrac{1}{2}h, u_n + \tfrac{1}{2}hk_2, v_n + \tfrac{1}{2}hl_2)$$
$$l_3 = g(x_n + \tfrac{1}{2}h, u_n + \tfrac{1}{2}hk_2, v_n + \tfrac{1}{2}hl_2)$$

$$k_4 = f(x_n + h, u_n + hk_3, v_n + hl_3)$$
$$l_4 = g(x_n + h, u_n + hk_3, v_n + hl_3)$$

$$u_{n+1} = u_n + \frac{h}{6}(k_1 + 2k_2 + 2k_3 + k_4)$$

$$v_{n+1} = v_n + \frac{h}{6}(l_1 + 2l_2 + 2l_3 + l_4)$$

■ **Example 7.13**
Consider the solution of the problem specified in Example 7.11 using the 4-stage Runge–Kutta method with the same step length as before.

Taking $n = 0$,

$$k_1 = f(x_0, u_0, v_0) = f(0, 1, -1) = (1)^2 - (2)(1)(-1) = 3$$
$$l_1 = g(x_0, u_0, v_0) = g(0, 1, -1) = (0)(1) + 1^2 \sin(-1) = -0.841\,47$$
$$k_2 = f(x_0 + \tfrac{1}{2}h, u_0 + \tfrac{1}{2}hk_1, v_0 + \tfrac{1}{2}hl_1) = f(0.025, 1.075, -1.021\,04)$$
$$= (1.075)^2 - (2)(1.075)(-1.021\,04) = 3.350\,86$$
$$l_2 = g(x_0 + \tfrac{1}{2}h, u_0 + \tfrac{1}{2}hk_1, v_0 + \tfrac{1}{2}hl_1) = g(0.025, 1.075, -1.021\,04)$$
$$= (0.025)(1.075) + (1.075)^2 \sin(-1.021\,04) = -0.958\,47$$

Continuing in this way gives

$$k_3 = 3.394\,03, \qquad l_3 = -0.976\,18, \qquad k_4 = 3.821\,78, \qquad l_4 = -1.127\,51$$

Hence

$$u_1 = u_0 + \frac{h}{6}(k_1 + 2k_2 + 2k_3 + k_4) = 1.169\,26$$

$$v_1 = v_0 + \frac{h}{6}(l_1 + 2l_2 + 2l_3 + l_4) = -1.048\,65$$

The values $u_2 = 1.388\,30$ and $v_2 = -1.115\,62$ can be calculated in a similar manner.   □

Finally, we show how a higher order differential equation can be written, and hence solved, as a system of first order equations. Consider the pth order equation

$$y^{(p)} = f(x, y, y', \ldots, y^{(p-1)}) \tag{7.22}$$

The general solution contains p arbitrary constants and so p additional conditions are needed to pin down the solution. The particular numerical approach used to solve higher order problems depends crucially on the form of these conditions. We shall consider the initial conditions

$$y(x_0) = y_0, \qquad y'(x_0) = y'_0, \qquad \ldots, \qquad y^{(p-1)}(x_0) = y_0^{(p-1)} \tag{7.23}$$

for given numbers $x_0, y_0, y'_0, \ldots, y_0^{(p-1)}$. (Methods for solving problems involving other types of conditions are described in Chapter 8.) Any pth order equation can be reduced to a system of p first order equations by introducing the variables

$$y_1 = y, \qquad y_2 = y', \qquad y_3 = y'', \qquad \ldots, \qquad y_p = y^{(p-1)}$$

since it follows that

$$y_1' = y_2$$
$$y_2' = y_3$$
$$\vdots$$
$$y_{p-1}' = y_p$$

and, from (7.22),

$$y_p' = f(x, y_1, y_2, \ldots, y_p)$$

Additionally, conditions (7.23) give

$$y_1(x_0) = y_0$$
$$y_2(x_0) = y_0'$$
$$\vdots$$
$$y_p(x_0) = y_0^{(p-1)}$$

For example, the second order equation

$$y'' + 3y' + 2y = e^x$$

with initial conditions $y(0) = 1$, $y'(0) = 2$ can be converted to the system

$$y_1' = y_2 \qquad\qquad ; \qquad y_1(0) = 1$$
$$y_2' = e^x - 2y_1 - 3y_2; \qquad y_2(0) = 2$$

The problem may then be solved using any of the techniques described in this chapter.

EXERCISE 7.4

1. Find the approximate solution of

$$u' = -uv^2; \qquad u(0) = 1$$
$$v' = -v - e^u; \qquad v(0) = 1$$

at $x = 0.2$, using

(a) Euler's method
(b) The Euler-trapezoidal predictor-corrector pair applying two corrections per step
(c) The classical 4-stage Runge–Kutta method

with a step length $h = 0.1$.

2. Construct the Taylor polynomials of degree 2 about $x = 0$ for the functions $u(x)$ and $v(x)$ which satisfy the initial value problem given in Question 1. Hence estimate $u(0.2)$ and $v(0.2)$.

3. Reduce the second order equation

$$y'' - 10y' - 11y = 0$$

to a system of two first order equations.

Hence estimate $y(1)$ in the case when the initial conditions are

(a) $y(0) = 1,$ $y'(0) = -1$
(b) $y(0) = 1.000\,01,$ $y'(0) = -0.999\,99$

by applying the fourth order Adams–Bashforth, Adams–Moulton predictor-corrector pair with a single correction. Use the classical 4-stage Runge–Kutta method as a starting procedure and take $h = 0.1$. Comment on your results.

4. Convert the following system of two second order equations into a system of four first order equations:

$$y'' + (y')^2 + y = \ln x; \qquad y(1) = 0 \qquad y'(1) = 0.5$$
$$z'' + 2y'z' + z = 0; \qquad z(1) = 0, \qquad z'(1) = 1$$

Hence estimate $y(2)$ and $z(2)$ by applying Euler's method with $h = 0.1$.

5. Construct the Taylor polynomial of degree 3 about $x = 0$ for the function $y(x)$ which satisfies

$$y'' + 3y' + 2y = x; \qquad y(0) = 0, \qquad y'(0) = -2$$

Hence estimate $y(0.1)$.

Use the finite difference approximations

$$y'(x_n) \approx (y_{n+1} - y_{n-1})/2h$$

and

$$y''(x_n) \approx (y_{n+1} - 2y_n + y_{n-1})/h^2$$

to derive the recurrence relation

$$(1 + \tfrac{3}{2}h)y_{n+1} + (2h^2 - 2)y_n + (1 - \tfrac{3}{2}h)y_{n-1} = h^2 x_n$$

for the step-by-step solution of this initial value problem. Hence estimate $y(0.2)$ by taking $h = 0.1$.

8
Ordinary differential equations: boundary value problems

Consider the second order differential equation

$$y'' = f(x, y, y') \tag{8.1}$$

defined on an interval $a \leqslant x \leqslant b$. Here, f is a given function of three real variables and y is an unknown function of the independent variable x. The general solution of (8.1) contains two arbitrary constants, so in order to determine it uniquely it is necessary to impose two additional conditions on y. When one of these is given at $x = a$ and the other at $x = b$, the problem is called a boundary value problem and the associated conditions are called *boundary conditions*. The simplest type of boundary conditions are

$$y(a) = \alpha \tag{8.2a}$$

$$y(b) = \beta \tag{8.2b}$$

for given numbers α and β. However, more general conditions such as

$$\lambda_1 y(a) + \lambda_2 y'(a) = \alpha_1 \tag{8.3a}$$

$$\mu_1 y(b) + \mu_2 y'(b) = \alpha_2 \tag{8.3b}$$

for given numbers λ_i, μ_i and $\alpha_i, (i = 1, 2)$ are sometimes imposed. Sufficient conditions for the existence and uniqueness of such problems may be found in Keller (1968).

In this chapter two numerical techniques are described for the solution of boundary value problems. The first of these, based on finite differences, reduces a boundary value problem to a system of algebraic equations which can be solved using the methods described in Chapters 2 and 3. The second technique, known as the shooting method, reduces a boundary value problem to a sequence of initial value problems which can be solved using the methods described in Chapter 7.

8.1 THE FINITE DIFFERENCE METHOD

We begin by considering the numerical solution of the linear second order differential equation

$$y'' = p(x)y' + q(x)y + r(x) \qquad (8.4)$$

subject to the boundary conditions (8.2a) and (8.2b). The interval $[a, b]$ is divided into N subintervals of length h, so that the mesh points are given by $x_i = x_0 + ih$, with $x_0 = a, x_N = b$ and $h = (b - a)/N$. The numerical approximation to the exact solution $y(x_i)$ is denoted by y_i. From (8.2a) and (8.2b), we may take $y_0 = \alpha$ and $y_N = \beta$. The remaining values $y_1, y_2, \ldots, y_{N-1}$, corresponding to the internal mesh points of the interval, are unknowns which need to be determined. This is illustrated as follows:

At $x = x_n$, the differential equation becomes

$$y''(x_n) = p(x_n)y'(x_n) + q(x_n)y(x_n) + r(x_n)$$

The simplest way of approximating this equation is to replace the derivatives $y'(x_n)$ and $y''(x_n)$ by their central difference approximations

$$y'(x_n) \approx \frac{1}{2h}(y_{n+1} - y_{n-1})$$

$$y''(x_n) \approx \frac{1}{h^2}(y_{n+1} - 2y_n + y_{n-1})$$

derived in Section 6.1. This gives

$$\frac{1}{h^2}(y_{n+1} - 2y_n + y_{n-1}) = p_n \frac{1}{2h}(y_{n+1} - y_{n-1}) + q_n y_n + r_n$$

where $p_n = p(x_n), q_n = q(x_n)$ and $r_n = r(x_n)$. This can be rearranged as

$$\left(1 + \frac{hp_n}{2}\right)y_{n-1} - (2 + h^2 q_n)y_n + \left(1 - \frac{hp_n}{2}\right)y_{n+1} = h^2 r_n \qquad (8.5)$$

Equation (8.5) can be applied at each of the internal mesh points by taking $n = 1, 2, \ldots, N - 1$ in turn. Furthermore, since the values of p_n, q_n and r_n are known, it represents a linear algebraic equation involving y_{n-1}, y_n and y_{n+1}. The totality of equations produces a system of $N - 1$ linear equations for the $N - 1$ unknowns $y_1, y_2, \ldots, y_{N-1}$. The first equation, corresponding to $n = 1$, simplifies to

$$-(2+h^2q_1)y_1 + \left(1 - \frac{hp_1}{2}\right)y_2 = h^2r_1 - \left(1 + \frac{hp_1}{2}\right)\alpha$$

because $y_0 = \alpha$. The last equation, corresponding to $n = N - 1$, reduces to

$$\left(1 + \frac{hp_{N-1}}{2}\right)y_{N-2} - (2 + h^2q_{N-1})y_{N-1} = h^2r_{N-1} - \left(1 - \frac{hp_{N-1}}{2}\right)\beta$$

because $y_N = \beta$. The values of $y_n (n = 1, 2, \ldots, N-1)$ can therefore be found simultaneously by solving the tridiagonal system

$$Ay = c$$

where

$$A = \begin{pmatrix} -(2+h^2q_1) & \left(1 - \frac{hp_1}{2}\right) & & & 0 \\ \left(1 + \frac{hp_2}{2}\right) & -(2+h^2q_2) & \left(1 - \frac{hp_2}{2}\right) & & \\ & \ddots & \ddots & \ddots & \\ & & & & \left(1 - \frac{hp_{N-2}}{2}\right) \\ 0 & & & \left(1 + \frac{hp_{N-1}}{2}\right) & -(2+h^2q_{N-1}) \end{pmatrix}$$

$$y = \begin{pmatrix} y_1 \\ y_2 \\ \vdots \\ y_{N-2} \\ y_{N-1} \end{pmatrix} \qquad c = \begin{pmatrix} h^2r_1 - \left(1 + \frac{hp_1}{2}\right)\alpha \\ h^2r_2 \\ \vdots \\ h^2r_{N-2} \\ h^2r_{N-1} - \left(1 - \frac{hp_{N-1}}{2}\right)\beta \end{pmatrix}$$

■ **Example 8.1**

As an illustration of the method, consider the solution of the boundary value problem

$$y'' = xy' - 3y + e^x; \qquad y(0) = 1, \qquad y(1) = 2$$

using $h = 0.2$. For this example we have:

$y_0 = 1$	$y_1 = ?$	$y_2 = ?$	$y_3 = ?$	$y_4 = ?$	$y_5 = 2$	y
$x_0 = 0$	$x_1 = 0.2$	$x_2 = 0.4$	$x_3 = 0.6$	$x_4 = 0.8$	$x_5 = 1$	x

Replacing the first and second derivatives at $x = x_n$ by their central difference approximations produces

$$\frac{1}{0.04}(y_{n+1} - 2y_n + y_{n-1}) = x_n \frac{1}{0.4}(y_{n+1} - y_{n-1}) - 3y_n + e^{x_n}$$

or equivalently

$$(1 + 0.1 x_n)y_{n-1} - 1.88 y_n + (1 - 0.1 x_n)y_{n+1} = 0.04 e^{x_n}$$

Taking $n = 1, 2, 3$ and 4 in turn gives

$$
\begin{aligned}
-1.88 y_1 + 0.98 y_2 &= 0.04 e^{0.2} - 1.02 \\
1.04 y_1 - 1.88 y_2 + 0.96 y_3 &= 0.04 e^{0.4} \\
1.06 y_2 - 1.88 y_3 + 0.94 y_4 &= 0.04 e^{0.6} \\
1.08 y_3 - 1.88 y_4 &= 0.04 e^{0.8} - 1.84
\end{aligned}
$$

This system may be solved using any of the techniques described in Chapter 2 to give

$$y_1 = 1.4651, \qquad y_2 = 1.8196, \qquad y_3 = 2.0383, \qquad y_4 = 2.1023 \qquad \square$$

Suppose now that the boundary conditions are given by (8.3a) and (8.3b) with $\lambda_2, \mu_2 \neq 0$. In this case the values of y_0 and y_N are unknown. Consequently two extra equations are needed so that the final system has the same number of equations as unknowns. These can be constructed by setting $n = 0$ and $n = N$ in (8.5) to obtain

$$\left(1 + \frac{hp_0}{2}\right)y_{-1} - (2 + h^2 q_0)y_0 + \left(1 - \frac{hp_0}{2}\right)y_1 = h^2 r_0 \qquad (8.6)$$

and

$$\left(1 + \frac{hp_N}{2}\right)y_{N-1} - (2 + h^2 q_N)y_N + \left(1 - \frac{hp_N}{2}\right)y_{N+1} = h^2 r_N \qquad (8.7)$$

The values of y_{-1} and y_{N+1} are approximations to $y(x)$ at $x = x_0 - h$ and $x = x_N + h$ which, unfortunately, lie outside the interval $[a, b]$. However, they can be eliminated from (8.6) and (8.7) by applying a central difference approximation to the boundary conditions. For example, we may approximate (8.3a) by

$$\lambda_1 y_0 + \lambda_2 \frac{1}{2h}(y_1 - y_{-1}) = \alpha_1$$

and so replace y_{-1} by $(2h/\lambda_2)(\lambda_1 y_0 - \alpha_1) + y_1$. If this is substituted into (8.6), then

$$\left(\frac{2h\lambda_1}{\lambda_2} + \frac{h^2 p_0 \lambda_1}{\lambda_2} - 2 - h^2 q_0\right)y_0 + 2y_1 = h^2 r_0 + \left(\frac{2h}{\lambda_2} + \frac{h^2 p_0}{\lambda_2}\right)\alpha_1 \qquad (8.8)$$

Similarly, we may approximate (8.3b) by

$$\mu_1 y_N + \mu_2 \frac{1}{2h}(y_{N+1} - y_{N-1}) = \alpha_2$$

so that equation (8.7) becomes

$$2y_{N-1} + \left(\frac{-2h\mu_1}{\mu_2} + \frac{h^2 p_N \mu_1}{\mu_2} - 2 - h^2 q_N\right) y_N = h^2 r_N + \left(\frac{-2h}{\mu_2} + \frac{h^2 p_N}{\mu_2}\right)\alpha_2$$

$$(8.9)$$

Equations (8.5), (8.8) and (8.9) comprise a system of $N + 1$ linear equations for the $N + 1$ unknowns $y_0, y_1, \ldots, y_N$. Moreover, if (8.8) is taken as the first equation and (8.9) as the last, the coefficient matrix is tridiagonal and the system is easily solved.

So far we have only considered the numerical solution of linear boundary value problems. The general non-linear equation

$$y'' = f(x, y, y')$$

can be solved in an analogous way. It is approximated by

$$\frac{1}{h^2}(y_{n+1} - 2y_n + y_{n-1}) = f\left(x_n, y_n, \frac{1}{2h}(y_{n+1} - y_{n-1})\right)$$

which is used to set up a system of algebraic equations as before. However, the system is non-linear and must be solved using one of the iterative techniques described in Chapter 3, such as Newton's method. This can be done fairly easily because each equation involves at most three unknowns and the corresponding Jacobian matrix is tridiagonal.

■ **Example 8.2**
Consider solving the boundary value problem

$$y'' = y^3 - yy'; \qquad y'(0) = -1, \qquad y(1) = 0.5$$

using $h = 0.25$, so that $x_0 = 0, x_1 = 0.25, x_2 = 0.5, x_3 = 0.75$ and $x_4 = 1$. From the boundary conditions, we have

$$\tfrac{1}{0.5}(y_1 - y_{-1}) = -1, \qquad y_4 = 0.5$$

The differential equation is replaced by

$$\tfrac{1}{0.0625}(y_{n+1} - 2y_n + y_{n-1}) = y_n^3 - y_n \tfrac{1}{0.5}(y_{n+1} - y_{n-1})$$

or equivalently

$$y_{n+1} - 2y_n + y_{n-1} - 0.0625 y_n^3 + 0.125 y_n(y_{n+1} - y_{n-1}) = 0$$

Taking $n = 0, 1, 2, 3$ and substituting for y_{-1} and y_4 gives

$$-2.0625y_0 + 2y_1 - 0.0625y_0^3 + 0.5 = 0$$
$$y_2 - 2y_1 + y_0 - 0.0625y_1^3 + 0.125y_1(y_2 - y_0) = 0$$
$$y_3 - 2y_2 + y_1 - 0.0625y_2^3 + 0.125y_2(y_3 - y_1) = 0$$
$$-1.9375y_3 + y_2 - 0.0625y_3^3 - 0.125y_2y_3 + 0.5 = 0$$

In order to solve these equations using Newton's method we need to supply initial estimates of y_i ($i = 0, 1, 2$ and 3). The simplest way of doing this is to use linear interpolation of the boundary conditions. If $P(x) = mx + d$, then the conditions $P'(0) = -1$ and $P(1) = 0.5$ lead to $m = -1$ and $d = 1.5$. Taking the starting values $y_i^{[0]}$ to be $P(x_i)$, we have

$$y_0^{[0]} = 1.5, \qquad y_1^{[0]} = 1.25, \qquad y_2^{[0]} = 1.0, \qquad y_3^{[0]} = 0.75$$

The results obtained using Newton's method with these starting values are given in Example 3.8. The final iterates are

$$y_0 = 0.9789, \qquad y_1 = 0.7888, \qquad y_2 = 0.6608, \qquad y_3 = 0.5989 \qquad \square$$

EXERCISE 8.1

1. Find the analytical solution of the boundary value problem

$$y'' + \omega^2 y = \omega^2 x; \qquad y(0) = 0, \qquad y(1) = 1$$

   where ω is a parameter.
   Show that this problem is ill-conditioned whenever ω is close to a multiple of π.

2. Consider the boundary value problem

$$y'' = p(x)y' + q(x)y + r(x); \qquad y(a) = \alpha, \qquad y(b) = \beta$$

   where $q(x) \geq 0$ and $|p(x)| \leq K$ on $[a, b]$.
   Show that the coefficient matrix of the approximating finite difference equations is diagonally dominant provided $h < 2/K$.

3. Use the finite difference method to estimate the solution of

$$y'' - 3y' + 2y = 0; \qquad y(0) = 0, \qquad y(1) = 2$$

   with (a) $h = 1/4$ (b) $h = 1/8$.
   Find the analytical solution of this problem and hence write down the corresponding errors at $x = 0.5$.
   Deduce that this method is second order convergent and explain why this might have been expected.

4. Use the finite difference method to estimate the solution of

$$x^2 y'' + xy' - y = 8x^3; \qquad y'(1) - 3y(1) = 0, \qquad 2y'(2) - y(2) = 15$$

with $h = 1/4$.

5. Use the finite difference method to estimate the solution of

$$y'' = -2yy'; \qquad y(0) = 0, \qquad y(1) = 1$$

with $h = 1/5$. Solve the resulting non-linear system using Newton's method. Linearly interpolate the boundary conditions to obtain the starting values, and stop the iteration when $|y_i^{[m+1]} - y_i^{[m]}| < 0.000\,1$ for all i.

8.2 THE SHOOTING METHOD

We now describe a second way of solving boundary value problems of the form

$$y'' = f(x, y, y'); \qquad y(a) = \alpha, \qquad y(b) = \beta$$

In the previous chapter several methods were considered for the numerical solution of second order differential equations given the initial values of y and y'. In the present context, it is known that $y(a) = \alpha$ but the value of $y'(a)$ is undetermined. One way of finding this is to make a guess, λ say, of $y'(a)$ and then to solve the initial value problem

$$y'' = f(x, y, y'); \qquad y(a) = \alpha, \qquad y'(a) = \lambda \qquad (8.10)$$

from $x = a$ to $x = b$. If the corresponding solution happens to satisfy the right hand boundary condition $y(b) = \beta$, then no further action is necessary since the solutions of the initial and boundary value problems are the same. On the other hand, if $y(b) \neq \beta$, then we need to repeat the exercise with a (hopefully) better guess.

Techniques based on this idea are known as shooting methods since the problem is analogous to that of the artilleryman trying to fire a cannonball at a fixed target. The cannon and target have coordinates (a, α) and (b, β), respectively. The artilleryman has to choose the elevation of the cannon so that the cannonball hits the target. Suppose that the trajectory of the first shot is as shown in Fig. 8.1(a). Clearly the cannonball has undershot the target, and it is necessary to increase the elevation for the second shot. If this then passes over the top of the target, as is shown in Fig. 8.1(b), it makes sense to try an elevation somewhere between the first two. The process is repeated until the target is hit.

■ **Example 8.3**

To illustrate the shooting method, consider the boundary value problem

$$y'' = -2yy'; \qquad y(0) = 0, \qquad y(1) - 1$$

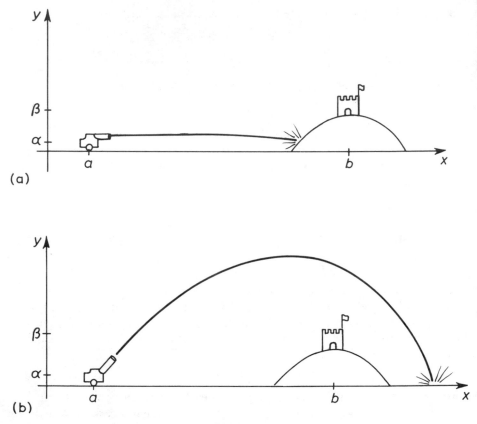

Figure 8.1

The corresponding initial value problem is

$$y'' = -2yy'; \qquad y(0) = 0, \qquad y'(0) = \lambda$$

which reduces to the first order system

$$y_1' = y_2; \qquad\qquad y_1(0) = 0$$
$$y_2' = -2y_1y_2; \qquad y_2(0) = \lambda$$

where $y_1 = y$ and $y_2 = y'$. The classical 4-stage Runge–Kutta method is used as the initial value solver, with $h = 0.1$. Taking $\lambda = 1$ and $\lambda = 2$ as the first two guesses of $y'(0)$ produces $y(1) = 0.761\,593\,2 < 1$ and $y(1) = 1.256\,361\,7 > 1$, respectively. The exact value of $y'(1)$ must therefore lie in the interval $[1, 2]$. Bisecting the interval gives 1.5 as the next guess, for which the corresponding value of $y(1)$ is $1.030\,067\,1$. This process can be repeated with the new interval $[1, 1.5]$. This and further steps of the bisection method are given in Table 8.1.

Table 8.1

λ	$y(1)$
1	0.761 593 2
2	1.256 361 7
1.5	1.030 067 1
1.25	0.902 122 1
1.375	0.967 519 5
1.437 5	0.999 133 1
1.468 75	1.014 683 1
1.453 125	1.006 929 1
1.445 312 5	1.003 036 4
1.441 406 3	1.001 086 1
1.439 453 1	1.000 109 9
1.438 476 6	0.999 621 6
1.438 964 8	0.999 865 7
1.439 209 0	0.999 987 9

The intermediate values of y associated with $\lambda = 1.439\ 209\ 0$ are

$$y(0.1) \approx 0.143\ 232\ 9, \qquad y(0.2) \approx 0.282\ 439\ 9, \qquad y(0.3) \approx 0.414\ 035\ 2$$
$$y(0.4) \approx 0.535\ 215\ 0, \qquad y(0.5) \approx 0.644\ 138\ 3, \qquad y(0.6) \approx 0.739\ 937\ 8$$
$$y(0.7) \approx 0.822\ 595\ 5, \qquad y(0.8) \approx 0.892\ 743\ 3, \qquad y(0.9) \approx 0.951\ 443\ 2$$

Although $|y(1) - 1| < (1/2) \times 10^{-4}$, it does not necessarily follow that the errors in the intermediate values are also bounded by $(1/2) \times 10^{-4}$. The method can only be as accurate as the initial value solver that is used. $\square$

Notice that the solution y of the initial value problem (8.10) depends not only on x but also on λ. Therefore y is a function of two variables x and λ, i.e. $y = y(x, \lambda)$. The right hand boundary condition is satisfied when

$$F(\lambda) = y(b, \lambda) - \beta = 0$$

In general this represents a non-linear algebraic equation for λ and may be solved using any of the techniques considered in Chapter 3. Example 8.3 is a special case of this in which the bisection method is used.

One of the most successful techniques for solving non-linear equations is Newton's method, and it is worth while to see how it may be used in the present context. Newton's method is defined by

$$\lambda_{n+1} = \lambda_n - \frac{F(\lambda_n)}{F'(\lambda_n)}$$

$$= \lambda_n - \frac{[y(b, \lambda_n) - \beta]}{z(b, \lambda_n)} \tag{8.11}$$

where $z = \partial y/\partial \lambda$. In order to use this formula we need to be able to calculate $z(b, \lambda_n)$. Unfortunately no explicit representation of the function $z(x, \lambda)$ exists. However, it is possible to construct an initial value problem involving z by partially differentiating both sides of the equation $y'' = f(x, y, y')$ with respect to λ. From the chain rule,

$$\frac{\partial}{\partial \lambda}(y'') = \frac{\partial f}{\partial \lambda} = \frac{\partial f}{\partial y}\frac{\partial y}{\partial \lambda} + \frac{\partial f}{\partial y'}\frac{\partial y'}{\partial \lambda}$$

Now

$$\frac{\partial y'}{\partial \lambda} = \frac{\partial}{\partial \lambda}\left(\frac{\partial y}{\partial x}\right) = \frac{\partial}{\partial x}\left(\frac{\partial y}{\partial \lambda}\right) = \frac{\partial z}{\partial x} = z'$$

assuming that the order of differentiation with respect to x and λ can be reversed. Similarly,

$$\frac{\partial}{\partial \lambda}(y'') = z''$$

and so

$$z'' = z\frac{\partial f}{\partial y} + z'\frac{\partial f}{\partial y'}$$

If the initial conditions in (8.10) are also differentiated with respect to λ, then

$$z(0) = 0, \qquad z'(0) = 1$$

Hence z satisfies the initial value problem

$$z'' = z\frac{\partial f}{\partial y} + z'\frac{\partial f}{\partial y'}; \qquad z(0) = 0, \qquad z'(0) = 1 \qquad (8.12)$$

Now (8.12) cannot be solved independently of (8.10) because f is a function of x, y and y', so $\partial f/\partial y$ and $\partial f/\partial y'$ are also functions of x, y and y'. If we write $y_1 = y, y_2 = y', y_3 = z$ and $y_4 = z'$, then (8.12) and (8.10) reduce to a system of four first order equations. The values of $y(b, \lambda_n)$ and $z(b, \lambda_n)$ are therefore found simultaneously and can be substituted into (8.11) to calculate λ_{n+1}.

Newton's method converges rapidly when the first guess λ_0 is sufficiently close to the exact value. In the absence of any prior knowledge about the true solution of the boundary value problem, λ_0 is taken to be the slope of the straight line interpolating the boundary conditions.

■ Example 8.4
For the problem given in Example 8.3 we have

$$y'' = -2yy'; \qquad\qquad y(0) = 0, \qquad y'(0) = \lambda$$
$$z'' = -2zy' - 2yz'; \qquad z(0) = 0, \qquad z'(0) = 1$$

Table 8.2

λ	$y(1, \lambda)$	$z(1, \lambda)$
1	0.761 593 2	0.590 782 1
1.403 544 4	0.982 044 1	0.506 277 3
1.439 010 9	0.999 888 8	0.500 034 0
1.439 233 3	1.000 000 0	0.499 995 4

Taking $y_1 = y, y_2 = y', y_3 = z, y_4 = z'$ gives

$$
\begin{aligned}
y_1' &= y_2 & ; \quad & y_1(0) = 0 \\
y_2' &= -2y_1 y_2 & ; \quad & y_2(0) = \lambda \\
y_3' &= y_4 & ; \quad & y_3(0) = 0 \\
y_4' &= -2y_3 y_2 - 2y_1 y_4; & & y_4(0) = 1
\end{aligned}
$$

The line passing through the end points $(0, 0)$ and $(1, 1)$ is $y = x$ which has slope $\lambda_0 = 1$. If the classical 4-stage Runge–Kutta method is used as the initial value solver with $h = 0.1$, then the numerical values of $y(1, \lambda_0)$ and $z(1, \lambda_0)$ are 0.761 593 2 and 0.590 782 1, respectively.

From (8.11),

$$
\lambda_1 = \lambda_0 - \frac{[y(1, \lambda_0) - 1]}{z(1, \lambda_0)} = 1 - \frac{[0.761\,593\,2 - 1]}{0.590\,782\,1}
$$

$$
= 1.403\,544\,4
$$

This and further steps of Newton's method are given in Table 8.2. Clearly convergence is much more rapid than with the bisection method used in Example 8.3, although we have had to solve an extra initial value problem at each stage. □

One disadvantage of the shooting method compared with the finite difference method is that it can be sensitive to rounding errors (see Exercise 8.2, Question 4). For a discussion of this problem and the incorporation of more general boundary conditions, see Keller (1968).

EXERCISE 8.2

1. Use the secant method to solve the problem given in Example 8.3 with $\lambda_0 = 1$ and $\lambda_1 = 2$. Use the same initial value solver and step size. Stop shooting when $|y(1, \lambda_n) - 1| < 10^{-6}$.

2. Use the shooting method in conjunction with Newton's method to solve the

boundary value problem

$$y'' = y^3 - yy'; \qquad y(0) = 1, \qquad y(2) = 1/3$$

with $\lambda_0 = -1/3$. Use the classical 4-stage Runge–Kutta method as the initial value solver with $h = 0.1$. Stop shooting when

$$|y(2, \lambda_n) - \tfrac{1}{3}| < \tfrac{1}{2} \times 10^{-6}$$

Verify that the analytical solution is $y = (1 + x)^{-1}$. Are the errors at the intermediate values of x also less than $1/2 \times 10^{-6}$ in magnitude?

3. Use the shooting method in conjunction with (a) the bisection (b) the secant (c) Newton's method to solve the boundary value problem

$$y'' - 3y' + 2y = 0; \qquad y(0) = 0, \qquad y(1) = 2$$

applying the classical 4-stage Runge–Kutta method as the initial value solver with $h = 0.1$. Take $\lambda_0 = 0$ and $\lambda_1 = 2$ in (a) and (b) and $\lambda_0 = 2$ in (c). Stop shooting when $|y(1, \lambda_n) - 2| < 0.0005$.
   Account for the rapid convergence in (b) and (c).

4. Attempt to use the shooting method in conjunction with Newton's method to solve the boundary value problem

$$y'' - 26y' + 25y = 0; \qquad y(0) = 0, \qquad y(1) = 1$$

applying an initial value solver, step size and starting values of your choice. What difficulties do you encounter, and how might they be resolved?

Appendix

BINOMIAL THEOREM

If a and b are any two numbers and n is a positive integer, then

$$(a+b)^n = \sum_{i=0}^{n} \binom{n}{i} a^{n-i} b^i$$

where

$$\binom{n}{i} = \frac{n!}{i!(n-i)!}$$

FUNDAMENTAL THEOREM OF ALGEBRA

A polynomial of degree n has n (possibly complex) zeros, including multiplicities.

INTERMEDIATE VALUE THEOREM

If f is continuous on $[a,b]$ with $f(a) \neq f(b)$ and c lies between $f(a)$ and $f(b)$, then there is a number $\xi \in (a,b)$ such that $f(\xi) = c$.

ROLLE'S THEOREM

If f is continuous on $[a,b]$ and differentiable on (a,b) with $f(a) = f(b)$, then there is a number $\xi \in (a,b)$ such that $f'(\xi) = 0$.

MEAN VALUE THEOREM

If f is continuous on $[a,b]$ and differentiable on (a,b), then there is a number $\xi \in (a,b)$ such that

$$f'(\xi) = \frac{f(b) - f(a)}{b-a}$$

INTEGRAL MEAN VALUE THEOREM

If f is continuous on $[a, b]$ and g is integrable and does not change sign on $[a, b]$, then there is a number $\xi \in (a, b)$ such that

$$\int_a^b f(x)g(x)\,dx = f(\xi) \int_a^b g(x)\,dx$$

TAYLOR'S THEOREM

If $f, f', \ldots, f^{(n)}$ exist and are continuous on $[a, a + h]$ and $f^{(n+1)}$ exists on $(a, a + h)$, then there is a number $\theta \in (0, 1)$ such that

$$f(a + h) = f(a) + hf'(a) + \frac{h^2}{2!} f''(a) + \cdots + \frac{h^n}{n!} f^{(n)}(a)$$

$$+ \frac{h^{n+1}}{(n+1)!} f^{(n+1)}(a + \theta h)$$

TAYLOR'S THEOREM FOR FUNCTIONS OF TWO VARIABLES

If $f(x, y)$ and its partial derivatives of order $n + 1$ exist and are continuous in a neighbourhood of (a, b) which contains the line joining (a, b) to $(a + h, b + k)$, then there is a number $\theta \in (0, 1)$ such that

$$f(a + h, b + k) = f(a, b) + \left(h\frac{\partial}{\partial x} + k\frac{\partial}{\partial y} \right) f(a, b)$$

$$+ \frac{1}{2!} \left(h\frac{\partial}{\partial x} + k\frac{\partial}{\partial y} \right)^2 f(a, b) + \cdots$$

$$+ \frac{1}{n!} \left(h\frac{\partial}{\partial x} + k\frac{\partial}{\partial y} \right)^n f(a, b)$$

$$+ \frac{1}{(n+1)!} \left(h\frac{\partial}{\partial x} + k\frac{\partial}{\partial y} \right)^{n+1} f(a + \theta h, b + \theta k)$$

References

Bell, W.W. (1968) *Special Functions for Scientists and Engineers*, Van Nostrand, London.

Cheney, E.W. (1966) *Introduction to Approximation Theory*, McGraw-Hill, New York.

Davis, P.J. (1975) *Interpolation and Approximation*, Dover, New York.

Davis, P.J. and Rabinowitz, P. (1984) *Methods of Numerical Integration*, Academic Press, New York.

Fox, L. (1964) *An Introduction to Numerical Linear Algebra*, Clarendon Press, Oxford.

Golub, G.H. and Van Loan, C.F. (1983) *Matrix Computations*, North Oxford Academic, Oxford.

Goult, R.J., Hoskins, R.K., Milner, J.A. and Pratt, J.A. (1974) *Computational Methods in Linear Algebra*, Stanley Thornes, London.

Henrici, P. (1962) *Discrete Variable Methods in Ordinary Differential Equations*, Wiley, New York.

Hildebrand, F.B. (1974) *Introduction to Numerical Analysis*, McGraw-Hill, New York.

Householder, A.S. (1970) *The Numerical Treatment of a Single Non-linear Equation*, McGraw-Hill, New York.

Isaacson, E. and Keller, H.B. (1966) *Analysis of Numerical Methods*, Wiley, New York.

Keller, H.B. (1968) *Numerical Methods for Two-Point Boundary-Value Problems*, Blaisdell, Waltham, Mass.

Lambert, J.D. (1973) *Computational Methods in Ordinary Differential Equations*, Wiley, New York.

Lapidus, L. and Seinfeld, J.H. (1971) *Numerical Solution of Ordinary Differential Equations*, Academic Press, New York.

Noble, B. and Daniel, J.W. (1977) *Applied Linear Algebra*, Prentice-Hall, Englewood Cliffs, New Jersey.

Ortega, J.M. (1972) *Numerical Analysis – A Second Course*, Academic Press, New York.

Ortega, J.M. and Rheinboldt, W.C. (1970) *Iterative Solution of Nonlinear Equations in Several Variables*, Academic Press, New York.

Ostrowski, A.M. (1966) *Solution of Equations and Systems of Equations*, Academic Press, New York.

Powell, M.J.D. (1981) *Approximation Theory and Methods*, Cambridge University Press, Cambridge.

Ralston, A. (1965) *A First Course in Numerical Analysis*, McGraw-Hill, New York.

Ralston, A. and Rabinowitz, P. (1978) *A First Course in Numerical Analysis*, McGraw-Hill, New York.

Rao, M.R.M. (1981) *Ordinary Differential Equations*, Edward Arnold, London.

Rice, J.R. (1964) *The Approximation of Functions. Volume 1: Linear Theory*, Addison-Wesley, Reading, Mass.

Traub, J.F. (1964) *Iterative Methods for the Solution of Equations*, Prentice-Hall, Englewood Cliffs, New Jersey.

Varga, R.S. (1962) *Matrix Iterative Analysis*, Prentice-Hall, Englewood Cliffs, New Jersey.

Watson, G.A. (1980) *Approximation Theory and Numerical Methods*, Wiley, New York.

Wilkinson, J.H. (1963) *Rounding Errors in Algebraic Processes*, HMSO, London.

Wilkinson, J.H. (1965) *The Algebraic Eigenvalue Problem*, Clarendon Press, Oxford.

Wilkinson, J.H. and Reinsch, C. (1971) *Handbook for Automatic Computation. Volume II: Linear Algebra*, Springer-Verlag, Berlin.

Young, D.M. (1971) *Iterative Solution of Large Linear Systems*, Academic Press, New York.

Solutions

to

exercises

EXERCISE 2.1

1. (i) (a) $(0, 2.00, 0)$
    (b) Fails because $a_{22}^{(2)} = 0$.
    (c) $(0.563, -0.723, -0.345, 0.743)$
 (ii) (a) $(1.03, 0.982, 1.01)$
    (b) $(1.00, 1.94, 2.99)$
    (c) $(0.578, -0.726, -0.344, 0.743)$
 In (i) (a) the multipliers are very large, so results are very different from (ii) (a).

2. $(0.0597, 0.359, 6.67)$; $(0.616, 0.736, 3.33)$
 The system is ill-conditioned. The value of $a_{33}^{(3)}$ is 0.003, much smaller than original data. Back substitution involves $x_3 = b_3^{(3)}/a_{33}^{(3)}$, so x_3 is sensitive to changes in $b_3^{(3)}$.

3. $\det A = -96.82$

$$A^{-1} = \begin{pmatrix} 0.041\,26 & 0.188\,7 & 0.176\,9 \\ -0.066\,82 & 0.239\,3 & 0.017\,17 \\ 0.220\,4 & 0.003\,921 & -0.074\,10 \end{pmatrix}$$

4. $L = \begin{pmatrix} 1 & 0 & 0 \\ 1 & 1 & 0 \\ 1/2 & -4 & 1 \end{pmatrix}$ $U = \begin{pmatrix} 4 & -2 & 2 \\ 0 & -1 & -4 \\ 0 & 0 & -18 \end{pmatrix}$

 (See Example 2.2 for an alternative derivation of L and U.)

5.

s.f.	x_1	x_2	x_3	x_4
3	1.83	-9.88	29.3	-18.3
4	1.258	-1.741	7.391	-3.073
5	1.0115	0.878\,80	1.2815	0.821\,05

 The coefficient matrix is a segment of the infinite Hilbert matrix with (i,j)th entry $1/(i+j-1)$. This matrix is notoriously ill-conditioned.

6. Step i uses 1 division and 2 multiplications, so the complete elimination requires $3(n-1)$ operations. The calculation of x_n uses 1 division. The calculation of $x_i(i \neq n)$ uses 1 division and 1 multiplication, so back substitution requires $2(n-1)+1$ operations. $3(n-1)+2(n-1)+1 = 5n-4$.

EXERCISE 2.2

1. (i) (a) $(1, 1, -1)^{\text{T}}$
   (b) Decomposition fails because $u_{22} = 0$.
   (c) Back substitution fails because $u_{33} = 0$.
   (ii) (a) $(1, 1, -1)^{\text{T}}$
   (b) $(1, 2, 3)^{\text{T}}$
   (c) $(23/11,\ 5/11,\ -14/11,\ 8/11)^{\text{T}}$

2. (a) $L = \begin{pmatrix} 1 & 0 & 0 \\ 1/2 & 1 & 0 \\ 1/2 & -1/2 & 1 \end{pmatrix}$ $U = \begin{pmatrix} 2 & 2 & -4 \\ 0 & 6 & 8 \\ 0 & 0 & 5 \end{pmatrix}$

$\det = (-1)^2 \times 2 \times 6 \times 5 = 60$

$A^{-1} = \begin{pmatrix} 1/6 & 2/3 & 1/12 \\ 1/30 & -4/15 & 7/60 \\ 1/10 & 1/5 & -3/20 \end{pmatrix}$

(b) Matrix is singular because $u_{33} = 0$. So $\det A = 0$ and the inverse does not exist.

3. (b) and (c) are positive definite.

4. $L_2^{-1}L_1$ and $U_2 U_1^{-1}$ are unit lower triangular and upper triangular, respectively.

$$L_1 U_1 = L_2 U_2 \Rightarrow L_2^{-1} L_1 U_1 U_1^{-1} = L_2^{-1} L_2 U_2 U_1^{-1} \Rightarrow L_2^{-1} L_1 = U_2 U_1^{-1}$$

The only matrix which is both unit lower triangular and upper triangular is I. So $L_2^{-1} L_1 = I = U_2 U_1^{-1}$, and hence $L_1 = L_2, U_1 = U_2$.

5. $A = LL^{\text{T}} \Rightarrow \det A = \det L \det L^{\text{T}} = \prod_{i=1}^{n} l_{ii}^2$

$$\Rightarrow |l_{jj}| \geqslant (\det A)^{1/2} \qquad \text{since } |l_{ii}| \leqslant 1$$

(This shows that l_{jj} cannot be small unless $\det A$ is pathologically close to zero, leading to ill-conditioning.)

EXERCISE 2.3

1. (i) (a) and (c) converge in 8 and 4 iterations, respectively.
   (ii) (a) and (b) converge in 6 and 9 iterations, respectively.

2.

	(a)	(b)	(c)
Jacobi	0.405	1.000	0
Gauss–Seidel	0.163	0.354	2

3. (a) is diagonally dominant. No; (b) and (c) are both counterexamples.

4. The system

$$4x_1 - x_2 - x_3 = 2$$
$$x_1 - 6x_2 + 2x_3 = 3$$
$$3x_1 + x_2 + 5x_3 = 1$$

is diagonally dominant.

$(0.41, -0.42, 0.04)^{\mathrm{T}}$

5. $0 < \omega < 1.3$;　　0.9

6. Equations (2.30) can be written as

$$a_{11}x_1^{[m+1]} \qquad\qquad = \omega b_1 + (1-\omega)a_{11}x_1^{[m]} - \omega a_{12}x_2^{[m]} \qquad - \omega a_{13}x_3^{[m]}$$
$$\omega a_{21}x_1^{[m+1]} + a_{22}x_2^{[m+1]} \qquad = \omega b_2 \qquad\qquad + (1-\omega)a_{22}x_2^{[m]} \qquad - \omega a_{23}x_3^{[m]}$$
$$\omega a_{31}x_1^{[m+1]} + \omega a_{32}x_2^{[m+1]} + a_{33}x_3^{[m+1]} = \omega b_3 \qquad\qquad\qquad\qquad + (1-\omega)a_{33}x_3^{[m]}$$

i.e.

$$(D + \omega L)x^{[m+1]} = \omega b + [(1-\omega)D - \omega U]x^{[m]}$$

i.e.

$$x^{[m+1]} = (D + \omega L)^{-1}\omega b + (D + \omega L)^{-1}[(1-\omega)D - \omega U]x^{[m]}$$

The iteration matrix is therefore

$$(D + \omega L)^{-1}[(1-\omega)D - \omega U]$$

EXERCISE 3.1

1. (a) See Fig. S.1.
   (b) See Fig. S.2.
   (c) See Fig. S.3.
   In each case, $f(1/2)\, f(1) < 0$
   After 7 bisections: (a) 0.74 (b) 0.65 (c) 0.57

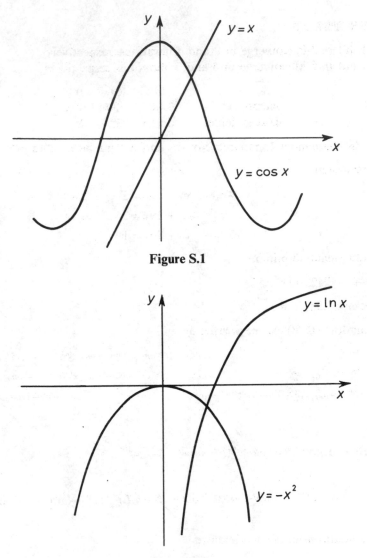

Figure S.1

Figure S.2

2. Same results as Question 1 but with (a) 2 (b) 1 (c) 4 steps.

3. If $f(a)f(c) < 0$ then the new interval is $[a, c]$ with length

$$c - a = \frac{af(b) - bf(a)}{f(b) - f(a)} - a = \frac{f(a)}{f(b) - f(a)}(a - b)$$

Similarly for $f(b)f(c) < 0$.

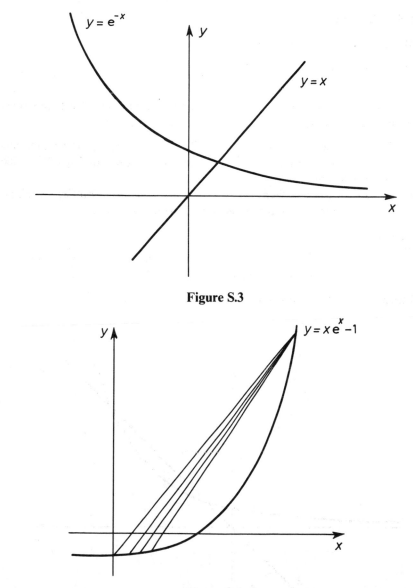

Figure S.3

Figure S.4

4. (a) 11 (b) 31.
   Five iterations for bisection in both cases.
   See Fig. S.4. Slow, since $|f(b)| \gg |f(a)|$ and the root is not close to a. This is also apparent from the results of Question 3, since the length of the new interval will only be slightly less than the original.

EXERCISE 3.2

1. (a) $x_{n+1} = \cos x_n;$ $x_{13} = 0.74$

   (b) $x_{n+1} = e^{-x_n^2};$ $x_{30} = 0.66$

2. $x_9 = 0.57$
   For any $x_0, e^{-x_0} > 0$, so x_1 and all subsequent iterates are strictly positive. In this region, $|g'(x)| = e^{-x} < 1$, hence convergence.

3. (a) If x_n converges to α, then $\alpha = (\alpha^2 + 2)/3$, i.e. $\alpha^2 - 3\alpha + 2 = 0$, which has roots 1 and 2. $g'(x) = 2x/3$, so $|g'(1)| < 1$ and $|g'(2)| > 1$, hence convergence to $\alpha = 1$.
   (b) If x_n converges to α, then $\alpha = 3 - 2/\alpha$, i.e. $\alpha^2 - 3\alpha + 2 = 0$. $g'(x) = 2/x^2$, so $|g'(2)| < 1$ and $|g'(1)| > 1$, hence convergence to $\alpha = 2$.

4. See Fig. S.5.

$$x_{n+1} = \tfrac{1}{5}e^{x_n}; \qquad x_7 = 0.26$$
$$x_{n+1} = \ln 5x_n; \qquad x_5 = 2.54$$

5. Writing $x_n = \alpha + (x_n - \alpha)$ gives result.

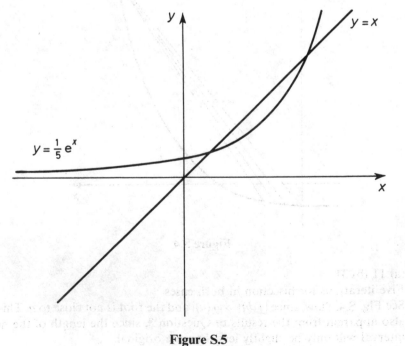

Figure S.5

If $g(x_n) = x_{n+1}, g(\alpha) = \alpha$ and $g'(\alpha) = 0$, then

$$x_{n+1} - \alpha = \frac{(x_n - \alpha)^2}{2} g''(\xi_n)$$

Hence $x_{n+1} - \alpha \approx C(x_n - \alpha)^2$ with $C = g''(\alpha)/2 \neq 0$.

6. If x_n converges to α then $\alpha = \alpha + \omega[h(\alpha) - \alpha]$, so $h(\alpha) = \alpha$.
$g'(\alpha) = 1 + \omega[h'(\alpha) - 1] = 0$. Hence $\omega = [1 - h'(\alpha)]^{-1}$.
If $h(x) = \cos x$ then $\omega = [1 + \sin(1)]^{-1} = 0.5430$, i.e.

$$x_{n+1} = x_n + 0.5430\,(\cos x_n - x_n)$$

$x_3 = 0.74$

EXERCISE 3.3

1. (a) $x_0 > 0$ (b) $x_0 < 0$ (c) $x_0 = 0$
$x_5 = 1.414\,214$

2. (a) (i) $x_4 = 0.739\,085$ (ii) $x_4 = 0.652\,919$ (iii) $x_7 = -3.000\,000$
(b) Same results as (a) but with 6, 6 and 19 iterations, respectively.

3. $x_{n+1} = x_n - \dfrac{x_n^k e^{x_n}}{k x_n^{k-1} e^{x_n} + x_n^k e^{x_n}} = \dfrac{(k-1)x_n + x_n^2}{k + x_n}$

(a) $x_5 = 0.000\,000$ (6D)
(b) $x_5 = 0.072\,003$ (6D)
If $k = 1$ and x_n is small then $x_{n+1} \approx x_n^2$; quadratic convergence.
If $k = 2$ and x_n is small then $x_{n+1} \approx (1/2)x_n$; linear convergence.
Note In (b), $f'(\alpha) = 0$.

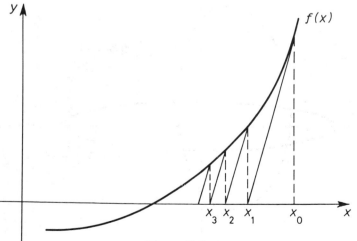

Figure S.6

4. See Fig. S.6. Lines giving $x_1, x_2, x_3, \ldots$ are parallel to the tangent at x_0.

$$x_{n+1} = x_n - \frac{x_n - \cos x_n}{1 + \sin(1)}$$

$x_6 = 0.739\,085$

5. (a) $g(\alpha) = \alpha - \dfrac{f(\alpha)}{f'(\alpha)} = \alpha$

   (b) $g'(\alpha) = \dfrac{f(\alpha)f''(\alpha)}{(f'(\alpha))^2} = 0$

   (c) $g''(\alpha) = \dfrac{f''(\alpha)}{f'(\alpha)} \neq 0$

6. (a) 1 (b) no convergence (c) 0 (d) 2

EXERCISE 3.4

1. $\dfrac{(x + 3/2)^2}{19/3} + \dfrac{(y - 1/2)^2}{19} = 1; \qquad \dfrac{x^2}{36} + y^2 = 1$

See Fig. S.7.

$$\begin{pmatrix} 1.009\,210 \\ 0.985\,987 \end{pmatrix} \quad \begin{pmatrix} 0.865\,110 \\ -0.989\,551 \end{pmatrix} \quad \begin{pmatrix} -4.012\,683 \\ 0.743\,460 \end{pmatrix} \quad \begin{pmatrix} -3.909\,423 \\ -0.758\,588 \end{pmatrix}$$

in 3, 4, 5 and 5 iterations, respectively.

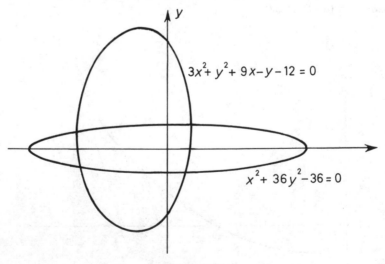

$3x^2 + y^2 + 9x - y - 12 = 0$

$x^2 + 36y^2 - 36 = 0$

Figure S.7

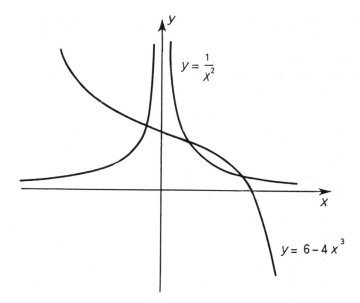

$$y = \frac{1}{x^2}$$

$$y = 6 - 4x^3$$

Figure S.8

2. $y = 6 - 4x^3$, $y = 1/x^2$
   See Fig. S.8.

$$\begin{pmatrix} 1.088\,282 \\ 0.844\,340 \end{pmatrix} \quad \begin{pmatrix} 0.418\,613 \\ 5.706\,575 \end{pmatrix} \quad \begin{pmatrix} -0.399\,819 \\ 6.255\,653 \end{pmatrix}$$

   in 5, 4 and 5 iterations, respectively.

3. (a) 8 (b) 12 (c) 25

4. $f = Ax - b \Rightarrow J = A \Rightarrow x^{[1]} = x^{[0]} - A^{-1}(Ax^{[0]} - b) = A^{-1}b$

5. Take $f = F_x$ and $g = F_y$ in (3.14).

$$F_x = 2 - x + 3x^2/y \qquad F_y = -1 - x^3/y^2 \qquad F_{xx} = -1 + 6x/y$$
$$F_{xy} = -3x^2/y^2 \qquad F_{yy} = 2x^3/y^3$$

$(-1, -1)$, $(-4, -8)$ If $x^{[0]} = y^{[0]} = a$ then

$$\begin{pmatrix} x^{[1]} \\ y^{[1]} \end{pmatrix} = \begin{pmatrix} a \\ a \end{pmatrix} - \begin{pmatrix} 5 & -3 \\ -3 & 2 \end{pmatrix}^{-1} \begin{pmatrix} 2 + 2a \\ -1 - a \end{pmatrix} = \begin{pmatrix} -1 \\ -1 \end{pmatrix}$$

EXERCISE 4.1

1. | *Matrix* | *p* | *Eigenvector* | *Eigenvalue* | *Iterations* |
   |---|---|---|---|---|
   | (a) | 0 | $(1.000\,000, -1.000\,000, -0.500\,000)^{\mathrm{T}}$ | 18.000\,004 | 14 |
   | (a) | 18 | $(1.000\,000, 0.500\,002, 0.999\,997)^{\mathrm{T}}$ | 3.000\,004 | 51 |

(b) $0\,(-0.499\,999,\quad 0.500\,000,\quad 1.000\,000)^{\mathrm{T}}\qquad 3.000\,004\quad 31$
(b) $3\quad(1.000\,000, -1.000\,000,\quad 0.000\,001)^{\mathrm{T}}\qquad 0.999\,998\quad 23$

From trace, (a), (b) have eigenvalues 6, 2 (approximately).
Rate of convergence for (a) and (b) depends on $(6/18)^s$ and $(2/3)^s$ respectively, so (a) is faster than (b).

2. $p = 1.5 = (\lambda_2 + \lambda_3)/2$ for $\boldsymbol{u}_s \to \boldsymbol{x}_1$
   $p = 2.5 = (\lambda_1 + \lambda_2)/2$ for $\boldsymbol{u}_s \to \boldsymbol{x}_3$

Quite generally,
If $\boldsymbol{u}_s \to \boldsymbol{x}_1$, optimal $p = (\lambda_2 + \lambda_n)/2$
If $\boldsymbol{u}_s \to \boldsymbol{x}_n$, optimal $p = (\lambda_1 + \lambda_{n-1})/2$

3. (a) $\lambda_1 = \bar{\lambda}_2$ (b) $\lambda_1 = -\lambda_2$

Shift will cure (b) giving $\lambda_1 = 3$.

4. (i) $A - 2.01I = \begin{pmatrix} -1.01 & 0 & -1 \\ 1 & -0.01 & 1 \\ 2 & 2 & 0.99 \end{pmatrix}$

$L = \begin{pmatrix} 1 & 0 & 0 \\ 0.500 & 1 & 0 \\ -0.505 & -1.000 & 1 \end{pmatrix} \qquad U = \begin{pmatrix} 2.000 & 2.000 & 0.990 \\ 0 & -1.010 & 0.505 \\ 0 & 0 & 0.004\,95 \end{pmatrix}$

$\boldsymbol{u}_1 = \begin{pmatrix} -0.993 \\ 0.499 \\ 1 \end{pmatrix}$

5.

Matrix	Eigenvalue	Eigenvector	Comment
(a)	3	$(1, 1)^{\mathrm{T}}$	
	6	$(1, -1/2)^{\mathrm{T}}$	
(b)	10	$(1, -1, 1)^{\mathrm{T}}$	Other two eigenvalues complex conjugates
(c)	-3	$(1, -1, 1/2)^{\mathrm{T}}$	For dominant eigenvector $\lim \boldsymbol{u}_s$
	6		depends on $\boldsymbol{u}_0$, so $\lambda_1 = \lambda_2$
(d)	1	$(2/5, 1, 2/5, 2/5)^{\mathrm{T}}$	
	4	$(1, 1, 1, 1)^{\mathrm{T}}$	

6.

s	$\boldsymbol{u}_s^{\mathrm{T}}$	$\boldsymbol{v}_{s+1}^{\mathrm{T}}$	$\max(\boldsymbol{v}_{s+1})$
0	$(1, -1/4, -1/4)$	$(0, -27/4, 27/4)$	27/4
1	$(0, -1, 1)$	$(12, -3, -3)$	12
2	$(1, -1/4, -1/4)$		

$u_2 = u_0$. Now

$$u_2 = \tfrac{1}{12}v_2 = \tfrac{1}{12}Au_1 = \frac{1}{12 \times 27/4}Av_1 = \tfrac{1}{81}A^2 u_0$$

i.e. $A^2 u_0 = 81 u_0$, so A^2 has eigenvalue 81 and A has eigenvalues ± 9. From the trace, third eigenvalue $= 18 \neq 0$.

Let $\quad u_0 = \alpha_1 x_1 + \alpha_2 x_2 + \alpha_3 x_3$ $\qquad\qquad\qquad\qquad$ (1)

So $\quad u_1 = \tfrac{1}{27/4}(\alpha_1 18 x_1 + \alpha_2 9 x_2 + \alpha_3(-9)x_3)$ $\qquad\quad$ (2)

and $\quad u_0 = u_2 = \tfrac{1}{81}[\alpha_1(18)^2 x_1 + \alpha_2(9)^2 x_2 + \alpha_3(-9)^2 x_3]$

i.e. $\quad u_0 = 4\alpha_1 x_1 + \alpha_2 x_2 + \alpha_3 x_3$ $\qquad\qquad\qquad\qquad$ (3)

(1) and (3) give $\alpha_1 = 0$, so from (1) and (2)

$$u_0 = \alpha_2 x_2 + \alpha_3 x_3$$
$$u_1 = \tfrac{4}{3}(\alpha_2 x_2 - \alpha_3 x_3)$$

Hence x_1 is a multiple of $u_0 + \tfrac{3}{4}u_1 = \begin{pmatrix} 1 \\ -1 \\ 1/2 \end{pmatrix}$

and x_2 is a multiple of $u_0 - \tfrac{3}{4}u_1 = \begin{pmatrix} 1 \\ 1/2 \\ -1 \end{pmatrix}$

EXERCISE 4.2

1. $Q = \begin{pmatrix} 0 & 1 & 0 \\ 1 & 0 & 0 \\ 0 & 0 & 1 \end{pmatrix} \qquad E = \begin{pmatrix} 2 & -1 & 1 \\ 1 & 1 & -2 \\ 1 & 0 & -1 \end{pmatrix}$

$P = \begin{pmatrix} 1 & 0 & 0 \\ 1/3 & 1 & 0 \\ 1/3 & 0 & 1 \end{pmatrix} \qquad B = P^{-1}EP = \begin{pmatrix} 2 & -1 & 1 \\ 0 & 4/3 & -7/3 \\ 0 & 1/3 & -4/3 \end{pmatrix}$

$C = \begin{pmatrix} 4/3 & -7/3 \\ 1/3 & -4/3 \end{pmatrix}$

C has eigenvalues $1, -1$ with eigenvectors $(7, 1)^T$, $(1, 1)^T$, respectively. So B has eigenvectors $(6, 7, 1)^T$, $(0, 1, 1)^T$
$\quad E$ has eigenvectors $(6, 9, 3)^T$, $(0, 1, 1)^T$
$\quad A$ has eigenvectors $(9, 6, 3)^T$, $(1, 0, 1)^T$

2. $P = \begin{pmatrix} 1 & 0 & 0 & 0 \\ 1 & 1 & 0 & 0 \\ 1 & 0 & 1 & 0 \\ 1 & 0 & 0 & 1 \end{pmatrix}$ $B = P^{-1}AP = \begin{pmatrix} 4 & -2 & -3 & 1 \\ 0 & 1 & 0 & 0 \\ 0 & 0 & 2 & 0 \\ 0 & 0 & 0 & 3 \end{pmatrix}$

$C = \begin{pmatrix} 1 & 0 & 0 \\ 0 & 2 & 0 \\ 0 & 0 & 3 \end{pmatrix}$

C has eigenvalues 1, 2, 3 with eigenvectors $(1,0,0)^T$, $(0,1,0)^T$, $(0,0,1)^T$, respectively.

So B has eigenvectors $(2/3,1,0,0)^T$, $(3/2,0,1,0)^T$, $(-1,0,0,1)^T$

A has eigenvectors $(2/3,5/3,2/3,2/3)^T$, $(3/2,3/2,5/2,3/2)^T$, $(-1,-1,-1,0)^T$

i.e. $(2,5,2,2)^T$, $(3,3,5,3)^T$, $(1,1,1,0)^T$

3.

Matrix	Deflated matrix	Eigenvalue	Comment
(a)	$\begin{pmatrix} 2 & -1/2 \\ 0 & 1 \end{pmatrix}$	7, 2, 1	
(b)	$\begin{pmatrix} 1 & -2 \\ 3 & 1 \end{pmatrix}$	10	Eigenvalues of deflated matrix are $1 \pm (\sqrt{6})i$
(c)	$\frac{1}{11}\begin{pmatrix} -6 & -36 \\ -54 & 39 \end{pmatrix}$	6, 6, -3	
(d)	$\frac{1}{3}\begin{pmatrix} 9 & 6 & -12 \\ 1 & 4 & -2 \\ 2 & 2 & -1 \end{pmatrix}$		

$$3, 2, 1, 1$$

$$\begin{pmatrix} 1 & 0 \\ 0 & 1 \end{pmatrix}$$

4. (a) 1.000,   1.997, -1.001
   (b) 2.617, -1.360,   0.843

Small changes to the coefficients have produced a large change in the eigenvalues, so ill-conditioned.

5. $P = \begin{pmatrix} 1 & 0 & 0 \\ -0.99 & 1 & 0 \\ -0.49 & 0 & 1 \end{pmatrix}$ $B = P^{-1}AP = \begin{pmatrix} 17.9 & -6 & -4 \\ -0.149 & 5.06 & -1.96 \\ -0.149 & -0.94 & 4.04 \end{pmatrix}$

$C = \begin{pmatrix} 5.06 & -1.96 \\ -0.94 & 4.04 \end{pmatrix}$

C has eigenvalues 3.1, 6 with eigenvectors $(1, 1)^T$, $(1, -0.4796)^T$
B has eigenvectors $(0.6757, 1, 1)^T$, $(0.3430, 1, -0.4796)^T$
A has eigenvectors $(0.6757, 0.3311, 0.6689)^T$, $(0.3430, 0.6604, -0.6477)^T$

i.e. $\qquad\qquad (1, 0.4900, 0.9899)^T, \qquad (0.5194, 1, -0.9808)^T$
Hence only slight loss of accuracy.

EXERCISE 4.3

1. $P^T A P = \begin{pmatrix} 5c^2 + 2s^2 - 2sc & -0.1s & c^2 - s^2 + 3sc \\ \cdot & -3 & 0.1c \\ \cdot & \cdot & 5s^2 + 2c^2 + 2sc \end{pmatrix}$

Require $c^2 - s^2 = -3sc$, and result follows from double angle formulas.
Approximate eigenvalues are 5.3028, 1.6972, -3 with approximate eigenvectors

$$\begin{pmatrix} 0.9571 \\ 0 \\ 0.2898 \end{pmatrix} \begin{pmatrix} -0.2898 \\ 0 \\ 0.9571 \end{pmatrix} \begin{pmatrix} 0 \\ 1 \\ 0 \end{pmatrix}$$

Maximum errors in eigenvalues are 0.0290, 0.0957, 0.1247

2. $\quad b_{ii}^2 = a_{ii}^2 \qquad i \neq p, q$
$b_{pp}^2 + b_{qq}^2 = (c^2 a_{pp} - 2sc a_{pq} + s^2 a_{qq})^2 + (s^2 a_{pp} + 2sc a_{pq} + c^2 a_{qq})^2$
$\qquad = (c^4 + s^4)(a_{pp}^2 + a_{qq}^2) + 4sc a_{pq}(c^2 - s^2)(a_{qq} - a_{pp})$
$\qquad\quad + 8s^2 c^2 a_{pq}^2 + 4s^2 c^2 a_{pp} a_{qq}$
$\qquad = a_{pp}^2 + a_{qq}^2 - 2s^2 c^2(a_{pp} - a_{qq})^2 + 4sc a_{pq}(c^2 - s^2)(a_{qq} - a_{pp})$
$\qquad\quad + 8s^2 c^2 a_{pq}^2$

since $c^4 + s^4 = (c^2 + s^2)^2 - 2s^2 c^2 = 1 - 2s^2 c^2$.
Now by choice of θ, $sc(a_{qq} - a_{pp}) = (c^2 - s^2)a_{pq}$, so

$$b_{pp}^2 + b_{qq}^2 = a_{pp}^2 + a_{qq}^2 + 2(c^2 - s^2)^2 a_{pq}^2 + 8s^2 c^2 a_{pq}^2$$
$$= a_{pp}^2 + a_{qq}^2 + 2(c^2 + s^2)^2 a_{pq}^2$$
$$= a_{pp}^2 + a_{qq}^2 + 2a_{pq}^2$$

3.

Matrix	Eigenvalues	Number of steps (i)	(ii)	(iii)
(a)	3.0000	8	7	8
	18.0000			
	6.0000			

(b)   9.0000   8   7   8
729.0000
81.0000

(c)   6.0000   3   2   2
− 3.0000
6.0000

(d)   1170.6102   18   12   12
233.9973
45.3082
− 1088.9157

4. $P^{-1}AP = \begin{pmatrix} 2.30 & 0.04 & 0.02k & -0.05 \\ 0.02 & 7.60 & 0.01k & 0.02 \\ -0.03/k & 0.02/k & 0.80 & 0.02/k \\ 0.01 & -0.01 & -0.02k & -5.20 \end{pmatrix}$

Circles corresponding to rows 1 and 3 touch when

$$0.8 + 0.07/k = 2.3 - 0.09 - 0.02k$$

i.e. when $k \approx 70$. The circles are then given by

Centre	Radius
2.30	1.49
7.60	0.74
0.80	0.001
− 5.20	1.42

Complex eigenvalues occur in conjugate pairs, both of which lie in the same circle. Since the third circle is isolated there is only one eigenvalue in this circle. Hence A has a real eigenvalue in the interval 0.8 ± 0.001.

5. $P^T AP = \begin{pmatrix} 2c^2 - 4sc + 6s^2 & 0.1(c-s) & -4sc + 2(c^2 - s^2) & 0.1c - 0.05s \\ \cdot & 3 & 0.1(c+s) & 0.2 \\ \cdot & \cdot & 2s^2 + 4sc + 6c^2 & 0.1s + 0.05c \\ \cdot & \cdot & \cdot & 1 \end{pmatrix}$

If $\theta = (1/2)\tan^{-1} 1 = (\pi/8)$ then

$$P^T AP = \begin{pmatrix} 1.17 & 0.05 & 0 & 0.07 \\ 0.05 & 3 & 0.13 & 0.2 \\ 0 & 0.13 & 6.83 & 0.08 \\ 0.07 & 0.2 & 0.08 & 1 \end{pmatrix}$$

The determinant of a matrix is equal to the product of its eigenvalues. Gerschgorin's theorem applied to $P^T A P$ indicates that all of the eigenvalues of A are non-zero. Hence $\det A \neq 0$.

EXERCISE 4.4

1. $f_r(1)$: $1, 0, -4, -20, 144$ has 2 agreements.
   $f_r(2)$: $1, -1, -1, 5, -9$ has 1 agreement.
   Hence only one eigenvalue in $[1, 2]$.

2. $f_r(0)$: $1, 1, 1, 1, 1$ has 5 agreements. Hence all of the eigenvalues are strictly positive.

3. $2.4375, 2.0625, -1.6875$

4. (a) $-0.1926, 1.3819, 3.6180, 5.1926$
   (b) Work with the two 3×3 submatrices separately to get $-1.0995, 1.2793, 7.8201$ and $-4.6458, 0.6458, 11.0000$.

5. Differentiate expressions for $f_0(p), f_1(p)$ and $f_r(p)$ given in text to get result. Newton's method is

$$p_{m+1} = p_m - \frac{f_n(p_m)}{f'_n(p_m)}$$

$p_0 = 1.5$, $f_0(p_0) = 1$, $f_1(p_0) = -1.5$, $f_2(p_0) = 2.75$, $f_3(p_0) = 0.125$
$f'_0(p_0) = 0$, $f'_1(p_0) = -1$, $f'_2(p_0) = 4$, $f'_3(p_0) = -3.75$

$$p_1 = 1.5 - \frac{0.125}{-3.75} = 1.533$$

$p_1 = 1.533$, $f_0(p_1) = 1$, $f_1(p_1) = -1.533$, $f_2(p_1) = 2.883$, $f_3(p_0) = -0.004$
$f'_0(p_1) = 0$, $f'_1(p_1) = -1$, $f'_2(p_1) = 4.066$, $f'_3(p_0) = -4.050$

$$p_2 = 1.533 - \frac{-0.004}{-4.050} = 1.532$$

Therefore $\lambda_1 = 1.53$ (2D).

6.

i	x_i
1	1.0000
2	0.7462
3	0.3030
4	0.0859
5	0.0188
6	0.0033
7	0.0002

8	0.0020
9	0.0153
10	0.1318
11	1.2692
12	13.5073
13	157.3903
14	1992.6209
⋮	⋮

Method is unstable, so not recommended. Wilkinson (1965) shows that these poor results are mainly due to the inexact value of λ. (The preferred method is inverse iteration, which is particularly easy to use if C is tridiagonal.)

EXERCISE 4.5

1. The formulas follow from the right-angled triangle shown in Fig. S.9. These results show how the rotation matrices can be constructed avoiding the use of the function $\tan^{-1} x$.

2. $w^T w = (1/3)^2 + (2/3)^2 + (2/3)^2 = 1$

$$Q = \frac{1}{9} \begin{pmatrix} 7 & -4 & -4 \\ -4 & 1 & -8 \\ -4 & -8 & 1 \end{pmatrix}$$

3. $C = \begin{pmatrix} 3.0000 & 3.0000 & 0 & 0 \\ \cdot & 4.0000 & 3.1623 & 0 \\ \cdot & \cdot & 2.0000 & -1.0000 \\ \cdot & \cdot & \cdot & -2.0000 \end{pmatrix}$

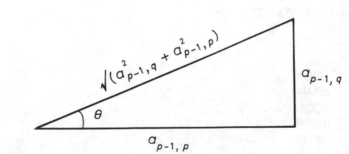

Figure S.9

4. $Q_1 = \dfrac{1}{15}\begin{pmatrix} -10 & -5 & -10 \\ -5 & 14 & -2 \\ -10 & -2 & 11 \end{pmatrix}$

$Q_2 = \dfrac{1}{10+3\sqrt{10}}\begin{pmatrix} -(9+3\sqrt{10}) & -(3+\sqrt{10}) \\ -(3+\sqrt{10}) & 9+3\sqrt{10} \end{pmatrix}$

$C = \begin{pmatrix} 3.0000 & -3.0000 & 0 & 0 \\ \cdot & 4.0000 & 3.1623 & 0 \\ \cdot & \cdot & 2.0000 & 1.0000 \\ \cdot & \cdot & \cdot & -2.0000 \end{pmatrix}$

5. $C = \begin{pmatrix} 10.0000 & -7.2111 & 0 \\ \cdot & 11.3077 & -1.5385 \\ \cdot & \cdot & 5.6923 \end{pmatrix}$

which has eigenvalues 18.0, 6.0, 3.0, with eigenvectors

$$\begin{pmatrix} -0.9014 \\ 1 \\ -0.1250 \end{pmatrix} \qquad \begin{pmatrix} -0.3606 \\ -0.2000 \\ 1 \end{pmatrix} \qquad \begin{pmatrix} 1 \\ 0.9707 \\ 0.5547 \end{pmatrix}$$

Pre-multiplication by

$$X_{23} = \begin{pmatrix} 1 & 0 & 0 \\ 0 & 0.8321 & -0.5547 \\ 0 & 0.5547 & 0.8321 \end{pmatrix}$$

gives the eigenvectors of A as

$$\begin{pmatrix} -0.9014 \\ 0.9014 \\ 0.4507 \end{pmatrix} \qquad \begin{pmatrix} -0.3606 \\ -0.7211 \\ 0.7212 \end{pmatrix} \qquad \begin{pmatrix} 1.0000 \\ 0.5000 \\ 1.0000 \end{pmatrix}$$

EXERCISE 4.6

1. $Q = \dfrac{1}{3}\begin{pmatrix} 1 & 2 & 2 \\ 2 & -2 & 1 \\ 2 & 1 & -2 \end{pmatrix} \qquad R = \begin{pmatrix} 3 & 1 & 1 \\ 0 & 1 & 1 \\ 0 & 0 & 3 \end{pmatrix}$

2.

Matrix	Eigenvalues	Number of iterations		Comment
		LR	QR	
(a)	7, 2, 1	14	14	Faster convergence for (b)
(b)	729, 81, 9	8	7	than (a) since λ_{i+1}/λ_i smaller in (b)

(c)	$10, 1 \pm (\sqrt{6})\text{i}$	10	9
(d)	$18, 9, -9$	24	20

(d) Oscillation, since $|\lambda_2| = |\lambda_3|$. Can still calculate eigenvalues, however, using 2×2 submatrices

(e)	$4.7453, 3.1773,$	37	28
	$1.8227, 0.2547$		
(f)	$1 \pm (\sqrt{6})\text{i}, 1 \pm 2\text{i}$	55	50

3. Without interchanges, method converges satisfactorily producing $\lambda_1 = 3$, $\lambda_2 = -2$. However, with interchanges,

$$A_2 = \begin{pmatrix} 1 & 2 \\ 3 & 0 \end{pmatrix} \qquad A_3 = \begin{pmatrix} 1 & 3 \\ 2 & 0 \end{pmatrix} = A_1$$

and so fails to converge.

4. $Q_1^T Q_2$ is orthogonal and $R_1 R_2^{-1}$ is upper triangular.
If $Q_1 R_1 = Q_2 R_2$ then $Q_1^T Q_1 R_1 = Q_1^T Q_2 R_2$, i.e. $R_1 = Q_1^T Q_2 R_2$ since $Q_1^T Q_1 = I$. Hence $R_1 R_2^{-1} = Q_1^T Q_2 R_2 R_2^{-1} = Q_1^T Q_2$.
The only matrix which is both orthogonal and upper triangular is

$$D = \text{diag}(\pm 1, \pm 1, \ldots, \pm 1)$$

Hence $R_1 R_2^{-1} = D = Q_1^T Q_2$ and $R_1 = D R_2$ and $Q_2 = Q_1 D$.

5. $A_{s+1} = Q_s^T A_s Q_s$

(a) This is a similarity transformation because $Q_s^T = Q_s^{-1}$.
(b) $A_{s+1}^T = (Q_s^T A_s Q_s)^T = Q_s^T A_s^T (Q_s^T)^T = Q_s^T A_s Q_s = A_{s+1}$

$Q_s = X_{12} X_{23} X_{34}$

$$= \begin{pmatrix} x & x & 0 & 0 \\ x & x & 0 & 0 \\ 0 & 0 & x & 0 \\ 0 & 0 & 0 & x \end{pmatrix} \begin{pmatrix} x & 0 & 0 & 0 \\ 0 & x & x & 0 \\ 0 & x & x & 0 \\ 0 & 0 & 0 & x \end{pmatrix} \begin{pmatrix} x & 0 & 0 & 0 \\ 0 & x & 0 & 0 \\ 0 & 0 & x & x \\ 0 & 0 & x & x \end{pmatrix}$$

$$= \begin{pmatrix} x & x & x & x \\ x & x & x & x \\ 0 & x & x & x \\ 0 & 0 & x & x \end{pmatrix}$$

$$R_s = \begin{pmatrix} x & x & 0 & 0 \\ 0 & x & x & 0 \\ 0 & 0 & x & x \\ 0 & 0 & 0 & x \end{pmatrix}$$

Hence

$$R_sQ_s = \begin{pmatrix} x & x & 0 & 0 \\ 0 & x & x & 0 \\ 0 & 0 & x & x \\ 0 & 0 & 0 & x \end{pmatrix} \begin{pmatrix} x & x & x & x \\ x & x & x & x \\ 0 & x & x & x \\ 0 & 0 & x & x \end{pmatrix} = \begin{pmatrix} x & x & x & x \\ x & x & x & x \\ 0 & x & x & x \\ 0 & 0 & x & x \end{pmatrix}$$

From (b), $A_{s+1} = R_sQ_s$ is symmetric, i.e.

$$R_sQ_s = \begin{pmatrix} x & x & 0 & 0 \\ x & x & x & 0 \\ 0 & x & x & x \\ 0 & 0 & x & x \end{pmatrix}$$

6. If $Ax = b$ then $Q(Rx) = b$, i.e. $Qc = b$ and $c = Q^{T}b$.

$$c = Q^{T}b = \frac{1}{3}\begin{pmatrix} 1 & 2 & 2 \\ 2 & -2 & 1 \\ 2 & 1 & -2 \end{pmatrix}\begin{pmatrix} 3 \\ 3 \\ 0 \end{pmatrix} = \begin{pmatrix} 3 \\ 0 \\ 3 \end{pmatrix}$$

Since $Rx = c$,

$$\begin{pmatrix} 3 & 1 & 1 \\ 0 & 1 & 1 \\ 0 & 0 & 3 \end{pmatrix}\begin{pmatrix} x_1 \\ x_2 \\ x_3 \end{pmatrix} = \begin{pmatrix} 3 \\ 0 \\ 3 \end{pmatrix}, \qquad \text{i.e.} \quad \begin{pmatrix} x_1 \\ x_2 \\ x_3 \end{pmatrix} = \begin{pmatrix} 1 \\ -1 \\ 1 \end{pmatrix}$$

EXERCISE 4.7

1. $\begin{pmatrix} 3 & -1 & 1 & -4 \\ 4 & 2 & -6 & 5 \\ 0 & 0 & 1 & 1 \\ 0 & 0 & -3 & 2 \end{pmatrix}$; $\frac{5}{2} \pm \frac{(\sqrt{15})i}{2}$, $\frac{3}{2} \pm \frac{(\sqrt{11})i}{2}$

2. (a) $\begin{pmatrix} 3 & 1 & 0 \\ 4 & 6 & 2 \\ 0 & 0 & 1 \end{pmatrix}$;   1, 2, 7

(b) $\begin{pmatrix} 8.0000 & -3.8571 & -1.8125 & -3.0000 \\ 7.0000 & -2.8571 & -1.8125 & -3.0000 \\ 0 & 0.3265 & 2.4821 & -0.8571 \\ 0 & 0 & -0.7109 & 2.3750 \end{pmatrix}$;   1, 2, 3, 4

3. 4.7453, 3.1773, 1.8227, 0.2547

Note that the symmetric tridiagonal form is retained.

4. Without interchanges, $M_2^{-1}M_1^{-1}AM_1M_2 = H$. Hence $AL = LH$ with

$$L = M_1M_2 = \begin{pmatrix} 1 & 0 & 0 & 0 \\ 0 & 1 & 0 & 0 \\ 0 & m_{31} & 1 & 0 \\ 0 & m_{41} & 0 & 1 \end{pmatrix} \begin{pmatrix} 1 & 0 & 0 & 0 \\ 0 & 1 & 0 & 0 \\ 0 & 0 & 1 & 0 \\ 0 & 0 & m_{42} & 1 \end{pmatrix} = \begin{pmatrix} 1 & 0 & 0 & 0 \\ 0 & 1 & 0 & 0 \\ 0 & m_{31} & 0 & 0 \\ 0 & m_{41} & m_{42} & 1 \end{pmatrix}$$

Writing

$$H = \begin{pmatrix} h_{11} & h_{12} & h_{13} & h_{14} \\ h_{21} & h_{22} & h_{23} & h_{24} \\ 0 & h_{32} & h_{33} & h_{34} \\ 0 & 0 & h_{43} & h_{44} \end{pmatrix}$$

and equating entries in columns 1, 2, 3 and 4 in turn gives

$$H = \begin{pmatrix} 8.00 & -3.86 & -1.93 & 1.00 \\ 7.00 & -2.86 & -1.93 & 1.00 \\ 0 & 0.31 & 1.72 & 0.14 \\ 0 & 0 & 0.81 & 3.14 \end{pmatrix} \qquad L = \begin{pmatrix} 1 & 0 & 0 & 0 \\ 0 & 1 & 0 & 0 \\ 0 & 0.86 & 1 & 0 \\ 0 & 0.71 & 1.07 & 1 \end{pmatrix}$$

5. $1, 2, -3, -7$

EXERCISE 5.1

1. $p_3(x) = \dfrac{x(x-3)(x-4)}{-60}5 + \dfrac{(x+2)(x-3)(x-4)}{24}1$

$\qquad + \dfrac{(x+2)x(x-4)}{-15}55 + \dfrac{(x+2)x(x-3)}{24}209$

$\quad p_3(1) = -19$

2. $p_1(0.55) = 0.522\,033\,9$
$f(x) = \sin x$, so $f''(x) = -\sin x$, which is continuous on $I = [0.5, 0.6]$.

Hence

$$f(x) - p_1(x) = -(x - 0.5)(x - 0.6)\sin \xi; \qquad \xi \in I$$

Now

$$\max_{\xi \in I} |\sin \xi| = \sin 0.6$$

so

$$|f(0.55) - p_1(0.55)| \leqslant |(0.55 - 0.5)(0.55 - 0.6)\sin 0.6|$$
$$= 0.0014$$

Actual error is 0.000 65.

3. (a) $p_2(1.2) = 0.052\,120$
   (b) $p_4(1.2) = 0.092\,256$
   (c) $p_8(1.2) = 0.090\,718$

4. Similar to Fig. 5.2(c) but with more violent oscillations at the end of the range.

5. Error term involves $(n+1)$th derivative, which will vanish for any polynomial of degree at most n. In particular, taking $f(x) = 1$ gives

$$1 = \sum_{i=0}^{n} L_i(x)y_i = \sum_{i=0}^{n} L_i(x)$$

6. $p_5(7) = -2.317\,734$
   Using piecewise linear interpolation, $f(7) \approx 0.710\,664$.
   The exact value is 0.75, so piecewise linear interpolation is more accurate. This is also apparent from a sketch of the two approximating functions.

EXERCISE 5.2

1.
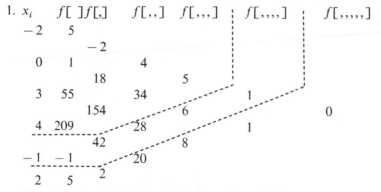

(Dashed lines indicate additions to the table.)
$p_3(1) = -19,\qquad p_4(1) = -1,\qquad p_5(1) = -1$
Since $p_4(x) = p_5(x)$, for all x, this suggests that $f(x)$ is a quartic polynomial.

2. If data is taken in the order given,
   $p_1(2.6) = 1.093\,85,\qquad p_2(2.6) = 0.914\,99,\qquad p_3(2.6) = 0.965\,81$
   $p_4(2.6) = 0.954\,99,\qquad p_5(2.6) = 0.954\,85,\qquad p_6(2.6) = 0.955\,65$
   $p_7(2.6) = 0.955\,53,\qquad p_8(2.6) = 0.955\,52$
   $|p_7(2.6) - p_8(2.6)| \leqslant 0.000\,05$
   However, if data is reordered so that 2.6 is roughly in the centre of the interpolating points at each stage, i.e.

   $$x_0 = 2.5,\qquad x_1 = 2.7,\qquad x_2 = 2.3,\qquad x_3 = 2.9,\dots$$
   then
   $$p_1(2.6) = 0.954\,77,\qquad p_2(2.6) = 0.955\,57,\qquad p_3(2.6) = 0.955\,52$$

3. All three divided differences can be expanded as

$$\frac{(x_2 - x_1)f_0 - (x_2 - x_0)f_1 + (x_1 - x_0)f_2}{(x_2 - x_1)(x_2 - x_0)(x_1 - x_0)}$$

4. $p_2(x_0) = f_0 + 0 + 0 = f_0$

$$p_2(x_1) = f_0 + \frac{f_1 - f_0}{x_1 - x_0}(x_1 - x_0) + 0 = f_1$$

$$p_2(x_2) = f_0 + \frac{f_1 - f_0}{x_1 - x_0}(x_2 - x_0)$$

$$+ \frac{(x_2 - x_1)f_0 - (x_2 - x_0)f_1 + (x_1 - x_0)f_2}{(x_2 - x_1)(x_2 - x_0)(x_1 - x_0)}(x_2 - x_0)(x_2 - x_1)$$

$$= f_2$$

EXERCISE 5.3

1. 1.0   0.183 939 7

                        − 175 042

    1.1   0.166 435 5   16 658

                        − 158 384   − 1 586

    1.2   0.150 597 1   15 072   152

                        − 143 312   − 1 434

    1.3   0.136 265 9   13 638

                        − 129 674

    1.4   0.123 298 5

    Forward difference: take $x_0 = 1.0$, $s = 2.5$
    Backward difference: take $x_4 = 1.4$, $s = -1.5$
    $p_4(1.25) = 0.143 252 4$

2. The sixth differences are zero, so q has degree 5. Using forward difference form:

$$p_5(x_0 + sh) = 6 + 0.090\,02s + 0.0594\frac{s(s-1)}{2}$$

$$+ 0.057 \frac{s(s-1)(s-2)}{6} - 0.0048 \frac{s(s-1)(s-2)(s-3)}{24}$$

$$- 0.0024 \frac{s(s-1)(s-2)(s-3)(s-4)}{120}$$

$s = x/0.1$ since $x = x_0 + sh$

Hence $\qquad q(x) = p_5(x) = -2x^5 + 10x^3 - x + 6$

3. Forward difference: $x_0 = 1.0$, $s = 0.5$ gives $p_3(1.05) = 0.0212$

   Backward difference: $x_6 = 1.6$, $s = -0.5$ gives $p_2(1.55) = 0.1903$

   Central difference: $x_3 = 1.3$, $s = 0.5$ gives $p_4(1.35) = 0.1303$

4. Differences initially decrease but then start increasing from the fifth difference onwards. This is due to propagation of rounding errors. The errors in the tabulated values of f_i do not exceed 0.0005. Hence the error in $\Delta f_i = f_{i+1} - f_i$ could be as large as $0.0005 + 0.0005 = 0.001$. Similarly, the errors in $\Delta^2 f_i$, $\Delta^3 f_i, \ldots$, could be as large as $0.002, 0.004, \ldots$.

5. (a) $f_{i+4} - 4f_{i+3} + 6f_{i+2} - 4f_{i+1} + f_i$
   (b) $f_i - 3f_{i-1} + 3f_{i-2} - f_{i-3}$
   (c) $f_{i+1} - 2f_i + f_{i-1}$

6. $p_4(x_2 + sh) = f_2 + \dfrac{s}{2}(f_3 - f_1) + \dfrac{s^2}{2}(f_3 - 2f_2 + f_1)$

$$+ \frac{s(s^2-1)}{12}(f_4 - 2f_3 + 2f_1 - f_0)$$

$$+ \frac{s^2(s^2-1)}{24}(f_4 - 4f_3 + 6f_2 - 4f_1 + f_0)$$

Now substitute $s = -2, -1, 0, 1, 2$ to get result.

EXERCISE 5.4

1. $L_0(x) = (-x^3 + 7x^2 - 12x)/60$ $L_0'(x) = (-3x^2 + 14x - 12)/60$

   $L_1(x) = (x^3 - 5x^2 - 2x + 24)/24$ $L_1'(x) = (3x^2 - 10x - 2)/24$

   $L_2(x) = (-x^3 + 2x^2 + 8x)/15$ $L_2'(x) = (-3x^2 + 4x + 8)/15$

   $L_3(x) = (x^3 - x^2 - 6x)/24$ $L_3'(x) = (3x^2 - 2x - 6)/24$

   $s_0(x) = (x + 2)[L_0(x)]^2$ $r_0(x) = 1 - 2(x + 2)L_0'(-2)[L_0(x)]^2$

   $s_1(x) = x[L_1(x)]^2$ $r_1(x) = 1 - 2xL_1'(0)[L_1(x)]^2$

   $s_2(x) = (x - 3)[L_2(x)]^2$ $r_2(x) = 1 - 2(x - 3)L_2'(3)[L_2(x)]^2$

   $s_3(x) = (x - 4)[L_3(x)]^2$ $r_3(x) = 1 - 2(x - 4)L_3'(4)[L_3(x)]^2$

   $H_7(1) = -3.84$

2. $H_5(1.2) = 0.090\,233$

3. Proof is similar to that of Theorem 5.1. Define

$$g(t) = f(t) - H_{2n+1}(t) - C(t - x_0)^2 (t - x_1)^2 \cdots (t - x_n)^2$$

Take

$$C = \frac{f(x) - H_{2n+1}(x)}{(x - x_0)^2 \cdots (x - x_n)^2}$$

so that g vanishes at $x_0, x_1, \ldots, x_n$ and x. Now g' vanishes at $x_0, x_1, \ldots, x_n$, and by Rolle's theorem it has an additional $n + 1$ zeros between the points $x_0, x_1, \ldots, x_n$ and x, i.e. g' has $2n + 2$ zeros. By repeated application of Rolle's theorem, $g^{(2n+2)}(\xi) = 0$ for some $\xi \in (a, b)$. Hence

$$0 = f^{2n+2}(\xi) - 0 - (2n + 2)!C$$

as required.

4. $e^x - H_5(x) = x^2(x - 0.5)^2(x - 1)^2 \dfrac{e^\xi}{6!} \qquad \xi \in (0, 1)$

Now $\displaystyle \max_{\xi \in (0,1)} |e^\xi| = e = 2.718\,281\,8$

Hence $|e^{0.75} - H_5(0.75)| \leq (0.75)^2 (0.25)^2 (-0.25)^2 \dfrac{2.718\,281\,8}{720}$

$$= 0.000\,008\,3$$

In fact $H_3(0.75) = 2.116\,994\,8$, so actual error is $0.000\,005\,2$.

5. Spurious oscillations towards end of range. Approximation to a large number of equally spaced data points is as much a problem in Hermite interpolation as in Lagrange interpolation because the polynomials $L_i(x)$ are used in both cases.

6. $H_{3,0} = 3x^3 - 5x^2 + x + 2$

   $H_{3,1} = (-x^2 + 2x + 3)/4$

EXERCISE 5.5

1. $s_0(1) = 0 = s_1(1)$; $s_0'(1) = 5 = s_1'(1)$; $s_0''(1) = 11 = s_1''(1)$

   $s_1(2) = 25/2 = s_2(2)$; $s_1'(2) = 22 = s_2'(2)$; $s_1''(2) = 23 = s_2''(2)$

2. Symmetry is obvious. Eigenvalues > 0 by Gerschgorin's first theorem.

3.

x	(a)	(b)
0.95	0.8356	0.8144
0.125	0.0004	0.0013

(b) More accurate than (a) since extra information about the function is used. Moreover $f''(-1) = 12 = f''(1)$, which is very different from the value 0 taken in (a).

4.

x	(a)	(b)	(c)
-3	-7	-5.6	-7
1	5	3.6	5
5	209	204.3	209

$f(x) = x^3 + 3x^2 + 2x - 1$

(a) and (c) reproduce this polynomial exactly. However, (b) cannot be exact because $f''(-6) \neq 0$ and $f''(7) \neq 0$.

5. No spurious oscillations present towards the end of range.

6.

If $s_i(x_i) = y_i$	then $a_i = y_i$	(1)
If $s_i(x_{i+1}) = y_{i+1}$	then $a_i + b_i h + c_i h^2 + d_i h^3 = y_{i+1}$	(2)
If $s_i''(x_i) = z_i$	then $2c_i = z_i$, i.e. $c_i = z_i/2$	(3)
If $s_i''(x_{i+1}) = z_{i+1}$	then $2c_i + 6hd_i = z_{i+1}$, i.e. $d_i = (z_{i+1} - z_i)/6h$	(4)

Substitute (1), (3), (4) into (2) to get

$$b_i = [(y_{i+1} - y_i)/h] - [h(2z_i + z_{i+1})/6] \tag{5}$$

Now, if $\quad s_{i-1}'(x_i) = s_i'(x_i) \quad$ then $\quad b_{i-1} + 2c_{i-1}h + 3d_{i-1}h^2 = b_i$

Substitute (3), (4), (5) to get

$$z_{i+1} + 4z_i + z_{i-1} = \frac{6}{h^2}(y_{i+1} - 2y_i + y_{i-1})$$

Free boundary conditions give $z_0 = 0 = z_3$.

$$\left. \begin{array}{l} 4z_1 + z_2 = (2/3)(-3.40) \\ z_1 + 4z_2 = (2/3)(0.30) \end{array} \right\} \Rightarrow \begin{array}{l} z_1 = -0.618 \\ z_2 = 0.204 \end{array}$$

From (1), (3), (4), (5),

$a_0 = 4.50,\quad a_1 = 7.00, a_2 = 6.10;\qquad c_0 = 0,\quad c_1 = -0.31, c_2 = 0.10$
$d_0 = -0.03, d_1 = 0.05, d_2 = -0.01;\qquad b_0 = 1.14, b_1 = 0.22,\quad b_2 = -0.40$

EXERCISE 5.6

1. $y = -1.1289 + 3.5585x$
   $y = -0.3884 + 2.6931x + 0.1562x^2$
   $y = \quad 1.7609 - 2.3007x + 2.1377x^2 - 0.2065x^3$

Do not necessarily get more accurate results at any particular point as degree increases.

2. (a) $y = -0.6500 + 3.1333x$
   (b) $y = -0.9551 + 2.7266x$
   (c) $y = -1.0334 + 2.6222x$
   Increasing values of α produce more accurate approximations at $x = 5$, as expected.

3. $a \sum_{i=1}^{n} w_i e^{2x_i} + b \sum_{i=1}^{n} w_i = \sum_{i=1}^{n} w_i y_i e^{x_i}$

   $a \sum_{i=1}^{n} w_i + b \sum_{i=1}^{n} w_i e^{-2x_i} = \sum_{i=1}^{n} w_i y_i e^{-x_i}$

   $a = 2.0202, \qquad b = 3.0144$

4. $\sum_{i=1}^{n} (y_i - ax_i^b)x_i^b = 0$

   $\sum_{i=1}^{n} (y_i - ax_i^b)ax_i^b \ln x_i = 0$

   System is non-linear.

   $A = 1.0976, \qquad B = 0.5076$
   $y = 2.9970x^{0.5076}$

5. $m = 3; (2.5, 103.5)$ should be $(2.5, 10.35)$

6. Minimize

   $$\sum_{i=1}^{4} [y_i - (a_0 + a_1 x_i + a_2 x_i^2)]^2$$

   subject to

   $a_0 - a_1 + a_2 = 1$

   $-2 \sum_{i=1}^{4} [y_i - (a_0 + a_1 x_i + a_2 x_i^2)] + \lambda = 0$

   $-2 \sum_{i=1}^{4} [y_i - (a_0 + a_1 x_i + a_2 x_i^2)]x_i - \lambda = 0$

   $-2 \sum_{i=1}^{4} [y_i - (a_0 + a_1 x_i + a_2 x_i^2)]x_i^2 + \lambda = 0$

   $a_0 = 2.3566, \qquad a_1 = 1.8316, \qquad a_2 = 0.4750$

   Otherwise: introduce the extra point $(-1, 1)$ and assign a large weight to it.

EXERCISE 5.7

1. $p_1(x) = a_0 + a_1 x$

$$\left.\begin{array}{l} \dfrac{\pi^2}{8}a_0 + \dfrac{\pi^3}{24}a_1 = 1 \\[3mm] \dfrac{\pi^3}{24}a_0 + \dfrac{\pi^4}{64}a_1 = \pi - 2 \end{array}\right\} \quad a_0 = -\dfrac{120}{\pi^2} + \dfrac{384}{\pi^3}, \qquad a_1 = \dfrac{384}{\pi^3} - \dfrac{1152}{\pi^4}$$

Three alternating extrema of opposite signs.

2. $\varphi_0(x) = 1, \qquad \varphi_1(x) = x - \pi/3$

$p_1(x) = a_0 \varphi_0(x) + a_1 \varphi_1(x)$

$$a_0 = \dfrac{8}{\pi^2}, \qquad a_1 = \dfrac{384}{\pi^3} - \dfrac{1152}{\pi^4}$$

3. $1, \qquad x, \qquad x^2 - 1/3, \qquad x^3 - 3x/5$

4. $p_2(x) = a_0 \varphi_0(x) + a_1 \varphi_1(x) + a_2 \varphi_2(x)$

$$a_0 = \frac{\displaystyle\int_{-1}^{1} f(t)\,dt}{\displaystyle\int_{-1}^{1} 1^2\,dt} = \frac{1}{2}\int_{-1}^{1} f(t)\,dt$$

$$a_1 = \frac{\displaystyle\int_{-1}^{1} t f(t)\,dt}{\displaystyle\int_{-1}^{1} t^2\,dt} = \frac{3}{2}\int_{-1}^{1} t f(t)\,dt$$

$$a_2 = \frac{\displaystyle\int_{-1}^{1} (t^2 - 1/3) f(t)\,dt}{\displaystyle\int_{-1}^{1} (t^2 - 1/3)^2\,dt} = \frac{45}{8}\int_{-1}^{1} (t^2 - 1/3) f(t)\,dt$$

5. If φ_i is multiplied by a constant α_i, then we require

$$\int_a^b w(x)\alpha_i^2\, \varphi_i^2(x)\,dx = 1, \qquad \text{i.e. } \alpha_i = \left[\int_a^b w(x)\varphi_i^2(x)\,dx\right]^{-1/2}$$

$1, \qquad \dfrac{1}{\sqrt{12}}(x - \tfrac{1}{2}), \qquad \dfrac{1}{\sqrt{180}}(x^2 - x + \tfrac{1}{6}), \qquad \dfrac{1}{\sqrt{2800}}(x^3 - \tfrac{3}{2}x^2 + \tfrac{3}{5}x - \tfrac{1}{20})$

6. From Leibnitz's rule,

$$\frac{d^n}{dx^n}[x^n e^{-x}] = x^n \frac{d^n}{dx^n}(e^{-x}) + \binom{n}{1}\frac{d}{dx}(x^n)\frac{d^{n-1}}{dx^{n-1}}(e^{-x})$$

$$+ \binom{n}{2}\frac{d^2}{dx^2}(x^n)\frac{d^{n-2}}{dx^{n-2}}(e^{-x}) + \cdots + \frac{d^n}{dx^n}(x^n)e^{-x}$$

Result then follows since

$$\frac{d^i}{dx^i}(e^{-x}) = (-1)^i e^{-x}$$

and

$$\frac{d^i}{dx^i}(x^n) = n(n-1)\cdots(n-i+1)x^{n-i}$$

If $m > n$, then

$$\int_0^\infty e^{-x}P_n(x)P_m(x)\,dx = (-1)^n \int_0^\infty \frac{d^n}{dx^n}(x^n e^{-x})P_m(x)\,dx$$

$$= (-1)^n \left[\frac{d^{n-1}}{dx^{n-1}}(x^n e^{-x})P_m(x)\right]_0^\infty$$

$$+ (-1)^{n+1}\int_0^\infty \frac{d^{n-1}}{dx^{n-1}}(x^n e^{-x})P_m'(x)\,dx$$

The first term is zero, since by Leibnitz's rule $(d^{n-1}/dx^{n-1})(x^n e^{-x})$ contains a factor of xe^{-x}. This process can be repeated a further m times to obtain

$$\int_0^\infty e^{-x}P_n(x)P_m(x)\,dx = (-1)^{n+m+1}\int_0^\infty \frac{d^{n-m-1}}{dx^{n-m-1}}(x^n e^{-x})P_m^{(m+1)}(x)\,dx = 0$$

since P_m is a polynomial of degree m.

$$x - 1, \qquad x^2 - 4x + 2, \qquad x^3 - 9x^2 + 18x - 6,$$
$$x^4 - 16x^3 + 72x^2 - 96x + 24$$

EXERCISE 6.1

1. Put $k = 4$ in (6.3) to derive approximation.
   From (6.4),

$$|e_4'(x_2)| = \left|\frac{2(2-1)(2-3)(2-4)}{5!}h^4 f^{(5)}(x_0 + \theta h)\right| \leqslant \frac{h^4 M}{30}$$

$$|\text{rounding error}| \leqslant \frac{\varepsilon + 8\varepsilon + 8\varepsilon + \varepsilon}{12h} = \frac{3\varepsilon}{2h}$$

2. $f_1 = f(x_0 + h) = f_0 + hf'_0 + \dfrac{h^2}{2!} f''_0 + O(h^3)$

$\quad f_2 = f(x_0 + 2h) = f_0 + 2hf'_0 + \dfrac{(2h)^2}{2!} f''_0 + O(h^3)$

Hence

$$4f_1 - f_2 = 3f_0 + 2hf'_0 + O(h^3)$$

$$\text{i.e. } (4f_1 - f_2 - 3f_0)/2h = f'_0 + O(h^2)$$

The other two formulas can be derived in the same way by expanding about x_1 and x_2, respectively.

Using each of these formulas in turn gives

$f'(0.01) \approx 4.316\,30, \qquad f'(0.05) \approx 1.631\,35, \qquad f'(0.1) \approx 0.770\,70$

3. $12.064\,00,\ 12.081\,25,\ 12.088\,89,\ 12.075\,00,\ 12.000\,00$

4. Bound follows from equation (6.9), which is minimized by taking $h = (48\varepsilon/M)^{1/4}$.

$\varepsilon = (1/2) \times 10^{-5}, \qquad M = 256|\sin 4.2| \qquad h = 0.032$

5. $\qquad p'_2(x_1) = \dfrac{(x_1 - x_2)}{(x_0 - x_1)(x_0 - x_2)} y_0 + \dfrac{(2x_1 - x_0 - x_2)}{(x_1 - x_0)(x_1 - x_2)} y_1$

$\qquad\qquad + \dfrac{(x_1 - x_0)}{(x_2 - x_0)(x_2 - x_1)} y_2$

$\quad p'_2(0.6) = 0.652\,62$

$f(x) - p_2(x) = (x - x_0)(x - x_1)(x - x_2)\dfrac{f'''(\xi)}{3!}; \qquad \xi \in (x_0, x_2)$

$f'(x_1) - p'_2(x_1) = (x_1 - x_0)(x_1 - x_2)\dfrac{f'''(\xi)}{6}$

$f'''(\xi) = \dfrac{3}{8}\xi^{-5/2} \leqslant \dfrac{3}{8}\left(\dfrac{1}{2}\right)^{-5/2} = \dfrac{6}{\sqrt{8}}$

Hence $|f'(0.6) - p'_2(0.6)| \leqslant \dfrac{0.04}{\sqrt{8}}$

Actual error $= 0.007\,12$

EXERCISE 6.2

1. Put $k = 3$ in (6.12) to obtain

$$\int_a^b p_3(x)\,dx = h\left[sf_0 + (f_1 - f_0)\frac{s^2}{2} + (f_2 - 2f_1 + f_0)\frac{1}{2}\left(\frac{s^3}{3} - \frac{s^2}{2}\right) \right.$$

$$\left. + (f_3 - 3f_2 + 3f_1 - f_0)\frac{1}{6}\left(\frac{s^4}{4} - s^3 + s^2\right) \right]_0^3$$

$$= \frac{3h}{8}(f_0 + 3f_1 + 3f_2 + f_3)$$

For error term, put $k = 3$ in (6.16) to obtain

$$\frac{h^5}{4!}f^{(4)}(x_0 + \xi h)\left[\frac{s^5}{5} - \frac{3s^4}{2} + \frac{11s^3}{3} - 3s^2\right]_0^3 = \frac{-3h^5}{80}f^{(4)}(x_0 + \xi h)$$

2. Closed:

k	(a)	(b)
0	0.000 000 0	1.000 000 0
1	0.077 543 0	1.859 140 9
2	0.095 383 0	1.475 730 6
3	0.095 388 9	1.468 713 7

Open:

k	(a)	(b)
0	0.104 303 0	1.284 025 4
1	0.101 337 6	1.338 571 3
2	0.095 403 0	1.451 691 0
3	0.095 400 1	1.454 877 0

3. It is easy to check that

$$\frac{b-a}{6}\left[a^i + 4\left(\frac{a+b}{2}\right)^i + b^i\right] = \frac{b^{i+1} - a^{i+1}}{i+1}$$

for $i = 0, 1, 2$ and 3.
Deduction follows from linearity of integration.
Error term contains a factor $f^{(4)}$, which is zero when f is a cubic.

4. (a) 0.965 926   (b) 0.982 963
   Exact value is 0.989 043.
   Error in (a) is

$$-\frac{(1)^3}{12}f''(\eta_0); \qquad \eta_0 \in (0.5, 1.5)$$

Error in (b) is

$$-\frac{(1/2)^3}{12}[f''(\eta_1) + f''(\eta_2)]; \qquad \eta_1 \in (0.5, 1.0), \qquad \eta_2 \in (1.0, 1.5)$$

So would expect error in (a) to be about four times that of (b).

5. Put $n = 1$, $x_0 = 0$ and $x_1 = 1$ in Theorem 5.1. Multiply both sides by x^α and integrate to get

$$\int_0^1 x^\alpha f(x)\,dx = f(1)\int_0^1 x^{\alpha+1}\,dx - f(0)\int_0^1 (x^{\alpha+1} - x^\alpha)\,dx$$

$$+ \tfrac{1}{2}f''(\eta)\int_0^1 (x^{\alpha+2} - x^{\alpha+1})\,dx$$

$$= \frac{1}{2+\alpha}\left[f(1) + \frac{1}{1+\alpha}f(0) \right] - \frac{f''(\eta)}{2(3+\alpha)(2+\alpha)}$$

(a) Error $= -e^{\eta/2}/30$; $1/30 \le |\text{error}| \le e^{1/2}/30$
(b) Error $= -(1+\eta^2)e^{\eta^2/2}/6$; $1/6 \le |\text{error}| \le e^{1/2}/3$
The error ranges do not overlap.

6. $\displaystyle\int_{x_0}^{x_1} f(x)\,dx = h\int_0^1 f_0 + s(f_1 - f_0) + \frac{s(s-1)}{2!}(f_2 - 2f_1 + f_0)\,ds$

$$+ h^4 \frac{f'''(\eta)}{3!}\int_0^1 s(s-1)(s-2)\,ds$$

$$\frac{0.5}{12}[5\sin(0) + 8\sin(0.5) - \sin(1)] = 0.1247$$

$$|\text{error}| = \frac{(0.5)^4}{24}|-\cos\eta| \le \frac{(0.5)^4}{24} = 0.0026$$

EXERCISE 6.3

1. $\displaystyle\int_a^b f(x)\,dx = h\sum_{i=0}^{n-1} f_i + \frac{h^2}{2}\sum_{i=0}^{n-1} f'(x_i + \xi_i h)$

By the intermediate value theorem,

$$\frac{1}{n}\sum_{i=0}^{n-1} f'(x_i + \xi_i h) = f'(x_0 + \eta h) \qquad \eta \in (0, n)$$

Result then follows since $nh = b - a$.
0,   0.202 732 5,   0.297 056 1,   0.342 322 2
Exact value is 0.386 294 3 (7D), so errors reduce by a factor of about two at each stage.

2. (a) 0.166 39   (b) 4.850 70   (c) −8.000 00   (d) 0.163 23
Trapezium rule:

nD	(a)	(b)	(c)	(d)
1	8	7	12	1
2	27	32	37	4
5	≈ 1050	≈ 850	≈ 1250	47

Simpson's rule:

nD	(a)	(b)	(c)	(d)
1	6	2	4	2
2	8	4	8	2
5	48	18	40	6

3. Rounding error $\leqslant \dfrac{h}{3}(2\varepsilon + 4n\varepsilon + 2(n-1)\varepsilon) = (b-a)\varepsilon$

4. (a) 46   (b) 196   (c) 1152   (d) impossible

5. $\displaystyle\int_0^1 x^2(1-x)^2 f^{(4)}(x)\,dx$

$$= [x^2(1-x)^2 f'''(x)]_0^1 - \int_0^1 (4x^3 - 6x^2 + 2x)f'''(x)\,dx$$

$$= -[(4x^3 - 6x^2 + 2x)f''(x)]_0^1 + \int_0^1 (12x^2 - 12x + 2)f''(x)\,dx$$

$$= [(12x^2 - 12x + 2)f'(x)]_0^1 - \int_0^1 (24x - 12)f'(x)\,dx$$

$$= 2\{f'(1) - f'(0)\} - [(24x - 12)f(x)]_0^1 + 24\int_0^1 f(x)\,dx$$

$$= 2\{f'(1) - f'(0)\} - 12\{f(1) + f(0)\} + 24\int_0^1 f(x)\,dx$$

However,

$$\int_0^1 x^2(1-x)^2 f^{(4)}(x)\,dx = f^{(4)}(\xi)\int_0^1 x^2(1-x)^2\,dx = \frac{f^{(4)}(\xi)}{30}$$

Now,

$$\int_a^b f(x)\,dx = \sum_{i=0}^{n-1}\int_{x_i}^{x_{i+1}} f(x)\,dx = h\sum_{i=0}^{n-1}\int_0^1 f(x_i + sh)\,ds$$

$$= \frac{h}{2}\sum_{i=0}^{n-1}(f_i + f_{i+1}) - \frac{h^2}{12}\sum_{i=0}^{n-1}(f'_{i+1} - f'_i) + \frac{h^5}{720}\sum_{i=0}^{n-1}f^{(4)}(\xi_i)$$

$$= \frac{h}{2}\sum_{i=0}^{n-1}(f_i + f_{i+1}) + \frac{h^2}{12}\{f'(a) - f'(b)\} + \frac{nh^5}{720}f^{(4)}(\eta)$$

6.259 370 7,   6.380 388 5,   6.388 504 8,   6.389 021 5

6. From (6.21),

$$\int_a^b f(x)\,dx = \sum_{i=0}^{n-1} \int_{x_{2i}}^{x_{2i+2}} f(x)\,dx = 2h \sum_{i=0}^{n-1} f_{2i+1} + \frac{h^3}{3} \sum_{i=0}^{n-1} f''(x_{2i} + \xi_i h)$$

$$= 2h \sum_{i=0}^{n-1} f_{2i+1} + \frac{(b-a)h^2}{6} f''(x_0 + \eta h)$$

for some $\eta \in (0, 2n)$ by the intermediate value theorem, since $(b-a) = 2nh$.
0.405 465 1,   0.391 379 7,   0.387 588 3

$$|\text{error}| \leqslant \frac{h^2}{6(x_0 + \eta h)^2} \leqslant \frac{h^2}{6} \quad \text{since} \quad x_0 + \eta h \in (1, 2)$$

If $h^2/6 \leqslant (1/2) \times 10^{-8}$ then $h \leqslant (\sqrt{3}) \times 10^{-4}$, i.e. 5774 subintervals.

EXERCISE 6.4

1. If $n = 2^{k-1}$ and $h = (b-a)/2n$,

$$T_{k,1} = (4T_{k,0} - T_{k-1,0})/3$$

$$= \frac{1}{3}\left\{ 4\left(\frac{h}{2}\right)(f_0 + 2f_1 + 2f_2 + \cdots + 2f_{2n-1} + f_{2n}) \right.$$

$$\left. -\frac{(2h)}{2}(f_0 + 2f_2 + 2f_4 + \cdots + 2f_{2n-2} + f_{2n}) \right\}$$

$$= \frac{h}{3}(f_0 + 4f_1 + 2f_2 + \cdots + 2f_{2n-2} + 4f_{2n-1} + f_{2n})$$

2. 0.346 573 5
   0.376 019 3   0.385 834 6
   0.383 699 5   0.386 259 6   0.386 287 9
   0.385 643 9   0.386 292 0   0.386 294 2   0.386 294 3

3.

nD	(a)	(b)	(c)	(d)
1	2	1	2	0
2	3	2	3	1
5	4	3	4	2

4. (a) 5   (b) 8   (c) 10

5. $I = R_{k,0} + \alpha_1 h_k + \alpha_2 h_k^2 + \alpha_3 h_k^3 + \cdots$

$$I = R_{k+1,0} + \alpha_1 \frac{h_k}{2} + \alpha_2 \left(\frac{h_k}{2}\right)^2 + \alpha_3 \left(\frac{h_k}{2}\right)^3 + \cdots$$

Multiply second equation by two and subtract from first equation to get

$$I = (2R_{k+1,0} - R_{k,0}) - \frac{\alpha_2}{2} h_k^2 - \frac{3\alpha_3}{4} h_k^3 - \cdots$$

i.e. $I - R_{k+1,1} = O(h^2)$

$R_{k+1,2}$ and $R_{k+1,3}$ similarly.

$R_{3,3} = 0.386\,273\,0$

6. If $n = 2^{k-1}$ and $h = (b-a)/2n$,

$$R_{k,1} = 2h(f_0 + f_1 + f_2 + \cdots + f_{2n-2} + f_{2n-1})$$
$$- (2h)(f_0 + f_2 + \cdots + f_{2n-2})$$
$$= 2h(f_1 + f_3 + \cdots + f_{2n-1})$$

EXERCISE 6.5

1. (a) $[0,2]; S_1 = 1.804\,74, S_2 = 1.856\,94; |S_1 - S_2| < 15\varepsilon$
    So $I \approx 1.856\,94$

    (b) $[1,2];$ $S_1 = 1.218\,87, S_2 = 1.218\,95; |S_1 - S_2| < 15\varepsilon/2$
    $[0,1];$ $S_1 = 0.638\,07, S_2 = 0.656\,53; |S_1 - S_2| \nless 15\varepsilon/2$
    $[1/2,1];$ $S_1 = 0.430\,93, S_2 = 0.430\,96; |S_1 - S_2| < 15\varepsilon/4$
    $[0,1/2];$ $S_1 = 0.225\,59, S_2 = 0.232\,12; |S_1 - S_2| \nless 15\varepsilon/4$
    $[1/4,1/2]; S_1 = 0.152\,36; S_2 = 0.152\,37; |S_1 - S_2| < 15\varepsilon/8$
    $[0,1/4];$ $S_1 = 0.079\,76; S_2 = 0.082\,07; |S_1 - S_2| \nless 15\varepsilon/8$
    $[1/8,1/4]; S_1 = 0.053\,87; S_2 = 0.053\,87; |S_1 - S_2| < 15\varepsilon/16$
    $[0,1/8];$ $S_1 = 0.028\,20; S_2 = 0.029\,01; |S_1 - S_2| < 15\varepsilon/16$

    $I \approx 1.218\,95 + 0.430\,96 + 0.152\,37 + 0.053\,87 + 0.029\,01 = 1.885\,16$

    Exact value is $1.885\,62$.
    In part (a), error > 0.005. This discrepancy is due to the assumption that the values of η_1 and η_2 in (6.32) and (6.33) are the same. In practice it may be better to 'play safe' and test on 10ε, say, rather than 15ε.

2.

	(a)	(b)	(c)	(d)
0.05	2	2	4	2
0.005	4	2	6	2
0.000\,005	22	12	30	6

3. (a) 10 (b) 30 (c) 49

4. $I - T_1 = -\dfrac{h^3}{12} f''(a + \eta_1 h)$

$$I - T_2 = -\frac{h}{12}\left(\frac{h}{2}\right)^2 f''(a + \eta_2 h)$$

If

$$f''(a + \eta_1 h) \approx f''(a + \eta_2 h) \approx f''(\eta)$$

then

$$T_2 - T_1 \approx \frac{-3h^3}{48} f''(\eta) \approx 3(I - T_2)$$

Error $\approx \frac{1}{3}|T_2 - T_1| = \frac{1}{3}|1.075\,13 - 1.091\,75| = 0.005\,54$

5.
n	Integral	Subintervals
1	0.367 880	8
7	0.112 384	22
20	0.045 545	30
60	0.016 133	30

6.
m	k	$\Gamma(m)$	Subintervals
1.5	11	0.886 175	48
3	15	1.999 926	26
6	26	119.999 941	84

EXERCISE 6.6

1. Put $k = 1, m = 3, a = -1$ and $b = 1$ in (6.36)

$$v(c_0 + c_1) + u(c_0 x_0 + c_1 x_1) + c_0 x_0^2 + c_1 x_1^2 = 2v + 2/3$$

i.e. $c_0(x_0^2 + ux_0 + v) + c_1(x_1^2 + ux_1 + v) = 2v + 2/3$
i.e. $0 = 2v + 2/3$ (since x_0, x_1 are roots of $x^2 + ux + v = 0$)
i.e. $v = -1/3$

Similarly for u.
If $x^2 - 1/3 = 0$ then $x = \pm 1/\sqrt{3}$.
Solving first two equations for c_0, c_1 gives $c_0 = c_1 = 1$.

2.
	$k = 0$	$k = 1$	$k = 2$	$k = 3$
(a)	0.014 05	0.259 80	0.180 84	0.165 55
(b)	2.828 43	4.840 65	4.849 78	4.850 70
(c)	0.000 00	− 8.975 80	− 7.967 51	− 8.000 50
(d)	0.155 71	0.162 84	0.163 22	0.163 23

3. Newton's method gives $x_0 = 0.112\,70, x_1 = 0.500\,00, x_2 = 0.887\,30$.
   Hence $c_0 = c_2 = 0.277\,78, c_1 = 0.444\,44$.

$$\int_0^1 e^x \, dx \approx 1.718\,28$$

4. Now p_3 is a polynomial of degree 3, for which the formula is exact, so

$$E = \int_{-1}^{1} e_3(x)\,dx - e_3\left(\frac{-1}{\sqrt{3}}\right) - e_3\left(\frac{1}{\sqrt{3}}\right)$$

$$\Rightarrow |E| \leqslant \frac{M}{24}\left\{\int_{-1}^{1} x^4\,dx + \frac{1}{9} + \frac{1}{9}\right\} = \frac{7M}{270}$$

5. $x_0 = \frac{3}{7} - \frac{2}{7}\sqrt{\frac{6}{5}}, \qquad x_1 = \frac{3}{7} + \frac{2}{7}\sqrt{\frac{6}{5}}$

$c_0 + c_1 = 2$
$c_0 x_0 + c_1 x_1 = 2/3$
$c_0 x_0^2 + c_1 x_1^2 = 2/5$
$c_0 x_0^3 + c_1 x_1^3 = 2/7$

Solving the first two equations for c_0, c_1 gives

$$c_0 = 1 + \frac{1}{3}\sqrt{\frac{5}{6}}, \qquad c_1 = 1 - \frac{1}{3}\sqrt{\frac{5}{6}}$$

It is easy to check that the last two equations are satisfied with these values.

$$\int_0^1 x^{-1/2}\,e^{x/2}\,dx \approx 2.3899$$

6. $c_0 + c_1 = 1, \qquad c_1 x_1 = 1/2, \qquad c_1 x_1^2 = 1/3$

have solution $x_1 = 2/3, c_1 = 3/4, c_0 = 1/4$.

Degree two.

EXERCISE 7.1

1. (a) 0.606 516 4
   (b) 0.584 262 3
   (c) 0.582 203 6

$$y' = -\frac{(e^x - 1)}{(e^x - x)^2} = \frac{(1-x)}{(e^x - x)^2} - \frac{(e^x - x)}{(e^x - x)^2} = (1-x)y^2 - y$$

$$y(0) = \frac{1}{e^0 - 0} = 1$$

   (a) 0.024 539 7
   (b) 0.002 285 6
   (c) 0.000 226 9

Since errors decrease by ≈ 10 at each stage, $p = 1$.
Now if error $= Ah$ then, from (c), $A \approx 0.23$.
For 8D accuracy, error $\leqslant (1/2) \times 10^{-8}$ and so $h \leqslant (1/2) \times 10^{-8}/0.23 \approx 2 \times 10^{-8}$.

2. (a) 0.566 698 5
   (b) 0.590 814 1
   (c) 0.589 348 1
   (d) 0.589 441 1

One correction produces most accurate estimate.

3. From the mean value theorem

$$f(x_{n+1}, y_{n+1}) - f(x_{n+1}, y_{n+1}^{[m]}) = (y_{n+1} - y_{n+1}^{[m]}) \frac{\partial f}{\partial y}$$

where f_y is evaluated at (x_{n+1}, ξ) for some number ξ which depends on m and n and lies between y_{n+1} and $y_{n+1}^{[m]}$.
If we let $e^{[m]} = y_{n+1} - y_{n+1}^{[m]}$ and subtract (2) from (1)

then $e^{[m+1]} = h\beta_0 e^{[m]} \dfrac{\partial f}{\partial y}$

Hence $|e^{[m+1]}| \leqslant h|\beta_0| |e^{[m]}| K$

and $|e^{[m]}| \leqslant (h|\beta_0|K)^m |e^{[0]}| \to 0$

provided $h|\beta_0|K < 1$

(This result also follows from Theorem 3.1.)

4. (a) 1
   (b) 3
   (c) 15

This is to be expected since, from Question 3, the smaller the value of h the faster the convergence.

5. Apply Simpson's rule to (7.8)

6. $y(x_r) = 1 - rh + \dfrac{(rh)^2}{2!} - \dfrac{(rh)^3}{3!} + \dfrac{(rh)^4}{4!} - \dfrac{(rh)^5}{5!} + \dfrac{(rh)^6}{6!} - \dfrac{(rh)^7}{7!} + \cdots$

$y(0.1) = 0.904\,837, \qquad y(0.2) = 0.818\,731, \qquad y(0.3) = 0.740\,818,$
$y_4^{[0]} = 0.670\,323, \qquad y_4^{[1]} = 0.670\,319, \qquad y_4^{[2]} = 0.670\,320$

EXERCISE 7.2

1. If $h = 0$ then $y_{n+1} = y_n$, i.e. $y_n = y_0$ for all n. Hence bounded.

2. (a) A-stable
   (b) $h \leqslant 6/\lambda$
   (c) Stable only for $h = 0$. (Methods with this property are said to be *weakly stable* and are not recommended for general use.)

3. (a) $\dfrac{h^2}{2} y''(x_n)$

   (b) $-\dfrac{h^3}{12} y'''(x_n)$

   (c) $-\dfrac{h^4}{24} y^{(4)}(x_n)$

   (d) $\dfrac{3h^4}{8} y^{(4)}(x_n)$

4. $y_{n+1}^{[0]} = (1 - \lambda h) y_n$

   $$y_{n+1} = y_{n+1}^{[1]} = y_n + \frac{h}{2} [-\lambda y_{n+1}^{[0]} - \lambda y_n]$$

   $$= \left(1 - \lambda h + \frac{\lambda^2 h^2}{2}\right) y_n$$

   Hence stable if and only if $-1 \leqslant 1 - \lambda h + (\lambda^2 h^2/2) \leqslant 1$.
   The right hand inequality gives $h \leqslant 2/\lambda$.
   The left hand inequality can be rearranged as $(3/2) + (1/2)(\lambda h - 1)^2 \geqslant 0$
   which is always true.
   Hence stable if and only if $h \leqslant 2/\lambda$.

5. If $h = 0$ then the auxiliary equation is $m^2 - (1/2)m - (1/2) = 0$ and has roots
   $1, -1/2$.

   (a) $\beta_0 = 0,$ $\beta_1 = 7/4,$ $\beta_2 = -1/4;$ $h \leqslant 1/2\lambda$
   (b) $\beta_0 = 3/8,$ $\beta_1 = 1,$ $\beta_2 = 1/8;$ $h \leqslant 2/\lambda$

6. (a) 0.5 (b) 0.3 (c) 3

EXERCISE 7.3

1.   *Estimates*   *Errors*
   (a) 0.581 991 5   0.000 014 8
   (b) 0.581 977 7   0.000 001 0

   On halving h the error has been reduced by 15 which is roughly in
   agreement with an $O(h^4)$ error.
   Now if error $\approx Ah^4$ then, from (b), $A \approx 0.01$.
   For 8D accuracy, error $\leqslant (1/2) \times 10^{-8}$ and so $h^4 \leqslant (1/2) \times 10^{-8}/0.01$, i.e.
   $h \leqslant 0.03$.

2., (a) $h \leqslant 2/\lambda$ (b) $\dfrac{h^3}{6}(f_x f_y + f_y^2 f)$

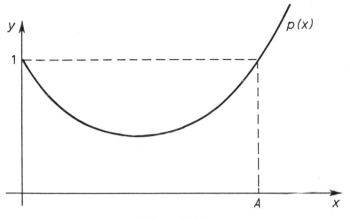

Figure S.10

3. $y_{n+1} = y_n + hf + \dfrac{h^2}{2!}(f_x + ff_y) + \dfrac{h^3}{3!}(f_{xx} + 2ff_{xy} + f_x f_y + ff_y^2 + f^2 f_{yy})$

$k_1 = f$

$k_2 = f + \dfrac{h}{2}(f_x + ff_y) + \dfrac{h^2}{8}(f_{xx} + 2ff_{xy} + f^2 f_{yy})$

$k_3 = f + h(f_x + ff_y) + \dfrac{h^2}{2}(f_{xx} + 2ff_{xy} + f^2 f_{yy} + 2ff_y^2 + 2f_x f_y)$

$$
\left.
\begin{array}{ll}
O(h): & c_1 + c_2 + c_3 & = 1 \\
O(h^2): & (1/2)c_2 + c_3 & = 1/2 \\
O(h^3): & (1/8)c_2 + (1/2)c_3 = 1/6 \quad \text{and} \quad c_3 = 1/6
\end{array}
\right\} \quad c_1 = c_3 = 1/6, \quad c_2 = 2/3.
$$

4. $p''(x) = 1 - x + \tfrac{1}{2}x^2 = \tfrac{1}{2}(x-1)^2 + \tfrac{1}{2} > 0 \Rightarrow$ convex

$p(0) = 1$

See Fig. S.10. Simple bisection gives $A = 2.8$ (1D).

EXERCISE 7.4

1.

	(a)	(b)	(c)
$u(0.2)$	0.864 49	0.904 09	0.909 49
$v(0.2)$	0.319 39	0.353 31	0.355 01

2. $u'' = -u'v^2 - 2uvv'$; $\quad v'' = -v' - e^u u'$
   $u(h) \approx 1 - h + (3 + 2e)h^2/2$; $\quad v(h) \approx 1 - (1 + e)h + (1 + 2e)h^2/2$
   Hence $u(0.2) \approx 0.968\,73$, $\quad v(0.2) \approx 0.385\,07$

3. $y_1' = y_2$

$y_2' = 11y_1 + 10y_2$

(a) 0.367 88   (b) 0.453 60

Ill-conditioned (see Example 1.2).

4. $y_1 = y$, $y_2 = y'$, $y_3 = z$, $y_4 = z'$ gives

$y_1' = y_2$; $y_1(1) = 0$

$y_2' = -y_1 - y_2^2 + \ln x$; $y_2(1) = 0.5$

$y_3' = y_4$; $y_3(1) = 0$

$y_4' = -y_3 - 2y_2 y_4$; $y_4(1) = 1$

$y(2) \approx 0.448\,26$, $z(2) = 0.586\,40$

5. $y(h) \approx -2h + \dfrac{6h^2}{2!} - \dfrac{13h^3}{3!}$

Hence $y(0.1) \approx -0.172\,17$

$[(y_{n+1} - 2y_n + y_{n-1})/h^2] + [3(y_{n+1} - y_{n-1})/2h] + 2y_n = x_n$

can be rearranged to get result.

Hence

$y_2 = \frac{1}{1.15}[0.01x_1 + 1.98y_1 - 0.85\,y_0]$

$= \frac{1}{1.15}[0.01(0.1) + 1.98(-0.172\,17) - 0.85\,(0)]$

$= -0.295\,56$

EXERCISE 8.1

1. General solution is $A \sin \omega x + B \cos \omega x + x$.
   The particular solution which satisfies the boundary conditions is

$$y(x) = \begin{cases} x & \omega \neq k\pi \\ x + A \sin \omega x & \omega = k\pi \end{cases} \quad k = \cdots -2, -1, 0, 1, 2, \ldots$$

If $y(0) = \delta$ and $y(1) = 1 + \varepsilon$

then $y(x) = \left(\dfrac{\varepsilon - \delta \cos \omega}{\sin \omega} \right) \sin \omega x + \delta \cos \omega x + x$

Change in solution is large whenever $\sin \omega \approx 0$, i.e. $\omega \approx k\pi$.

2. If $q(x) > 0$ then $|-(2 + h^2 q_i)| = 2 + h^2 q_i$

   If $h < 2/K$ then $|1 \pm hp_i/2| = 1 \pm hp_i/2$

   and $2 + h^2 q_i > \left(1 + \dfrac{hp_i}{2}\right) + \left(1 - \dfrac{hp_i}{2}\right)$

3. $y(x) = \dfrac{2}{e - e^2}(e^x - e^{2x})$

At $x = 0.5$,

	Estimate	Error
(a)	0.4348	0.0232
(b)	0.4525	0.0055

Halving h reduces error by 1/4.
To be expected since both of the finite difference approximations used have errors $O(h^2)$.

4.
$$-54y_0 + 32y_1 \qquad\qquad\qquad\qquad\qquad = 8$$
$$22.5y_0 - 51y_1 + 27.5y_2 \qquad\qquad\qquad = 15.625$$
$$33y_1 \;\; -73y_2 + 39y_3 \qquad\qquad = 27$$
$$45.5y_2 - 99y_3 + 52.5y_4 = 42.875$$
$$128y_3 - 112y_4 = -191$$

$$y_0 = 0.15, \qquad y_1 = 0.42, \qquad y_2 = 1.22, \qquad y_3 = 2.61, \qquad y_4 = 4.70$$

5.
$$5y_2 - 10y_1 + y_1y_2 \qquad\qquad = 0$$
$$5y_3 - 10y_2 + 5y_1 + y_2(y_3 - y_1) = 0$$
$$5y_4 - 10y_3 + 5y_2 + y_3(y_4 - y_2) = 0$$
$$-9y_4 + 5y_3 - y_3y_4 + 5 \qquad\qquad = 0$$

$$y^{[0]} = (0.2, 0.4, 0.6, 0.8)^{\mathrm{T}}$$
$$y^{[6]} = (0.2845, 0.5383, 0.7428, 0.8944)^{\mathrm{T}}$$

EXERCISE 8.2

1. $\lambda_{n+1} = \dfrac{\lambda_{n-1}F(\lambda_n) - \lambda_n F(\lambda_{n-1})}{F(\lambda_n) - F(\lambda_{n-1})}; \qquad F(\lambda) = y(1, \lambda) - 1$

$\lambda_5 = 1.439\,233\,3, \qquad y(1, \lambda_5) = 1.000\,000\,0$

2. $\lambda_4 = -1.000\,013\,7, \qquad y(2, \lambda_4) = 0.333\,333\,3$

$$y'' = \frac{2}{(1 + x)^3} = \frac{1}{(1 + x)^3} + \frac{1}{(1 + x)^3} = y^3 - yy'$$

$$y(0) = \frac{1}{1 + 0} = 1, \qquad y(2) = \frac{1}{1 + 2} = \frac{1}{3}$$

No, since $|y_1 - y(0.1)| = |0.909\,093\,0 - 0.909\,090\,9| = 2.1 \times 10^{-6}$

3. $\lambda_n = 0.4282$ for (a) $n = 13$ (b) $n = 2$ (c) $n = 1$
Secant and Newton's methods converge in one step for linear equations.

4. λ_n does not converge. Exact solution of initial value problem

$$y'' - 26y' + 25y = 0; \qquad y(0) = 0, \qquad y'(0) = \lambda$$

is

$$y(x, \lambda) = \frac{\lambda}{24} (e^{25x} - e^x)$$

and so $y(1, \lambda)$ is sensitive to small changes in λ.
This difficulty can be avoided by shooting from 'right to left', i.e. by solving the initial value problem

$$y'' - 26y' + 25y = 0 \qquad y(1) = 1 \qquad y'(1) = \lambda$$

from $x = 1$ to $x = 0$ and choosing λ so that $y(0, \lambda) = 0$. No difficulty this time because $y(x, \lambda)$ decays with decreasing x.

Index

MONDAY'S CHILD

Louise Bagshawe

KT-377-507

headline

First published in Great Britain in 2004
by HEADLINE BOOK PUBLISHING

5

Cataloguing in Publication Data is available from the British Library

ISBN 0 7553 0423 3 (hardback)
ISBN 0 7553 0865 4 (trade paperback)

Typeset in Meridien by Avon DataSet Ltd,
Bidford-on-Avon, Warwickshire

Printed and bound in Great Britain by
Clays Ltd, St Ives plc

Headline's policy is to use papers that are natural, renewable and
recyclable products and made from wood grown in sustainable
forests. The logging and manufacturing processes are expected to
conform to the environmental regulations of the country of origin.

HEADLINE BOOK PUBLISHING
A division of Hodder Headline
338 Euston Road
London NW1 3BH

www.headline.co.uk
www.hodderheadline.com

MONDAY'S CHILD

Monday's child is fair of face . . .

This novel is dedicated to my readers; I hope you have half as much fun reading it as I had writing it. Please come and visit me at www.louise-bagshawe.com (don't forget the hyphen!). I hope you'll email me there; I promise to read and answer all of them.

My thanks are, as ever, due to my wonderful agent Michael Sissons and his whole team at PFD, especially Jim Gill and Tim Corrie; to my editor, the peerless Rosie de Courcy, to Alexandra Mason, and everybody at Headline; and to the boys and girls on the Board, and here's hoping I make it to a Freelance Friday.

1

'And so movies, now more than ever, represent the UK's creativity. And its rich cultural diversity...'

I stare at the MC, trying not to seem too bored. It's hard going. I'm tired. It's been a long day at work, and it's not going to end anytime soon. My boss, Kitty Simpson, has been invited to this bash, and naturally she has to show off by taking an assistant. Because it costs two thousand pounds a head. It's a film industry night for 'charidee', where overpaid actors, directors, agents and producers supposedly raise money for AIDS awareness, but actually just want to see and be seen. The opulence is everywhere: mounds of caviar, circus performers walking around swallowing fire, expensive appetizers, women in dazzling gowns, men with watches that cost six figures. There's at least twenty grand's worth of out-of-season floral arrangements. You wonder why they don't just write a cheque to the Terrence Higgins Trust and save the overhead, but of course that would be no fun.

'Anna,' Kitty hisses at me. 'I *said* to go fetch my bag.'

'Sorry,' I whisper back.

'You're here to *assist* me. You should be paying attention,' she says, tossing her immaculately styled hair. Tonight she is wearing a long-sleeved, opaque dress with a mandarin collar to hide the wrinkles on her neck; I think it's Armani. She has teamed this with diamond chandelier earrings and an AIDS ribbon, but naturally her AIDS ribbon is a platinum brooch studded with rubies.

'You don't need to just stand there like a huge great sack of potatoes,' she snaps, obviously annoyed. 'Really, you could

1

have at least made an effort. At least *tried* to find a flattering dress.'

My face falls. I thought I'd done OK with this dress. It's navy velvet with sheer sleeves. Comes right down to my flats.

'But I suppose there's not much point, is there?' Kitty sighs, losing interest.

I am five-eleven. I have a bit of a tummy, strong arms and hands, and an unfortunate beak of a nose. There are some things I like about myself; I've got a decent bottom, not too flabby, and good legs, but I have to cover those up because I'm so tall.

I thought I didn't look too bad tonight, but evidently I was wrong. Kitty is wasp-waisted, a woman of a certain age, with a great plastic surgeon so nobody knows what that age is.

'I thought it was an all right dress,' I mutter.

Kitty ignores me. 'My bag? Before I die, please.'

'OK,' I sigh. 'Do you have the cloakroom ticket?'

She shrugs. 'Lost it. Just describe it to them.'

'But that'll take them forever to find it,' I protest. There are at least a thousand people here. 'And Mark Swan's going to be speaking soon!'

Watching him was going to be the one bright spot in this nightmare of an evening. Somehow they'd actually managed to get Swan, England's hottest director, to give the speech. Mark Swan has won three Oscars for best picture and he's still in his thirties. He's so talented he makes Sam Mendes seem like an amateur. But he's also reclusive, never appears in the press, no parties in Cannes, no photo shoots in *Hello!* I'm a producer, and I don't understand the fuss they make about directors. But even I want to hear what Mark Swan has to say. (Well, OK, I'm not technically a producer. All I do is read scripts and act as Kitty's general dogsbody. But I work for producers, which is almost the same thing.)

'Just describe it to them,' Kitty says again. 'You know my handbag. It's the black one.'

The black one. Great.

'Could you be a little more—'

'Ssh!' hisses Kitty, her beady eyes fixed slavishly on the stage. 'They're introducing him.'

So they are.

'. . . third Academy Award for *King Harald* . . . Ladies and gentlemen, Mr Mark Swan.'

I crane my neck towards the stage too. But Kitty has the good seat. My view is obscured by a fat bloke with a huge bald head. I daren't ask him to move. He paid two grand for his seat which makes him a lot more important than me.

'Thanks,' Mark Swan says. He's got a rich, sexy baritone and he's very tall. But I still can't see what he looks like. 'When you put it like that it sounds impressive.'

A ripple of sycophantic laughter from the crowd.

'You know, the first thing I think when I see these affairs is why couldn't we all just write a cheque,' Swan says. 'How much do those fire-swallowers go for, anyway?'

More laughter, a bit more uncomfortable. I grin and shift in my seat, trying to see. I love the guy already.

'Anna,' Kitty hisses, eyes narrowing. 'Are you deaf?'

I get up reluctantly. 'I'm going, I'm going . . .'

I thread my way through the tables, getting tsk-tsked at when I impede somebody's view of Swan for a second (he is now telling a very politically incorrect joke about a studio head's wife and a pool cleaner) and rush as quickly as I can out to the lobby. Hurry! I really would like to hear this speech. It's a huge opportunity. He doesn't speak, doesn't give interviews . . .

'Yes? Can I help you?' inquires the cloakroom attendant with a plastic smile.

'Um, yes. I need to retrieve my boss's handbag.'

'Ticket?'

'She lost it.'

'Then I can't help you.'

'It's a black one,' I say pathetically. She treats this with the scorn it deserves. 'Mine was number three sixty. It might be near that.'

'There are about fifty thousand black bags near three sixty, love.'

'Look,' I say desperately. 'I'm missing Mark Swan's speech—'

'He came by here earlier,' she says, softening. 'Isn't he handsome?'

'I wouldn't know. I'm missing the speech.'

'He is,' she says dreamily. 'Handsome. Tall, dark and handsome. He said he liked me hair,' she adds, primping a bit.

'He sounds gorgeous,' I say. 'And I really don't want to miss him. D'you think I could go in there and have a look? I know what it looks like, roughly.'

'Suit yourself,' she says, shrugging.

I plunge into the vast cloakroom, hopelessly rooting through fur coats (real ones) and leather jackets, trying to spot Kitty's nondescript Prada clutch. All the bags look identical. I understand why the cloakroom attendant was so insistent on the ticket. It's no use. I spend a good twenty minutes searching hopelessly, all the while imagining Swan being so warm and funny and teasing all the overblown executives.

There aren't that many really good film guys in Britain and I'm missing the best one out there. And, naturally, Kitty will do nothing but yell at me and tell me I've 'lost' her stupid bloody bag.

'Excuse me.'

I turn round in the gloom of the closet to see a tall, bearded man in black tie trying to squeeze past me.

'I didn't expect to see anyone else in here,' he says.

'Oh. I hope it's OK, the other attendant said I could look for my bag.'

'I'm not the attendant, I'm doing the same thing,' he says sympathetically. 'Lost your ticket?'

'My boss lost hers.'

'Millions of black coats,' he says, sighing. 'Why did I choose black?'

'It's a bugger, isn't it?'

'Sure is,' he says, looking down at me, amused. He's very attractive, from what I can see in the gloom. Rugged, muscular. 'How long have you been here?'

'Almost half an hour.' I sigh. 'I've totally missed Mark Swan's speech now.'

He pauses. 'I'm afraid you have. Wasn't worth hearing, anyway.'

'He sounded pretty funny when I left,' I tell him. 'Not all puffed up and luvvie like you'd expect. Not even pretentious.'

'Really.'

'Which is strange, considering he's a recluse.'

'Why's that?' he asks, thumbing through the coats.

'Well,' I say, warming to my subject, 'being a recluse is a bit wanky, isn't it? Like you're so important, you have to hide all the time. It's the sign of an unhealthy self-preoccupation. Look at Stanley Kubrick.'

'Maybe he just doesn't want to be hounded,' he says mildly. 'He came here tonight, after all.'

I snort. 'Someone should tell him he's not Tom Cruise.'

'I expect he knows,' he says, stroking his beard thoughtfully.

'He sounded funny though. I liked him,' I say. 'But I'll never get to hear him now. I wanted to see what he was like, you know. And now I won't.'

'Found it,' he says triumphantly, pulling a huge and beautifully made black wool coat from the racks. 'What's the bag you're looking for like?'

'It's a Prada clutch. Should be somewhere in the three hundreds. But you get going,' I say. 'No reason both of us should be stuck in this hole.'

'My ex-girlfriend had one of those,' he says. 'Is this it, by any chance?'

It's a miracle. He reached into a pile of bags and unerringly pulled out Kitty's. I quickly open it. Yep, it's hers all right. There are the business cards, the fags, the Eve Lom appointment card.

'Oh, you are brilliant,' I say. 'Thanks so much.'

'Anything to help out a pretty girl,' he says, with a slight bow.

Pretty girl! That's a good one. It must be gloomier in here than I thought.

'Who's in here? You can't all come in here!'

It's the cloakroom attendant. She storms in towards us.

'I take a five-minute break and everybody's – oh,' she says, stopping dead. 'Excuse me! You didn't have to come in here, I'd have found it for you,' she says to him, simpering.

'I lost my ticket,' he apologizes.

'I remember *your* coat, Mr Swan,' she says.

I'm so surprised I lose my balance and fall over backwards, upsetting an entire rail of coats.

'Bloody hell!' she roars.

'There now, no harm done,' says Swan, reaching down and pulling me to my feet. I thank God for the gloom now. He can't see my flushed red cheeks. 'That's easily fixed.' He reaches over and pulls up the entire rail, coats and all, with one hand.

'Oh, well,' she says, de-fanged. 'Well.'

'I'm sorry,' I say.

'Be more careful next time,' she snaps. But I'm looking at Swan.

'Relax, it was fun,' he says. 'More fun than the speech.' And he winks at me.

I lift the clutch. Yes, I believe that even I have had smoother evenings than this.

'I'd better be going,' I say. 'Take care.' Ugh. Why did that come out of my mouth? Take care? Take care?

'Nice to meet you,' he says. 'What's your name?'

'Anna. Anna Brown,' I say. 'OK, well, goodbye, Mr Swan.'

'It's Mark,' he says, grinning. 'Bye, Anna.'

'Where the bloody hell have you been?' Kitty asks, snatching her bag from my flustered hand. 'You missed Mark Swan, you know.'

'Ahm – yes.'

'Too bad for you. Fetch me another champagne,' she says, dismissively.

'Sure,' I say, glad of the opportunity to escape.

Oh well. At least she'll never know. Right?

My flatmate Lily is sitting hunched over her computer. Her

long legs are revealed in a short, sleek skirt in buttercup leather, showing off her deep golden tan. Her top is a white halter neck in clingy jersey that drapes her perfectly formed, perfectly fake boobs, and shimmering blond hair tumbles down her back.

I'm wearing my best summer outfit too. Black jeans, black T-shirt, Doc Martens. Apparently going monochrome is 'really slimming', or so Lily always tells me.

So far I don't seem to have shrunk very much.

Horizontally or vertically.

My flatmates are both models. Not catwalk, mostly just photo shoots. Lily's five six, and Janet's five seven, but they are both so slim and petite, especially Lily, that they just seem tiny. And I'm just the right height to be a cat-walk model and out-do both of them, but that's only going to happen in an alternative universe. When people are feeling kind, they describe me as 'strapping'. I'm a big girl. All over. I'm very tall, and not thin and gangly; I have useful, farm-labourer hands, a bit of a tummy, huge boobs, not much of a waist, my bum isn't bad – it's a nice size – but it's sort of flat and squashy from sitting down all day, and my legs are no help. I can't show them off, can I? You look like me, you don't want to draw attention to yourself. I wear flat heels and baggy jeans, mostly. Camouflage.

And then there's my face. I'm the daughter of a handsome father, with masculine Yorkshire features, and a beautiful mother. She's still beautiful now, at fifty. She has those Michelle Pfeiffer cheekbones you can hang towels on. She's also petite, five five, with an elfin bone structure, huge blue eyes and gorgeous, naturally raven hair. She looks as if she should be Janet's mother, not mine.

But I take after Dad. Do I have my mother's slender torso? Nope. Do I have her raven hair? Nope. Do I have her tiny, petite, retrousse nose, so winsome and feminine? I do not. I am mousy-haired, freckled, hearty, strong, and I have a big nose to go with my big face. Dad always told me I was beautiful, growing up. And I only gradually found out that I wasn't. Being stood up at school by Jack Lafferty, for example, who was supposed to take me to the fourth-year disco, and everybody giggling and laughing. And the next year, at the St John's School dance, the big one, I found out why he'd done it. Fifth years got to celebrate taking their GCSEs with a big dance at the grammar school across town. We prepared for it all year, and most girls didn't want to be taken by somebody, because the St John's boys were fresh meat and you wanted to see what talent was out there. So I suppose I didn't notice that I had no date, most of the girls didn't. I prepped myself like everybody else. Spent the previous Saturday afternoon at Supercuts, had a free makeover courtesy of the No. 7 counter at Boots (blue eyeshadow was big back then), picked out a dress, a black velvet Laura Ashley thing with a bow on the back. I knew I was tall and had a big nose, but it didn't bother me, not back then. I thought I was beautiful. Princess Diana was tall and had a big nose and people thought she was the most beautiful woman in the world. I hadn't really had too many boyfriends, but I put that down to the fact that they were shy.

I don't think I'll ever forget what it was like once we got inside that hall. I was so excited, and everybody else seemed to be, all the St John's boys giggling and laughing and whispering, and after five minutes of hanging nonchalantly around the drinks table, cradling my glass of non-alcoholic punch, one of them actually came up to me. Swaggered, even. He was handsome and muscular, he looked great in his dinner jacket, and he was flanked by a couple of friends. As I looked down at them smirking up at me I felt great. All the Feldstone Comprehensive girls just standing around, and these guys were asking *me*.

'All right?' said the vision. I thought he was a vision, despite his spots and slightly too greasy hair.

'Hi,' I said flirtatiously.

'D'you wanna dance?' he asked. His friends grinned. I smiled at all of them.

'Sure, why not?' I said casually.

'What's your name?'

'Anna,' I said.

'Mine's Gary,' he said. 'And yours can't be Anna. It's got to be Beanstalk,' he added, sniggering.

His two friends nudged him, cackling.

'Here, love,' one of them said. 'What's the weather like up there, eh?'

'She should come with lights,' said the other one. 'Warn the low-flying aircraft!'

All three of them laughed heartily, right in my face, and then turned and walked back over to their side of the room as I stood there, stonily, my cheeks burning. I wanted to put a good face on it, I really did, but when I heard some of the girls next to me snigger meanly too, I couldn't take it. I burst into tears, right in front of everybody. I was so humiliated, and my tears were big, scalding ones that trickled down my cheek and gave me a runny nose and made my face even redder, with streaks in the foundation.

I grabbed a napkin off the pile next to the plastic plates and cubes of cheese speared with pineapple and dabbed at my eyes, but it was too late. I can still remember running to the loo, through that crowd of girls and slouching aren't-we-cool boys, all giggling and whispering. Seeing my face in the dirty mirror in the guest bathroom that stank of urine and disinfectant, looking at it all streaked and red-eyed. Trying to repair it, but still crying, so I just made it worse. My face was such a mess, so smudged and stained with mascara, my nose all red, my eyes watery, that in the end I just splashed water on myself and washed it off – washed off the semi-professional make-up job it had taken two hours to put on, and watched it swirl down the sink in little rivulets of tan and black.

I didn't go back out there. Maybe I should have, maybe I should have gone and found one of my friends and hung out with them defiantly and made witty, cutting remarks about everybody else. But I couldn't. I couldn't do anything except crawl inside one of the stalls, sit down on the loo, and cry.

I stayed inside that loo for four hours, hearing the disco booming in the dance hall, listening to other girls gossiping as they came in to refresh their blusher and giggle about the boys they were pulling. When it was five minutes before ten and the coach was ready to leave, I walked out straight to the car park and got on the bus. I was the first girl on board and when my best friend Clara Bryant came on, all breathless and smiling, she took one look at me and didn't say anything because she didn't know what to say.

That was the night I learned my lesson. I wasn't pretty, I wasn't striking, I didn't look anything like Princess Diana. It had been so easy to believe my parents, especially my daddy who always told me I was beautiful and that all the boys would be fighting over me. And I believed them, despite the evidence of the mirror. I believed them through all the boys who didn't ask me out and the few who did who dumped me. It was always them, never me. I believed Daddy right up until the night of that dance.

Never again, though. I adjusted immediately. I shoved all my fancy clothes straight to the back of my wardrobe and wore the plainest things I had. I threw away all my high heels and fire-engine red lipsticks and little tubs of body glitter. I didn't live near a Gap when I was a teenager, but if I had, I'd have shopped there permanently, for plain tees and basic jeans and khakis, dull, straight-up clothes that nobody would ever notice. But I did my best at Marks and Sparks and Top Shop. When I was older I found a few classic 'don't notice me' dresses and bought four in different colours, and so I go from the office to dinners to the occasional disastrous date, all in clothes expertly designed to conceal me.

I will absolutely never look like Janet or Lily, not even after major plastic surgery, which I can't afford. I've got a Nose Job Fund box, and I had a look in there recently. I've saved a massive ninety-eight quid and thirty-four p. There used to be more, but I did take that weekend CityBreak in Bruges six months ago, when it rained the whole time.

'Anna, what's your date of birth?' Lily asks suddenly, not looking around from her screen.

'Why?' I ask.

'Just for fun,' she persuades. 'It tells you what day of the week you were born.'

'July the third, nineteen seventy-one.'

Lily's long talons peck over the computer, one letter at a time.

'Monday!' she says.

Janet looks over at me. Janet is wearing a teeny, body-hugging pale pink dress that looks incredible against her olive skin. Filigree gold bangles run up her arms and make a chinking sound as she moves.

'Monday's child is fair of face,' Janet says.

Lily sniggers. 'Sorry,' she says, clasping her bony hand over her mouth. 'It slipped out! Honest!'

'Anna can take a joke, can't you, Anna?' Janet says.

I sigh. 'No problem.'

And it really isn't. I mean, it's not like I haven't heard it before. Primary school is probably the worst; after the first ten years, you get used to it.

Being ugly, that is.

I don't deceive myself about my looks. Even my granny stopped saying, 'It's a just a phase. You'll see,' when I reached fifteen. And now I would prefer not to talk about it. I've tried various things. Flat shoes, plain colours. Make-up tips to 'minimize' my nose, shading and highlighting, but nothing helps. It's still huge. It's still there. And just in case I might be tempted to forget about it, I tend to get helpful reminders from male members of the public.

'Bloody 'ell! Packed yer trunk, Nellie?'

That was delivered last week by a drunk teenager as I walked home from Tesco's. I thought it was rather imaginative of him.

'Don't worry, Anna, *I* think you're, like, totally beautiful,' says Janet.

'Because real beauty is on the inside,' lies Lily. 'And that's what counts.'

'What's that on the end of your nose?' asks Janet.

'What?' Lily asks, concerned.

'It's a bit red. It might be the start of a spot,' Janet says, ominously.

'Oh my God!' Lily shrieks in horror. She rushes to the mirror over the mantelpiece and examines the infinitesimally small pinkish area. 'Oh my God! I'm hideous! And I'm working tomorrow!'

'You look fine,' I offer.

'What would you know?' Lily wails, tossing her fountain of flaxen strands about like a Timotei commercial.

Janet shakes her head sagely. 'I told her not to go with that foundation, I really think it clogs the pores.'

'What day were you born, Lily?' I ask, in a vain attempt to distract her.

'I'll look it up,' Janet says. She strolls over to the computer, sticking out her perfectly firm, rounded bottom. Janet has a J-Lo fixation. She watches all her videos religiously and favours lots of gold jewellery and real fur, which she steals from shoots. She also has an unfortunate tendency to say 'Bling bling' and likes to be called 'Jay-Me', which she prefers to her real name, Janet Meeks.

'Wednesday,' Janet says.

'Wednesday's child is full of woe,' I say, a touch more cheerfully. Who knows? Anything could happen. Lily could come down with a disfiguring bout of adult chickenpox, or have a terrible reaction to a collagen implant. That would be fantastic!

'It's a stupid rhyme,' Lily says dismissively. 'It doesn't mean anything.'

'Not like astrology then,' I suggest.

She looks at me disdainfully. 'Astrology is totally proven.'

'By who?'

'By everyone,' she says, triumphantly closing down my argument.

'I have to get back to work,' I say, reaching for another red paper-bound script. My glasses are heavy on my nose and my eyes are watering, but it's Sunday already and I have to get coverage typed up for five more screenplays, all of which will, without a doubt, be as dreadful as the last sixteen I've read this weekend.

'Oh, take a break, Anna. Live a little!' says Janet brightly. Janet works maybe two days a week and makes three times what I do. She looks languidly into the camera for three hours and goes to look-sees and auditions. I type my fingers to the bone all weekend writing 'coverage' for crappy scripts that will never get made, then during the week I run errands, type letters, answer phones, make copies, walk dogs, and generally act as an office slave for a total bitch named Kitty.

Janet and Lily make about forty grand a year.

I make sixteen.

They are twenty-eight and twenty-three.

I am thirty-two.

And yet, and I know this is stupid, I keep thinking things are going to change for me. I mean, I'm in the right industry. It took me four years to find a job as a script reader, and now I have one at a proper production company with fancy Covent Garden offices. I even have a pension plan and health insurance.

It could happen, right? I could find that needle in the haystack, that one great script I could recommend. I could get made into a development executive like Kitty. I could be a producer and make millions and win an Oscar . . .

It has happened before. It happens all the time. Sometimes I think I should take stock of my life, try and get a better-paying job somewhere else, but doing what? It wouldn't be in the film business. They pay us bugger all because they know they can. There are forty little Annas out there, fresh-faced from film school, who would kill for my job. And anyway, I don't have time to take stock of my life because I'm too busy. I save a bunch of money by paying only three hundred a month for my room in this flat. Instead of money I give Lily and Janet all the invitations and tickets that come my way – the movie premieres, the industry parties, the VIP passes to the members-only clubs.

That works out well because I wouldn't bother with those parties anyway. The people who go to them are all varying combinations of rich, beautiful and successful. I'm none of the above, so I just stay home and read more bad scripts.

'We could try a makeover on you,' Janet says.

'I don't think so, but thanks.'

'Come on. I know I could do *something*,' Janet says, encouragingly.

'It's not all about looks, OK?' I tell her. 'I'm fine like I am!'

'I wonder if Brian thinks that,' Lily says archly.

Brian is my boyfriend. He works for Barclays as a teller. He's a bit skinny and has some problems with sweaty armpits and bad breath, but I've been working on those, subtly. Brian's always telling me he 'doesn't care' about my looks because 'real beauty is on the inside'.

I hear that one a lot.

Brian is no prize himself though. He's all skinny and he breaks out from time to time, doing a good impression of a pepperoni pizza. Plus, he's shorter than me – but who isn't? Still, you have to have a boyfriend, don't you? I mean, especially if you're ugly. A boyfriend is great camouflage. He stops people from making pitying comments, and stuff like that. So I cling to him.

'Brian likes me for who I am,' I tell Janet defiantly.

'Sure,' she says, brightly. 'Whatever you say.'

The buzzer rings and she leaps to her feet. Lily tears herself away from the mirror above the mantel. They adore the phone, the buzzer, anything. Always expecting something fabulous. And why not? For pretty girls, it mostly is something fabulous.

'Hello?'

'Hi, 's Brian,' Brian's voice slurs.

'Speak of the devil!' says Janet, laughing lightly. She can't help it, she flirts with everything in trousers, even men she wouldn't be seen dead with. 'We were just talking about you. Come on up.'

'OK,' he says, dully.

I get up and check my own face in the mirror.

'Well, it's too late to do anything about it *now*,' Lily hisses.

A second later I hear the ping of the lift. Our flat is located above a feminist bookshop in Tottenham Court Road, one of those old Victorian buildings with an ancient, narrow lift. It fits one normal person, or two models, and feels like a coffin without the velvet.

Brian opens the lift door and steps out. He's wearing a white polyester short-sleeved shirt and saggy chinos, but I'm still glad to see him. He's my boyfriend, after all. As in, I have one!

'Hi, honey,' I say, kissing him on the cheek. 'Come in.'

'Hi, Brian,' coo Lily and Janet together, shaking their hair and smoothing their already tight clothes tighter round their bodies.

'Hi,' he says, staring. I do wish he wouldn't drool like that. I mean, I am standing right here.

'Would you like some coffee?' I ask pointedly.

'Oh. No,' he says, continuing to stare at my flatmates.

I cough. 'Are we going out to dinner?'

'I thought you had all that work to do,' says Janet, innocently. She smiles cosmetically whitened teeth at him. 'You know Anna, always work, work, work.'

'I can take a break,' I say. 'For you,' I add to Brian.

He looks a bit awkward. 'No, it's not that . . . can we go in your room?'

'Oooh!' says Lily, widening her blue eyes and pretending to be shocked.

'No rumpy-pumpy!' says Janet, shaking her finger at him.

Brian giggles. Which makes my skin go all bristly in distaste. 'It's not that either,' he says. 'I just need some privacy.'

'Oh, don't mind us,' Lily reassures him. 'We don't have any secrets, do we, Anna? We're like sisters.' She bats thick black lashes at him.

'Come on,' I say, heading towards my 'bedroom'. It's really more of a walk-in closet with a bunk in it, but what can you expect for three hundred a month?

'Oh no, don't worry, we'll go into the kitchen,' says Janet. 'Come on, Lily.' And they head into the kitchen, shutting the door. I can hear the scraping and frantic muffled whispering instantly. They're probably arguing over who gets first turn with an ear at the keyhole.

'What is it, darling?' I ask, with self-conscious tenderness.

'Our relationship,' Brian says. 'I have to express myself on our relationship.'

15

Oh God. He's been at the self-help books again.

'Could you do it over dinner?' I ask hopefully. 'We could always go to Pizza Express.' Brian is a touch cheap.

'Dutch treat,' I hasten to reassure him, like it's ever been anything else.

'That wouldn't fit my moral paradigm at this time,' Brian says heavily.

'Sorry, I don't speak weirdo,' I say and instantly regret it.

'Oh yeah. That's just typical,' Brian snaps. 'You've always stood in the way of my self-actualization!'

I swallow. 'Sorry. What was it you wanted to say?'

'We've been in each other's lives for some time now,' he begins.

'Three months.'

His sandy eyebrows beetle together. Brian hates being interrupted. 'Yes, well, and I think we have both gained a lot from the uniqueness of the experience,' he says, offering mc a brisk smile.

Hey! Maybe he's going to ask me to move in with him. Brian owns his own flat, a one-bedroom ex-council in Camden with its own scrap of patio. I could sit outside all summer and read my scripts in a deck chair.

'I know I have,' I say, smiling back at him encouragingly.

'Of course *you* have,' says Brian as though this were perfectly obvious. 'But my personal boundaries have been trammelled and I'm at a place in my life where I need fresh stimulus, which is not to say that we haven't offered each other very real positivity.'

I digest this for a second.

Then light dawns.

'You're breaking up with me,' I say slowly. I drink him in, all eight and a half stone, lanky ginger hair, pizza face, tell-tale grey underarm patch (I did warn him to at least wear dark shirts, but he wasn't having it). Brian is, in short, one of the least attractive men I've ever seen. '*You're* breaking up with *me*.'

'Dunno why you have to be so crude about it,' he says prissily. 'I think of it as a growing experience.'

Oh boy. This is a new depth of humiliation. 'I don't believe this,' I say, trying for incredulity, but my voice wobbles and it comes out all plaintive.

'Out of suffering comes strength, Anna,' says Brian wisely.

'You really are a total tosser,' I blurt. 'Just go away, Brian, would you?'

But he stays rooted to the spot.

'It's not how you look,' he says virtuously. 'Well, that did have *something* to do with it. I need to be honest with myself. Looks are important to me because the body is a reflection of the spirit.' He gestures at my pudgy tummy. 'I think you should work on that. Just a friendly comment.'

'A reflection of the spirit?' I retort. 'Let's hope not, because looking at your face, I guess your spirit has measles.'

He flushes a dull red. 'Mildred says I'm very handsome, *actually.*'

'Who's that? Your new girlfriend?'

He doesn't answer.

'Well, Mildred is either a liar or blind, and if she hasn't mentioned your halitosis, her sense of smell can't be all that great either.' I stand up threateningly. 'Now *get out.*'

I have two and a half stone on him. Brian turns and flees.

The kitchen door opens.

'Everything OK?' Lily asks carelessly.

'Fine,' I lie.

I walk over to the window and look out. There is a girl loitering below, outside the bookshop. She has blonde hair and seems ordinary looking, which puts her out of my league, and Brian's too. But of course, he does own that flat in Camden . . .

Brian exits our building and greets the blonde. Then he high-fives her in another pathetic attempt to look American.

'Well,' says Lily sweetly, 'I'm glad he values you for who you are!'

2

God. I hate the tube.

Every single day I tell myself I'm not going to take it any more. I could get up an hour earlier and walk. Or buy a bike and ride. That would be good for my bum as well as my bank balance. Imagine if I didn't have to buy a Travelcard. I'd practically be rich.

Yet somehow it's Monday morning, eight thirty, and here I am again, squashed in between a fourteen-year-old boy who I think has an erection – anyway, he keeps suspiciously bumping into the girl in front of us – and a fifty-year-old drunk who reeks of stale alcohol and BO. The pathetic thing is that I feel almost affronted that the fourteen year old never tries to rub up against me. They aren't *supposed* to be discriminating at that age, are they? They aren't even supposed to care.

Maybe I'm a little sensitive because of the Brian incident. After they had had their fun, Lily and Janet tried to be nice to me.

'He's not worth it, anyway.'

'Surely you can do better,' Janet said, without conviction.

'He was too young for you,' Lily said.

I looked at her. 'He was thirty-six.'

'Aren't you thirty-eight?' asked Lily, innocently.

'I might as well be,' I said morosely.

So anyway, it's another dull Monday like any other except that I am now boyfriendless. I had the worst boyfriend in the world, and he dumped me.

'Thank *God*,' cries Vanna (her real name) when she hears.

19

Vanna is my best friend. We met back when we were in college together and we've stayed best friends ever since, even though our lives have gone in slightly different directions: me, reader of bad scripts and dogsbody for no money, her, senior commissioning editor at one of London's top publishing houses for about a hundred and fifty grand a year; me, now dumped by ugly, bad-breath reeking loser, her, blissfully married to Rupert, an investment banker, with two small children.

All men adore her. I can't understand why we still get on.

'He was vile, hon.'

'I know. But he dumped *me*.'

'Who else would have him?' Vanna scoffs.

'Some girl,' I say glumly. 'I saw her.'

'Was she pretty?'

'Yes,' I admit. Well, she was compared to me, let's face it.

'I bet she was a dog,' Vanna says. 'And you didn't want him anyway.'

'But it was so nice to have a boyfriend,' I say sadly.

'You'll get another one. A *better* one. You work at Winning Productions, after all. Just think of all the talent that walks in there! And I mean talent in the strictly trouser sense,' Vanna adds. 'You know I think all those actors and writers are just self-indulgent twats.'

'I know.'

'They really are.'

'I know you think that.'

'I don't know how you put up with them,' she says, as though I am hand-holding major Britpack stars every day.

'You handle authors,' I point out.

'Them? Bunch of self-important tossers, too. I let PR take care of them,' Vanna says confidently. 'I only suck up when absolutely necessary, for example at sales conferences.'

'I don't get much of a chance to socialize at work anyway.'

'You need to try,' Vanna says, ominously. 'You may be letting your best chance slip away.'

'Aren't office romances frowned upon?'

'Well, yes. If you get caught. But where else are you

supposed to meet your match, eh? A busy, professional woman like you.'

I'm not sure this is an exactly accurate depiction of me, but still.

'A harassed executive,' she continues firmly. 'On the creative side. With very little free time in the evenings. No, it's got to be work, babes. You have to take this Brian thing as a wake-up call.'

'What do you—'

But when she's in full flow, there's no stopping her. This is one reason she is so successful. Nobody ever dares interrupt.

'A wake-up call that says, "I will not be ignored any more! I will not settle for the dregs in life! I will only go out with *authentically shaggable men* who can think themselves lucky to get a look-in with a hottie like me!" That's your new mantra, darling – *Anna is a hottie*!'

'Very noble,' I said, but at least now I'm laughing.

'I wasn't joking,' she says earnestly. Vanna is totally blind where I am concerned.

Anyway, at least she got me thinking that being shot of Brian wasn't so bad. I mean, I couldn't stand him. Of course, that doesn't mean that I am somehow going to replace him with somebody better. I know reality by now. But perhaps I could think more about my job, try and make a bit more of an impression. Get a raise. Something.

They say that there's someone out there for everybody, but we all know that's a load of old bollocks, don't we? Maybe some of us should focus on trying to be happy alone.

I look down at the fourteen-year-old kid, for example. Concentrating very hard on the pretty slip of a thing in front of me.

He looks at me.

'Whatcher staring at?' he demands, flushing.

'Don't you think you should give her some room?'

'Fuck off, fatso,' he says, charmingly.

Apparently he didn't get Vanna's 'Anna is a hottie' memo. I lift up my flat shoe and bring the heel down hard on his toes.

'Owwww!' he yells. Everybody is staring now. The pretty girl turns round.

'He was rubbing against you,' I tell her.

She stares at him. 'You little perv!'

'I wasn't. She's making it up, the tart,' he says nastily.

'How dare you! That lady's old enough to be your mother,' she says, nodding gratefully at me.

The train heaves to a stop and I get out. The day is boiling hot, and when I finally emerge from the tube station it's like going from an oven into a sauna.

Ah yes, I think. A perfect start to another perfect week!

'Hi,' I say to Sharon and John, my fellow readers and slaves. We all have a cubicle here on the west side of the office, next to the secretaries and right in front of Kitty Simpson's office. Technically speaking, we are not support staff like the secretaries but, as Vanna keeps spinning it, executives. However, we get paid less and are expected to do what they say.

Sharon and John greet me with an equal lack of enthusiasm. Sharon is a pert 22-year-old who is only doing this job as an alternative to waitressing while she hones her acting skills. She has evidently decided that if she flirts with enough male executives here, one of them might give her a part (not that kind of part) or get her an agent or something. She would not need a pep talk about looking for romance in the office. Sharon is a pro. Her light brown curls are always bouncy, her freckled, creamy skin always glowing. In summer she favours teensy little dresses with cute cardigans embroidered with flowers; in winter she likes very tight pants and body-skimming jackets. All year round she likes kitten heels and dangly earrings.

John is twenty-eight and regards himself as an utter failure, but unlike me he sees this as a choice. John believes the noble art of the cinema is being bastardized by Hollywood, and only a really serious *auteur* like himself can rescue it. He wants to direct. Surprise. In the meantime, he gets perverse pleasure from reading so many bad scripts and passing on them all. He always wears brown corduroys and an orange or a plum-based print shirt, because he's all about the seventies (except the good bits, like the Wombles). John likes jazz, beat literature and French cinema. He also likes Kitty, who is an utter bitch, which seems to turn him on.

John keeps his job because he is a world-class suck-up and fawner.

Sharon keeps her job because the men in the office won't let Kitty fire her.

I keep my job because I do all the work.

'How was your weekend?' Sharon asks. She is playing Expert Minesweeper with a serious frown of concentration. Sharon has never won Expert Minesweeper but it is not for want of trying. 'Meet any hot guys?'

'Not really.'

'Great,' she says absently. 'Oh, SHITEHOLE.'

'Maybe you should stick to Intermediate,' I suggest.

Sharon looks at me pityingly. 'I've moved on . . . *I* met someone,' she adds, ringlets bouncing.

'Oh? Who?'

'I met him at the *Legally Blonde 2* premiere party,' she says, determined to spin it out.

'Was he handsome?' I ask, 'Sexy? Funny?'

Sharon waves a hand as if to brush away such minor considerations. 'He works for *MGM*,' she says.

'Exciting,' I concede.

'In *LA*,' she adds triumphantly. It is Sharon's lifelong dream to get out to LA and get discovered, so she can be the next Catherine Zeta-Jones. Unfortunately for her Sharon is not talented and only a bit pretty.

'You'd do well out there,' I say supportively.

She shrugs. 'Well, of course I would. I've got what it takes. I'm talented.'

'You're hungry.'

'That's right.'

'Committed. Passionate,' I suggest.

Sharon's smile broadens. She crosses her long, lean legs

under today's white mini dress. 'Exactly,' she agrees. 'You can really spot talent, Anna.'

'Do you think I should go to LA?' I ask. Maybe that's what's missing in my career. Maybe I'm just in the wrong place.

Sharon takes a long, assessing look at me. I'm wearing my plain camel cotton skirt and long-sleeved white shirt with a pair of slides. It doesn't attract too much attention and I think it's businesslike.

'LA's not for you, is it, really,' she says. 'I'm only being honest.'

I sigh.

'You're doing really well here, anyway,' she lies. 'Kitty relies on you.'

'Good morning, team,' Mike Watson says.

We all look up. Oh joy, it's Mike Watson, also a development executive, and a total pig. He hates Kitty, but that's about all you can put in the plus column. Mike is a deeply sad man. He loves American slang, working out at the gym and putting women down. Every actress is 'too fat' or 'too old'. Mike has one reader only: Rob Stanford. He's blond, nineteen, and upper class; strictly window dressing. More importantly, Rob is the nephew of a big agent, Max Stanford, whom Mike sucks up to. I don't think I've ever actually seen Rob read a script. He just marks them all 'pass'.

Mike gets his scripts direct from agents he likes and ignores all the rest.

Mike hates me in particular. I once was asked by Max Withers, the head of Winning, in a meeting, about a script Mike was championing and I told him I thought it sucked. They passed on the project, and Mike has never forgiven me.

'Hi,' I say. John nods bleakly. Sharon tosses her ringlets and smiles at Mike engagingly.

'Hey, Mike,' she says, in her best breathy little-girl voice. 'Can I get you some iced tea? I know it's your favourite.'

'We don't have that,' Mike says, giving Sharon's shoulder a friendly squeeze.

'I could run round to Starbucks for you,' she breathes. 'It's no trouble.'

'Thanks, babe,' he says, 'but Rob's already fixed me up.'

'Oh,' says Sharon, crestfallen.

'But you can bring me some biscuits,' he says, grinning as though this is a great treat.

'Sure thing, Mike,' says Sharon, fluttering her eyelashes.

I think I'm going to be sick.

'What's the matter with you?' Mike demands, seeing my expression. 'Monday blues?'

'Something like that,' I mumble.

'At Winning we look *forward* to Mondays,' says Mike sternly. 'We dread Fridays.'

The phone trills. Sharon jumps on it.

'Winning Productions,' she says. 'Hello, you're with a Winner. What's that? Now? OK. OK, Kitty. Yes, right away.' She stands up dramatically, giving Mike a nice flash of tanned thigh. 'That's Kitty, she wants us all in her office right away.'

'About what?' Mike demands instantly.

'I'm afraid she didn't say, Mike,' says Sharon, pouting.

'That's a hold on the biccies then,' Mike concedes. 'You can bring them to me later.' Yeah, and tell him all about our meeting with Kitty, no doubt.

Sharon smiles at him regretfully.

John gets to his feet. 'I believe Kitty said now,' he insists, opening her office door. He looks daggers at Mike. 'Excuse us, Mike.'

'I got to go brief my own *support* staff,' says Mike, looking at him disdainfully. 'See you all later.'

I head into Kitty's interior-designed office. With all the originality of a new burger joint, Kitty has chosen to do up her gorgeous Victorian corner space, complete with intricate leaf mouldings on the ceiling, as a 1950s American diner. She has a couch, two red plastic-covered stools, a non-working soda fountain and posters for James Dean and Rock Hudson movies. And then of course there's the Oscar.

Kitty once actually won an Oscar for producing Best Foreign Film, *Questa Sera*, back in the seventies, and she's been trading

25

on it ever since. Rumour has it she was shagging the director to get that credit, but I don't believe it. Who on earth would shag Kitty? Other than John, of course.

Kitty's scary.

Nobody knows her age. Forty-nine? Fifty-one? She's been Botoxed to death, so she can't frown or smile properly. But she can still yell. She's small and wiry, with the dress sense of Coco Chanel and the warm fluffy personality of Mussolini. And even though she's five foot two, she manages to always make me feel smaller.

Despite being a pitbull in the office, Kitty is a social butterfly. Put her in a room with a major actor, director, or agent, and her personality transforms, super-hero like, immediately. You can almost see the total-bitch exterior ripping away to reveal a charming, witty woman, absolutely fascinated by the other person and what they are saying. She takes long, languid lunches at all the right clubs, sends flowers on birthdays, calls everyone important in her Rolodex twice a month without fail just to catch up. She's a presence on the scene. She's known. And she has especially good links with the older crowd of stars. They might not be box-office sizzle, but they're all impressive names: Judy, Helen, Sean . . .

Kitty likes success and despises failure.

She keeps me on for only two reasons. One, somebody has to give her good coverage on scripts. Two, it annoys Mike.

We slide onto the tiny, uncomfortably hard couch (Kitty hates her guests to be comfortable) and wait nervously. This is not like my boss. Normally she comes in, walks into her office and slams the door. Then her mousy, terrified secretary, Claire, brings her the day's call sheet. Then she tells us to prepare for a coverage meeting where she cuts us off and berates us for not finding her the next *Titanic* or *Harry Potter* at the very least.

After that she usually has me run errands for most of the afternoon, while she lunches with somebody fabulous, and John takes her notes while she's in meetings. Sharon floats about pretending to read more scripts but mostly offering male executives cups of coffee and gossiping in the kitchen.

It's a fairly well-honed routine.

Why is Kitty calling us in here now? Is one of us going to be sacked? I don't think I could take it. My mind runs to my bank account. I've got one hundred and three pounds forty-seven in there right now. And to my CV. What would I say? Thirty-something script reader seeks lucrative position as development executive. Film credits: none. Recent promotions: none. Scripts recommended for development: none.

I glance at Sharon and John. Neither of them seem to be concerned.

'What do you think she's up to?'

'I don't think Kitty's "up to" anything,' says John, all tight-mouthed. 'I think as her team we all owe it to her to wait and see what she has to say.'

'Maybe she's resigned,' Sharon says hopefully.

'Why would she do that?' he snaps.

Sharon smiles fulsomely at John, but he is immune. Sharon does nothing for him. 'Maybe she's been poached,' Sharon adds, placatingly.

'Poached. Yes. Well, that's possible. Kitty *is* incredibly talented,' says John, looking at me and daring me to disagree.

'Coming through, coming through,' bellows a voice in the hallway. I turn to see Kitty striding through the office belligerently. The effect is spoiled by the fact there's nobody there. She doesn't seem to care. Her mobile is pressed to her ear.

'Yeah, yeah. Yeah, yeah. Right, absolutely. Yes, *devastated*,' she purrs. I perk up. I can tell it's good news.

Kitty snaps her phone shut, walks into her office and slams the door loudly. John looks up at her approvingly, like an adoring dog. And indeed Kitty does look particularly designer bitch from hell today. Yellow Dolce suit with trademark DG buttons, check; Louis Vuitton baguette in spring colours, check; enormous canary yellow diamond ring flashing ostentatiously, check. She wears her hair up in her trademark French plait, probably to show off the two-carat diamond studs in her ears.

Kitty believes in labels. If you can't see who it's by, why bother?

'There are going to be some changes around here,' she says dramatically.

I panic. I *am* going to get sacked.

'We're going to raise our game. Find that key project that's out there.'

Wow. She sounds like Mike.

'Somebody from Winning has to be the first to deliver, and I intend the Kitty Simpson team to be that somebody,' she says. 'It's absolutely vital that we get ourselves noticed.'

'I totally agree,' fawns John. 'You've got such a compelling vision, Kitty.'

She favours him with a wintry smile. 'Yes,' she muses. 'It needs to be the whole team's vision – the "Kitty Simpson" vision,' she says, drawing a circle in the air with her hand, diamond flashing. 'As I make my mark!'

'As *we* make our mark, you mean,' says Sharon, smiling winsomely.

Kitty's eyes narrow. She hates Sharon, which is silly. Sharon's just Sharon. You know exactly what you're getting when she walks into an office.

'We can start the team effort with you getting us some coffee,' she says to Sharon. 'You seem to be awfully good at that.'

'Of course,' Sharon says demurely. 'I'll make it for you specially.'

This is obvious code for 'I'll spit in it'.

'Actually, why don't you go, Anna?' says Kitty, after a second.

'I'll have a cappuccino,' says Sharon, smiling broadly. 'Non-fat milk.'

'I'll have a chai,' says John, looking adoringly at Kitty. 'For the energy! Sounds like I'm going to need it!'

'You know how I like mine by now,' Kitty says dismissively. 'Espresso. Double. Lemon peel twist.'

I walk out gloomily. It's a reflection of our luvvie production house status that we actually have a kitchen with the ability to make all these things. If I do get fired as a script reader, there could always be a job for me at one of those fancy Soho coffee shops. Of course, Kitty correctly assumes that I will not spit in her coffee. I suppose I lack the killer instinct.

I stand over the machines, frothing milk and carving off a twist of fresh lemon peel. I even sprinkle cocoa powder over Sharon's foam. I know I'm far too nice, but I can't really help it. I think it has something to do with being ugly. Pretty girls are sometimes nice, but often they don't need to be. They just toss their hair and smile and everyone's enchanted, men and women. Non-pretty girls can't get away with that. We need to be extra sweet and accommodating all the time.

I pile everything onto a tray and head back to Kitty's office. Suddenly, out of nowhere, Rob Stanford materializes. He manages to plant himself squarely in my way.

'Getting the coffee again?' he asks, in that horrible plummy voice of his. I bet he puts it on. I bet his mother was a dinner lady or something. 'They really do work you too hard.'

'It's heavy, so I'd better get on,' I say.

'Mike's wondering what the "top secret" meeting is all about,' Rob says which causes my skin to go all creepy with dislike.

'Is he?' I ask sweetly. Two can play this game.

Rob purses his lips. 'We're all on the same team, you know,' he hisses. 'Mike has thought for some time you're not a team player, Anna. *Actually*.'

'If Mike wants to know what the meeting's about, why doesn't he ask Kitty?'

'I'm asking you,' Rob says.

'I can't tell you.' This is perfectly true, as I have no idea myself.

'Well!' he says huffily. 'I won't forget this, you know.'

'I will,' I say. 'Bye, Rob,' and swoop the groaning tray past him with an agility that will do me credit in my future coffee-shop career.

Kitty looks up sharply when I come in. She is tapping her

pencil impatiently against her desk. I really don't think she needs any more caffeine.

'Get lost?' she inquires acidly.

'No, Rob Stanford stopped me,' I say, attempting innocence. 'He wanted to know what this meeting was about.'

'Did he indeed,' Kitty hisses darkly. Her heavily lined eyes sweep the three of us. 'For Mike Watson, no doubt. What I am about to tell you remains on this team. Understand?'

'Absolutely! Of course,' John gushes. 'I would never betray your confidence, Kitty, I hope you know that . . .'

She cuts him off with a wave of her blood-red talons. 'Anna?'

'Sure,' I shrug. I can't stand Mike anyway.

Kitty gives Sharon a piercing gaze. 'If it leaks, Sharon, I'll know it was you. This information is to be absolutely private.'

'OK,' says Sharon, sulkily.

Kitty takes a sip of her espresso just to draw out the suspense.

'The company is being taken over,' she says. 'My sources tell me there's interest from New York, there's interest from LA. *Serious* players. Looking to take over our projects. Our people and our talent.'

'Somebody from LA?' asks Sharon, her eyes gleaming now. Visions of being discovered dancing in her head.

'Of course, if any bids *are* made, they're going to look at all the resources of Winning,' says Kitty sternly. 'Bringing in a new, fresh approach. Cutting the dead wood.'

'Winnowing out the chaff,' says John, adoringly.

'And I – we – are not going to be seen as dead wood. I want a *project*. Something big and brash that I can bring to the table. Something with "hit" all over it,' Kitty says. 'Find me something I can package. I want to attach Hugh. Or Catherine. Even Jude, at a pinch. And I want a jump on—'

'Your colleagues?' asks Sharon.

'The competition,' Kitty snaps. 'You do know, Sharon, that the first thing new management does is fire people? People that aren't performing, people that get *bad evaluations*?'

Sharon swallows drily.

'Was there anything in the weekend read?'

We all shake our heads.

'Nothing?' Kitty seems very put out. 'Don't give me that. Where is your coverage?'

John and Sharon look sheepish.

'I've done a page of coverage notes,' John blusters.

'For twenty scripts?'

'It was all they deserved, Kitty, I assure you,' he says defensively.

She sighs. 'Get it on my desk. Sharon?'

Sharon blushes. 'My notes were mainly in my head . . .'

'You mean you didn't read any of the scripts,' Kitty says. 'You're a waste of space. You'd better come up with something fast, Sharon. Or you won't be here long enough to meet the new bidders.'

'OK,' Sharon says meekly.

'Anna?'

'There was nothing,' I agree. 'But I have two pages on each script. I'll bring them to you.'

'God, no,' says Kitty, losing interest immediately. 'How dull. Just *get* me something, Anna. Get me something. There's something in it for the first person who does.'

For the first time all day I feel a flicker of something unusual. Hope. That's what it is.

'What would that be?' I ask, trying to sound all casual, like I don't care.

'Advancement,' Kitty says in her most serious voice. 'Your big break.'

Ooh. Why not?

It's not impossible, I think to myself as I sit in my cubicle. I could be the reader that finds that one gem and champions it. Kitty backs me up, and together we find ourselves in development meetings, where I will, naturally, speak so eloquently and passionately that they'll *have* to take on my project. And Kitty gets made into a Vice-President of Production, while I move up to Development Executive. With a transfer to Los Angeles. And readers of my own, and

31

an apartment in a complex with gated security, manicured lawns, and a pool . . .

Maybe not a pool. My tummy has not seen a swimsuit since I was fourteen and it was compulsory. But I could have a nice car. A convertible. I bet Brian would love to date me then.

I look over at Sharon and John. They have already been down to the mail room for their mail. Sharon has actually swiped more scripts than usual – that's not something you see every day. They both have a nice neat pile stacked on their desks, and are eagerly flicking through them.

My phone buzzes.

'Hi.' It's Kitty. 'Come back in.'

'Do you want the others?'

'Did I say I wanted the others? Just you. Come back in,' she snaps. 'And don't let them see you.'

Gosh, this is all very exciting today. It's almost like *Wall Street* or something. I gingerly get up and go to the kitchen for a plate of biscuits. Low-carb oatmeal, Kitty's favourite. Though I think she's kidding herself, because how can biscuits be low-carb?

I sneak back towards her office carrying the plate of cookies for cover, but neither Sharon nor John even looks at me. They are too busy straining their eyes with the unaccustomed task of actually reading something.

I shut the door behind me and sit on the couch.

'Sit down,' Kitty says.

'Um, thanks,' I say. 'Cookie?'

'What? I didn't ask for cookies.'

'It was just an excuse,' I say.

'Oh. Yes, very clever. No thank you, *I'm* watching my weight,' she says with heavy emphasis, letting me know she is not fooled by the forgiving drape of my skirt, nor my long-sleeved shirt that is meant to disguise my flabby arms. 'Now, Anna, you know you're the only one I can trust and that I have *absolute confidence* in you.'

This is news. 'You do?' I ask hopefully.

'Oh, of course. You're my go-to girl,' she says. 'And I want you to know the real skinny.'

I bite my cheeks to stop from saying, 'Sorry, I don't speak American.'

'Great,' I offer weakly.

'The bidder for the company,' she says, 'is *Eli Roth*.'

I sit a bit straighter. 'Eli Roth? Of Red Crest Productions?'

'You know who he is?' demands Kitty, her eyes narrowing. 'How?'

'I read the trades,' I say. Like, only every day. How could I not know who Eli Roth is? He founded Red Crest as a bright Stanford MBA and built it into a mini studio, the West Coast's answer to Miramax. He's mostly known for two things: hit movies and cost-cutting. Red Crest are constantly buying out smaller production houses, taking their best talent and projects, and firing everyone else.

I can see why Kitty's anxious. We need a hit movie soon, or we're dead.

'Anna. Anna.' Kitty snaps her fingers in front of my face. 'Are you paying attention to me? Eli Roth is a big ideas man. We need to be ready when he comes. I'm relying on you to help me find the right project.'

'OK,' I say. How I'm going to do it I have no idea. It's not as if I wasn't already looking, is it?

'And I have some good news for you,' Kitty says, lowering her voice to a paranoid whisper, though the door remains shut. 'This will be a Greta Gordon film.'

I perk up. Greta Gordon. Oscar-winning actress, former Hollywood uber-babe, huge in the eighties. Greta went through a My Drugs Hell and, after getting bounced off six major movies, her career in rags, finally did a stint at the Betty Ford to clean up. And then she quit films altogether, moving to England to be a recluse and escape the paparazzi.

'She wants to get back into films?'

'This is classified,' hisses Kitty, as though she had handed me a folder from MI6. 'But yes. I've been working on her,' she adds proudly.

'Is she still clean?'

'Nothing but mineral water and the odd vitamin,' says Kitty.

I nod. It would certainly be big publicity if Greta returned. 'But isn't she a touch old?'

Kitty draws herself up. At her, undetermined, age, she refuses to accept that any woman is too old.

'She's as vibrant and lovely as ever.'

'Of course.'

'And she's looking for a lead. A romantic lead.' Kitty draws another circle in the air, diamond glittering. 'Something light, she wants. A comedy.'

'Riiiiiiiight,' I say.

A romantic comedy lead for a 45-year-old former reclusive actress. Well, that should be easy to find, huh? In an industry where thirty equals past it?

I want to speak up, tell her it's impossible. But instead I amaze myself by saying, in a very small voice, 'And then can I have a promotion?'

'What?' barks Kitty.

'You know, if I find you the script,' I say. 'Can I be made a development executive?'

Kitty stares at me.

'Of course I'd still be working for you,' I reassure her, 'because you would have got promoted.'

'I would,' she purrs.

'You'd be a vice-president. Out in LA. And you'd need a development girl you can trust.'

Kitty hums, pressing her skinny lips together.

'You'd need to stay in England,' she says eventually.

Man. She's really agreeing? If she were just blowing me off she'd yes me to death.

'You have expertise in the UK market. It's a good source Eli'd want to keep using,' she says, dreamily. Already he's 'Eli' to her and she's his right-hand woman.

'That's fine,' I say, with a pang of regret for the tan and the hibiscus flowers. But hey. What would a girl like me do with a tan anyway? 'So, that's a deal, then? If I can get you the right script, you'll give me a promotion?'

'Absolutely.' Kitty looks deep into my eyes. 'You have my word on it.'

Then she spins her chair away and looks out of the window, building her empire as she stares over London.

'Just get me that script, Anna.'

3

Why am I here?

I'm standing on the stone steps that lead up to Vanna's porch, having just blown twelve quid on a taxi fare down to leafy Barnes. Vanna has one of those fabulous houses, all Georgian, with grey columns around the porch, high yew hedges to shield her from the road, and pebbles all over her drive. Her Land Rover is parked next to Rupert's racing-green Aston Martin, and behind the house is a small but glorious walled garden.

Vanna has carved out a little slice of the country in said walled garden. The hedges are thick with dogroses, and she has wisteria planted all along the walls, a few apple trees, and bulbs planted around them, so that there are flowers all year round. For winter, she has holly bushes and Christmas roses and winter jasmine.

My finger hovers over the buzzer. I could still run away.

I love Vanna to bits, and I even love coming here, especially on late summer evenings – when Rupert is away, doing business in New York or Tokyo, and Vanna's two little angels have been packed off to bed by the live-in nanny. We go out in the garden and drink chilled white wine and eat organic strawberries, or whatever I want, really. It's like holidaying in a life you could never afford. And I try not to feel too guilty about it, because we really are best friends and there's just no way I could possibly pay her back. I can't even afford to bring a bottle of champagne with me every time. Though I do have one tonight.

However, I feel totally different about evenings like this

one. Rupert is here, and he's an ass. Well meaning, but an ass. And so is some other bloke. His name's Charles, but at least he's gay, so Vanna won't start one of her wretched matchmaking evenings again. But still, that's dinner for four, and I always feel completely out of place with Vanna's brilliant, glittering friends, many of whom I can regularly read about in Nigel Dempster or *Hello!*

I so don't want to do this, but Vanna said she was desperate for a fourth, so here I am. I press the button.

'Hi! Darling!' Vanna is there in two seconds, hugging me, pressing me into a cloud of expensive scent. 'Winston! Get down! He's just trying to be friendly,' she says to me, ineffectually tugging at bloody Winston's collar.

'Yes. I know. Hi, Winston,' I say. Winston is a huge and ancient golden retriever who could shed dog hairs for England. Once again, I have somehow forgotten this fact and worn navy, and my neat little sheath dress is now covered in yellow hairs. 'Get down, Winston. Good boy.'

Winston enthusiastically jumps up on me and gives my entire face a huge lick, smudging my mascara and getting dog spit on my lipstick. I think the Koreans have the right idea when it comes to dogs. Though who'd want to eat this lumpen thing I have no idea.

'Oh, look!' Vanna laughs. 'How hysterical. Look, darling . . .'

Rupert emerges from the kitchen, a martini glass in hand. 'Ha ha ha, great look for you, Anna! Winston, you're such a bad dog,' he adds, giving Winston a friendly shake of the head. 'He's only friendly, you know, Anna.'

'I know,' I say, trying to get my mouth to twist into some sort of rictus smile. 'I'll just go and mend my face.'

A man appears by Rupert's side. He's five foot four, slender, wiry, and wearing an expensive-looking rather dandified grey suit and a lilac silk tie. He has a neatly clipped goatee beard. And a horrified look on his face. He murmurs something to Rupert.

'Oh yes, that's Anna,' Rupert says jovially. 'But she scrubs up a lot better than that! Don't you, Anna? Just got the old Winston treatment! Anna, this is Charles Dawson!'

'Excuse me a sec,' I murmur, and flee to the downstairs loo.

Oh hell. It's a trick, isn't it? Vanna has sunk to new depths this time. She *promised* me Charles was gay. She promised me interesting and cultured conversation, and expensive Indian food catered from their local high-class Indian restaurant.

Instead, she's trapped me. Again.

Vanna is always trying to matchmake me. She has made my happiness her mission in life, which is fine, but there are other ways to make me happy than to stick me next to a bunch of uptight, self-absorbed wankers, I mean bankers, and make me suffer through an excruciating meal making small talk as they all try to get away. I've thought to myself in the past that I should carry a little black book to record all the excuses. Some of them are highly creative and I could probably recycle them at work. 'Have to run, got that video conference with Koyoto.' 'Sorry, Vanna, have to pick up my dog – I have custody this weekend.' 'Is that the time? Must dash, got an operation tomorrow morning, better get some rest . . .' At which point, a crestfallen Vanna will plead with them to stay, see them to the door, and then come back to me and hiss, 'That went great, I think he really likes you,' very loudly. And Rupert will snort.

I dab my face with some wet loo paper and get rid of most of the mascara smudge. I suppose I shouldn't be so harsh on Winston; he's the only male I know who really values me.

I take a deep breath. Face repair, check. Neutral make-up, check. I'm wearing my invariable palette: foundation, the slightest bit of brown blusher, mascara and liner and that's it. Dress, check. It's navy. It's a shift. It's lined. Discreet string of small white (fake) pearls, check. Touch of perfume, check. Chanel No. 5. Very expensive, but I have made the same bottle last since Christmas. Mum and Dad always give me Chanel No. 5. I wouldn't mind something else, like a cool mobile phone or, say, a fifty-pound note. Or even a different, lighter scent. But it's the same every year and I always tell them it's my *absolute* favourite present, Christmas wouldn't be the same without it.

Maybe they're on to something, too. When you are not pretty, you have different dress requirements from everybody else. Gorgeous girls spend ages in shops, feeling the fabric of this exquisite scarf, trying on that mini dress, or selecting something long and clingy. They dress to stand out, to get noticed. They like to change their looks. They are always trying to look like a different pretty girl, they want the must-have print or the must-grab handbag.

Ugly girls have totally different needs. We dress to *not* be noticed. We want to be wallpaper, invisible. Don't notice me. I want to just blend in.

I have shopping down to a fine art. I am usually in and out in twenty minutes or less.

The key is to be appropriate. Like tonight. Dinner with posh friends equals nondescript navy dress in heavy and therefore forgiving fabric, control-everything tights, a little bit of sheer to give it class, a string of cheap pearls (ditto), and Chanel No. 5 is a standard scent. I have three or four dresses much like this one and they're great. Nobody ever gives you a second glance.

I head out of the loo, bracing myself, and instantly run into a small throng of people. Vanna is moving amongst them, her neat brown cap of hair bouncing jauntily, kissing the air on the side of cheeks, and passing out glasses of champagne (not the stuff I brought; vintage Veuve Cliquot).

'Vanna.' I grab her toned arm with a pincer-like grip. 'What is this?'

'Well, originally it was going to be just the four of us,' she lies weakly, 'but these are all *such nice people* and I thought you would have a wonderful time with everybody . . .'

Vanna also believes I need a social life. This is a nightmare. I could be at home, ploughing through scripts. Watching paint dry. Anything.

'Vanna—'

'You say you don't like it, but you always have a great time. And you meet such nice men!'

Grrr. 'Give me one of those,' I hiss, snatching a crystal flute of champagne from her. I would down it in one go but there's just too many of these people here to get away with that. I have to content myself with a huge gulp.

'Charles,' Vanna says, as he passes. 'Come and meet Anna Brown. She'll be sitting next to you at dinner.'

I smile faintly at Charles, who is already wearing that harried, hunted look of a man trapped by Vanna's inexorable will.

'We're best friends,' Vanna tells him. 'And Charles is the brother of Crispin, Rupert's friend from work. You remember Crispin, Anna?'

Sure do. A total tosser.

'Hello,' Charles says, smiling through gritted teeth.

'Charles is a very literary writer,' says Vanna. 'And he has a new novel that he's sending to agents.'

'Yes,' says Charles, visibly preening. 'And you're going to publish it, Vanna! If you're lucky!'

'Would *love* to,' Vanna lies. 'But *unfortunately* our house isn't taking on any literary fiction at the mo'. My hands are tied. *Such* a shame.'

'Hello,' I say back, just to remind him that I am still standing here.

'Vanna and Anna,' Charles says suddenly, as though he's just got the joke. 'You practically have the same name. And you look like sisters,' he says, with a stiff little bow.

'Thank you,' says Vanna, which is sweet of her.

'Of course, Vanna is Cinderella,' he adds, with a high-pitched laugh. 'Ha ha ha, just my little joke, Anna!'

I glare at my hostess who in turn is glaring at Charles.

'No offence, you look marvellous,' he says faintly.

'Anna works in the film industry,' says Vanna determinedly. She will force us to like each other if it kills her.

'Oh?' says Charles, slightly more interested.

'Come on, everyone!' Rupert's bass tones boom. He strikes on his beloved little brass gong, which I detest. 'Chop chop! Time for dinner!'

I know what hell will be like. An endless dinner party, just like this. We could start with the food. Is it delicious curry and prawn kormas from Vanna's local, expensive and fantastic Indian, the Star of Bhopal? No, it is not. It is French, from Rupert's favourite local, expensive and pretentious bistro, Le Coq D'Argent. First course, grilled escargots in sickly garlic butter.

Everybody is eating them as though they were Pringles. But I haven't got the taste for slime-trailing garden pests yet.

'Not fond of snails, Anna?' Rupert says, arching an eyebrow as he sees me pushing them around my plate.

'I'm on a diet,' I say, 'though they look divine.'

'Good idea,' says Charles, supportively. 'A journey of a thousand miles starts with just one step! Or even of thirty pounds!'

'Have you ever had weight issues, Charles?' asks Priscilla, Rupert's banker colleague who is sitting the other side of him, next to her mousy house husband, Justin. Vanna hates Priscilla. She wears tiny little twinsets and Alice bands, and plays little girl lost whilst being an utterly ruthless wheeler-dealer. Vanna is sure she's after Rupert.

'No,' says Charles, admiring her tiny waist. 'I believe in self-discipline. Just like you. Presentation is so important in today's world, I always think.'

I can't take it.

'So when are you going to see someone about your bald patch?' I inquire pleasantly. 'It might give you the wrong image, unless you want to be a tonsured monk.'

Charles's face flushes, and Vanna winks at me. As well she might, she got me into this.

'Next course!' Rupert announces, as two uniformed waitresses enter and whisk away our plates. 'An *amuse-guele* of smoked salmon and foie gras.'

This is more like it. I hate most fish but I love smoked salmon. I try a bite of my foie gras and that's delicious too.

'You may find it a bit too fattening on your diet,' Charles says to me, in hushed tones. '*Gras* means fatty in French.'

I want to ask what 'Fuck off, baldie' is in French but content myself with spearing an extra-large chunk and eating it right in front of him.

'Now this,' says Rupert, waving his bone-handled silver fork at the dish in front of us, 'is the real stuff. None of this modern diluted *foie* for me. They get it from a special place that makes it the old-fashioned way.'

'And what's that?' asks somebody from his end of the table.

'You know, get the goose, hold it steady, force-feed it until the liver bursts,' Rupert says.

I stop chewing. I feel ill.

'You know, Charles,' says Vanna, seeing me turn pale, 'Anna reads scripts for Winning Productions.'

'Never heard of them,' Charles says dismissively.

'Oh, but you must have heard of their movies,' Vanna insists. 'They did *Twickenham* last year, the rugby comedy, and that adaptation of *Bleak House* that won all the Baftas.'

Charles has heard of these. He deigns to turn to me properly.

'So you are interested in quality material?' he asks.

'Certainly,' I say, though our last two movies had nothing to do with me. Kitty claims part credit for *Bleak House*, but Kitty likes to claim part credit for everything.

'That's very commendable,' he says, 'because these days the studios are only interested in commercial crap, aren't they?'

'They do like films to sell tickets,' I concur.

'Crap!' he declaims. 'And the so-called British cinema – double crap!'

'I do agree,' purrs Priscilla.

Vanna gives me a pleading look, so I swallow hard and merely say politely that perhaps Charles's book could be adapted for the screen.

'Well,' he says, as though considering it. 'It could only be done by a company with real taste, and delicate sensitivity. The exccutive in charge would need to be cultured, and understanding of a great work. Too many people ignore subtle tones and shades of emotion.' He looks at me suspiciously. 'Do you have a refined sensibility, Anna? Would you be capable of handling true literature?'

'I just evaluate scripts and source material.'

'Oh.' He immediately loses interest in me. 'You don't have

43

the power to green-light? Perhaps I should be talking to your boss. Who is he? I think I'll go straight to him.'

'Her name's Kitty,' I say, smiling sweetly, 'but she doesn't accept unrecommended projects.'

'This has literary merit to recommend it,' he counters. 'What else does it need? Plus, of course, I can use your name.'

'I'd need to know something about it first,' I say, 'other than its great literary merit.'

'I don't see why,' he sniffs.

I look round the table for help, but Vanna, having secured her aim in getting the two of us to talk, is busy in conversation with somebody else, and Rupert is drooling over Priscilla. In fact there is a low hum of conversation everywhere. Nobody seems fascinated by Charles's masterpiece. I am trapped.

'Well,' I say slowly, 'films have different needs from books. Sometimes what makes a good book doesn't do well on screen . . . especially literature.'

'I see,' Charles sniffs.

'We look for a great idea. Something that people will come to see just based on the idea alone. For example *Jurassic Park* or *Fatal Attraction*.'

'Junk!' he says.

'Great books can make lousy box office,' I tell him. 'The *Les Miserables* film was a bomb.'

'A travesty!' he sneers. 'The viewing public are too *dumb* to know what's good for them!'

Wow. I should really introduce Prince Charming here to John. They'd get on like a house on fire. Not that John would ever recommend his book. John would hate to see somebody else be successful.

'Sounds like the kind of trash Trish writes,' he says, upper lip curling in disgust.

'Who's Trish?' I ask.

'Trish Evans,' he says. 'My sister's nanny. She writes scripts. Always telling me about her ridiculous ideas.'

'Like what?'

'The last one was about a wedding that goes wrong.'

'Oh,' I say, losing interest myself. Done to death.

'Yes, inane. It's called *Mother of the Bride*,' he snorts. 'She's jealous of her daughter and trying to interfere and control the wedding, and then she falls in love with the groom's uncle.'

I laugh. 'Actually it sounds like fun.' And perfect for Greta Gordon! *Mother of the Bride*. I love it!

'Fun?' he demands crossly. 'It's totally forgettable.'

'You remembered it.'

'My novel—' he begins.

'Charles,' I say, 'what might be a good way to get in touch with your sister's nanny?'

He stiffens. 'I really don't think you should be bothering her,' he says. 'Now, about *my* book—'

'Here's the thing,' I say, determinedly. 'I need to present a mix of ideas to Kitty. If I had something low-brow, something pulp—'

'Commercial and crass,' he adds.

'Exactly. Then I could manage to get her to also consider a literary work.'

He muses. 'So you need to get Trish's work in to smooth the way for mine?'

'Exactly,' I agree.

'I see.' He pulls a thick, gold-embossed business card out of his pocket and writes a number down on it. 'My sister is Lady Cartwright. Wife of the eminent podiatrist Sir Richard Cartwright.'

'Very impressive,' I say.

'And does this Kitty actually listen to you?' he asks suspiciously.

'I'm her go-to girl,' I tell him.

Charles lifts his red wine glass to me in a toast. 'To Anna,' he says, 'whose ship has just come in!'

I chink glasses with him and wonder what Trish is like.

'I certainly hope so,' I tell him.

The rest of the evening passes in a dull blur of conversation and vile food, and by the time the coffee comes I can't wait to get away.

'Wasn't it *wonderful*?' Vanna insists, pressing my arm. She looks so hopeful I think it would be cruel to tell her the truth. 'You and Charles! You two were engrossed!'

'He's very nice,' I fib weakly.

'I hope you two see each other again. He's very suitable. The book's dire, of course, but he's got pots of money,' Vanna promises. 'Trust fund. Flat in Eaton Square with a sixty-year lease.'

'He's a real catch,' I agree.

'Charles, darling,' says Vanna, grabbing his arm and threading it through mine. 'See Anna into a taxi, won't you?'

'Of course,' he says graciously. 'One can't trust a dear lady on the streets by herself.'

'I can get the tube,' I say, trying to wrestle myself free.

'Allow me to take you,' he says, bowing low and practically kissing my hand.

In the taxi on the way home I wonder what I must have done in a previous life to deserve this. On the one hand, Charles is an improvement on Brian. He does not, for example, suggest the cab fare should be my treat, nor does his breath reek of bad fish marinated in old beer. On the other, Charles does not even pretend to be interested in me. He is endlessly entertained by discussing, in no particular order, his own book, his brilliance, and his attractiveness. I soon learn that the quickest way out of it is to say 'Uh-huh' and 'Mmm' and nod as though I am equally enthralled.

'Of course,' he says, as we turn by Leicester Square (almost home, thank God), 'women just don't understand me, Anna.'

'Don't they? That's a shame.'

'Would you believe I haven't had a steady girlfriend for three years?'

Yes. 'No.'

'It's true,' he says, bitterly. 'They can't cope with the rigours of living with a creative genius. All they ever say is perhaps I should get a "real" job. But what is more real than Art?'

'What indeed?' I ask.

'Of course, they don't object to spending my money,' he says. 'And staying in my flat uninvited, then claiming to have headaches and so-called woman's trouble.'

'Oh dear,' I say, struggling for composure.

'Woman's trouble doesn't last for two weeks every month, does it?' he demands.

'Not typically,' I say.

'Because I looked it up on the internet,' he adds.

My building is coming up. Hoorah!

'I have so much to offer,' he says, dramatically. 'Yet nobody is prepared to see the real me!'

'I'm sure you'll find somebody soon,' I tell him.

'Are you seeing anybody?' he asks.

'Not right now.'

'No, of course not,' he agrees.

Hey, thanks. 'Well, this is me,' I say gratefully. 'Thanks very much for the lift. And I'll be waiting for your book.'

'Anna,' he says, as I step out of the taxi. 'I like you. You have a wonderful way of listening. Very feminine, so many women want to talk all the time, banging on about themselves.'

'Thanks,' I murmur.

'So maybe I'll give you a call and we can go out?' he asks. 'After all, you are a friend of Vanna and Rupert's. We must have something in common.'

'Ahm . . .' Help. *How* do I get out of this? *Why* did I say I wasn't seeing someone?

'That's settled then!' Charles exults. 'I'll call you. It's a date. My, but you *have* been lucky this evening, Anna. Good night.'

'Have you heard the news?' Sharon asks, as soon as I get to my cubicle. I haven't even dumped my bag and already she's hovering like a hawk, which means it must be something really big.

I speculate. 'John's proposed to Kitty.'

'No.'

'You've read the greatest script ever.'

'No,' says Sharon, shaking her bouncy curls as though to wonder how I can be so stupid. Of course, that would mean she'd actually read a script.

'Kitty's been fired.'

'Did you hear that?' she asks, eyes brightening.

'No. I'm just trying to guess your big secret.'

'I've got a transfer,' she says exultantly. 'I've been promoted.'

My mouth falls open. 'What?'

This can't be true, can it? There are Pet Rocks out there brighter than Sharon. She has only just avoided being fired by the skin of her teeth for the last six months.

'Mike Watson saw my potential,' she says triumphantly. 'I'm now officially working for him as a junior development executive.'

I feel faint. 'But you work for Kitty.'

In fact, I am clearly not the only one to be aggrieved. At that moment Kitty storms out of her office, a mini dynamo clad in circulation-killing Azzedine Alaia, hands on her bony hips.

'What the fuck is this?' she screams, brandishing a memo in Sharon's face. 'Is this some kind of a joke?'

I look round as Mike materializes from nowhere, smiling his big fake smile at Kitty. The two of them square off, pacing around each other like leopards about to fight to the death.

'Is there a prob, Kitty?' he asks coolly.

'Yes there is,' she snaps, 'actually. There's a memo from Personnel on my desk saying that Sharon reports to you now.'

'That's right,' Mike says. 'I needed somebody else on my team, and you seem well served with these two.' He indicates John and me dismissively.

'And you didn't think to ask me?' Kitty demands.

'You're so busy, Kitty,' Mike says smoothly. 'Preparing for the buyout of the company. Eli Roth and all that.'

Kitty shoots a look of loathing at Sharon, who tosses her curls triumphantly.

'I see you got a promotion, too,' she says to Sharon. 'No need to ask what for!'

'For my talent,' Sharon says blithely. 'Of course.'

'You do realize,' Kitty says to Mike, 'that having sponsored her for that position, your ass is on the line if she fails to deliver? Which she will.'

'My goodness, Kitty,' says Mike. 'Maybe you should have some faith in your team.' But he looks a little less smug. Mike knows that Sharon is about as effective as Cherie Blair's astrologer.

'Go fuck yourself, Mike,' snaps Kitty.

'Ladylike as ever,' says Mike, grinning. 'Come on, Sharon.'

'You two, in my office,' says Kitty to us, and John and I proceed into her office, where she slams the door and screams about disloyalty for twenty minutes.

John spends most of the time agreeing. 'Of course, Kitty . . . she isn't worth it . . . you can't let her get to you . . .'

I just sit there. Trying to deal with it. Sharon Conrad, Junior Development Executive. Sharon has just got herself the position I have been slaving to achieve for the last six months, Sharon, a girl who has no brains, no drive, and no sense. All she had to do was spill the beans on Kitty's secret to Mike Watson, and there you go.

Of course, that wasn't all of it.

If I had done that, for example, gone and told Mike about Kitty's knowledge of our impending buyout, what would have happened? Transfer? Promotion? Not bloody likely. He'd have said, 'There's a good girl, Anna,' and smirked at me. Maybe. And then told me to go and make him some coffee. Whereas Sharon, with her need to please and her bright smile and general cute as a button-ness, Sharon he rescues. He actually promotes.

It's all because she's pretty.

Well, I've had it. From now on I vow never to trust a pretty woman again. I hate them all!

Except Vanna, obviously.

Pretty women are proof that God is a man. They do no work and get other people's promotions. They laze around the flat all day making tons of money just for having their picture taken. Everybody loves them. They get into clubs for free, they float to the head of the queue, men give them their seats on the tube. And for what? A set of perfect features they did absolutely nothing to deserve.

Bloody hell. Enough of sisterhood and solidarity. I hate *all* of them. And I hope that Rob spends all day long spitting in Sharon's coffee.

'Anna?' Kitty is talking to me. 'Did you come across anything yet?'

I think of Charles Dawson's card in my pocket.

'Might have done,' I say morosely.

'Well, get to it,' Kitty says intently. 'Now Mike knows, everybody will. There's no time to waste! He could be here tomorrow.'

'Eli Roth is coming here?' John asks.

'Anytime,' Kitty says. 'So let's get to it. Our franchise is out there, people!'

For the record, even though she's not strictly pretty, I also hate Kitty.

'Hello? Is that Lady Cartwright?'

'No,' says a voice. Pure Albert Square. 'Her ladyship ain't in. Take a message?'

'I'm not actually looking for her,' I say. 'I'm trying to get in touch with her nanny, Trish Evans.'

'This is her, innit?' says Trish, her tones as soothing as a cheese-grater scraped across a blackboard. 'What do you want?'

This is so obviously a wild goose chase that I'm about to hang up.

'Are you from Nice Nannies?' Trish continues. 'Because I already told you I ain't switching again. They offered me another four grand to stay. And my own car. Besides, I'm tired of moving.'

'Not Nice Nannies as such,' I tell her.

'Or Mother's Help?' she asks, suspiciously.

'I'm not trying to poach you as a nanny,' I say. 'Though I hear it's a cut-throat world.'

'You don't know the half of it,' she says darkly.

'I met Lady Cartwright's brother the other day.'

'Prick,' she says, loudly and clearly. I can't help wondering where exactly the little Cartwrights are right now.

'Well, he recommended you as a writer,' I lie.

'Did he? Not like him. Thinks he's Charles Dickens.' She snorts. 'What did he say, then?'

'Only that you were writing a movie and I thought maybe we could meet for coffee.'

'Are you a serial killer?' she asks suddenly.

'No,' I say.

'Well, how do I know?'

'You could ask Charles.'

She treats this response with the contempt it deserves.

'Look,' I say desperately, 'I would just like to meet up with you for coffee in a public place. Absolutely no knives or acid baths. My name's Anna, Anna Brown, and I work for a production company. Looking for good scripts.'

'All right,' she concedes. 'I got lunch at one but you'll have to come here. I don't have time to muck about on the tube and that.'

'Give me the address,' I tell her. Well, it's easy enough – Lady C lives in Albany, Piccadilly. Just round the corner from me. And a million miles away.

This is totally futile.

I'm standing here talking to a snooty porter at the entrance of London's absolute grandest rental address (cars in front – Bentleys, Rolls-Royces, and one Lamborghini the others probably look down on as far too *nouveau*), trying to feel more than two feet high, and absolutely sure I'm wasting my lunch hour.

On the other hand, it gets me out of the office.

I need that.

I can't sit still for one more minute listening to John bitch about Mike, or Kitty screaming at her secretary just to have someone to bully. And most of all I can't sit there thinking about pretty Sharon and her totally undeserved promotion that I should have got bloody months ago. So maybe it's worth listening to this doorman telling me that if 7F doesn't answer soon he's going to have to ask me to leave the property as I can't 'loiter'. I can always go to a coffee shop here and sit and stew in my own bitterness. Or, I could just take the rest of the day off, go home and stew in my own bitterness there. What an enticing prospect!

'Look, miss, you'll have to – oh. Yes? Hello? Miss Evans? Yes, there's a young . . . *lady* waiting downstairs for you . . .'

He hangs up, clearly surprised that I am indeed expected.

'She'll be right down, madam,' he says.

And indeed she is, one minute later. I'm sure she's an absolute favourite with the staff here. She comes racing down the hall, her footsteps clattering loudly on the glorious old flagstones.

'Fucking hell!' she half shouts. 'I couldn't get the little bastards to sit still long enough for me to answer the bloody phone. Wotcher! You all right?'

'Fine,' I say through clenched teeth.

She extends a hand. 'I'm Trish. There's a coffee place down the street, Cook's got the little ones while I'm on my break.'

I shake, gloomily. Just my luck!

Trish is absolutely, cast-iron, triple-certified gorgeous. Let's run though the list, shall we? Long blond hair, bleached. Legs that would do credit to an Arabian racehorse. Smooth, pearly skin with a natural rosy bloom on high cheekbones that will ensure she'll look great at sixty. Big green eyes with thick, dark lashes. A full, pouty mouth. If Kate Moss were just a touch more upholstered, and a blonde, she might look a bit like Trish.

I want to cry off right here and now. No way Barbie here has written anything remotely interesting.

'I bet you're fucking glad to get out of the office, eh?' she asks with a disarming smile. 'When I worked as a temp I couldn't stand it. Wanna fag?'

'Um, no thanks.' I look around for a cafe, and there's Costa Coffee, rising to greet me like an angel of caffeinated mercy. 'We'll just go for a quick coffee and you can tell me all about your writing. I can't promise anything, though,' I hasten to add. 'It's really just to get to know you.'

'You don't need to know me,' she says. 'You just need to read my script, 'cause it's great.'

I smile thinly. 'I'm sure it is.'

She grins back at me. 'Wow! I can't believe I'm really talking to you. I'm so lucky!'

We sit down at Costa Coffee and order. I get a cappuccino and she gets a plain black decaf.

'Watching your weight, I suppose,' I say glumly.

'I have to,' she says. 'Girl like me. Not much going for me. Dropped out before university. Bunch of crappy jobs, waitressing and temping, then I found being a nanny. That pays well, but I hate it,' she says passionately. 'So I wanna get married. Old fashioned, right?' she asks, self-deprecatingly. 'But it's the only chance I got.'

'Why do you hate being a nanny?' I ask. 'If it really pays well?'

'It does,' she says, 'but they treat you like a maid. And make you give the kids all this vile stuff they don't want to eat. Macrobiotic. More like macro old bollocks.'

I laugh. She's actually not that bad. For a fox.

'Lady C don't care about them kids,' she says. 'Breaks your heart, watching them try to please her. Drawing little pictures and she just throws them out and they find them and start crying. I have them framed and say she did it.'

I grin. 'So you want to marry for money?'

'Can't afford a flat,' she says. 'Can't afford anything. You have to be practical, don't you?'

'I suppose.'

'Maybe you'll read my script and like it and offer me a million pounds,' she says hopefully. 'Then I wouldn't have to.'

'Would you stop dieting?'

'Just watch me,' she says. 'I could murder a McDonald's.'

I haven't had a McDonald's in months. My mouth starts watering. 'Oh, don't.'

'Big juicy burger and fries with three packets of ketchup,' she says temptingly.

'Um. You're not very likely to get a million pounds. You're not all that likely to get any pounds,' I say, and watch her face fall.

'Oh well.' She shrugs. 'Knew it was too good to be true, didn't I?'

'Sorry if I gave you the wrong impression,' I say, and to my surprise I am. I actually like her. 'I'm just looking for projects right now. I'm looking all over. For what it's worth, I'd like to read yours, but I should tell you, we reject almost every script we read.'

'Why's that, then?' she asks. Genuinely curious.

'Because they're crap.'

'Oh. Fair enough. Well, I'll be OK then because mine's not crap,' she says, reaching into her bag and pulling it out. She's done it up properly, good formatting, holes punched in the right places.

She's a fun girl and I don't want her to be disappointed.

'I just think you need to know that everybody believes their script is great,' I tell her, 'and almost everybody's wrong so please don't be too upset if it's not right for us. It might be right for somebody else . . .'

'Look at you,' she says, grinning away. 'Trying to let me down gently and that. I'm not thick, honestly. I was just lazy in school. If you don't like it, no harm done, right? It's hard to even get scripts to readers.'

'If I don't like it,' I say, because I know I won't, 'I'm going to call you and give you some pointers to help you with your next one. OK?'

'All right,' she agrees. 'Is there any money in script reading?'

'No. Fuck all.'

'Always the way, innit?' She sighs. 'The boring jobs pay the most, that's why there are so many lawyers.'

Amazingly enough I am in a slightly better mood when I get back to my desk. She was a funny girl. She can't help being stunningly beautiful. She would be, wouldn't she?

I put Trish's script to one side and make some phone calls, trying to sound very authoritarian.

'Yes, Kitty wants your best stuff. Right away. And nothing like that one you sent me last week about the two circus midgets.'

'Hi. Yes, I heard you have the galleys of *Permanent* in. Any chance you could slip them to us? Kitty's office, care of Anna Brown . . .'

I don't know if it'll work but at least it makes me feel useful. I get a few good responses to the urgency in my voice. When Kitty passes me, she gives me a thumbs up. To my astonishment, she even returns a minute later with a coffee for me.

'Good work,' she hisses, shooting a look of loathing over at Mike's office. 'Keep it up.'

I drink the coffee and feel all jittery. Two shots of caffeine in two hours and I'm looking for some chewing gum to stop my teeth grinding, but never mind. It helps the nervous energy. And you need it, to try and get your hands on something good.

I shoot a look across the hall at John. He's sitting crouched in his cubicle, hand covering his face, so I can't see what he's saying in those low, fast tones. Like I care. John thinks he's Jerry McGuire now. Mr Superagent.

'What's this?'

I look up to see Sharon standing by my desk, flicking through Trish's script.

'*Mother of the Bride*?' she asks. 'By Trish Evans. When did this come in?'

I snatch it back from her. 'That's mine.'

Sharon extends one hand, her nails glittering with frosted silver polish. 'Actually, it belongs to the company,' she says. 'Hand it over.'

I take it and lock it in my bottom drawer. Sharon's radiant cheeks redden. She scowls.

'I am a development executive now,' Sharon says. 'I'm *senior*.'

'I'm reading this for Kitty,' I say. 'Or do I report to you now? Because I haven't got *that* memo yet.'

Sharon tosses her curls. 'Maybe I'll call Personnel and see if I can arrange it,' she threatens. I take a gulp of coffee.

'Oh, will you?' asks Kitty in icy tones. She has seen this from her office and snuck up behind Sharon, walking as lightly and predatorily as a cat.

Sharon jumps out of her skin. 'Oh, hi, Kitty.' She recovers, then looks at Kitty defiantly. 'Maybe you can help me convince Anna that all script submissions are to the agency, and not just to Anna Brown.'

'Or Kitty Simpson?' Kitty asks with quiet menace.

Gosh, this is great. It's just like one of those BBC2 nature shows where the young lion squares off against the old lion.

'Exactly,' Sharon agrees, unfazed. 'We work as a team, we all need to share our leads.'

This is so ridiculous that I give a derisory snort, but unfortunately coffee spurts out of my nose and ruins the effect. Winning Productions is not a 'team'. It is a hothouse of fear and loathing. And greed.

'That's an excellent idea,' Kitty says smoothly. 'Why don't you run back to Mike's office, gather up copies of everything he's working on and deliver them all to me. Then we'll send you a copy of Anna's discovery.' She rests a proprietary bony hand on my plump shoulder; I can see her huge canary diamond flashing and glittering with my peripheral vision.

Sharon searches in vain for a good response.

'I can't take Mike's things,' she says. 'But I can send you everything *I've* come up with.'

'No thanks,' says Kitty immediately. 'I don't need any tips on nail-painting or eyeshadow application.'

Sharon flounces off, but not before muttering, 'Oh yes you do,' just loud enough so the whole office can hear it.

'What is that you've got?' demands Kitty, as soon as her tight little butt has minced out of sight. 'Anything good?'

'Doubt it,' I say. 'First try by someone's nanny.'

'Oh. Well. Don't give it to her anyway. And don't just sit there, Anna. Start dialling. Eli Roth will be here *tomorrow*.'

I finally make it home, weighed down with a huge overnight read. All I want is to go down to the offie and pick up a bottle of wine. Or possibly a couple of those Mixed Doubles instead. Three. Or four. And a Snowflake. And then nuke one of my Marks and Spencer ready meals. And have a hot bath with some Matey bubble bath, wrap myself in my big, frayed and,

OK, a bit grimy white towelling robe and just sit down and pig out.

Ooh. That sounds really nice. My muscles are just starting to unknot from the day's tension as I turn the key in the lock, and then I hear the unmistakable sound of sobbing.

I dump my stuff on the couch. It's Janet. She's curled up in floods of tears.

'What's the matter?' I ask.

'It's Gino,' she sobs.

Of course it's Gino. Her Euro-trash boyfriend, an Italian count or something. Inherited a bunch of money from a car-manufacturing daddy, and now dedicates his wastrel life to getting rid of it all as fast as possible. Gino is your typical Trustafarian. He espouses socialist and anarchist principles while hanging out at *very* exclusive clubs, dating models and sneering at the poor.

I'm not his biggest fan.

'What's he done now?'

'Du-du-dumped me,' she sobs. 'We were at Brown's and he suddenly says that he thinks it's time to see someone else. He's going with Katrina Pereshkova!'

'Who?'

'Oh, you have to know her. She was in *Company* last week. And she's just done German *Vogue*,' Janet wails. 'She's so hot right now! And when I asked him why, Gino shrugged and said that my butt was too big. He said that curves are going out . . . and he called me Janet instead of Jay-Me!'

I look at Janet's incredibly slim form, and her ludicrously high and tight bottom. Perky doesn't begin to do it justice.

'What curves?' I ask. 'You're 34B if you're lucky. That's not curvy.'

'It is for a model,' Janet says. 'I'm huge! I'm a size eight,' she whispers, ashamed, and then dissolves again.

'Oh. I see,' I say, trying really hard to be sympathetic.

Because the tough thing is, living with two models I do actually know she's right. Janet is just this side of anorexic looking, which in her world makes her a hefty girl. The more usual look is Lily's, all bony and angular, without an ounce of fat anywhere. And a jerk like Gino trades up with his models like he trades up his watches or his cars. The trends have been all about so-called curves, which means shots of Kate Hudson and Catherine Zeta-Jones, but that was last month. Now we're back to waifs and heroin chic, and Janet's tiny, perfect little bottom is too much.

So Prince Charming has dumped her.

'I told him I'd go on a diet,' she sobs. 'But he said I'd never be thin enough and I was getting too old, anyway.'

'Gino's a pig.'

'He was only being honest,' Janet sniffs.

'He was only being spiteful. Look, you can do so much better.'

'But he was a millionaire,' Janet says. 'And a count. I could have been a countess.'

'Those Italian titles are ten a penny. I bet he got his out of a magazine by sending away twenty Euros for a certificate.'

'You think?'

'I bet,' I tell her. 'And I don't think he's going to have any money left in a year or so. He'll be calling you and begging you to take him back, but you'll have moved on to Bill Gates.'

'He's already married,' says Janet mournfully. 'I checked.'

'You stay there,' I tell her, 'and I'll pop out to the offie and get us some booze and then I'll go down to the Golden Dragon for a Chinese takeaway.'

Janet's eyes round. 'Do you even know how many calories are in that stuff?'

'Chinese people eat it, don't they? And they're all skinny.'

'I suppose once couldn't hurt,' says Janet doubtfully. 'Just get me something from the diet menu, OK?'

'You got it.'

She fumbles in her little Prada purse and hands me a couple of twenties.

'That's all right, it can be my treat.'

'Don't be silly, Anna,' Janet says, wiping her eyes. 'Everybody knows you're poor.'

We stay up late drinking Mixed Doubles (rum and Coke for me, gin and tonic for her. I decanted it and told her it was diet) and eating dim sum and shrimp lo mein, which I also told her was diet. Janet eats most of it, and who can blame her? She probably hasn't had a decent meal in five years.

Eventually, she hugs me.

'I'm going out,' she announces.

I blink. 'Are you sure? You're a bit merry.'

Janet is swaying dangerously around the room and we haven't even got to putting on the J-Lo records yet. She likes to sing them, too. She has her own version of 'Jenny From the Block' cunningly re-titled 'Janet from the Block'. She can't rap and she can't sing, but she doesn't let that stop her.

'I'm fine,' she says. 'Jusht fine. The night ish young! D'you want to come to some clubs?'

I tap my pile of scripts. 'Got work to do.'

'You don' need to worry that they won't let you in,' she says, reassuringly. 'They will if you're with me. I'll jusht say, "She's my mate. She's with me." ' Janet waves one arm generously. 'I'll say, "Step off, fool, Jay-Me and Anna B are in the hizzouse!" '

'Well, I'd love to be in the hizzouse,' I say. 'But unfortunately I have to be in the flat. Working.'

Janet nods. 'OK,' she says, getting her coat, 'but you know, you're not as bad as you think you are, and if you are, it's your own fault, you know what I mean?'

'Um. Yes.'

'See ya,' she says, wafting out in a trail of Dune and gin and tonics.

I look at my watch. Quarter to midnight. The booze is making me sleepy, but that's no good. I have to read some of these scripts. At least a few.

I make coffee and flick through the first few on the pile. Immediately my headache comes back. Why do people write these things? Arty, pretentious indie flicks, formulaic romcoms, endless Britflick 'Lock Stock' rip-offs, somebody has kidnapped the President's daughter (yawn), a master thief is brought out of retirement for one last heist . . .

The euphoria of the rum and Cokes is slipping away and a bone weariness starts to seep through my system. It's no good. I'll never find anything because there isn't anything to find. There are just millions and millions of rubbishy scripts, and Kitty will fire me and I'll never make anything of myself.

In desperation, I pick up Trish's script. I have to read this one all the way through. Why, why did I promise to give her notes? I won't get to bed till two.

Idly I read the first page. And the second. And third. And then I slowly put my coffee to one side.

I can't believe it. It's funny, it's fast-paced, the characters are believable. I want to know what happens! I keep reading, keep flipping. It makes me laugh, it's a bit sexy, and sometimes, like the very best comedies, it's touching. Plus, there are no big sets or special effects needed. It could be a film you shot cheaply and made millions on.

In a daze I peel off my clothes, dump them on my bedroom floor and climb into bed. It's one thirty but I'm too excited to sleep. I just lie there staring into space.

This could be it. This could be my chance!

4

I wake up early. Or you could say I never went to sleep at all. I mean, I did a bit, but it was that sort of fitful, tossing and turning dozing that leaves you more tired than if you'd stayed up with coffee and a pack of Pro-Plus.

At any rate, I get to hear all the nocturnal sounds of Tottenham Court Road that I normally tune out. Drunks yelling. Clubbers screaming for a taxi. Janet coming back around half four and heaving up in the bathroom. All in all I'm quite glad when the sun puts its head up in the east and the first pink tendrils of dawn are spreading over the grimy London skyline. I jump out of bed and into the shower. I feel a bit punch-drunk, sure, but that feeling is still there, shining through the exhaustion. A sort of Christmassy feeling, as if I'm about to get a very big and shiny present.

I wash my hair and dry it. At 6 a.m., this would normally have the models screaming, but Lily hasn't been around for days (probably staying with her latest footballer boyfriend) and Janet is totally passed out. Anyway, I don't even care. The day feels big and bright and full of opportunity.

After doing my hair I go into the kitchen and steal some of Lily's vanilla hazelnut coffee, specially imported from Seattle, and brew up a delicious-smelling pot. Unfortunately her food is not worth stealing – calorie- and taste-free rubbish – so I content myself with the coffee and re-reading *Mother of the Bride*. At first I'm afraid that maybe it was the booze and the script has turned crap overnight. This happens to script readers the way beer goggles happen to men – you go to bed with a beauty and wake up with an absolute dog.

But not today. It's just as exciting this morning. It actually improves on a second read. I dress hastily, unable to believe my luck. Beige trousers, Gap white T-shirt, my greying sports bra – nobody's going to see my bra, are they, plus it sort of flattens my boobs a bit. When you're as tall and galumphing as me, you really don't want a couple of huge boulders drawing attention, do you? I add a large, cable-knit cardigan that belts round me and hides everything, quickly do my make-up – foundation, bronzer, pathetic bit of nasal shading – and I'm ready to go. I brave the mirror for a quick check. No lipstick on the teeth, no obvious rat's nests in the hair. It's as good as I'm going to get. I grab the script and hurry off to the tube.

When I get to the office, Kitty is already there.

'Wow,' I say. 'You look nice.'

She gives me a piercing look, as though 'nice' is totally inadequate.

'Well,' she sniffs, '*some* of us believe in making an effort.'

And how. Kitty has chosen an exquisitely cut, tomato-red dress, looks like Versace, with discreet pearl buttons down the front, tanned Woolford tights, and perfectly matched red shoes. It's sort of eighties power dressing but done with a modern twist: three-quarter length sleeves, very fitted, her skirt slightly flippy. Her bag is a tiny black thing in supple leather, and her make-up very subtle, shades of chocolate. She could pass for thirty-five.

'I hope you haven't forgotten that Eli Roth is coming in today,' she says, looking around the half-empty office. 'Apparently everybody else has.'

'I – no,' I say, smoothing down my cardigan in a vain attempt to look more presentable. 'It is only eight thirty.'

'Yes, Sharon and John should have been in *hours* ago,' Kitty snaps, magnificently ignoring the fact that we never drink our first cups of coffee until ten. 'But at least you're here. Why are you here?' she probes, eyes narrowing. 'Hoping to meet Mr Roth early, eh? Maybe get in there before everybody else? Make an impression?'

'Of course not,' I say. 'It's not very likely he'd be impressed by me, is it?'

Kitty looks me over, craning her neck as though she can hardly see high enough to look at me. 'I suppose not,' she says, mollified.

'I think I have something for you,' I tell her.

Kitty's eyes round in a greedy O. 'You do?'

'I think so.'

'The script? By the nanny?'

'It's actually pretty good,' I say. 'It could be perfect for Greta, anyway.'

Kitty fairly dances on the balls of her toes. 'What are you waiting for?' she demands. 'Get into my office. No, wait, bring me some coffee, then get into my office. You can get some for yourself, too,' she adds magnanimously.

I hand her the script and rush to the kitchen. That Christmas-morning feeling is still there as I load up the espresso machine and put the kettle on for my own PG Tips. Kitty's in two hours early. She must really think this buy-out is important. And nobody else is here!

I have the office all to myself. I can go in and pitch to her, no John or Sharon to try and sabotage me, no filing to get done, no errands to run. It's my chance. And as I carefully slice off a curl of lemon peel and place it artfully on her saucer, I find my hands are shaking.

'There you go,' I say, setting her coffee in front of her.

'Thanks.' Kitty's bony hands reach out, diamonds flashing, and claw delicately at the bone china cup, while my rough paws grip my chipped mug that says 'Old Dieters Never Die, They Just Waist Away'. 'Where's your coverage?'

'In my head,' I say, and then, because that sounds too much like Sharon, add hurriedly, 'I came in early to write it up, but you're already here, so maybe I could just give it to you in person?'

'Go on,' she says.

'Well, it's a romantic comedy,' I say. '*Mother of the Bride.*'

'I like the title,' Kitty says crisply.

'The heroine is jealous of her daughter's wedding and unconsciously sabotaging it. But then she falls in love with the groom's uncle. Only he's gone off her because she's been a total prima donna bitch, and then she has to save the wedding and her romance.'

'I don't know,' Kitty says doubtfully. 'I don't think Greta's going to want to play a total bitch. She's America's sweetheart.'

Well, she was, before she was revealed to have done more class A substances than Keith Richards.

'I think she would,' I say deferentially. 'Remember *Starlight* and *Outcasts*? She made her name in those. If she's going to come back, the part's funny, it's bitchy, it's poignant . . .'

'Poignant,' Kitty says, scornfully.

'She could be nominated,' I lie. This is rubbish, of course, but execs love to hear it. 'You could play a part in *two* Oscar-winning movies.'

'I do more than play a part, I drive the projects,' Kitty says, magnificently. 'And it's funny, you say?'

'I think it's everything *The First Wives Club* tried to be and wasn't, and that movie was still a hit,' I point out. 'Plus,' I move in for the kill, 'there aren't that many great romantic leads for older women actresses.'

'Greta is hardly older,' Kitty says.

'She's older than Kate Hudson,' I say, 'or Natalie Portman. You know how it is in casting, Kitty.'

'Mmm,' Kitty says non-committally, but at least she's not throwing me out.

'The mother, Diane, is the only really huge role, so all the rest would be character actors, and there aren't that many locations, it could be shot quite cheaply. It could be another *Full Monty*,' I say. 'Or *Four Weddings*.'

Kitty presses her bony fingers together. 'Get me coverage,' she says. 'Make it sizzle. Get me something to sell with. Make it appeal to Greta. Maybe we have something here.'

I want to ask if this means I get a promotion, but Kitty's eyes are flinty. 'Hurry up,' she barks. 'We've only got a matter of hours.'

I bring the stapled sheets to her forty minutes later, my fingers stiff from writers' cramp. By now, the office is crowded; it's still early for us, I guess nobody wants to be late when new management is here.

'Thanks,' Kitty says, as I hover in front of her desk. She covers her phone receiver with one hand. 'Something else?'

'I – no, but . . .'

'What?'

I suppose I was hoping she'd read it through, sit me down while she gets on the phone to Greta Gordon. Something.

'Nothing,' I say.

'Go to Starbucks, fetch me a fat-free frappucino,' Kitty says, dismissively, and that's it. I honestly don't know why I thought it would be any different. Luck and me just don't go together, do they?

'Don't sulk, Anna,' says Kitty acidly as I turn dejectedly towards the door.

I come back from Starbucks in a vile mood. I felt so miserable and down-hearted I ordered myself a raspberry ice tea and a cheese Danish and scarfed them both down on my way round the corner. They didn't help, just make me feel even more bloated. Which always happens when I stuff myself, and yet I keep doing it, which is the definition of insanity, isn't it? I should get my own version of the Verve song and call it 'The Cakes Don't Work'.

'Visitor for you in main reception,' mumbles Claire, keeping her voice down, as I hand her Kitty's frappucino. Kitty likes her secretaries to be seen and not heard.

Visitor? I never get visitors.

'Who is it?'

'Some bloke called Charles Dawson. He says he has that novel you wanted.'

'Oh . . . oh, right.' I sigh.

'He knows you're here because I told him,' she adds helpfully. 'Shall I send him up?'

'Go ahead,' I say, despondently. I can't very well refuse, can I? He did put me in touch with Trish. And he's Vanna's friend. And anyway, I sort – of promised to read his stupid book.

I can take it off him and send him on his way. Then he'll be happy and so will Vanna. Plus, it might be a good idea to be seen taking delivery of a manuscript today, since Kitty is yelling at me for 'more projects'. Nobody need know it's pants. It could be a slipped copy of the new John Grisham. Or something.

I plaster my professional smile back on as Charles steps out of the elevators. Oh dear, did he absolutely *have* to wear that three-piece tweed suit with the gold watch on a chain? All he needs is a monocle. People are staring in the halls, sniggering.

I stride forward with my most businesslike smile and shake Charles's hand with a firm, dry grip.

'Charles, good of you to come,' I say loudly. Maybe I can get people to think he's an eccentric agent. 'Thanks for slipping me this manuscript for a first look.'

'Slipping it to you?' he says, and unfortunately his voice is loud as well as high-pitched. 'I'm hardly doing that. It doesn't have a publisher, after all – yet.'

'We prefer our material fresh,' I say, suppressing the urge to blush. 'We like to surprise the markets. Anyway, thanks for dropping by. I'll get right back to you,' I say, putting a friendly arm round his shoulders and steering him back the way he came. There. That didn't go too badly, did it?

'Hell*oo*,' coos a voice.

Oh great.

'I'm Sharon,' Sharon says, giving him a pearly white smile. She's outdone herself today. She's wearing a dress that, while technically not slutty, still manages to give that impression. It's made of very fine and light navy wool, low-cut to show a hint of cleavage, cap sleeves, and sits right on the knee. It clings lovingly to every curve of her body, showing no signs of VPL, not even a thong. Suggesting she's not wearing any. Sharon's been under the tanning beds, too, and her long legs taper down to cute little white leather slides with a navy trim. She gives me a sidelong glance and extends her slim hand to him.

'Delighted,' Charles drools, eyes popping. 'Oh, delighted.'

Sharon retrieves her hand and wipes it surreptitiously on her dress. His palms *are* a bit clammy.

'You're bringing a goodie for Anna?' she asks.

This is amazing. She can't get over that I wouldn't give her

Trish's script, and now she wants to jump on this just because it's mine. I suppose it's every girl for herself this morning.

'I am,' Charles confirms. He puffs out his scrawny chest like a starving pigeon trying to attract a mate. 'It's an extremely exclusive "sneak peak" at a brand new novel.' He makes those little quote marks in the air while saying 'sneak peak', just like Rob, and I almost die of shame.

'Well, thanks, Charles, I'll just take it now,' I say.

'I think not,' says Sharon, lightly but distinctly, smiling away. 'You see,' she adds to Charles, 'I'm senior to her. She just reads material. It's people like me who make the decisions.'

'Is that true?' Charles asks me.

I swallow. 'Well, sort of.'

'Sort of?' Sharon demands. 'I *am* a junior development executive, and Anna here is only a reader. I was promoted,' she adds, tossing her curls.

It burns, it really does. I look down at Charles, towering over him, in my plain trousers and shapeless, baggy jumper, and there's Sharon, about his height, small-boned, delicate, in her cute little dress, curls bouncing; Sharon, with all the power. I wait for Charles to give her the sodding book.

Charles turns to Sharon, very politely, and says, 'I'm awfully sorry, but I do think I should give it to Anna.'

'And why's that?' she demands, scowling now.

'Because Anna and I have a very special relationship,' Charles says, stoutly.

I smile at him. I can't believe it. He's sticking up for me!

'I didn't know *Anna* had any special relationships with *talent*,' says Sharon, licking the outside of her lips provocatively.

Charles hesitates, but sticks firm.

'No,' he says, 'I promised to give the book to Anna and I must keep my word.'

'*Thank* you, Charles,' I say, triumphantly, swiping the book from him and clinging on to it tightly.

'And I'll see you soon,' he says, looking at me hopefully. 'How about tonight? Are you free tonight?'

My heart sinks.

'Oh, well, if I'd known he was your *boyfriend*,' Sharon says scornfully. 'I thought he was a *real* writer.'

'Charles is a very gifted writer,' I say, because one good turn deserves another. Charles is beaming at me. Sharon storms off, doubtless in search of someone else whose project ideas she can steal. I seize the opportunity. 'Look, Charles, about that date—'

'Seven thirty for eight,' he says, stepping back into the elevator and punching the door close button.

'But I—'

'No need to say anything, Anna,' Charles says, magnanimously. 'You *deserve* a date. See you tonight!'

The doors hiss shut on him. I stand there, gnashing my teeth. Why? Why do I deserve a date? What did I do wrong?

I'm reading Charles's book when Eli Roth arrives on our floor. You can tell it's him by the crackle of electricity that ripples through our normally moribund workspace: all the novels and tabloid papers are shoved into drawers, the secretaries are tossing their hair and checking their nails, sitting up very straight and typing busily. I put the manuscript aside, glad of a legitimate distraction.

Mind you, it hasn't taken much to distract me from this masterpiece. I've been finding all sorts of things infinitely more interesting, such as arranging my stored emails into folders, changing my desktop background six times, and tidying up my desk. I want to give him a chance, I really do, but it's just so dull. At one point I realized I'd just read the same ten pages over twice and nothing had sunk in either time. It's sort of like Proust but far, far duller. I'd rather read the Yellow Pages. At least from there I could order a pizza.

There's a little knot of people gathering around Roth. I can just about make out a tall bloke in a charcoal grey suit. Mike Watson is pumping his hand, there's Sharon simpering and flicking her hair about, Rob Stanford is hovering, John's kowtowing . . .

I glance over at Kitty's office. Magnificently, she has not come out yet. She's waiting for him to come to her. Kitty's the ice queen. I wish to be just like her, career-wise, instead of sitting here morosely feeling inadequate all the time. Where does Kitty get this insane self-confidence?

Roth walks over towards her office now, having pressed the flesh, and sticks his head inside. He's about five eleven. Same height as me. This is something I always notice about men – how many of them I tower over in an unattractive fashion. I feel sure that even if I were pretty, a man like Roth wouldn't be interested in a woman who could look him right in the eye.

Charles is five four, five five tops. I can't believe he actually *wants* to be seen in public with me. He's probably a pervert or something. I shiver with horror at the idea of actually going out with him, we'll look utterly ridiculous. Maybe he'll let me order in. But then I'd have to share the experience with Lily and Janet and their perfect bodies. Just shoot me now.

Kitty says something to Roth, and he steps inside her office. And shuts the door. I sneak a look over at Mike Watson. He's just standing there, looking all pissed off. You have to hand it to Kitty, she's such a smooth operator, she's not about to let anybody in. She has Roth all to herself.

The light on my phone flickers.

'Yes?'

'Anna?' It's Kitty. 'Would you step into my office, please? There's somebody I want you to meet.'

I blink. I can't believe it. Kitty never does things like this. I get up nervously and knock on the door of her office, feeling John's eyes boring jealously into my back.

'Come,' she says.

I open the door just a crack. Roth and Kitty are both sitting on the couch. Kitty looks all relaxed and at ease with the world. Is she on drugs?

'Eli,' she says, smiling broadly, 'I want you to say hi to Anna Brown, one of my readers. Anna's a *great help* to me refining the material I find.'

'Hey,' Roth says.

He's young and powerful-looking. Broad shoulders, nicely defined muscles without being all steroidy, dark Hugo Boss suit, gold Rolex (natch). In fact if you were to conjure up the image of a Hollywood executive, all you'd have to do would be to stick a pair of shades on him and he'd be a perfect match. 'Kitty tells me you're a real asset to her team,' Roth says, looking at me intently.

'Oh – I – yes,' I splutter. 'Thank you, Mr Roth.'

'When you call me that, sugar, I turn around and look for my dad,' Roth says. 'It's Eli. We don't stand on ceremony at Red Crest.'

Kitty looks at me. Obviously my moment in the sun is over now. That's her 'get out' look.

'Would – would you like some coffee?'

'That's OK,' Kitty says.

'I'm good for now, honey,' says Roth.

'Nice to meet you, Mr – Eli,' I say, quickly withdrawing and shutting the door. I walk back to my desk on a high. Kitty introduced me to the boss! She actually introduced me to the boss! I feel a surge of gratitude. She's finally taking me seriously.

Nothing happens after that until lunchtime. I flick through the book (crap, all the way through), type up more notes, and do important office things like playing Free Cell and Spider Solitaire, until finally it's one o'clock and I can buzz Claire.

'Want to go out?'

'I shouldn't. I'm not even a quarter of the way through Kitty's call sheet,' Claire says in mournful tones that make her sound like Eeyore.

'Just knock off all the really boring ones. She'll never notice.'

'I really could use a fag,' says Claire.

'Perfect, come on, then,' I say. 'We'll go out to Pret a Manger, my treat.'

'All right,' Claire says, perking up.

We go out to the front of the building where Claire quickly chain-smokes two Marlboro Lights, and then head off to the Pret down the road. I want a chicken curry sandwich, but it might make my breath smell, so I settle for a mozzarella, tomato and focaccia instead, along with some fresh squeezed orange and raspberry juice and some of those baked vegetable crisps. They can't be that fattening, right? They're not like real crisps, after all, they're *vegetables*. Eli Roth would probably eat them.

Claire settles for sushi and a bottle of mineral water. She's a cheap date.

We grab a couple of empty stools by the window and I'm about to tell her what happened with the script when she starts talking.

'Sharon's pretty cross, you know,' she says darkly.

'What do you mean?'

'With you. She thinks you deliberately set up that book guy coming in to make her look bad.'

'That's insane,' I say. 'Sharon's a moron, she never should have been promoted.'

Claire looks shocked. 'She *is* a development executive now. You should be careful how you talk about her.'

'Oh, that's all rubbish!' I exclaim, then add at her hurt look, 'Come on, Claire. They're just people. You're not going to get anywhere letting them walk all over you.'

'Well, *you* haven't been promoted,' she retorts, with a touch of spirit.

'I might be,' I say. 'I've found a good script. Kitty promised me that if it goes anywhere I get made a junior D-girl too.'

'Yeah,' says Claire, darkly. 'And she promised me a raise and an extra week's holiday.'

'Maybe things will be different now Eli Roth's here.'

'Or we could all get fired,' says Claire. 'But he's something, isn't he?' She sighs. 'Did you ever see anybody who looked like that? I mean in real life?'

'He looks like an actor in a soap opera,' I tell her.

'Naw, he's too sexy,' she says. 'Too masculine. He's all pretty, but he's not. It's the muscles. Or something about the eyes.' She primps her hair a little. I blink; this is *Claire Edwards*, the quietest piece of wallpaper in the world. 'D'you think . . . you know, if I maybe went to a salon? And I could get a dress like Sharon's?'

71

'Do I think Eli Roth would fancy you?'

'It could happen,' she says, blushing.

'Of course it could,' I agree hastily, 'Absolutely. And you're much prettier than Sharon,' I lie for good measure. 'But you know, he's probably got somebody at home in California. They don't leave the rich good-looking ones on the market, do they?'

'But he's not in California *now*,' says Claire slyly. 'When the cat's away the mice will play.'

They certainly will! Look at her! She's shaking out her hair and examining her face in her powder compact mirror (she's the kind of girl who still uses one of those). I wonder if tomorrow I'll see her in make-up and fuck-me heels.

'Well, good luck,' I say.

'Thanks,' Claire says. She lowers her voice. 'I bet Sharon wants him too.'

'I expect lots of the girls will,' I agree.

She looks despondent at the competition. 'Oh well, if it's meant to be it's meant to be. Que sera sera,' she says.

'Absolutely.'

'And you've got a new boyfriend too!' she says, helpfully. 'That book guy, Charles. That's nice, isn't it?'

'Mmm,' I say, non-committally.

'At least I know you won't be going after Eli Roth,' she says, giggling.

I blink. 'What's so funny?'

She covers her mouth. 'Oh. Nothing. I didn't—'

'You didn't mean anything by it.' I sigh. 'It's OK. Have you finished?' I add, looking at her empty plastic tray. 'Perhaps we should get back to the office. Eli Roth might be walking around our area right now.'

'He might,' she says, panicked. 'Come on, let's go.'

When we get back in there's a little Post-It note stuck to my computer from Rob, forced to give me my messages since Claire went out.

'Meeting at 2pm in conf room 3,' it says.

I check my watch. Fuck it! It's two fifteen right now. When do we ever have afternoon meetings until three?

Kitty is always out at some three-martini deal schmoozing somebody famous or powerful. And why conference room three? We always have initial story meetings in Kitty's office, then she takes our recommendations to the development meeting where all the powerful people sit and scrap over what to try and make. And I'm about as likely to be invited to one of those as I am to win Miss America.

I grab my *Mother* notes and scamper down there. The conference room door has a little window in it and I can see there are loads of people sitting round the table. Everybody is going to notice me coming in late. I get a momentary urge to run, just go home and say I had a stomach ache . . .

I open the door as quietly as possible and try to sneak into the room. There are four executives sitting at the table – Kitty, Mike, Carl Smith, and Paul Walker. Their readers are all down the other end, with Sharon, who is looking furious at the seating arrangement. And right at the head of the table is Eli Roth.

Carl Smith is in the middle of a presentation, pitching an idea about toys who come to life and go evil and start attacking everybody. He's sweating a bit and looking rather ruddy. He seems grateful for the interruption and stops, dramatically, as I take the last available seat, next to John and out of the way.

'Afternoon, Anna,' says Mike nastily. 'Glad you could join us.'

'Sorry,' I mutter. 'Lost track of time.'

'We're all usually very efficient,' says Carl Smith to Eli. Kitty scowls at me. 'Time is money,' he adds.

'I'm sure it's a one-off,' Eli Roth says kindly.

'I do apologize, Eli,' says Kitty, tightly.

'Carl, why don't you continue,' Eli says, turning his face towards the board as I blush richly.

'Well, I think this a great paradigm,' Carl says loudly. 'The script charts out like this . . .' He pulls up a diagram. A diagram! For a movie script! 'The Xs indicate where each plot point conforms to the Hero's Journey mythological structure as invented by Christopher Vogler, and—'

'It sounds like a fun project,' says Roth, cutting him off. 'But I think it's been done before.'

'Some ancient classic movie,' says Carl defensively. '*Gulliver's Travels* . . .'

'No, recently,' Roth says. '*Small Soldiers.*'

Carl looks blank, but clearly he isn't going to say, 'You what?' Instead, he nods and clears his throat. 'Maybe we can translate the elements to a new premise,' he says, sitting down. I glance round the rest of the table. Everybody is looking blank. Of course. This is because they only notice movies that are on at their local multiplex, if then. They don't bother keeping up with what happens in America.

'You guys remember *Small Soldiers*?' asks Roth. Apparently he is thinking the same thing. His eyes scan the room, and I notice that they have lost their friendly look and gone quite sharp.

Everybody looks down and makes notes on their pads so as not to catch his eye. There is a horrible silence.

'Of course we do,' I say, defensively. Somebody's got to say something!

Roth's head comes up and he looks at me. A slight grin plays across his mouth.

'Anna Brown,' he says. 'We met this morning, right?'

'Right,' I mutter, going scarlet.

Roth leans back in his chair and crosses his arms. 'Tell me about that movie.'

'It was an Elliott/Rossio thing,' I say. 'Toys that came alive. Evil Barbie dolls.'

'I don't know those directors,' says Carl sharply.

'They're not directors, they're writers,' I say.

'Oh well, *writers*,' he says scornfully. 'No wonder I haven't heard of them!' He looks smugly around the table for support. A bunch of the executives snigger. And it's true, writers don't have much clout in Hollywood, usually. There are a whole legion of writer jokes: 'Have you heard the one about the blonde? She was so dumb, she went to Hollywood and slept with the writer.'

I go red. Why does he have to be so mean? Our movies

74

would be a bit better if we cared about the writers! You can have the biggest star and the biggest budget and if you've got a lousy script you've got a flop. Just look at Arnie in *Last Action Hero*.

'I'm a little surprised,' says Eli Roth evenly, to Carl. 'Anna, you know some of their other movies?'

'Yes,' I say. '*Shrek*, *Aladdin*, *The Mask of Zorro*—'

'And *Pirates of the Caribbean*,' Roth finishes for me. 'One of the biggest films of all time.'

People stop sniggering and Carl opens his mouth like a landed fish, then thinks better of it.

'Other ideas?' Roth asks, and the meeting moves on. I sit there, listening to pitch after pitch, and watching the shutters coming down over Roth's eyes. He's obviously not impressed. Mike Watson pitches a couple of big budget sci-fi movies with muddy plots and a remake of a cult sixties TV show, and you can see him stifle a yawn.

'Kitty?' Eli says.

Kitty is the last development executive to go. She stands up and minces towards the presentation board in the front of the room, gold bracelets jingling.

Ooh. I sit up straighter. This will be fascinating, I've never actually seen her pitch. I'm usually the one pitching to her.

She runs through a couple of bog-standard ideas John offered her and there's a general shaking of heads. Kitty shrugs her bony shoulders in her Chanel, unaffected.

'I do have one more idea,' Kitty says. 'Something I found myself. It's called *Mother of the Bride*.'

Found herself? What's going on? I look at her, but she refuses to meet my eye.

'It's an interesting premise well executed,' she says confidently, 'and it won't cost much to make.'

She's quoting my coverage!

'The movie's about a socialite,' Kitty says, 'who's ruining her own daughter's wedding . . .'

And I just sit there while she tells the story in a nutshell. She even whips out the script and quotes a few of the funnier lines I had highlighted in yellow pen.

'That's interesting,' Eli says. His eyes are sparkling now. 'That's really funny. I haven't seen a good older lead flick in a long time. Where did you find this?'

Kitty waves one hand airily, her diamond flashing away.

'Personal contacts,' she says. 'I gave it to Anna to write the coverage on.'

Eli glances at me. I sit there, dumbstruck. Should I say something? Kitty's eyes are like chips of ice.

'Right, Anna?' she asks pleasantly.

I nod miserably. 'Right.'

Kitty relaxes perceptibly and flashes me a smile.

'Well, this has got potential,' Roth says. 'Make some copies of the script and the coverage and send it round to everybody, could you?'

'No problem,' Kitty purrs.

'Great job,' he says to her. 'OK, I think I've got a lot to work on here. Thank you all.'

Kitty puts her hand on my shoulder as we're walking out of the meeting and squeezes it tight. 'Go straight to my office,' she says.

I walk right there, sit on the sofa and wait. I feel all nervous and sick, little prickles of adrenaline crawling all over my skin like spiders. Kitty comes in a second later and shuts the door.

'I hope you understand what went on in there,' she says.

'Not really.' I cough. 'I found the script . . .'

Kitty shrugs impatiently. 'We work as a team. I want to get your movie *made*, Anna, and that means it needs the right backing. If Eli thinks it comes from an executive, he'll listen longer, he'll consider it more carefully – and who cares how we get it done, as long as we get it done?'

'But I won't get any of the credit,' I say in a small voice.

'You don't *need* to,' she says slowly, as if explaining things to a very small child. 'I know. I'm your champion in this firm. Mike Watson tried to have you sacked last month, did you know that? And I stood firm.'

'But how will it advance my career?'

'Everybody will read your excellent coverage, for a start,'

Kitty says. 'Why, even Eli Roth will notice you. I introduced you, didn't I?'

'Yes,' I admit.

'And that was only the start,' Kitty says. 'I'll be making sure you climb up the ladder. We're a team.'

She hands me a typed sheet of paper. It's a memo to Personnel asking that I get a raise.

I have to read it three times, I can't quite believe Kitty's done this. I know she promised, but I never trusted her. Obviously I am just a manipulative person who reads the worst into everyone else.

'That's great,' I say, stunned. 'Th-thank you.'

'You're welcome,' she says, smugly. 'Of course you realize you've now agreed that I found the script. It's very important we stick to that. So as not to look foolish.'

'Oh. Sure.'

'And bring the writer in for a meeting with me,' Kitty says. 'Better still, just give me her phone number. I should make contact.'

I write it down for her.

'Thanks, Anna,' Kitty says warmly. 'This is going to be a very exciting time for you. You're going to see this project succeed.'

She turns back to her desk, to show the meeting is now over.

I hover around the door.

'Yes?' she says, a little impatiently.

'The raise is great,' I say. 'Wonderful, really, but what about the promotion? I'm still going to get that, aren't I?'

'Absolutely,' Kitty says. 'But right now isn't a good time, not until Eli settles on whom he's going to hire. I don't think we should push him *just* yet. If this project is a success, then the sky's the limit. As long as you and I stick together.'

'OK.'

'I've been looking for a younger woman to mentor,' Kitty says, musingly. 'There aren't enough of us in this business, Anna, and I think you might be the one.'

'Thanks, Kitty,' I say, basking in her praise. Maybe I've misjudged her, all this time. After all, she did win an Oscar.

'I want you to be intimately involved in developing this with me,' she continues. 'Do you have ideas for casting, other than Greta?'

'Oh, sure,' I say, fairly stunned. 'I've got *loads* of ideas. *Masses.*'

'You watch quite a lot of these sort of movies, don't you?' she says, consideringly. By 'these sort' I take it she means 'successful and popular'.

'Yes,' I agree. Has she only just noticed this?

'So get me some suggestion lists,' she says. 'Directors, actors, cinematographers – you know the score.'

I'm glowing with happiness, I can't believe she's actually going to trust me with this. This is real producer stuff!

'No problem,' I say confidently.

'Get me everything by tomorrow morning. First thing,' she instructs, then picks up the phone, dismissing me.

'You got it.' I open the door to her office to withdraw.

Kitty covers the receiver with one hand. 'You can take the rest of the afternoon off,' she hisses. 'Well done.'

Bloody hell. It's a miracle!

I open the door to the flat to find Lily sitting on the floor cross-legged, doing her Tantric yoga thing. I hate Lily's Tantric yoga. It means she gets to sit there in one position chanting loudly and we can't watch any TV. Or make coffee or toast or anything.

'Om om om om om om,' says Lily loudly, pretending she hasn't noticed me come in.

'So you're back,' I say.

Lily opens one eye. 'Really, Anna, I'm trying to concentrate here. It's very important I clear my mind.'

'Shouldn't take long,' I remark. 'Not much in there to begin with.'

'All that spite stems from the fact your body isn't cleansed,' says Lily. 'You should try yoga. I use it to control my appetite.'

'Funny, I thought you used cocaine,' I say.

Lily's eyes widen. And honestly, I don't know what's got into me today.

'Don't be ridiculous,' she snaps. 'I never ingest harmful substances.'

'Tobacco?'

Lily waves a hand. 'All that nonsense is overrated. In France *everybody* smokes. It's practically compulsory. And they're very healthy.'

'Don't you have a party to go to or something?' I ask. 'I'm going to watch *EastEnders*, I warn you.'

'Well, my concentration's shot now anyway,' Lily says, standing up and stretching. 'And yes, I have tons of invitations, but I'm actually going to stay home and relax too.' She glances at the phone.

'Expecting a call?'

'No!' she snaps. 'Of course not. I don't wait by the phone.'

'Who from?'

The door opens and Janet comes in, looking a bit down.

'What's the prob?' says Lily, a look of fake concern on her face. 'Work-related?' She obviously can't wait to hear Janet's tale of modelling woe so she can feel all superior.

'My bloody agent,' Janet complains. 'Told me the shoot was for *Heat* but when I got there it was for *Good Housekeeping*. They made me wear these horrible shirts. And there were three other girls there, and when they did the close-up shots they told me they didn't need me, only for the group work.'

'Oh my,' says Lily, pressing her hand to her mouth and widening her eyes. 'Not needed for close-ups. That *is* bad.'

I really would like to slap Lily sometimes. Janet's eyes redden as though she might be about to cry again.

'Especially after Gino dumped you,' Lily says. 'That's awful. How humiliating.'

'Hey, Janet,' I say. 'Lily has to stay in and wait for the phone to ring.'

'Nonsense,' says Lily sharply.

'Is it?' I ask innocently. 'Then I'll just jump on the internet, OK? I may be a couple of hours.'

'No, you can't do that,' Lily says at once. 'I'm not waiting in, but I do have a friend who may be calling.'

'Who's that?' asks Janet.

Lily tosses her blond hair, plaited down her back in sexy sixth-former fashion.

'Actually, he's a very important man,' she says. 'Claude Ranier.'

Janet's mouth drops open. 'Not Claude Ranier, the financier? Not Claude Ranier, the one with the huge private yacht?'

'The *Trixabelle*, yes.'

'Not Claude Ranier . . . the one who's ninety years old?' I ask.

I mean, I can't believe it. You know Claude Ranier. He's the one who's always in *Hello!* and Nigel Dempster. Franco-Greek shipping millionaire, switched all his money into real estate when shipping started to slide. Has a vast yacht, a house in Cannes, a palazzo in Venice, and a walled-off estate in Notting Hill. Plus, a reputation as a real old goat. You always see vile pictures of the old fat bastard, looking like a leathery wrinkled prune, sitting on the deck of his bloody yacht with a bevy of bikini-clad twenty year olds. Claude's man-boobs are often bigger than theirs, too.

'He's no such thing,' snaps Lily.

'How old is he then? Sixty?' asks Janet.

'At least,' I tell her.

'You two are so superficial,' Lily says. 'Claude is a fascinating man. All that—'

'Money?' I ask.

'Wisdom,' she retorts. 'Age is nothing but a number, you know. Anyway, I think he's going to invite me to head down to Cannes for the film festival.'

'No way,' says Janet, enviously.

'We're going to sail around the Côte d'Azur on the *Trixabelle*,' says Lily happily, 'and we might pop into Monte Carlo for a spot of gambling and the Grand Prix. Of course he has a box. And then the film festival, mingling with the stars, I mean actors. I may give my CV to a few agents myself. It might be time for me to consider a career change,' she muses. 'Claude says I have a lot of potential. He says he can help me find the right project. Of course, he's got plenty of money to bankroll a film, you know,' she says, looking at me. '*All* the indie producers want to be in bed with him.'

'Yes,' I say, 'but do *you* want to be in bed with him?'

'What do you mean?' she asks, aggrieved. 'It's not like that.'

'The hell it's not,' I say. 'He wants you on his boat as arm-candy, Lily. Trophy girl.'

'Well,' Lily admits, loosening her plait and shaking out her fountain of blond hair. 'Gentlemen do like to look at attractive women, but that's just nature, isn't it? It's not like I'm going to do anything.'

'You think he invited you along just to be decorative?' I ask.

'You're wrong,' Lily snaps. 'Anyway, I don't want to discuss it with you any more. He's just a friend. An admirer, actually. Something you wouldn't know anything about!'

The phone rings and Lily snatches it up. 'Yes?'

Her face falls. 'It's for you,' she says, with a disbelieving air.

I take the receiver. 'Hello?'

'Hi, Anna,' says a cheery voice. Oh hell. I'd forgotten. 'It's Charles. All ready for the big date?'

I think fast. I've forgotten to call him and cry off!

'I'll be round there in five minutes,' he says. 'Top buzzer, right?'

'Right,' I agree. There's nothing I can do, is there? 'I'll come down when you buzz.'

So much for *EastEnders* and pizza.

'Who was that?' Lily demands. 'Somebody from work, I suppose.'

'It was a date, actually,' I tell her.

'Ooh. Who is it?' asks Janet, encouragingly. 'Somebody special?'

'Don't be stupid,' snaps Lily, still cross. 'You know the kind of person Anna dates. Poor as a mouse with no personality, probably a social worker who lives with his mother.'

'He doesn't work at all,' I tell her.

'There you go. Unemployed. Anna will probably have to pay for dinner,' says Lily viciously.

'Actually, Charles has a private income,' I tell her. 'He owns a flat in Eaton Square. And his sister has a title,' I add, watching Lily sulk.

'Really,' says Janet eagerly. 'Anna, that's wonderful! Is he very upper-class?'

'Went to Eton with Vanna's husband.' I shrug.

Lily says, 'Well, he must be mad.'

'At least he isn't senile,' I respond.

'Well!' she says huffily, and storms into her room, banging the door.

'Don't mind Lily,' Janet says supportively. 'She just can't believe you could get a man like that . . . I mean, no offence . . .' She's floundering. 'Anyway, well done!'

And you know, the sick thing is I do feel a bit of pride. Not that I fancy Charles. Or find him even remotely interesting. But Lily and Janet, the pretty girls, are forever comparing the men they're dating starting with the bank balance, then moving on to social position. Everything else is a long way down the list. Charles may be a bit of a dandy and a midget, but he's better looking than Claude Ranier any day. And he's loaded. This is the kind of boyfriend that can secure a woman's future, or so they think. And Lily is thoroughly rattled, while Janet is looking at me with admiration for once instead of pity.

And . . . it's nice.

It's much nicer than when Brian was my so-called boyfriend. Because Brian was a loser. Charles is rich. In their world, that means he counts. And I'm feeling just a tiny bit pleased that I have a rich date. I feel all glowy and well disposed towards Charles.

'Aren't you going to get ready?' Janet asks.

'He'll be here any second,' I say airily. 'He can take me as he finds me.'

'That won't do!' Janet says. 'You're not going to hook him like that.'

'Maybe I don't want to "hook" him,' I protest.

'Come here.' She leans over me and fusses with my hair, then dives in her make-up bag and starts to brush something on my cheekbones.

'Get off,' I say.

'No –' the buzzer goes. 'Too late.' She sighs. 'Oh well,' and reaches for her scent, spraying it all over me, asphyxiating me.

I press the buzzer, coughing. 'Coming right down,' I say. Then to Janet, 'I smell like a tart's knickers now!'

'It wears off,' she says knowingly. 'Leaves a woody under-note. I had it custom-blended for me in Paris. It was a present from Gino,' and her face falls. 'Oh well,' she says bravely. 'Get out there and have a great time!'

I grab my bag and head downstairs. It doesn't feel quite like marching to the guillotine the way I'd expected. Maybe this will be fun!

'Hi,' says Charles expansively to me as I emerge from the building. 'Got a parking spot right in front.'

We walk to his car. It's a sleek black Rolls-Royce. Of course, what else would he drive? I wonder.

'Blimey,' Charles says, gesturing at the feminist bookshop. 'Rather you than me. Get harassed by all those lesbians and feminists, do you?'

'Not really,' I say.

'All those lesbians asking you for dates,' he says.

'I don't get asked out by lesbians,' I tell him, or anyone else come to that, I don't add.

Charles hurries to the car and opens the door for me to get in. 'I expect they'll come out and picket me now,' he says, 'holding open a door for a lady. Ha ha ha.'

I try for a dutiful titter as he slides into the driver's seat. He's wearing a dark suit but has spoiled it a bit with a pink shirt. At first I thought he'd grown, but looking down discreetly I see it's just stack heels. I still tower over him, but doesn't everybody?

I feel the familiar clutch of shame in my belly. I'm so bloody huge Charles actually decided to wear *stack heels*. Oh well. I give myself a little shake to make the feeling go away. Let's get stuck in to this date. The sooner I start it, the sooner it's over with.

'So where are we going?' I ask. Trying for enthusiasm.

He was nice to me in the office today, the poor sod.

'I thought we'd try Mock Turtle,' Charles suggests. '*Fabulous* new place off Kensington High Street. Dreadful waiting list, but I got right in,' he adds smugly. 'I know a few people.'

'Sounds great. What does it serve?'

'All fish,' he says.

Fish. Ugh. I hate fish. Unless it's battered and comes with chips in newspaper. I find it bland and clammy and it reminds me of dead people.

'Great,' I say again uncertainly.

'They do the most wonderful lobsters,' he tells me, steering smoothly through the London traffic. 'Not only can you pick your own, but you can watch as they cook them! They have this big glass wall and you can watch them trying to climb out of the pots. It's awfully funny.'

He catches sight of my face.

'You're not one of those liberal loonies, are you, Anna?' he demands.

'I'm not a vegetarian or anything, but . . . cooking them alive . . .'

'They're only bloody lobsters,' he says crossly. 'Not like they know what's going on.'

'I can't watch that,' I say. 'I'm sorry, but I'll be sick.'

He looks over at me, exasperated. 'Bloody hell,' he says. 'I pulled strings to get that reservation.'

'I know a nice Chinese place,' I suggest. 'Very reasonable prices.'

'Reasonable prices?' Charles repeats, as though he doesn't know what I'm talking about. 'Good Lord, no. I know, we'll just pop down to the Savoy. They know me there. We'll get a table.'

And, when we get to the hotel and the car is valet-parked, they do. Charles is welcomed in by discreetly bowing, perfectly dressed staff.

'Good evening, Mr Dawson.'

'How nice to see you again, sir.'

'Good evening, madam,' to me.

Ooh. It's all very flash. I'm a bit nervous, I wonder if it'll cost me to breathe the air in here.

'I'm afraid your usual table is taken,' says the maître d', 'but we'll make one up for you, of course.'

'That's fine,' says Charles, with the air of one suffering indignity patiently. He turns to me while we wait. 'Have you come here before?'

The Savoy? I'm lucky my budget can stretch to Bella Pasta.

'Not as such,' I admit.

'I lunch here every day,' he says. 'Wonderful food, and very obliging staff. Ah, they're ready.'

He leads me through the throng of well-dressed people making polite conversation in a quiet murmur, and the waiters sit us at a table for two and leave us menus.

'Mine's wrong,' I say.

Charles looks at me. 'What do you mean?'

'It doesn't have any prices on it,' I say.

Charles blinks. 'My dear girl, of course a lady's menu has no prices. Where have you been eating?'

'Oh,' I say, feeling small.

'Shall I order for both of us?' he says, and a waiter instantly materializes. 'My guest will start with the quails' eggs – they're not on the menu, Anna, but they're divine – and then . . .'

He rattles off a list he obviously knows by heart and I don't say a word. I wouldn't dare.

'There!' Charles finishes proudly. 'That's you all taken care of. Now tell me, Anna, have you been reading my book?'

Oh fuckity fuck. What am I supposed to say? I can't just tell him it sucks like a vacuum cleaner, can I?

'Ahm,' I begin, going bright red.

'Ah, say no more,' says Charles, seeing my reaction. 'Dreadful manners. Excuse me. I should never discuss work with . . . a *beautiful* young woman,' he adds after a pause. 'And I don't want you to think I'm only asking you out for professional reasons.'

I smile weakly at him.

'You don't need to worry about that with me,' he says. 'I

can see that it might have occurred to you, given . . .' His voice trails off.

'Given what?' I ask.

'Oh, nothing,' he says hastily, looking at his napkin.

Given that I have a nose that would do credit to Gonzo from the Muppet show and a bottom with its own post code? And that the only woman taller than me is the Statue of Liberty?

'Why did you ask me out?' I ask him, hoping vainly for a confidence boost.

'Well . . . you were such a good listener,' he says. 'And, you know, you didn't ask *me* out.'

'Excuse me?'

'I don't trust the ones who ask me out,' he says, suddenly, bitterly.

I look at him. He's only five six even in the stacked heels, and the goatee is so neatly trimmed, and the shirt's pink, and he's a bit balding . . .

'Does that happen a lot?' I ask, taking a sip of wine to mask my disbelief.

'All the time,' he says. 'Girls get introduced to me at parties, you know. And then they take my card, and then they suggest we get together over a meal. And they want to go out with me.'

'Ah,' I say, mystified. Maybe it's pheromones. Those things they advertise in the back of dodgy magazines and *Private Eye* – you spray them on yourself and you suddenly become a chick magnet. I sniff the air. Nothing, except the lingering scent of Janet's perfume.

'They want to go to the best restaurants,' he says. 'And they all love the flat. Of course it is a marvellous flat. And then they stay overnight without being asked. Just turn up with overnight cases!' he splutters. 'And they stay . . .'

'Maybe some of them are just keen,' I suggest.

He's twirling his wine glass now, his fingers all tight on the stem. 'They aren't keen until somebody tells them about Chester House.'

'Chester House,' I repeat, but the waiter is serving us and

he clams up until he's gone. Charles is eating something heavenly smelling and there are a pile of tiny boiled eggs with grey salt by my plate.

'What's that?' I ask.

'Goat's cheese and caramelized onion tart,' he says, without offering me any. 'Try your quails' eggs.'

I pick one up gingerly and lower my voice. 'I think their salt's a bit manky.'

'Manky?' asks Charles, horrified. 'That's celery salt. Surely you've had quails' eggs before?'

'Oh yes. Millions of times,' I say, gingerly dipping one in the salt and eating it. It tastes all right. Like a boiled egg, only smaller. I could have boiled my own eggs at home.

'Yes, Chester House,' Charles says significantly. 'They find out I stand to inherit, and then . . . well. You can't get rid of them.'

'I don't know what Chester House is,' I admit.

'It's the family seat,' Charles says. 'Rather special, I suppose. Eighteenth century. Nice little park surrounding it. In Gloucestershire.'

Light dawns. Charles is like Mr Darcy from *Pride and Prejudice* and lives in a huge mansion with servants and deer grazing in his grounds.

'But aren't these girls pretty?' I ask.

'Some of them,' he agrees. 'But it doesn't matter. They won't . . .' He looks at me and trails off. 'Or very rarely, anyway. They seem to want to just when I get up enough courage to kick them out.'

'I see,' I say.

I'm feeling a bit sorry for him now. He's a pompous ass, and all that, but he deserves better than this.

'Not all girls are like that, you know,' I tell him. 'Some are very nice.'

'You didn't ask me out,' he points out.

'Can't you find a nice, rich girl with lots of money of her own?' I ask him.

'D'you have lots of money?' he asks, interested.

'Not a bean,' I say cheerfully, and it's a great weight off my mind. Now I won't have to pretend I know all about vintage champagne and things.

He slumps a bit.

'Still, I asked you,' he points out, as though this is a great novelty. 'And you said yes. Did Vanna tell you about Chester House?'

I shake my head.

'Perhaps if I take *you* back to my flat, we can . . .?' he asks hopefully.

'Charles! I only just met you,' I say. 'And I don't want to move in with you. Honestly.'

He smiles broadly at me.

'I like you,' he says. 'How are you fixed for tomorrow night?'

Janet and Lily had waited up for me. Obviously curiosity had got the better of Lily's fit of pique. They were sitting on the sofa together, curled up with the telly and steaming mugs of something when I got home.

'Ooh, what's that? Hot chocolate?'

'Don't be silly,' Lily sniffs. 'It's boiling water with lemon.'

'Makes your stomach feel full,' says Janet. 'It's great!' she adds, uncertainly. 'Anyway, how did it go with Charles Dawson?'

'It was fine,' I say. 'He was nice. He wants to go out again.'

'Where did he take you?'

I shrug. 'The Savoy.'

Janet nudges Lily. '*Told* you. Anyway, he's from a really good family. His grandfather was an earl.'

'How do you know that?' I ask.

'Made some phone calls,' Janet says blandly. Of course. Janet makes it her business to know every two-bit aristo in the British Isles and every minor count on the Continent.

'He's got an *enormous* country house,' Janet says enthusiastically. 'It's *huge*. He's, like, one of the most eligible bachelors.'

'I don't care about all that sort of thing,' I say, and it's mostly true. I mean, obviously I can't totally ignore the vast mansion and the sacks of money. But I'm going to do my best.

'Bullshit,' says Lily, tossing her hair. 'You're so lucky, Anna,' she adds, jealously. 'Does he have any single friends?'

'I expect he has loads.'

'Anyway, Claude called me,' she says, flicking her blond hair. 'And Claude's *really* loaded.'

'So I gather,' I say.

'You know you truly can't see him,' Janet says to Lily.

She pouts. 'You're both just jealous.'

'Is he really handsome?' Janet asks me, supportively.

I think about Charles. 'Um, no.'

'See?' Lily demands. 'You're no different from – I mean,' she says, correcting herself, 'you're a hypocrite! You tell me I shouldn't see Claude just because of a little thing like age, but you're quite happy to go out with an ugly bastard who's got a country estate worth millions of pounds.' She sounds quite wistful.

'And a flat in Eaton Square,' Janet reminds her. Lily scowls.

'I can't afford to be fussy about looks,' I say defensively. 'I didn't know he had all that money, not really. I just think he's a bit vulnerable.'

And, obviously, I want to be with someone, and I know the top totty is probably out of my reach. That's the brutal truth, isn't it? You've got to date in your own attractiveness range. Which is fine for all the pretty girls, they get to date men like Eli Roth. Girls like me have to grin and bear it. Charles is pretentious and obnoxious, but I think it's mostly because he's sad and lonely. A defence mechanism. Anyway, why not give him a try? He's the only thing on offer and he's better than nothing.

A lot of good marriages have been built that way.

Does that sound cynical? Think about it. You can wait forever for Prince Charming to come along with the white horse, or you can get out there and try to find yourself someone. Mostly, that seems to mean not being too fussy. If you fuss over every single bloke, you're liable to wind up fifty-five and alone forever, consoling yourself with some form of small domesticated animal.

No thanks. I'm a practical girl, I tell myself. I want a man.

I can't get too many, so I'll take what comes along.

'Yeah, right,' says Lily. 'Vulnerable, but fortunately also rolling in it. I can't understand it,' she says, shaking her head. 'He could have had anybody.'

I flush. 'Thanks a lot, Lily.'

'I'm only being honest,' she says, for the millionth time. 'As a friend.'

'Well, as a friend,' I say, 'he wouldn't pick a girl like you. He's dated loads of modelly types and they're all after him for his money. You know, men aren't as stupid as you think. They know what you're doing.'

'I don't know what you're talking about,' Lily snaps. 'You've been acting very strangely lately, Anna, with all these silly ideas.'

'I think it's great,' says Janet. 'Anyway, Lil, wouldn't you rather be introduced to one of Charles's friends than hang out with Claude Ranier?'

'Maybe,' says Lily, with a long-suffering air.

'I would,' says Janet eagerly.

'Um, I'll ask him,' I say, and head into the bathroom to brush my teeth.

5

The next morning, when I wake up, I still feel jazzed. It's wonderful; I emerge from sleep with that great feeling that something wonderful has happened, if only I could remember what it is.

The script. The movie.

I jump out of bed – well, it takes me less than five minutes to push the duvet off, and that's the same thing – and head to the shower. I'm thinking about actors and directors while I'm brushing my hair. Rachel Weisz, would she be interested? Or Sadie Frost, maybe, for the bride. And unknowns for the male leads, the whole point about this is that we should do it cheaply . . .

It's so exciting. Kitty actually valuing my opinion. Letting me suggest talent, be part of this. I nuke my hair dry, brush my teeth, stealing some of Lily's Rembrandt, so much better than my Aquafresh, and carefully pick out an outfit. Flats (course), black low-rise H&M jeans, and one of those little Bon Jovi style T-shirts from '86 with the three-quarter length sleeves in a different colour. I even make myself up, this time stealing some of Lily's Shu Umera. Well, she's got so much make-up she'll never notice the difference, and I really want to look good today.

Not 'good', obviously. More like . . . acceptable.

I grab my bag and head out the door, stopping only to drop a two-pound coin into the Nose Job box.

Perhaps I'll get a huge bonus when *Mother of the Bride* gets made, and instead of frittering it away on buying my own flat or paying off my credit cards, I'll wiscly invest it in a tiny, Michelle Pfeiffer-like protuberance. Unfortunately, they haven't yet invented the surgery that can make tall, strapping girls slim-boned and petite. But I'll be waiting!

John is already hovering by my desk when I step out of the lifts.

'Kitty wants to see you,' he says, importantly.

'I'll be right there.'

'She needs those casting suggestions from you right away,' he says, threateningly. I don't know why John feels he has to be Kitty's personal bouncer. 'I hope you've been working on them.'

'Of course I have,' I say shortly. 'And I hope *you've* been working on finding Kitty a project, too.'

'Nothing worth making so far,' says John, and sniffs. 'I have rather higher standards for the projects I show her. She is an Oscar-winner, you know, Anna.'

'Yes, I know.' As if any of us could forget.

'And she's none too happy about the flowers,' John adds, spitefully.

'Flowers?'

He indicates my desk. 'You should try to keep your personal life out of the office. It can get very distracting.'

I follow his gesture and my mouth drops open. What on earth is that on my desk? There's the hugest, almost obscenely large bunch of roses perched on the corner of my desk, filling my entire cubicle. There must be at least three dozen of them. Yellow and pink roses, twined round with ivy and twigs with berries on them, very designer florist. Everybody walking past is looking at them, and no wonder. You can smell the heavenly scent from here.

I can't quite believe it. This has never happened to me before. Girls like Lily and Janet and Vanna get flowers, not girls like me. I think the sum total of my previous flowers were some Shell station, dyed-red carnations Brian got me at the last minute when he forgot Valentine's Day this year.

Sharon spies me from across the floor and saunters over.

'Got an admirer, Anna?' she asks. 'Better check the card, I suppose you'll find it's printed in Braille.'

'Very funny,' I say, and go over to my desk, fishing around in the huge swathe of blooms for the little envelope. Here it is, very thick and creamy, with the name of a Chelsea florist embossed on the cover.

'Lila Sturgeon,' says Sharon sneeringly, but you can tell she's unwillingly impressed. I know she's doing exactly the same calculation I am, namely what these flowers cost. I'm saying, trendy London prices, at least two hundred quid. Maybe more. I scan the card.

Thank you for a wonderful evening. Call you tonight. Love, Charles.

Well. That's certainly very nice of him. He's not so bad, really.

Kitty sticks her head out the door. 'Sharon,' she snaps. 'Don't you have to fetch Mike some coffee? And Anna. Are those flowers for me? You can bring them in.'

'Actually, they're for me,' I say, blushing slightly.

'For you?' Kitty demands, rounding her eyes. 'Well, do tell your boyfriend not to do it again. It's hardly suitable for an office, some of us have allergies,' she adds waspishly, though she didn't seem to mind when she thought they were for her, did she?

'Charles?' asks Sharon, peering over my shoulder. I instantly shield the card from her.

'None of your business,' I say, blushing.

'Oh, that's too funny,' Sharon says meanly. 'Little and Large! Of course, he's the book guy. Probably just trying to bribe you,' she adds.

'No he's not,' I say, feeling protective. 'He's very nice, and you're just jealous.'

'Oh yeah, I'm green with envy,' Sharon says. 'I really wish *I* could get roses from a midget.'

'He's not a midget,' I say. 'He's a millionaire.'

Sharon laughs scornfully. 'Anna, you have to stop fantasizing! You're not going to bag a millionaire, get over it!'

'And why not?' I ask her. Although I know the answer, don't I? Sharon shrugs, as though she doesn't want to hurt my feelings.

'Well,' she says, after a pause. 'If you can't figure that out for yourself . . .'

'Well, Charles fancies me,' I say defiantly. And I look at my roses and know that it's true, and it feels good, it really does. He's certainly a step up from Brian. Or nothing. Which were my two previous choices.

'Anna,' Kitty says, sticking her bony neck out of her office door again. 'Stop mooning over those ridiculous roses and bring your lists in here.'

I nod and gather my papers together, shuffling them loudly. 'Sorry, Sharon,' I say. 'Got to go. Some of us have work to do.'

She tosses her curls and walks off, nodding at the roses.

'It won't last, you know,' she says.

She's probably right. But I don't care if it lasts, I just want to have it for a little bit. Being courted, being made a fuss of, just like a normal-height girl with a small nose. Because flowers at the office is a lot of fun. I make a note to myself. I'll call Charles later, he definitely deserves another date.

Kitty sticks her head out for the third time. 'Anna. I don't have all day. And bring some coffee with you, I'm dying of thirst.'

I sigh and head towards the kitchen. Another fabulous day in the glamorous world of films!

'Yes . . . yes,' Kitty says, approvingly, making notes by the names of various actresses I've suggested for the bride. 'Get on it. Call their agents – no, wait. I'll call the agents. I've got the magic touch with talent, after all,' she says, smugly. 'Greta agreed to do it.'

'Really? That's fantastic news. Oh, well done,' I tell her.

Kitty drums her diamond-encrusted fingers on her desk, sending light sparkling around the room. 'It was, rather,' she agrees. 'Now, directors.'

I hand her my next list. It's a short one: Roger Michell, who did *Notting Hill* and *Changing Lanes*; Mike Newell, who did *Four Weddings* and *Harry Potter and the Goblet of Fire*; Peter Cattaneo, *The Full Monty*; and the Weitz brothers, *American Wedding* and *About a Boy*.

'Hmm, yes,' she drawls. 'Very nice, but mostly unavailable. Who else?'

What does she mean who else? That's it, that's my list!

'There is one other name,' I tell her. 'Though I suppose there's no way . . .'

'Who?' Kitty demands.

'Well, Mark Swan.'

Kitty stares at me. 'Mark Swan?'

'Yes.'

She gives a short, barking laugh. 'Don't be ridiculous, Anna! We can't get *Mark Swan*. What are you thinking?'

I don't really know. What *am* I thinking? Mark Swan is known for gritty drama movies. Why would he agree to direct a romantic comedy? Just because I met him in the cloakroom and he was nice. OK, very nice. But it still doesn't mean I can get him attached to a comedy. I must be losing it.

'At least let me call his agent,' I hear myself plead.

Kitty shrugs. 'Knock yourself out, darling. Just don't expect anything. OK?'

She's right, of course. Swan's agent, a terribly busy terrier of a woman called Carly Smith, gives me ten seconds before hanging up on me ('Not his sort of thing. Thanks for thinking of us. Bye'). I should have known, really.

Only I can't stop thinking about Mark Swan.

I so want to get this movie made. Eli Roth will want to do it, with Greta on board, but I don't think she'll be enough for financing, not by herself. Roth likes to have studios put up at least half the money, and a studio will want more of a package. A name like Mark Swan would tip it over the edge. For the first time in what you might laughingly call my 'career', I feel I'm close to something.

I check out Kitty's office. She's in there, blinds drawn. Probably talking to Greta. She won't miss me if I go out for an hour or two. I pick up my bag from the desk.

'Early lunch?'

It's John. Standing there hovering by my desk, eyes narrowed. John just loves to sneak on people to Kitty. It used to be all Sharon, but as Mike's stolen her, he only has me now.

'No,' I say. 'I'm just going out.'

'Going *out*?' he asks, with fake shock, rounding his eyes. 'You can't do that, Anna. I think Kitty's made it very clear that she needs us in the office right now. Finding projects. It's all hands on deck.'

'This is to do with a project,' I say.

'Oh really?' he asks, folding his arms. 'How so?'

I sigh. 'I've got a meeting with Mark Swan about *Mother of the Bride*,' I tell him. 'Be right back!' and I walk towards the elevators, leaving John standing there with his mouth open, gaping after me.

By the time the lift doors have closed I'm sweating bullets, of course, but it had to be done. I can't have John running to Kitty. Of course, now I have to actually get to talk to Mark Swan. Which is where I'm headed.

It probably won't be as bad as all that. What are they going to do, throw me off his set? He's only a bloody director, in the end. I'm sure he'll be perfectly reasonable.

'Step back please.'

A beefy man in a nylon windbreaker throws one muscled arm against my chest, crushing my boobs under their cotton shirt.

'But if I could just—'

'You're not on the list.'

This is maddening. I can see the shoot going on up the heath. And I can't get anywhere near it. I've been standing in the rain in bloody Hampstead for forty minutes. A perfect English summer's day.

'Just ten seconds,' I say.

'You're not on the list.' He looks bored.

'But I'm from Carly Smith's office,' I say, in a burst of inspiration.

He looks at me with pity. 'No you ain't,' he says, flatly.

No. I ain't.

'OK, OK,' I say, dejectedly. 'I'm going.'

I look at him to see if this has softened him up and made him want to let me on set, after all. But it hasn't.

I turn away and walk slowly up the street. He doesn't call me back. That's OK, I wasn't really expecting it. I turn the corner and head into the newsagent's for a quick fix of something. Maybe a Bounty Bar. The taste of Paradise. Refined sugar is obviously going to be as close as I'll come to it today. The shop is full of all sorts of tempting things. I grab a *Sun*, a *Mail*, a copy of *Heat*, a family-size pack of Quavers, a Bounty, a Snowflake, a Creme Egg and a Diet Coke and march up to the counter, where a very tall bloke is buying a packet of cigarettes.

'Excuse me.' The *Heat* is slipping out of my grip. 'Can I just put these down while you . . .'

'Sure.' He grabs the *Heat* for me on its inexorable way to the floor. Wow. What a good-looking man. All craggy and masculine in a younger Ted Hughes sort of way. I didn't think they made them like that any more.

'Thanks,' I mutter.

'Diet Coke,' he says, amused.

'Excuse me?' I demand again. Bloody cheek! Just because I have some crisps. And a few sweets. I blush bright red.

'Four twenty,' says the checkout girl. I sullenly fork over a fiver.

'I hope you realize you've ruined it now,' I tell him. 'With your *comment*.'

'Oh,' he says. 'You mean . . .' He gestures at my pile of loot.

'Yes,' I say. 'And I hope you know those fags are going to give you cancer. Why don't you think of that when you light up, eh? With every puff, think, this is giving me cancer,' I say, triumphantly. 'Try to enjoy them when you're forced to think about the consequences!'

He has the good grace to chuckle. 'You're right, I'm sorry. Stress relief?'

'Yup,' I say, pocketing my change. 'I can't get on to the set over there.'

'Johnny Depp fan?' he asks, sympathetically.

'No. Well, yes. Of course. Loved *Pirates of the Caribbean*. His Keith Richards impersonation was spot-on, wasn't it?'

'Sure was.'

'I wanted to speak to the director.' I sigh.

'You an actress?'

'Oh. No. Nothing like that. I work in the film business, I've got a script I wanted him to read. And his agent wasn't interested.'

'So you thought you'd try the direct approach.'

'They won't let me on set, though,' I say. 'You know, Steven Spielberg started his career by sneaking onto the Paramount lot, but they must have had rubbish security back then.'

He stares at me. Gorgeous dark eyes, long lashes, but they are regarding me as though I'm some sort of circus freak. I step back, self-consciously.

'Are you for real?' he asks.

I'm getting a nasty feeling. There's something familiar about him. His voice, his eyes. 'What do you mean, am I for real? You're not with the security, are you? I only asked, you know, I haven't done anything illegal.'

Not so far, anyway. My next move is to sneak onto the set. But I haven't tried that yet.

'You mean you don't recognize me?' he asks.

And then of course I do. With a sickening lurch of horror. I gasp and relax my grip and all the rest of my purchases slither to the floor, and I'm on my knees, face flaming, scrambling to pick them up.

He bends down to help me, not unkindly. I stagger back up.

'No – I – I didn't. I do now, Mr Swan,' I say, miserably. 'It was the beard. You've shaved your beard,' I cry. Not fair! Why should men be allowed to shave their beards, step out of gloomy cloakrooms and look completely different? 'Um, look, this was obviously a really bad idea. Please just forget about it. OK? Um, goodbye.'

'Wait a minute,' Swan says. 'Don't I know you?'

'Not really,' I mutter.

'Yes I do. Yes, I do,' he says, insistently. 'Yes, I've got it. You're the cloakroom girl. Aren't you?'

I don't want to tell him. What if he calls Kitty and complains or something?

'Yes,' I admit.

'The one with the evil boss with the boring Prada clutch.'

I smile slightly. 'Yes. Er – no. No. I mean, she's not evil. And the bag, I suppose you'd call it classic.'

He steps back. 'I'd call it boring. But at least she isn't a wanky recluse, eh?'

The shop attendant's head is zipping back and forth between us like a Wimbeldon spectator.

'You know,' I say with dignity, drawing myself up to my full height, 'all you had to say was no, Mr Swan.'

'Mark,' he says. 'And you're Anna Brown. Right?'

I blink with surprise. He remembered?

'That's right.'

'And what company are you with?'

'Winning Productions.'

He looks dubious. 'They did *Starlight Dance*? Couple of years ago now.'

'Yes, but we've just got bought out by Red Crest Productions,' I say. 'Eli Roth.'

'A big name,' Swan concedes.

I have to do it. I snap open the magnetic clasp to my battered old Prada bag that I spent five hundred quid on five years ago and so can't stop using even though it's all scruffed-up and battered, and fish out my *Mother of the Bride* script.

'Here,' I say, shoving it at him. 'Just read the first ten pages. Please?'

'And why do I want to do this?' Swan asks. He makes no move to take the script and it hangs there, limp and pathetic, in my hand.

'Well.' I take a breath, then my words come tumbling out, falling all over themselves. 'It's a good script. Funny. Like nothing you've ever done before, but I thought, you know, he could try something different . . . it's a romantic comedy for an older actress . . . *Mother of the Bride* . . . she's ruining her daughter's wedding . . . Greta Gordon is attached . . .'

'Is she?' Swan asks, eyes glinting. 'Thought she'd retired.'

'She wants to come back.'

'I'll think about it,' he says, taking the script. 'No promises. Your phone number on here?'

'Yes,' I say, gratefully.

'It really doesn't sound like my kind of story,' he says, gently. 'Don't wait by the phone, Anna.'

'Like Bud Fox in *Wall Street*,' I say, laughing nervously.

Swan grins. 'Yes. I love that scene. Anyway, don't wait like that.' He turns to go out of the shop.

'I'm surprised,' I call after him. '*Wall Street*'s not very arty.'

Swan turns back to look at me and lifts his eyebrows. 'Who cares about arty?' he says. 'Don't you think *Star Wars* is the best film ever made?'

And then he's gone, and I'm just standing there, ecstatic, clutching my Quavers.

'Do you think,' I ask John dreamily when I get back to the office, 'that *Star Wars* is the best film ever made?'

He gives me a look of withering contempt. 'What?'

'*Star Wars*.'

'Yes, I thought that's what you said,' he replies, acidly. 'I just assumed you were joking. Either that, or you haven't heard of *Citizen Kane* or *Casablanca*. Or *The Bicycle Thief*.'

'I thought *Star Wars* was brilliant,' I say.

'Yes, well,' sneers John. 'I believe you were the one who said she liked *Speed*.'

'Yes.'

'And *Pretty Woman*.'

I want to hug myself. I wonder if Mike Swan liked *Pretty Woman*. I wonder if he likes all my favourite films. Films that aspiring producers aren't supposed to like. You know, *Die Hard*, *Goodfellas*, *Trading Places* . . .

There's a real stigma to liking mainstream movies. And I can't get enough of them, while *Citizen Kane* put me to sleep. I mean I got to the bit where she's doing the jigsaw on the floor and I switched off. When was the story going to start? No, we're all supposed to be like John and be into classic movies and film noir and things that win the Palme D'Or at Cannes and nobody ever goes to see. You know what I call a classic? *Raiders of the Lost Ark*, that's what. I never wish I'd written worthy movies. Instead I wish I'd had the idea for

Shallow Hal. Or *The Sixth Sense.* I've tried with the worthy movies, I really have. David Puttnam once told me that *Citizen Kane* changed his life. But what can I say? I just don't get it. And I couldn't get through the first twenty minutes of *Breakfast at Tiffany's,* either.

I'd love to write. The best thing about being a scout is reading, getting to sift through scripts. Even though so many of them are so awful. I write such scathing things about most of them in the coverage I draw up, and I really feel bad about it, but you've got no idea just how terrible most of them really are. When will people stop trying to write bad pastiches of the film they saw last week? I truly don't want to know about a retired thief/bounty hunter/private detective called out of retirement for one last job. I could not care less if somebody wants to rob Fort Knox. And please don't send me an action script featuring a leather-clad girl martial arts expert with magical powers . . .

Sometimes I think I could do better. But then I remember I'm only Anna Brown, who am I kidding?

Kitty got the company to pay for me to go to one of those screenwriting workshops once. She wanted me to learn three act structure and common clichés, so I could give her better coverage. I wasn't supposed to try and write something myself, so I didn't. I mean, who do I think I am? I'm not a writer of movies. I'm just a fan.

But you know, sometimes I do wonder if what I'd churn out could be any worse than the stuff I have to read.

I figured out long ago that what I love about movies is the stories. You can have a bunch of unknown actors, no special effects, even an ordinary director, and as long as you have a sparkling story it really doesn't matter. You know, *Four Weddings and a Funeral.* Or *Phonebooth.* In *Phonebooth* the lead character hardly ever gets out of the phone booth! Talk about low-budget. And it was still a great thriller.

And then I pull myself together and dismiss those kinds of thoughts. And go back to my slush pile.

'You wouldn't know a good film if it bit you in the arse,' says John. 'Oh, and Kitty wants to see you.'

'What for?'

'Probably so you can tell her all about your big meeting with Mark Swan,' he says, with heavy sarcasm.

'Oh.' I toss my hair, as if I was Sharon. 'No problem, it went pretty well, actually.'

John's eyes narrow. 'You did *not* have a meeting with Mark Swan.'

'Didn't I?' I ask innocently.

'You're making it up,' John hisses, looking sick. 'Why would someone like him want to meet someone like you?'

'Maybe to direct my movie?'

'Kitty's movie,' snaps John, but he's gone all pasty and his heart isn't in it.

I go to Kitty's office and knock timidly on her door.

'Come.'

I let myself in, shutting the door behind me, but can't resist winking at John's stricken face.

'What's all this I hear about you and Swan?' Kitty barks. 'Did you really have a meeting with him? And if you did,' she drums her talons on her mahogany desk, 'why wasn't I informed? *You* don't take meetings with talent, Anna, *I* take meetings with talent.'

'It wasn't like that,' I protest hastily. Kitty is scowling furiously at me. The thought of being shut out of a meeting with Mark Swan has driven her insane. 'I just went up to Hampstead Heath where he's shooting and asked to give him the script.'

Kitty's scowl switches from rage to horror. 'You did *what*? But that's so unprofessional, Anna! How could you? He doesn't know you work for Winning, does he?' she demands. 'He doesn't know you work for *me*?'

'Absolutely not,' I lie, hastily. 'He has no clue!'

'I don't suppose you managed to see him,' she says, slightly mollified.

'Actually, he did see me,' I say carefully. 'And he decided to take the script and have a look at it. But he said it probably wasn't his kind of thing.'

'Of course it isn't,' Kitty snaps, bad humour restored. 'If

you'd listened to me you'd have known that and saved us all some embarrassment.'

Her phone buzzes and she presses the speaker button with irritation. 'What the hell is it now, Claire?'

'Phone call for Anna,' Claire's disembodied voice says meekly.

'So bloody what?' snaps Kitty. 'She can take her phone calls on her own time.'

'He – he says his name is Mark Swan,' says Claire, nervously.

Kitty and I exchange looks.

'I'll just get back to my desk,' I suggest.

'No you don't,' hisses Kitty, shoving the receiver at me. 'You take it right here. Put him through,' she says to Claire.

I take the receiver, heart pounding, and Kitty picks up an extension, pressing the mute button so she can eavesdrop but Swan won't be able to hear her heavy breathing. Oh bugger. Please don't let him say anything to get me in trouble.

'Anna Brown,' I say.

'All right, Bud Fox,' says Mark Swan's voice, richly baritone and confident, and I can hear the grin in it. 'I want you to buy me twenty thousand shares of Bluestar . . .'

I can't speak. Kitty's eyes are popping out of her head. Clearly she doesn't get it.

'Mr Swan,' I say.

'If we're going to work together, don't you think you should call me Mark?' he says.

'You liked it,' I say. I can't breathe. My insides are melting. 'You liked it?'

'Very good, Sherlock,' he says. Oh man, he's sexy. 'My agent's having a fit, but I think this might be kind of fun. And I have some space in my schedule for the autumn. Are you developing this project?'

Kitty is making frenzied hand gestures at me.

'No,' I say. 'That's my boss, Kitty Simpson.'

'Well, have her call my assistant Michelle and set something up,' he says. 'And make sure you're in that meeting too, kid.'

Bless him. I love him! 'Whatever you say, boss.'

'I get final cut,' he warns. 'Non-negotiable.'

I look at Kitty, who nods frantically.

'No problem,' I say. 'And thank you so much.'

'Don't thank me,' he says. 'I own you now. I'm going to be working you so hard you're gonna throw up.'

'Sounds good to me,' I manage.

'Have her make the call,' he says. 'See you, Bud Fox.'

He hangs up and so do I. Kitty stares at me as if she can't believe it.

'Who the hell is Bud Fox?' she demands.

'*Wall Street*. You know, the movie?'

'Oh,' Kitty says, giving a little tinkling laugh. 'Wretched trashy film. Dear Mark, so ironic,' she says. 'And well done, Anna. I'll call Personnel, get them moving on that raise.'

Get them moving? I thought that had already gone through.

'Why don't you call the writer – that Trish person,' Kitty says. 'Have her ready to meet her director tomorrow morning. I'll call Eli Roth. And Carly Smith.' She smiles with satisfaction, like a cat, and then admires her canary diamond. 'My movie is really starting to come together,' she says, and nods at me as though she's the Queen and I've just won a gold medal in the Duke of Edinburgh's Award Scheme.

I exit her office to find John jumping away back to his desk. Had he actually had his ear pressed up against the door?

'So what happened in there?' he asks, casually.

I favour him with a smile. 'Oh, I was just chatting to Kitty and Mark,' I say, casually.

'Really,' says John, furiously. 'Congratulations.'

I mustn't let him make me any coffee for the foreseeable future.

I know it's childish, but I just feel wonderful. I'm going to go to the off-licence on the way home and buy some cheap champagne. And go to Blockbuster and rent a copy of *Wall Street* . . .

* * *

The phone is ringing as I walk through the door.

My arms are laden with bags. I have an Indian prawn korma from Marks and Spencer, a family-sized bar of Dairy Milk, a pack of tree-ripened peaches (so expensive, but what the hell), and two bottles of champagne (M&S's own, but it still counts). I don't care, it's a celebration, and Janet and Lily will be on the champagne too (Lily has convinced herself champagne is calorie-free) so there has to be enough. I also got a six-pack of crisps, all salt and vinegar. And I'm going to eat at least two packets. Maybe three!

Before I left the office today – after the tech guys turned all my files on to Kitty's computer – Personnel rang and told me my new salary and benefits.

I am going to be making thirty grand!

Thirty. Grand. A. Year!

And there's a bonus if we actually get *Mother of the Bride* made, but they weren't specific. Who cares? I just went up from sixteen grand a year to thirty grand a year in five minutes! And my own parking space in the underground lot!

Which would be nice if I had a car. But I can't afford one, with petrol and parking and congestion charge. Where would I park it round here? Monthly spaces are almost as much as the rent. Maybe I could rent out my new parking space. Rob is loaded, he drives a Porsche Boxster . . .

Course, he probably wouldn't rent from me, I think to myself. Too demeaning! Ha ha ha ha ha!

The phone is still ringing, so I drop my bags and pick it up.

'Hello?'

'Anna?'

'Oh, hi, Charles,' I say, with forced enthusiasm. It feels like my dentist calling. But I know I have to get over it. It's not his fault, after all. 'Um, thank you for the lovely flowers.'

And I do feel a bit guilty. Before all the drama started today he sent me those roses, and they made me feel fantastic, didn't they?

'Did you like them?' he asks, sounding all pleased.

'They were great. They're sitting on my desk right now.'

'Well, I hope they brightened up your morning,' he says, a

bit stiffly. But he sounds nervous. I know the feeling. I instantly want to put him at ease.

'They were the perfect start,' I tell him. 'And then I got a raise! I think they were my lucky flowers.'

'A raise?'

'I'm assisting on a project now,' I tell him. 'I can try to get films made.'

'Terrific,' Charles says happily. 'So you'll be able to put my novel into production!'

Oh bugger. The dentist feeling comes back. How the hell am I going to tell him his book sucks? I should be really brave and tell him the truth, like a real executive.

'I'm still working on the novel,' I lie brightly. 'I want to let it sink in.'

That was much easier. Obviously I'll tell him. Later. When I've had time to think of how.

'We're going to go out again this week?' he asks, gingerly.

'Oh yes, absolutely,' I say, deflating.

I wonder unhappily how long I can wait before I have to shag him. I hate sex. Most women hate sex, don't they? Real women, I mean, not pin-ups like Lily and Janet. It's just so embarrassing. Why do men insist on looking at you even though you've got a bit of a tummy and would prefer to do it in the dark?

And I personally don't think even Lily and Janet like it. They go out with such losers. It's probably as much of a chore for them as it is for me. Everybody pretends sex is so great, but it's dreadful. It's something you have to do to keep your boyfriend, as Brian kept reminding me. I wish I lived in Victorian times. And wasn't a poor prostitute, obviously.

It could be to do with the fact I don't care too much for any of the men I've been out with. None of them made me feel excited and edgy like Bruce Willis or Brad Pitt. Or even Mark Swan. But those aren't real men, those are fantasy figures, and not very likely to go out with me, are they? I know they're real, technically. But they might as well not be. The memory of Mark Swan in that shop comes back to me, standing there, all leonine, strong, hugely tall, mountain-craggy.

I shake it off, and remind myself how lucky I am that Charles is talking to me. Hell, Charles is a fantasy date for a girl like me! Even sent me nice, proper flowers to the office. Back in the Brian days I'd have been *dying* for that kind of attention. I pull myself together, try to pay attention.

'Well,' Charles says. 'Actually, I'm having a bit of a house party this weekend. Been planned for yonks. Up in the country.'

'At the mysterious Chester House?' I ask, then hearing his sulky silence move to cover. 'I mean, Chester House, that sounds lovely.'

'Everybody's going to be there,' he says. 'Binky and Jacob and Charlotte and Olivia.'

'I don't know those people, Charles,' I remind him gently. He means well, after all.

'Oh. Right. Course. Anyway, Vanna and Rupert will be there,' he says. 'Loads of people, actually. All staying over. Dancing. Kedgeree and champagne in the morning. Jolly nice to have you there too,' he adds, hopefully.

The door opens behind me and Lily walks in, going straight into her bedroom.

'Well, I – I suppose so,' I say. I can't get out of it, can I? Quick, what's a good excuse?

'Hello?' says a voice. It's Lily, picking up the extension in her bedroom. 'Who is this?'

'Excuse me,' I say coldly. 'Didn't you see I was on the phone?'

'No, sorry,' she lies coolly. 'Sorry to interrupt.'

'That's all right,' says Charles.

'Oooh, is this the *famous* Charles Dawson?' Lily asks. She's dropped her voice several octaves to that breathy, sexy smoker's throat thing she does.

'Hi. Ya,' says Charles, warmly. 'Who's this?'

'Do you mind?' I ask.

'This is Lily, Anna's flatmate,' says Lily. 'My friend Janet and I live here too. We can't wait to meet you. We've heard so much about you.'

'Oh, well. Come along on Saturday,' Charles says. 'Taking Anna to a house party. Plenty of room. Love to have you!'

'They're busy that day,' I say instantly.

'No we're not,' says Lily, equally instantly, 'and we'd *love* to come. Thanks so much for the invite, darling!'

Bastards!

'See you all on Saturday night, then,' Charles says, sounding pleased as punch. 'Drinks at seven, dinner at eight. Dancing starts at nine. Black tie, obviously.'

Obviously.

'Bye, Anna,' he says.

'Bye,' I say gloomily. 'See you Saturday.'

I slam down the phone and march into Lily's room. I am just about to commit physical violence on her when the door opens and Janet comes in.

'Hi,' Lily says. 'Great news! Charles Dawson just invited us to a super house party. At Chester House! All black tie. I bet there'll be loads of country gents there. Just swimming in money and no idea how to spend it.' She laughs. 'Well, we can help.'

Janet shrugs. 'I don't want to go.' She looks so down.

'Why not?' Lily demands.

'They sent me away from a booking today,' Janet says. 'Told me I had the wrong look. All the other girls were eighteen or nineteen and under a hundred pounds.'

'You really must do something about your weight,' says Lily severely. 'It's your own fault. No discipline.'

'My booker wouldn't take my calls this afternoon,' Janet says, tearily. 'They kept saying he'd have to call me back, but he never did. He thinks I'm a failure.'

'Have you considered a facelift?' Lily asks.

'Oh, shut up, Lily,' I say. 'Janet's only twenty-eight.'

'*Only*,' says Lily, scornfully. 'That's *ancient*.'

Janet rubs her eyes, then blows her nose to cover it.

'This is so stupid,' I say. 'Janet, you have to come to the party.' The two of them there is my worst nightmare, and now I'm trying to guilt her into coming? 'I need back-up,' I say firmly, 'and Lily's coming, so you can't leave me alone with her.'

'You've got Charles, at least he fancies you,' Janet says.

'Oh, come off it,' I say. 'When those . . .' I want to say chinless wonders but that's not very nice, so I settle for, 'country boys see you, their eyes will bug out of their heads.'

'There are going to be tons of society men there,' says Lily. By this, she means rich.

'And most of them are from *very good families*,' I add temptingly. By that, I mean titled.

'Maybe I'll come,' Janet says. 'If I can lose five pounds by the weekend. I can go on that watermelon diet again.'

'Too much sugar,' Lily says.

'You think a plain fast?' asks Janet, worried. 'Just the vitamin pills and water?'

Lily shrugs. 'Some of those vitamin pills are five calories each,' she warns. 'You'd better check the labels.'

'Oh, what a crock,' I say, impatiently. 'You'll eat like a normal person or you can't go. Same goes for you, Lily.'

'We're already invited,' says Lily, tossing her glossy platinum mane.

'And I can get you *un*invited,' I threaten. 'Don't think I won't.'

'OK,' Janet says. 'To be honest, I've had a rubbish day and I don't think I could take a fast.'

Lily looks at her pityingly, as if she has no self-control, but backs off when I give her my death stare. I pass her a bottle of champagne.

'Here, open this,' I say. 'Make yourself useful.'

'Oh, champers,' says Lily. 'Not vintage,' she adds, disappointed. 'But it'll do. What are we celebrating?'

'I got a raise,' I say proudly, and tell them all about it. Janet seems genuinely delighted, and Lily pretends to be. Which is about all I can expect. I eat the korma and they both decline (good, there isn't enough to go round) but they agree to eat the peaches. And one and a half bottles of booze later, all three of us are eating salt and vinegar crisps.

It's funny watching Lily struggle with herself over the crisps.

She wants another packet. I don't think she's eaten crisps in five years, every time she bites one she looks as if she's having an orgasm. Desperately she gets up and goes into the kitchen and returns with a tired-looking stick of celery.

'What the hell is that?' I ask.

'Celery's great,' says Janet earnestly. 'It takes more calories to eat it than it gives you.'

'It's weight loss in a stick!' Lily says, biting into it. 'You should try some,' she adds to me.

'Looks vile,' I say, eating another crisp. 'Dairy Milk?' I ask, unwrapping the bar and waving it under her nose.

Lily looks as though she might faint. But she's a strong-willed girl. 'See you two later,' she says, getting up and draining her champagne. 'I'm going out,' she adds, grabbing her coat.

'Where?' asks Janet.

'Anywhere,' says Lily, slamming the door.

'I'll have some Dairy Milk,' Janet says once the coast is clear. 'It's not like anybody's ever going to hire me again anyway.'

I break off two squares and give them to her and then, wrestling with myself, put the bar away. Because she's only going to eat it all. And then feel sick in the morning. Plus, for Janet, eating a family-sized bar of chocolate might cause a mental breakdown and then where would I be?

'So, the party,' I say, trying to cheer her up.

'What are you going to wear?' she asks me.

I shrug. I don't care what I'm going to wear, I only care what Janet's going to wear. I'm going to be humiliated whatever I put on, aren't I?

'My black dress with the pearls,' I say.

'You always wear that,' Janet says.

'It's suitable,' I explain. 'Anyway, what are *you* going to wear?' Maybe I can find somebody for Janet. Fix her up. She's actually a nice girl, under all that beauty.

'I haven't decided,' she says, then fixes me with a stare. 'But I'm taking you shopping on Friday. And to the hair-dresser. And we're going to get you made up.'

I feel a flash of anger. 'Give over, Janet. What's the point of that?'

'There's a point,' she says. 'You'll see.'

I shake my head and reach for another packet of crisps. Why can't she just leave me alone? I've been supportive to her, haven't I?

'Why don't you have another peach instead?' Janet says.

'Bloody hell, I can't believe this,' I snap. 'I'm being so nice and sympathetic to you . . .'

'It's not about beauty,' Janet lies. 'It's about health. I mean, you're an executive now. Maybe you should think about changing your diet just a *tiny* bit. For more energy.'

I put the crisps down sullenly. She's ruined them for me now, anyway.

'You like that new guy at work, right?'

'Eli Roth. Yes, he's nice,' I say, a bit lamely. 'I mean, obviously I don't like him in *that* way. My friend Claire really fancies him. But he's cool.'

'Well, you want to be like him, right?' Janet asks encouragingly. 'All successful and making loads of blockbuster movies with big stars?'

'Sure,' I say. I'd also like to be able to leap tall buildings at a single bound.

'Sooo, you should *model* him,' Janet says earnestly. 'Look.' She goes to our sparse bookshelf (all I ever read is scripts, and I wouldn't be surprised if Lily's illiterate, apart from fashion magazines). 'I adore this book. *Take Control! Take Over!*'

I look at it. It's very glossy and American, with a woman in workout gear giving a big thumbs up on the cover. As well as the workout gear, she's wearing a huge diamond necklace, like she forgot to take it off on her way to the gym.

She looks very rich and thin and healthy.

'What does it say?' I ask. You might as well give it a chance, mightn't you?

I'd better come clean. Despite my scorn for Brian, I do have a secret addiction to self-help books. I've got most of them. All the famous ones. I am an expert on feeling the fear and doing it anyway, the seven habits of highly successful people, and so forth.

Of course I never put any of their tips into practice, but I will. Just as soon as I have some time.

'Here,' Janet says, flipping through the pages. ' "The Seventh Success Secret, Model Your Mentor! By now you should have acquired your very own millionaire mentor," ' Janet reads slowly. She looks up. 'Is he a millionaire?'

'Several times over,' I say airily.

' "There is no need to spend years perfecting your own strategies," ' Janet continues. ' "You can piggyback by using hers or his! She or he has gotten to where you want to be! Do not be afraid to ask questions! Imitate your millionaire mentor in every way! Think like them! Dress like them! Be like them! By consistently following her or his actions, you too can gain her or his results!" '

'Hmm,' I say, all excited. I love discussing my career, now it looks as if I might have one. It makes me feel all giggly, like I was thirteen again.

Is it possible some of his success could rub off on me?

'Maybe there's something to this,' I exclaim.

'What does he look like?' Janet asks. 'Is he full of energy?'

'Oh yes,' I agree. 'He's tall. And he's got dark hair and dark eyelashes. And he wears these black suits which pick out his eyes.'

'Married?' asks Janet, interested.

'Girlfriend,' I say.

'Oh well,' she says. 'They all do. Anyway, what is his body like?'

I blush. 'What a question! How should I know?'

'You haven't shagged him in the broom closet yet?' she teases. I shake my head. 'Oh, sorry, you needn't look so po-faced about it. I know you're mad about Charles,' she says. 'I'm just asking, can't you see if he's fit and healthy?'

'Of course he is, he's Californian.'

'There you go, then.' Janet takes an exultant swig of champagne. 'You should try to be fit and healthy. Just to model him.'

I look regretfully at the salt 'n' vinegar crisps.

'But I'll never be skinny,' I say. 'I don't want to eat celery. And drink hot water with lemon in.'

'Baby steps,' Janet says. 'Slice the salami. You could just do a little. More peaches fewer crisps.'

'They are delicious, though,' I point out.

'They are,' agrees Janet. 'Tell you what, there are only two packets left. If we have one each we'll have eaten them up and then there won't be any temptation left.'

'It's the healthy thing to do,' I concur.

6

I wake up the next morning feeling a bit fragile. You wouldn't call it a full-blown hangover, but you wouldn't call it bounding out of bed to greet the day, either.

Groaning for some water, I stagger into the bathroom and cup my hands under the tap. Then I climb into the shower.

As the water is sluicing down over me, rinsing away the shampoo, yesterday comes back to me. The best day ever! I've got Mark Swan. And then there was what Janet said.

As I step out of the shower, swathing myself in my huge bath sheet – I bought it so I'd never have to see any part of my own naked anatomy first thing in the morning – I suddenly decide to do something different.

I actually examine myself.

This is too weird. I've spent years not looking at myself, except for a quick check in the mirror to see I've no lipstick on my teeth. In fact, I have elaborate and well-tested avoidance methods. And now here I am, facing the grim truth.

And it is quite grim.

There I am. Am I fat? Depends how you define it. I may not need reinforced floors to walk on or a crane to winch me from room to room, but there's that big tummy . . .

I look at my tall frame, my big, strong hands. You can't see my feet from here, but there's my nose, and then my hair which is plastered wetly around my face, ready to be combed through. My skin is really pale, but not in a porcelain way, in an 'I stay inside all day and never see the sun' type way.

I'm just not very attractive, all round.

115

I do have my good points. My bottom isn't bad and I have nice, strong legs. I don't hate my arms either – that's from lugging around all those scripts.

Part of me says what it always says. And I always listen. The part that says, well, since you'll never be pretty, you might as well eat whatever you like and dress invisibly. Nothing will ever make you look good. But this morning, there's another part of me that wants things to be a bit different. Because they *are* different. I've got more money. I've been promoted. Kitty believes in me.

I would like to model Eli Roth. I would like to be just like him, except a girl. And I bet he doesn't guzzle huge packets of Dairy Milk and salt 'n' vinegar crisps all day. He probably eats tofu and drinks wheatgrass juice.

Oh, it's ridiculous, says voice no. 1. *You're not going to eat tofu!*

And of course I'm not. But maybe I'll just . . . experiment, I think, guiltily. Guilt because I'm actually thinking I could make a change, which is obviously stupid. Anyway, for a laugh, perhaps I'll try to cut back a bit. Just slightly.

'Exciting, innit?' asks Trish, when I arrive to pick her up. She's got her long, blond hair braided into a sleek plait, and she's picked out a silvery shirt with a black pleated skirt, stacked Maryjanes, and spider-web tights. It's a very disturbing look, very hot sixth former. I can just imagine Kitty's reaction to seeing her like this. Or Mark Swan's. Next to her, I feel about as feminine as Lennox Lewis.

But Trish is genuinely thrilled. She's doing her patented Tigger impression, bouncing up and down like a child on the way to EuroDisney, making it impossible for me to hate her.

'I can't believe it,' she says. 'You're a bloody genius, you are.'

'The deal's not done yet,' I remind her.

'But it will be,' she says with total confidence. 'They've got Greta Gordon and Mark Swan. Did you see *Suspects*?'

'Yep.'

'Fuckin' *great* film. What's he like, then?' She breaks out a cigarette and strikes a match. 'Fag?'

'No thanks.'

'Great for appetite suppression,' she says, kindly. 'Is he a moody genius?'

I grin. 'I don't know about moody. But he's a genius all right.'

'That Kitty called me,' Trish says. 'Came round to see me yesterday an' all.'

I blink. 'What?'

'Yeah, in the afternoon.' When I was up in Hampstead Heath. 'Nice lady,' she says, doubtfully.

I wrestle with myself. 'I guess so,' I say, eventually.

What did Kitty mean by that? Going without me? I suppose she had asked for a meeting with the writer, and after all, I wasn't in the office. She could have told me, but maybe the news about Swan pushed it out of her mind.

Look, she gave you that raise, I tell myself. No need to be so bloody paranoid.

'She told me all about how she got Greta, and how she's going to be producing,' Trish says, looking sidelong at me under her dark lashes. 'She said she's the one I can trust.'

'Well, so you can,' I tell her, loyally.

'She didn't mention you,' Trish says.

'No need. She gave me a raise,' I feel compelled to add. 'She's letting me work on the project too. Casting, directors, everything.'

'No need to get all snippy,' Trish says, shrugging her slender shoulders. 'Long as you're cool with it.'

'I'm totally cool with it,' I say. 'Kitty's championed the script to Eli Roth, and she attached Greta.' I'm blushing and wondering why I sound like I'm trying to convince myself. I flag down a taxi.

'Jump in,' I tell her. 'Big meeting to get to.'

'Mmm,' Trish says happily, sliding herself onto the battered black leather. 'Moody genius Mark Swan and Trish Evans. D'you think he'll like me?'

I look at her and imagine her tiny, slim frame next to Mark Swan's huge one, like a ballerina draped over Arnold Schwarzenegger (in his *Total Recall* days).

'Oh yeah,' I say, a touch darkly. And who can blame me? 'I bet he'll love you.'

This meeting is obviously going to be big. Kitty is pouting because it's not going to be in her office; Eli Roth has insisted we all take it upstairs to the fourth floor. She is, however, determined to run the show anyway.

'Trish, *darling*,' she says, when I turn up with the writer in tow. 'What an amazing look, that's just fabulous. Eli will absolutely *love*.'

She herself is particularly resplendent today in a fitted suit in yellow wool, pearls the size of marbles, her canary diamond, and towering stiletto heels in lemon leather, with a white quilted Chanel bag slung over her shoulder. She looks like a terrifying, power-mad daisy.

'Greta will be here any minute,' Kitty promises, twitching to look over her shoulder at the lifts. She's like a girl in a club casting around for a richer man. Trish is nice, but she won't rate the full court press today. 'And Mark Swan.' I wonder, if they arrive together, whose ass will she kiss first? 'Anna, bring your notes up, would you? And John, bring my pad.'

'John?' I ask. What the hell is John doing in this meeting?

'Of course, Anna,' says Kitty with the kind of forced brightness that means if I open my mouth again she will kill me. 'John's on the team too!'

John smiles snidely at me. 'Right away, Kitty,' he says, obsequiously.

I try to smile. I suppose I'm just being paranoid. John does work for Kitty. Perhaps she's found him something to do. Maybe he's been contacting, I don't know, location scouts, costume designers . . .

'Trish, you go with Anna, darling. Get introduced to Eli,' says Kitty, anxiously. She obviously wants to lose us so she can be the only one welcoming Greta. And Mark.

I take Trish upstairs to Eli Roth's palatial fourth-floor suite. And for the first time, it dawns on me that I might actually be going to get a film made. Because this place seems a million miles from our chaotic little offices downstairs. It just reeks of corporate money. Already Roth has ripped out the Winning logo and installed art with Red Crest's logo on it. There is soft, white wall-to-wall carpeting that deadens your footfalls, a Japanese style flat stone wall with water cascading soothingly down it, and little topiary sculptures in terracotta pots. The phones are still buzzing, but they have a special, muted ring to them, and the secretaries are not frazzled and hunted like Claire but supremely confident thirty year olds in twinsets and neat little flats.

'So,' Roth says, welcoming us at the door. He has tinted floor-to-ceiling windows up here and LA-style southwestern furniture. He ignores John and me and smiles at Trish. 'This is the writer,' he says. 'You have a fabulous vision for the piece.'

Trish takes in his even-featured, rather anodyne good looks and grins.

'Should do, shouldn't I?' she asks. 'I bloody thought of it.'

'Absolutely,' Roth says, soothingly. 'And we only have a couple of notes . . .'

My heart sinks. He's going to rip it apart, and it's almost perfect just as it stands. Why do producers do this? Fall in love with a script, buy it and then change everything about it until it's unrecognizable and really sucky.

'No problem,' Trish says, looking trustingly at me. She smiles flirtatiously at Roth but he doesn't seem to notice. Business is business, I guess.

Kitty bustles in without knocking, shepherding a woman dressed head to toe in black and wearing dark wraparound glasses.

'Greta,' says Roth, with a delighted smile. He crosses the room at the speed of light and links his arm in hers, taking her towards the sofa. Kitty scowls. 'What a pleasure. What an honour.' He then kisses her hand.

'I'd like a coffee,' says the great lady, distantly.

'Of course,' Roth says. He snaps his fingers. 'Get Ms Gordon a coffee.'

Nobody moves.

'Don't just stand there, Anna,' says Kitty. 'You heard Eli, didn't you?'

I am just getting to my feet to go play dogsbody once more when the door opens again and Mark Swan walks through it.

'Sorry I'm late,' he says.

'You're right on time,' Kitty says warmly, darting towards him and thrusting out her jewel-encrusted claw. 'I'm Kitty Simpson, *Mother of the Bride* is my baby,' she says, smiling sweetly. 'This is Trish Evans, who wrote us such a *fabulous* script, of course we have just a few *teeny* changes, and this is the great Greta Gordon, who I'm sure you know . . .'

'We've not met,' says Swan, inclining his head down towards Greta. 'But I've always admired your work.' Gosh, he's tall, isn't he? He has to be at least six five. And so broad-shouldered. He's not pretty, like Eli Roth, but still, powerful. Strong-jawed, dark-eyed. Thick eyebrows. Yet I notice that Trish, me, Kitty, even Greta, we've all straightened our shoulders a bit since he walked in here. We're all standing a touch better, more self-consciously.

'And your script is fantastic,' Swan adds to Trish.

'Great to meet you, Mark,' says Roth, shaking his hand. I can see that Roth doesn't enjoy having Swan tower over him, but he handles it well. 'I love all your movies. Voted for you in the Academy.'

'Thanks very much,' says Swan. He casts a sidelong glance at me, winks. I stare at my shoes, trying not to smile. He winked at me! Mark Swan winked at me!

Kitty notices. Of course. She never misses a trick.

'You've met Anna, who works for me, and this is John, who also works for me,' she says, smiling. 'Anna was just about to go for coffee, would you like some?'

I blush. I'm being introduced as the dogsbody I am. It doesn't seem fair. How about 'This is Anna, who found the script'? But of course we can't say that. It needs senior backing, that's what Kitty told me.

'Well, since Anna's completely responsible for getting me to read the script and attach myself,' says Swan, easily enough, 'I think someone should be getting *her* coffee.'

I can't believe he just said that! Oh! I love him.

'That's OK,' I say, going scarlet. But now Eli and Kitty are falling over themselves to agree.

'Ha ha ha, of course,' says Kitty. 'Go and get everybody coffee, John.'

'Right away,' says John, fawningly. 'How do you like it, Mr Swan? I'm such an admirer. Really. Such a pleasure to get your coffee.'

'Black's fine,' says Swan. He looks at me again as if he's barely controlling an impulse to roll his eyes.

I stare at my pad. Am I mental, or is Mark Swan actually trying to bond with me? I can see both Eli and Kitty noticing, and the vibes coming my way aren't exactly appreciative.

'So, the good news is, I ran the package by Paramount this morning,' says Roth briskly, dragging Swan's focus back to him. 'And they're willing to go for it.'

'That's good of them,' says Swan, sarcastically. He's the hottest thing in the UK film industry and he knows it. Of course any studio would jump at the chance to bankroll the next Mark Swan film.

'What I thought would be useful was to get you and Greta together with us, and we can present our vision to the writer,' says Roth. 'Have her change the script.'

'Do you have a notepad?' Kitty asks Trish. 'Here's one, if not. You'll want to take very careful notes.'

'The part of Elsie needs to be beefed up,' says Greta. 'And in the first two acts made more sympathetic.'

'The script's too long. You'll need to chop off ten pages. And we should work on a zinger chart,' Roth says. 'A zinger scene every ten minutes, that's the formula. Ten pages, zinger. Ten pages, zinger,' he repeats, ignoring Trish's horrified face. 'You'll get the hang of it.'

'Elsie needs to be much nicer. And much more attractive,' Greta says. 'Perhaps she should be some sort of model for older women. Highly successful. And a philanthropist. And she needs a wonderful career. How about a judge? I know I look good in black,' she muses.

'Coffee,' sings out John. He has returned with a tray laden with cups and saucers and little jugs of milk. They all match, too, unlike the plastic spoons and chipped mugs we have downstairs bearing legends like 'If I Want Your Opinion I'll Give It To You' and 'Old Stockbrokers Never Die, They Just Freeze Their Assets'. 'Get it while it's hot,' he sings, arching his wrist to pour in the warm froth of milk and setting down some of those little cocktail sticks encrusted with sugar crystals. John is unbelievably camp.

'I don't know about all them changes,' says Trish dubiously. 'Anna said she liked the script how it was.'

'Whatever *Anna* thought, writers always have to make changes,' Roth says.

'Yeah, but all them changes to Elsie? She's the whole thing. It'll ruin it.'

'You'll make the changes, Trish,' says Kitty, smiling at her like a crocodile. 'Your job is to execute our vision.'

'I'm going to be playing the part. I think I know what's best for Elsie,' says Greta, majestically.

'But Anna said—'

'Anna's not in charge here. I am,' says Kitty, with soft menace that Trish takes no notice of. Eli Roth clears his throat. 'And Eli, of course.'

'Actually,' Mark Swan says, 'I'm in charge.'

Kitty, Eli and Greta all look over at him.

'Well, of course,' says Kitty placatingly, 'once filming starts, Mark.'

'No. Through the whole thing. Pre-production to final cut.' Swan shrugs. 'I don't work any other way. Any script changes will need to be approved by me; casting, crew hire, everything. I run my movies. It's my name on the film. When it says "A Mark Swan Film", that's the truth. I get the blame if it all goes wrong, and I take credit when it all goes right. Which means total control, or I don't attach. If you aren't up for that, it's been nice to meet you all, and I'll just go home.'

Greta shakes her head, frantically. She has no intention of losing the man who can single-handedly revive her career.

'That's fine with me,' she says instantly. 'You are the maestro!'

'That's right,' Kitty says at once. 'Whatever you say, Mark.'

She looks nervously at Roth, who gives a curt nod and a forced smile.

'No problem, no problem,' he says, spreading his manicured hands. 'Trish, let me rephrase. Your job is to execute *Mark's* vision.'

I don't know where to look. The atmosphere in here is as tense as a Florida election. You can almost feel the loathing crackle under the fake smiles pasted on Kitty and Eli's faces. You can almost smell the fear seeping out from Greta Gordon. And Trish is just bewildered and resentful.

The only person who seems completely relaxed about the whole thing is Swan.

'I signed up because I liked Trish's vision,' he says. 'Although why we all toss around the word "vision" is beyond me. It's only a bloody movie.'

Trish cackles with laughter, and I can see Kitty and Roth suppressing their winces.

'Fuckin' 'ell,' she says. 'You're all right, mate. You're almost normal!'

Swan chuckles. 'Well, cheers.'

'I thought you were a moody genius,' adds Trish. 'But Anna said not. She said you weren't moody, anyway.'

Swan turns round to look at me properly, and I have to force myself not to stare at my shoes. He's such a hulk of a man. And he's a bit of a legend. More than a bit.

'Did she?' he says. 'Well, I'm not a genius either, Bud Fox.'

I can't look at him.

'*Wall Street*,' says Kitty instantly. 'Such an amazing film! Such a powerful vis— er, movie.'

'The first two thirds are great,' says Swan. 'Which was your favourite scene?'

Kitty starts to flush. 'Ahm . . .' She looks helplessly around. She hasn't seen the movie. I suddenly wonder when was the last time she saw *any* movie that wasn't one of ours.

'You were telling me you loved the scene where Bud Fox walks into Gordon Gekko's office,' I pipe up.

'Yes,' she says, exhaling. 'Love that scene! So funny! Great sense of humour.'

'Funny?' Swan asks, bemused.

I shake my head imperceptibly at Kitty.

'The . . . drama of it . . . I just found funny,' she says, desperately.

'Kitty thought the underlying satire was really well observed,' I say.

'That's exactly right,' Kitty agrees.

'I see,' says Swan, and I wonder if he's going to wink at me again. Fortunately, he lays off. 'Anyway, I want the lead role, Elsie, to stay as written. I'm surprised you'd want it changed,' he adds to Greta. 'If you're coming back, it shouldn't be with the same cute character you've always played. Elsie's greedy, mean, selfish, pathetic – but that's what gets Oscars.'

'You really think I . . .?' asks Greta, simpering.

'Only if you work really hard at it, and take direction,' says Swan bluntly. 'You've got a bad rep, Greta. I'm the boss on my sets, and I don't like prima donnas. You won't get any special trailers or have somebody to pick out all the red M&Ms. Understand? It'll be written into your contract, and if you throw tantrums, or try to hold up filming, I'll sue.'

Greta looks stunned, but then shakes herself. I can see the little hamster wheel turning in her brain, humming *Oscar Oscar Oscar*.

'You're in charge, Mark,' she whispers.

Eli Roth and Kitty look impressed. As well they should be. And Trish is starting to smile again.

'Now, you,' Swan says to her. 'Same goes for you, sweetheart. I don't want too many changes, but no bitching about the ones I do want made. I'll listen to you, but I have final say and if you don't like it we can hire another writer. Fair enough?'

'Not really,' says Trish, grinning. 'But I s'pose I don't have much choice, do I?'

'None at all,' says Swan, cheerfully. 'And you,' he turns to Eli and Kitty. 'The budget's low. That's OK, I don't care too much about my fee. Labour of love.'

'I admire your passion for the work,' Eli Roth says smoothly.

'But the flipside is I get approval on the lot. And I probably won't want too many suggestions from Winning.'

'We're Red Crest now,' says Roth.

'Whatever.'

'But Mark,' says Kitty tremulously, 'surely you'll allow us to sit in on meetings. The producers have to be represented in the process.'

'You've done fine, Kitty,' says Swan easily. 'Found the script, the star and the director, all for peanuts. The rest of it you can leave up to me. I don't mind if you come to an occasional meeting,' he says. 'But no more than that.'

'I'm sorry, but I have to put my foot down here,' says Eli Roth, a bit anxiously. I can't believe the transformation I'm seeing. Roth, with his beautifully cut, loose suits, his LA tan and his gold and diamond Rolex, up until today I thought was a powerhouse. He made even Kitty buckle and scrape. He exuded self-confidence. His glossy good looks and perfect grooming were exactly what you'd expect from one of Hollywood's richest Young Turks. And now this scruffy, craggy man has him all nervous.

Eli Roth is buff and probably has a hundred-dollar-an-hour personal trainer, but you'd take Mark Swan in a fight any day.

'We have to be represented throughout the process of film-making,' Roth insists. 'I need to know what's going on.'

Swan pretends to consider it.

'Well,' he says eventually, 'you can send Anna.'

I blink.

'Anna!' Kitty explodes. 'But Anna's just a reader!'

'Anna found me,' Swan says simply, 'and I like her.'

'That was my idea,' says Kitty, instantly.

'No it wasn't,' says Swan. 'You'd have gone through my agent, a serious development executive like you. Trying to sneak onto the set is the sort of mental thing only juniors do. It was a ballsy move,' he says, turning directly to me. 'If you sit in on meetings you can learn something about movies in the real world. It'll make her a better producer,' he says to Kitty.

'She isn't a producer. She's just a reader,' says Kitty, venomously. 'And we don't feel she's ready to move on just yet. Nor does she, do you, Anna?'

Of course I do! But I can see by the look in her eye that it's not worth arguing the point.

'No,' I mumble, heart-sick.

'I have all the experience. I can sit in,' Kitty says, smiling winsomely. 'And report back daily to you, Eli,' she adds deferentially.

'I want Anna,' says Swan.

'Maybe you should take Kitty, Mr Swan,' I say, seeing the fury return to my boss's eyes. 'I mean, she won an Oscar,' I bleat. 'She's a really excellent producer. It's true I don't know anything about it.'

'You will when I'm done with you,' Swan says. He looks back at Kitty and Eli and shrugs. 'I'm about to sign on to do this film for scale. It has to be fun for me. I like Anna, so having her there to rep you guys makes it fun. If you're not cool with that, we can save ourselves a lot of trouble and I'll bugger off so you can call other directors.'

'That's fine, Mark,' says Roth, with a warning look at Kitty who has opened her mouth again. 'We're cool with that.'

'Excellent,' Swan says. 'You three, come over to my hotel at five – 47 Park Street.'

Trish and Greta say they'll be there, and I just nod my head. I don't dare even look up. My insides haven't churned up like this since I was in the front row of a Beastie Boys gig when I was sixteen and fancied Ad Rock. Kitty is going to blame me, I know it. Maybe Eli Roth too. But they can't do anything. Mark Swan wanted me, and that's all that counts.

He stands up and everybody jumps to their feet. Swan goes round the room, shaking hands, starting with Greta. Finally he gets to me.

'Nice seeing you again,' he says, clasping my hand in his giant one.

'See you this afternoon, Mr Swan,' I say, trying to sound businesslike.

'We agreed on Mark,' he reminds me. He gives my hand a gentle squeeze. 'This is gonna be fun,' he says. And then he's gone.

* * *

I'm sitting on Eli Roth's couch, twisting my hands nervously in my lap. Even though I've nothing to feel guilty about, I still do. I'm one of those people who start to cough and shuffle suspiciously whenever they see a copper or walk through the 'Nothing to Declare' channel at Customs.

Kitty and Roth seem none too pleased with my coup. This was to be expected, but they really are giving me the third degree. I feel like a Guantanamo Bay detainee.

'And you got this idea how?' Kitty asks acidly.

'I don't know,' I say lamely. 'It just came to me.'

'And you happened to bump into him,' Roth says. He's smiling crisply, but his body language leaves no doubt that he finds me highly suspicious. 'What were you doing again?'

'Just buying some sweets.'

'That figures,' says Kitty, meanly.

'And he agreed to read the script.'

'Yes,' I say. 'Honestly, it was just luck.'

'You're quite sure you had no other *contact* with him?' asks Roth, raising his eyebrows, which I now notice someone has shaped. A bit girly. I like Mark's bushy ones better.

'What do you mean?'

'You haven't had . . . *intimacy* with him?'

Intimacy? Not unless you count him catching my copy of *Heat*. Oh, wait a minute, he's speaking American, intimacy means sex.

'Me? And Mark Swan?' I'm so nervous I laugh out loud. The very idea is insane. I mean, Mark Swan could have anyone.

Kitty relaxes a little. 'I suppose that is rather fanciful,' she says, the cow. 'I hardly think Mr Swan would be having a relationship with *Anna*, Eli.'

'He seems to like you very much,' says Roth, as though this continues to be an inexplicable mystery that defies all rational explanation.

'You know directors,' I say meekly.

Roth nods curtly. 'Eccentricity goes with the territory when it comes to creative people,' he says. I take this as Hollywood code for 'Sure do, they're all a bunch of tossers'. I have

127

learned that the cardinal rule of the movie business is never to say anything bad about anybody or anything, in case you need them one day. Eli Roth would find something complimentary to say about *Gigli* if asked for a quote.

'Just remember, Anna,' he says. 'This is my project. And Kitty's,' he acknowledges, at her look. 'You're not the producer. You've just been delegated, at the director's request, to *report* to the producers.'

'I understand,' I say humbly.

'And you must be clear on your loyalties,' says Roth. He's warning me. 'You work for Red Crest Productions, not Mark Swan. You will account everything you see and hear fully to us. Especially any problems.'

'Especially any problems with Mark himself,' says Kitty, bitterly.

'I will,' I promise.

'You're gonna be my eyes and ears here,' Roth concludes. 'You've been shoved into this position even though you're only a reader. I expect you to use it fully. I expect to know everything, even things the actors and crew might not want me to know.'

'I'll definitely let you know every detail,' I promise. 'I'm hoping for a promotion,' I add, braving Kitty's scowl.

'Well, let's see how you handle this,' Roth says. 'Your future depends on it, Anna.'

'Thank you, Mr Roth,' I say earnestly. 'I won't let you down.'

'That's all,' Roth says. His face adds, 'You better not.'

I stand up, grab my bag and flee back downstairs.

John has piled a huge mound of scripts on my desk when I get back to it.

'What's this?'

'The weekend read,' he says.

'But . . .'

The elevator doors hiss open and Kitty emerges.

'But nothing,' John says.

'You'll have your normal duties, Anna, of course,' she says.

'I want you to realize that this whim of Mark's doesn't change anything – apart from possibly putting our production at risk, trusting somebody so inexperienced.'

I look at the huge pile. 'That's fine,' I say. 'Kitty, can I see you a minute?'

'Anything you have to say, you can say in front of John,' she snaps.

'Please,' I beg.

Kitty relents. 'Thirty seconds,' she says, gesturing to her office. We go in there and I shut the door while Kitty takes her bony frame back round her desk and sits down, and I'm standing there like a schoolgirl in front of the headmistress.

'Look,' I say. 'I just want to say that I didn't plan this. I didn't have anything to do with it – I had no idea he was going to ask for me.'

'Hmmm,' says Kitty.

'You can trust me,' I say, urgently. I want that promotion, and I know I'm not going to get it from Mark Swan. 'I'll be your right-hand girl. I'll report everything right back to you.'

Kitty looks at me, and her shoulders relax just a little.

'You know you can trust me, Anna,' she says. 'I gave you the raise.'

'Yes, thanks so much,' I say.

'And I did say you could be promoted when the time is right, so assuming you do a good job for us with Mark . . .' she waves her hand in the air. 'Perhaps after pre-production, then.'

I smile. She's actually putting a time frame on it. She means it, she's really going to promote me.

'And when I do move you up,' she continues, as though reading my mind, 'it won't be a fake promotion like Sharon's. You'll have your own office, your own readers, and we'll announce it in the trades.'

Fantastic! I want to jump up and down.

'Just make sure you don't go native on me,' Kitty says. 'You're on our side, not Mark's.'

'I know,' I say. 'Thanks so much, Kitty, you won't regret it.'

I float through the rest of the day. I flick through the new scripts (they're all dire) and make a few half-hearted notes. But mostly I'm just watching the clock, waiting for four thirty so I can get out of the office and get down to Park Street. I can't wait. I'll get a chance to thank Mark, and sit in on a real pre-production meeting. And prove myself to Kitty. I know which side my bread's buttered. I'm going to give them such exhaustive notes that Roth and Kitty will feel as if they were in there themselves.

It's like winning the Lottery! I'll get a chance to impress everybody. My bosses, Mark Swan, Trish, and Greta. I indulge in all these great career fantasies. In five or six years I could be a major Hollywood mogul. I could be as rich as Eli Roth!

Everybody in the office is looking at me. Sharon and John are standing by the kitchen, heads together, whispering. Mike Watson is staring at me from inside his office's glass walls. He scowls and looks away when I catch him at it. You can almost hear them all thinking, *not Anna*. But it *is* me, it is Anna.

And I have Mark Swan to thank for it!

Sharon's coming over. Quickly I close my game of Minesweeper, don't want her reporting me to Personnel or anything.

'Hi,' she says, smiling sweetly.

'Hello,' I say, suspiciously.

'Would you like some coffee?' Sharon asks winsomely. 'Or some herbal tea?'

'No thanks,' I say. What, with her spit in it?

'I could run round to Starbucks and get you a cappuccino or a hot chocolate if you'd like,' she offers. Something's definitely up.

'What do you want, Sharon?' I demand.

'Don't be like that,' she says. 'Why do I have to want anything? We're just friends, aren't we?'

We used to at least be civil, until she got promoted and decided her mission in life was to rag on me mercilessly. I don't say anything.

'That was quite a coup you pulled off with Mark Swan,' she says flatteringly. 'The whole office is talking about it.'

'I just bumped into him.'

'And apparently he really likes you.'

'He was very nice,' I admit.

'I heard what Eli Roth said. That was out of order,' Sharon commiserates, dropping her voice to a conspiratorial whisper. 'Of course you weren't *sleeping* with him.'

'No.'

'Mark Swan wouldn't do that,' she says. 'Obviously.'

'Obviously.' I sigh. I could wish I had a million pounds for plastic surgery to remake my entire body and face, but short of that . . .

'And anyway, you're taken,' she adds brightly. 'By the nice little man who sent you flowers.'

Of course I'm not taken, I'm just going out.

'He's not little,' I say defensively.

'No,' she agrees hastily. 'I'm sure there are lots of shorter men, and anyway, size doesn't matter, does it?'

I think of my own wretched height and how I tower over Charles.

'That's right, it doesn't,' I lie.

'You both have inner beauty,' she says consolingly. 'Anyway, how well do you know Mark?'

'I only just met him.'

'Does he actually have a girlfriend?' Sharon asks, casually. Ah. I should have guessed.

'I've got no idea,' I say. 'I expect so. Gorgeous millionaires normally do, don't they?'

'He's not gorgeous,' Sharon says, shocked. 'He's a great big beast. I expect he'd be lucky to get a girlfriend. I saw him when he came in. Of course,' she adds, 'I don't care about things like looks. I'm just attracted to his *talent*.'

'I'm sure he'd be very flattered,' I say.

'Maybe you could find out for me?' wheedles Sharon.

'I'll ask him,' I say reluctantly.

'Thanks.' Sharon beams at me. 'You're a real friend, Anna.'

'Mark Swan?' asks Claire, who's been listening in intently. 'Why would you bother with *him* when Eli Roth works right in this very building?' She sighs, dreamily.

Sharon eyes Claire's new, sexy self with disdain. She obviously doesn't think much of the home dye job and the short leather skirt.

'Eli Roth's not available,' she says curtly.

'And how do you know?' asks Claire, bristling.

Sharon tosses her curls confidently. 'I've taken him coffee a couple of times when he's been in to see Mike,' she says.

'So what? So have I,' says Claire.

'Well, he didn't ask me out,' says Sharon. 'So, you know. He's either got a girlfriend or he's gay.'

Claire snorts.

'He didn't ask you out either,' Sharon points out.

'Doesn't mean he won't. Maybe he's working up to it,' Claire retorts. They glare at each other and I grin slightly to myself. Pretty girls! What would it be like, I wonder, to be ignored by a man and actually conclude there was something wrong *with the man*? I'd love to have that self-confidence. Just once, just for five minutes.

'He's not going to ask you out,' says Sharon, cattily. 'He'd be just as likely to ask Anna.' She laughs. I feel all my happiness just draining out. I try to think of Mark Swan and my new honest to goodness career, but I can't. One comment, and I'm back to Anna Brown, the tall, big-boned, big-nosed girl.

A wave of sadness crashes over me. To my horror, I realize tears have started to prickle in my eyes. I may be the story of the day as a future success, but as a woman, I'm still just the office joke. I mean, I know Mark Swan and Eli Roth wouldn't date me, but does it have to be so bloody funny?

'I'm sorry,' says Sharon, catching my expression. 'Of course I didn't mean it like *that*.'

'I just have something in my eye,' I lie. My phone buzzes and I snatch it up.

'Winning Productions, you're with a Winner,' I say automatically, swallowing the thick lump in my throat.

'Anna,' says Charles.

'Oh, hi,' I say.

'How is the most beautiful girl in London?' Charles asks politely.

I smile, despite the tears that are trickling down my cheeks now. I quickly brush them away, although Sharon's already seen them. How does he know to say exactly the right thing when I need it most? I feel a rush of warmth and gratitude.

'How would I know? I've never met Kate Moss,' I joke.

'That bony thing?' Charles says scornfully. 'Looks like a golf club on a diet.'

That one actually makes me laugh.

'I'll be off then,' says Sharon, relieved. 'Just let me know what he says, OK?'

I nod and turn my attention back to the phone.

'Are we going out tonight?' Charles asks hopefully. Well, we weren't, but why not?

'Sure,' I say. 'I'd love to. I have an afternoon meeting, how about seven thirty?'

'Wonderful. I'll pick you up.'

'No, don't do that,' I say, panicking slightly. I have no wish to expose him to Lily before I have to. Even Janet, who's being so much nicer. She wouldn't be able to help herself, she'd have to ask really embarrassing questions about his sister and her title. 'I'll come to you. What's your address?'

'Forty-eight Eaton Square, flat twelve,' he says.

Eaton Square. Oh yes.

'Right. I'll see you there at seven thirty.'

'Brilliant. I'm really looking forward to it,' he says.

I find I've stopped crying. 'So am I,' I reply.

Forty-seven Park Street is a hotel. A very discreet hotel that looks like a private house, just off Marble Arch tube station, a stone's throw from Hyde Park. I've been here many times, mostly running errands for Kitty to some film star or another we'd put up here. I would leave little gift baskets at the front desk and wait to see if there was a reply, like some eighteenth-century butler. It wasn't very glamorous, unless you count that time I saw Hugh Grant all hung over. He's much fitter in real life, even bleary-eyed.

This will be the first time I've ever got further than the lobby.

'Mark Swan?' I ask nervously.

'Your name, please, madam?'

'Anna Brown,' I say.

He consults a list. 'Will you wait here, madam?' He lifts the phone and speaks discreetly into the receiver.

'Somebody is just coming to get you,' he tells me.

I wait, nervously, but within a minute a gorgeous, rather shocking young thing has appeared on the stairs. Of course. This is exactly the kind of girl who would wind up assisting Mark Swan. She's got the best red hair money can buy, a frighteningly low-cut pair of jeans, a safety pin through her nose, and no bottom or boobs. She's a film school graduate. Probably top of her class, now working for the great man.

'You Anna?' she demands.

'Guilty as charged.'

She doesn't smile. 'My name's Michelle Ross, I'm Mark's assistant. In the future, you want to talk to him, you can call me. OK?'

'OK.'

'Nice to meet you,' she says, shaking my hand briskly. 'Follow me.'

Swan has a suite on the fourth floor. When I get there Trish and Greta are already settled on the sofa, sipping drinks. Greta is on a mineral water and Trish is guzzling a huge gin and tonic. I must have a word with her about drinking in front of stars that have just come out of rehab.

Mark Swan is drinking a Heineken from the bottle. I must have a word with him, too. In some alternative universe where I would actually have the balls to criticize him.

'Hi,' I say.

Swan glances at his watch. 'You're late.'

I look guiltily at mine. It says five past. 'Oh, yeah, sorry, the tube was delayed.'

'Then leave extra time,' Swan says. 'I'm working two projects. I expect you to be exactly on time or early.'

I chuckle at his impression of a stuck-up movie mogul.

'What the fuck is so funny?'

'Oh, nothing,' I say, as it dawns on me he wasn't joking. 'It won't happen again.'

'It better not,' he says curtly.

Michelle sits in the corner, smirking and pretending to take notes.

I sit down on an empty chair, blushing. He was all over me this morning, does he have to be such a martinet? My face is burning like I've just downed eight of Trish's gin and tonics, but it's not the sexy, semi-enjoyable blushes his praise gave me this morning. Now I feel like a turbulent child called out in front of the class. I look reproachfully at Mark, but he just gazes evenly back at me.

Apparently he can't be manipulated.

I should be angry at the rebuke, but instead I'm feeling something odd. I feel . . . respect. Yes, that's what it is. I wasn't sure at first, it's been so long since I've actually met somebody I respect.

'The reason we're here is for me to give you notes,' he says, looking towards Trish and Greta. 'Greta, I'm going to tell you what I want to see in Elsie. Trish, I'm going to tell you what to rewrite. It's good for you to hear each other's notes, because knowing about the story will help Greta shape her performance, and knowing how I'm gonna direct the lead will help Trish with her rewrites. Everybody with me?'

Greta and Trish nod obediently.

'And why am I here?' I ask him.

Swan glances back at me, his bushy eyebrows knitting together, as though he doesn't like to be interrupted. 'You're here to listen.'

'I've some great production ideas,' I offer enthusiastically.

He shrugs. 'Well, keep 'em to yourself.'

'So I'm just going to sit here and say nothing?' I ask, my voice rising slightly.

'I see you've got the gist of it,' he says. 'Now can I proceed?' He's not giving way.

'Yes, of course,' I say. 'Sorry.'

Swan stares at me for a second, his lip twitching. Then he looks back at the other two women.

'We'll start with Greta,' he says. 'Now, my image of Elsie is . . .'

I stare at my yellow legal pad while he goes on about the character of Elsie and how he wants her acted. He refers to other films, other actresses, previous parts Greta's played, and adds some technical detail about shots and lighting. Greta nods every two seconds like one of those bobble-head dolls. I wish Swan wanted me to contribute something. Frustrated, I doodle words on my pad. 'Master shot', 'Complex', 'selfish', 'scared', 'haughty'. Just enough so he won't catch me writing the grown-up equivalent of 'I fancy Jason Connery' over and over like Miss Wilson did once in French.

'Yes, yes, I see,' Greta purrs once he's done. 'Fascinating, yes, I can bring all this to the role, I see her in a whole new way now.'

What a suck up. Everybody is staring at me. Oh hell, did I make that slurping noise out loud? I swallow, conspicuously. 'Mintoe,' I say. 'Stuck in my throat.'

'I don't want too many changes,' Swan says to Trish, after looking hard at me for a couple of seconds. 'But I do think you need to work on your second-act pacing, particularly in the scenes with the wedding planner.'

Trish nods. 'D'you think we should have more scenes where Gemma is fighting with the wedding planner?'

'Maybe,' Swan says. 'The dialogue's a bit wooden, though. I've seen this before. *Father of the Bride* with Steve Martin. This has to be different.'

I bite my lip. No, no, that's all wrong! You don't want to see more of Gemma, the bride. You want to see more of Elsie, her mother. I mean the film's about Elsie, not Gemma.

Swan continues to talk, telling Trish what he wants. This bit of the script does sag, it gets boring in the second act. And Trish keeps giving him her ideas. But I think they're all wrong.

I'd love to say something. I really would. But it's not my script, is it? And anyway, I'm supposed to be learning production, not screenwriting.

'Gemma could insist that if she doesn't get the flowers in blue and white the whole day will be ruined.' Trish clenches her fists and stamps her feet. '*Ruined*!'

I look dubiously at Mark. No, no, that's not it. Yeah, a petulant bride could be funny, but it's out of character. Gemma is sweet and passive while Elsie is the bitch queen mother from hell. If you do jokes that aren't consistent with the character, soon the whole story gets muddy. What would be far better here is if Elsie said . . .

I jot down a couple of funny lines, like doodling. You know, just what I would put if it *were* my script. But I'm careful to angle my body so she can't see what I'm doing. Everybody thinks they can write, don't they?

'That's great,' Swan says, finally. 'I'll look forward to talking to both of you in a couple of days.'

Trish and Greta both get to their feet. Greta is air-kissing the side of Swan's face, or trying to – even on tiptoe she only comes up to his chin – and Trish shakes his hand. I hang back, waiting politely till they've finished. I wonder how long it'll take me to get home and get changed for dinner, what I should wear. Thank God for Charles, I think to myself. It's been such a long day, and now I just feel drained. It was such a triumph to get here, but all I can think of is how frustrating it is not being able to do anything. There's not even any intrigue to report back to Kitty, no plots to go over-budget or anything, no diva-like tantrums from Greta. In fact she seems slavishly obedient to Mark Swan, which goes entirely against her reputation.

Perhaps she just knows she can't get away with it, with him.

I sneak another look at him. He's in his element, relaxed, enjoying himself. His whole body is engaged. When he talks to them he sits forward on his seat, his eyes locked on them; I feel I might as well not be there. Swan focuses totally on the people he's talking to, and his energy and enthusiasm are electric. *Mother of the Bride* is really just an above average, quite funny story that suited the actress Kitty wanted cast. But sitting here, I get the feeling that Mark Swan will make it something more. That he's going to mould it into some kind of comedy classic.

I wish it were me he was looking at like that . . .

Oh, don't be ridiculous, Anna! I cough, clear my throat. Those two are leaving. I try to compose myself. Don't want Mark Swan suspecting an idiotic schoolgirl crush or anything.

'Bye, Anna, see ya,' Trish says.

'Goodbye,' says Greta majestically to me. 'Wonderful to be working with you,' she lies, in a gracious I'm-Hollywood-royalty sort of way.

Swan glances at me, very quickly, a twinkle in his eye, then leaps forwards to hug them both goodbye.

'See you guys later,' he said. 'Michelle'll walk you out.'

It suddenly occurs to me I should be thrilled that these three beauties are leaving the room. It's my lot in life to hang around skinny, petite women, but I realize, with a jolt, that it hadn't actually bothered me this afternoon. I was so focused on what Mark Swan was saying that I didn't care. But now, as I watch Swan hug Trish goodbye, his tall, muscular frame enveloping her tiny, feminine one reality seeps back in. I don't really want to be alone with Swan. I feel vulnerable, exposed. Right now I just want to go home so I can hang around a skinny, petite man instead. Charles may not be my type, but at least he's dedicated to making me feel good, and that makes him pretty rare and special in my book.

Anyway, what do girls like me want with a type? Men aren't exactly falling over themselves to go out with me. It's insane to hold out for my type. My type has to be a male somewhere between puberty and death, doesn't it?

'OK,' I say briskly, in a nice impersonation of Eli Roth. 'That was a very valuable and, er, insightful meeting, and I'll be reporting back to the producers.'

Swan stares at me. I have the urge to look down and see if I've put my T-shirt on backwards or something. I ate a Walnut Whip on the way over here, I hope I haven't got chocolate all round my mouth. Surreptitiously I wipe my lips. But he's still staring. Maybe he needs some ass-kissing, that's what directors like, isn't it? You know the joke – how many directors does it take to change a lightbulb? One. He just holds the lightbulb and the world revolves around him . . .

'We're so delighted you're on board with this project,' I try. 'It's such an honour.'

'God, Anna, do you have to talk like a total wanker?' Swan says.

I blink. 'Excuse me?'

He waves his hand dismissively. 'Don't give me that. Just because you work for some producers doesn't mean you have to talk like a Hollywood executive. Or what you think they sound like.'

'I bet they do sound like that,' I say, stung.

Swan grins. 'Well, actually, a lot of them do. But you shouldn't. I liked you because you weren't like all the others.'

'What others?'

'What, you think you're the only one to come looking for me? I get two or three film students a week.'

'Is it the Steven Spielberg story?' I ask, crestfallen. I had thought it a bold and brilliant stroke.

'Yeah,' Swan says. 'Security guards everywhere curse the day that story started making the rounds.'

'So . . . why did you like me?' I ask.

He shrugs. 'I don't know. You took me on over that comment I made about the Diet Coke.' He grins. 'D'you know, I've stopped smoking? Whenever I try and enjoy a nice quiet smoke I keep seeing your face saying every time I light up I'm getting cancer.'

'Ha ha,' I say, triumphantly. 'Now you know how it feels.'

'Yes, well. You've ruined my cancer sticks for me.'

'You deserved it,' I say.

He inclines his head. 'I did. Yes.'

Then I remember who I'm talking to. I have to be careful, this is Mark Swan, as in, *Mark Swan*, all round film-making god and the only reason I've got this chance. He's got a way about him that makes you forget who he is.

'I expect the other film students don't say stuff like that to you,' I offer tentatively.

'No. Well, for one they recognize me.'

I blush.

'And for two, then they're dumbstruck. Like I was George Best or something. They sometimes ask to sit at my table when I'm in the pub, and if I say yes, they just sit there with their arms crossed, staring.' He gazes unblinkingly at me. 'But they don't say anything.'

'Like you're in the zoo,' I say, delighted.

'Exactly.' He grins, and then there's a pause. I wrench my gaze away. Shouldn't relish his company like this.

I glance at my watch. 'I'll be off, then.'

'Hold on,' Swan says, and his face turns serious again. 'What was that?'

'What?'

'That meeting. You sat there like a pudding. What's the matter, not glad to be here?'

'Oh,' I say, panicked. He's not going to kick me out, is he? 'No. I was glad to be here. So glad. I loved listening to you,' I say earnestly, then I blush. That came out wrong. I can't let him suspect my admiration.

'Then what?'

'I was listening,' I say.

'But you didn't say anything,' he explains, patiently.

'You were talking to Greta and Trish,' I point out.

'That doesn't mean I didn't want to hear from you. I didn't ask you to the meeting just so you could be decorative,' he says.

Decorative! Hah. 'But I'm just representing the production company.'

'Not as far as I'm concerned,' Swan says. 'Look, I didn't have to have anybody from Winning—'

'Red Crest.'

'Whatever. I could just have said no. Do you think they wouldn't have agreed to whatever I wanted?'

I shake my head. I know they would have.

'I value your opinion. That's why I asked you there.' He looks at me. 'When we were discussing those wedding planner scenes you looked like you'd bitten into a lemon, but you still didn't say anything.'

'I didn't think it was my place,' I say, blushing.

He smiles. 'But I'm telling you it is. You have a suggestion?'

'Uhm.' I feel rather stupid and exposed, so I blurt it out. 'I don't think those scenes should be about Gemma at all, it's not her story, it should be all about Elsie, and this is where she's really sabotaging the wedding so she could, you know, she could make sure to ask for things which will be impossible. She wants it to be chaos, or for the planner to quit.'

'Hmmm,' he says. 'Go on.'

'But Elsie needs it to sound reasonable. And you want it to be funny. So she needs to make all the outrageous demands with a very uptight little smile. Sounding really . . . saccharine,' I suggest.

He's silent for a beat. I daren't look up at him. I feel my heart thud-thudding in my chest.

'Can I see that?'

He reaches for the yellow pad on which I've doodled my dialogue. Flushing, I move it away.

'Oh no, that's nothing,' I protest.

'Hand it over,' he says, inexorably, swiping it. I shuffle my feet together while his eyes flicker over it, reading everything I've written. He seems taken aback.

'Did you come up with this?'

'What?' I ask, playing for time.

'These lines.'

'Yes, but I was listening, I swear.'

'Anna. These lines aren't bad. Not at all.' He casts his eye over them again, reading more carefully. 'That's closer to what I was looking for. The humour, the tone – the whole bit.'

'Oh.' I don't know what to say. I feel a wash of pleasure all over.

'You've got a great ear for dialogue,' he says. 'Ever thought about writing?'

'Who? Me?'

'There's nobody else here,' he points out, reasonably.

'I couldn't be a writer. I'm just a producer. Apprentice producer,' I say hastily.

'You're a lot more impressive as a writer,' Swan says. 'Anyway. Think about it.'

'You really think I could be a writer?' I ask, delighted.

'What, are you deaf? How many times do I have to say it?'

'Well. Thanks,' I say. 'Thank you *very much*.'

'You're welcome, sweetheart,' he says.

I turn away, spell broken. I had been gazing at him, so full of happiness and gratitude, and of course he's so gorgeous and he's being so nice. But then he said *sweetheart*, and I don't like the teasing. I swallow hard against the sudden lump in my throat. Which is totally ridiculous, of course.

'What happened to that smile?' he asks, grinning. 'Have you got plans for the evening? Maybe you'd like to come out for a drink? Michelle and I usually get a pint about now.'

Oh, absolutely. Love to, so I can sit there while much prettier women come up and fling themselves at him, and skinny Michelle sneers at me. No, no. Mark Swan is far too dangerous to hang out with on a social level.

'I've got plans,' I say, not looking at him.

'Cancel them,' Swan suggests.

'I've got a date,' I say, with a sudden surge of gratitude to Charles. Yes, I do have a date. Once again, Charles saves me without even knowing it. I glance at Swan. I'm waiting for the look of disbelief or mockery I'm sure will cross his face, only it doesn't.

'Oh. Who?'

'His name's Charles Dawson,' I say, as brightly as I can manage.

Now Swan does frown slightly. 'Charles Dawson, the one who tipped you off to the script? Trish's employer's brother?'

'How do you know that?'

'I called Trish this afternoon. I make it a rule to get to know the people I work with. It helps me get the best out of them if I can see inside their heads,' he says.

'Yes,' I say.

'Humph,' he says, non-committally.

'What, you think it'll interfere with the movie?' I ask, nervously. 'Him being Lady Cartwright's brother?'

'Oh no,' Swan says. 'Trish quit her job, anyway.'

'She did?' I ask. I feel guilty. I realize I really haven't taken the time to call Trish, bond with her, since I met up with her over the script. I've just been so busy. Mark Swan has known her two seconds and already he's her new best friend. 'Isn't that a bit risky for her? Nobody knows how the film will do, and she's only getting scale.'

'Her film will be a huge hit,' says Swan. 'I'm involved.'

'You're so modest.' I smile.

He shrugs. 'We both know it's the truth.' And of course I do. 'Anyway, you'll find that not risking anything is what you need to be afraid of.'

I look at my watch again. 'If it's OK with you, I really have to go,' I say. 'I have to get ready.'

'No problem,' Swan says. 'I'm sure Charles doesn't like to be kept waiting either. See you tomorrow. Ten o'clock, sharp.'

'I'll be there,' I say, gratefully. 'Thank you for the chance.'

He doesn't look at me, just nods. He's already gone back to his notes. I walk out of the room, trying not to look back at him.

I don't want to feel like this. So strongly. I mean, just because he's masculine and powerful but still funny and nice, that's no reason to fancy a man, is it?

OK, maybe it is.

But I can never *have* Mark Swan. He's out of my league. He's the Premiership and I'm five-a-side on Sundays.

Now that's an excellent reason *not* to fancy him. No point. And the fact that he said lovely things about me maybe being able to write is to do with my career.

I must think about my career.

I must not get a crush on a gorgeous, famous, powerful director.

Note to self. Do not fall for Mark Swan.

OK. I'm not going to think about him any more tonight.

But it's a real effort to shake the feeling as I emerge onto the street, already busy with harried office workers heading home. I mustn't get carried away with Mark Swan. I can't get that close to him. He's being so nice, but he's still in charge, and if I go too far and annoy him? Then what? I have to watch my step. Certainly not show stupid, unprofessional feelings like jealousy when he's close to Trish or other pretty women. I mean, he works with Michelle every day.

Women! I forgot to ask him about his girlfriend, for Sharon. Oh well, it'll have to wait. But there has to be some lucky cow, doesn't there? Fame and fortune and good looks, they don't go begging for company. Sharon doesn't think he's good-looking; she must be mad, either that or she likes pretty boys. Swan is a vast hulking mass of testosterone. He's completely attractive. Absolutely, totally attractive . . .

Anna. Stop it. You've got Charles, and that's a miracle.

I think about Charles and the lovely flowers until I reach the tube station and shove my way down into the mass of sweaty, tired humanity. At least I'm only two stops away. It's six o'clock. I should be able to do my hair and get Janet's advice on damage limitation with my make up and clothes. I'd like to look nice for Charles this evening, just for once. I'd like to feel pretty. Although I know that's not going to happen, so I'll settle for just feeling normal.

'Come in, come in,' Charles says, flinging open his door. 'So lovely to see you.'

He's beaming from ear to ear. I wait for a compliment, but none comes, which is a bit disappointing, considering how long I spent on this look. First I washed my hair and Frizz Ease'd it and then blew-dry it to within an inch of its life (what did we all do before Frizz Ease?). Janet helped me with my make-up; she used my neutrals, but she made my eyes really pop. I suppose it does distract from my nose a tiny bit and, more to the point, it at least makes me look put together, sort of elegant. My dress is my nice navy shift with pearls I wore to the dinner at Vanna's, the first time we met. Janet shakes her head over it.

'Makes your arse look huge,' she says, flatteringly. 'Which it isn't. And shows off your stomach,' she adds, grabbing a good three inches of love handle to illustrate her point. 'Meanwhile it flattens your tits and covers your legs.'

'So what?' I demand.

'You should show them off,' Janet says, her beautiful, olive-skinned face serious. 'And you've got no waist in that dress.'

'I've got no waist anyway.'

'We could create one,' Janet says judiciously.

'Who are you, God?' I scoff. 'Look, this is my body. Best thing for it is to wear something conservative.'

'Dowdy.'

'Classic,' I insist.

'Boring.'

'Neutral?'

'Vile,' Janet says pityingly. 'Anna, I am a fashion expert, you know. You're not doing yourself any favours. I'm taking you to Harvey Nicks this Friday to get you some nice clothes, tart you up a bit for the dance.'

'I can't be tarted up,' I say. 'I'm about as tarty as Ann Widdecombe.'

'Everybody can be improved,' says Janet. 'I'd *love* to get my hands on the Queen,' she adds, musingly. 'She's got great potential.'

I blink. 'You're mental.'

'Just say yes,' Janet pleads. I look at my watch; I don't have time for this.

'Fine, whatever. Just pass me my shoes.'

'If I must,' Janet sniffs, handing over my flat Hobbs pair with the white stitching, which I thought was a great match for navy and pearls. 'They make me want to puke.'

Thus encouraged I set off for Eaton Square in a taxi, even though it was twelve quid. I can't take another tube ride, not tonight. Plus, I'm trying not to be intimidated by the fabled flat. It's only a bloody flat at the end of the day, and so what if he's got loads of money? I've got a raise. I'm making thirty grand now.

And so here I am. The building is gorgeous. No porter or common lobby with glossy black leather sofas and grass growing in square white pots, none of the new London, Met Bar-style wealth on display. No, the building is old, with the paint slightly peeling, flagstones in the lobby, wide

145

proportions, and one of those old, very attractive and very terrifying elevators where you have to slide the iron cage shut after yourself before the damn thing will work.

To me it says rich more than any piece of abstract sculpture or haughty doorman. And not just rich, but mega-rich. Old, old money that doesn't need to shout about it.

No wonder Charles has been plagued by gold-diggers. I feel for him. Lots of men in his position wouldn't object. They'd just pick the sexiest chick and bed down with her, and trade her in ten years later for the next model.

'So this is home,' I say, stepping inside.

'Yes.' He looks around, half embarrassed, half proud. It's pretty much what I'd expected, but that doesn't stop it being insanely gorgeous: red damask wallpaper, prints of hunting scenes and Victorian cartoons, the odd oil here and there, some antiques crammed onto his shelves in a haphazard manner. A working fireplace with ashes and a brass guard. A mantelpiece crammed with stiff, creamy white invitations, most of them bearing crests. Books lining the walls, leather-bound and dusty looking. Sisal matting on the floor, the occasional threadbare Persian rug, deep, worn burgundy leather armchairs with little brass studs around their edges. Everything is upper class, lived in, valuable. The sole modern touches are the electronics. He has a huge flat-screen TV and a sexy-looking ultra-slim laptop on his desk, which is strewn with bits of paper, covered in red ink and underlining. With a pang I recognize it. It's part of his novel.

'I'm sorry it's so messy,' he says. 'The char doesn't come till tomorrow.'

'You should see my room,' I lie. I'd hate him to see my room. I am trying to bolster all my socialist, liberal feelings and tell myself that Charles represents the enemy. He's a parasite feeding off the backs of the workers and come the revolution he'll be first against the wall. So why should I care that I camp out in a room the size of a large closet, in a flat above a shop?

Only I do care and I don't resent Charles. He can't help being rich. Or having a gorgeous flat and ancestors with good taste. I wonder what it would be like to live somewhere like this, all the time.

'So this is where you work?' I say, and immediately wish I hadn't. What is wrong with me today?

'Yes, on my book.' He looks over at me. 'Any word on that yet?'

'I'm still reading it. I got a bit distracted by the *Mother of the Bride* thing, and I want to give it all the consideration it's due. It's very complex,' I say, truthfully enough. I was confused after two paragraphs.

'Ah, yes,' says Charles, stoically. 'It is complex, multi-layered. I can see you need time to fully appreciate it.'

'Yes,' I agree, smiling weakly. 'Time. Anyway, it's a lovely flat.'

Charles eyes me nervously. 'I do have a guest room,' he says. 'If you want to stay over?'

'Absolutely not. I hardly know you. I'm not about to move in,' I say indignantly, and then feel bad, because I know that soon I'll have to dump him. I ought to have done it before now, really. I don't fancy him, even though I like him now. I might tell myself that I'm going out with the poor sod to give him a chance, but in reality I just like how he treats me, that he tries to make me feel good about myself. I've had a bad day and I know that Charles will say nice hings over dinner, lie manfully, and call me pretty. He'd kiss my hand if I let him.

It's emotional dressing-up. I'm trying on Charles for a pretend relationship, the same way I might try on a tutu and pretend to be a ballerina, back when I was a little girl and thought of myself as feminine and fluffy like everyone else. Dating Charles, you can imagine exactly what ordinary girls get. Flowers, dinner, compliments, doors held open for them. I've even fantasized about bumping into someone from the office while at dinner with him. Sharon or Mike Watson or John . . .

It'd never happen though. Sharon and John can't afford the kinds of places Charles takes me. Frankly, I doubt Mike can either. But they all saw the roses.

I wouldn't mind if Charles did that again, with the flowers. Or maybe some chocolates or something. He could call me, too. Claire would put it through and then she'd tell everybody. She's as good as a full page ad in the *Standard*.

Stop that, I tell myself with a guilty start.

'Sorry,' Charles says meekly. 'I'm so used to girls wanting that.'

'Not me,' I say breezily. 'I'm a career girl.'

'I know,' he says admiringly. 'I think it's wonderful. Getting promoted and everything,' and I feel even worse. I mustn't use him, like all the other girls. Even if they were using him for money and I'm using him for compliments, it's really the same thing, isn't it? I should say something. Let him know it's not going to work.

'I've been boasting about you to all my friends,' he says. 'They can't wait to meet you.'

I start. 'Meet me? I thought it was just us for dinner.'

'Oh, it is,' he says. 'I mean at Chester House, this weekend.'

'Oh yes,' I say. 'Right.'

Well, I'm stuck now. I can't break up with him. Not if he's been telling all his friends. I look at him and feel a pang of protectiveness. I'd never expose Charles to some of the pain I've been through.

'I'm looking forward to meeting them this weekend,' I tell him.

'Are you?' His face lights up. 'That's marvellous. 'I know you'll *love* them.'

I don't think I'm going to love some clone of Rupert's called Binky or Crispin, but I nod and smile as though it was the best thing in the world and I'm looking forward to it as much as two weeks' free holiday in Mustique.

Charles smiles up at me. I smile back. It's so nice to see him happy. It's like watering a plant. I always think that the plant looks instantly better if you give it water when it's dying and dried-out. I like making him happy, it's the least I can do for all the trouble he takes for me.

'Where are we going to dinner?' I ask.

'Well, here,' he says.

'Here?' I glance around the flat. 'You cooked something?'

'Not exactly,' he admits. 'Bit of a dunce in the old egg-boiling department.'

My eyes narrow. 'You aren't expecting me to cook, are you?' My idea of cuisine is a Marmite sandwich, and besides, I am definitely not the Oxo Mum type. I hope Charles hasn't got some idea that, not being sexy, I'm going to come into his life and mother him.

'Absolutely not,' he hastens to reassure me. 'I called the caterers. I thought it'd be more romantic to eat here. More intimate.'

I smile but my skin's crawling. I do hope he doesn't mean intimate in the Eli Roth sense. I look around, trying to guess where the bedroom is. I'm just not ready to have sex with Charles and I don't know if I ever will be. Sex isn't something I enjoy; with Brian I had to get absolutely hammered and then would lie back, faking it desperately in the hope he'd hurry up and finish and I could get into the shower. It was boring, it was uncomfortable – trying to turn my head away from his breath whilst pretending to be in the throes of passion was tricky – and it felt a bit dirty, it felt wrong. Seeing what happened with Brian, I know why.

I don't know if I can face going through all that with Charles.

'Got anything to drink?'

'Of course,' he says, beaming, springing to his feet in his haste to be of service. 'Sherry? G and T? Wine? Scotch? Irish?'

Of course, Charles is a *much* nicer person. 'Gin and tonic would be lovely, thanks,' I say. Charles leaps into action, hurrying to the kitchen (Smeg appliances and Sub-Zero freezer meets terracotta tiles and ancient wooden counter top) and returning with a beautiful Waterford crystal tumbler filled with ice, slices of lime and a subtly fizzing drink. I take a good hit and immediately start to relax. It's been a long day, and the alcohol unknots my muscles like a massage on the back of the shoulders. That, plus I realize I won't actually have to endure another restaurant, won't have to see the sidelong glances and hear the hushed voices of people laughing at us. What was it Sharon said? Little and Large? Me with my flats and my please ignore clothes, Charles with his dandified suits, neatly trimmed goatee and stacked heels.

Maybe the flat is better after all. I refuse to think about what happens after dinner. I take another big slug of G&T.

'You've finished, let me get you another,' he says.

I shake my head. 'Don't want to get too tipsy before we start eating.'

'No,' he says, admiringly. 'Very proper. Shall we go in?'

He's so formal. I wonder if my dress is smart enough. Charles might be one of those weird Old Etonians who still insist on dressing for dinner, for all I know.

'Where's the butler? No "Dinner is served"?'

'I don't have one,' Charles responds, crestfallen. 'I can get one if you want.'

'Don't be silly, I was joking,' I say, aghast. 'Nobody has servants these days.'

He looks a bit sheepish.

'You have servants?'

'Just a couple,' he admits. 'At Chester House. But you know, it's a bare bones staff,' he excuses himself. 'Just a butler and a couple of maids. And a gardener. And a cook.'

'Oh, well, that's all right then,' I say.

'And my valet,' he admits.

'Charles, that must cost you a fortune.'

'It's a necessity when you live out of town,' he says.

No it isn't. 'Sure, I understand.'

'No, look, Anna,' he says, reading my expression and pleading his case. 'Don't think I'm just some rich egomaniac with servants.'

That's pretty much exactly what I was thinking, so I start guiltily. 'Hey, it's a free country, right? You can do what you like with your money.'

'They all worked for my father,' he says. 'Except the valet. But he found it hard to get other work and my butler recommended him. And the other staff are too old to just send them packing.'

I soften. 'So you continue to employ the old family retainers?'

'Pretty much,' Charles agrees. I like him for it, I like him immensely. I chide myself for being so bloody judgemental.

'What was the valet doing before?'

'Fifteen to twenty for GBH at Strangeways,' Charles says. 'Couldn't get a job after that. But Wilkins knew him, and he's an absolute genius at picking out ties.'

I laugh. 'You know what, Charles, I really like you.'

He glows. 'I like you, too.' He gestures to the dining room. 'Shall we?'

Fortified by my gin and tonic and Charles's dodgy valet, I manage to sit down in his dining room without feeling overwhelmed. I give myself a gold star for that, because, if anything, it's even lovelier than the rest of the place: oak panel walls, a gorgeous table to match, and chairs with his family crest carved on the back of each one. They're a little narrow, probably because they're obviously hundreds of years old. The china is plain white with silver piping round the edges, the cutlery is silver and antique, and there are some beautiful yellow and white roses arranged in low silver bowls dotted around the room. Candles everywhere, and a magnum of champagne chilling in an ice bucket. Krug. Very nice.

'I hope you like it,' Charles says, nervously.

'It's beautiful,' I say, smiling to reassure him.

'I thought we could start with caviar. Do you like caviar?' he asks, anxiously. 'Some people hate it.'

'I've never tried it, but I'm sure it's delicious.'

'And then there's roast guineafowl with stuffing and roast baby parsnips.'

Bloody hell, that sounds delicious. My mouth is watering in a very unladylike manner.

'And there's green tea sorbet, and then pudding is a bitter chocolate tart with ginger ice cream, and there's some petits-fours with coffee, or you could have cheese and fruit, I had them make up a plate in case you aren't the pudding type.'

I look ruefully down at my ample tummy. 'I am,' I inform him solemnly, 'the pudding type.'

'Champagne, and there's a very nice brandy afterwards, or you could have port. I have some other digestifs . . .'

'Charles,' I say, smiling with genuine warmth, 'this is absolutely brilliant, honestly. I think it's the nicest thing anyone's ever done for me.' And I kiss him on the cheek.

'Well,' he says, blushing scarlet. 'Well.'

He's absolutely at a loss for words, so I jump in. 'Let's sit down and eat,' I suggest. 'I really want to try some caviar.'

He offers me his arm. He actually offers me his arm, but I know how to rise to the moment. I take it, just like Elizabeth Bennett in *Pride and Prejudice* or any of the leading ladies in those old Sunday afternoon shows, and let him escort me the two feet into the dining room, where he sits at the head of the table and I sit in the chair to his right. I think briefly of Lily and Janet and their strings of glamorous escorts, the tribute life usually pays to beautiful women. Is this what it's like for them all the time, is this what good-looking girls expect on dates? Men who can't do enough for them, boyfriends desperately eager to please?

It has never happened to me before and I don't know how to handle it. Ever since the St John's dance, during the rare times in my life when I've actually been attached to a man, I've been the keener person, I've been the one who needed the approval. I see the same look in Charles's eyes that I've seen in my own.

And it gives me an incredible feeling of power, because I can put things right for him the way they weren't put right for me. I can make Charles feel good. I can compliment him, say nice things about him to all his friends, accept what he offers me with enthusiasm. I can protect him from all the snubs and bitter insults and jokes that were levied at me.

'I can't believe you've never had any caviar,' Charles says, heaping a little bone spoon with the glistening black pearls and handing it to me. 'You could put lemon juice on them or mix them up with chopped egg but I think the first time you taste them you have to do it straight.'

I try them. They look slimy but they're actually delicious. I tell him so.

'Don't look so surprised,' Charles says. 'Champagne?' he asks, expertly popping the cork. I pass over the cut-glass flute by my plate.

'Why the hell not,' I say.

I have a damn good time. Can't actually remember the last time I had a better one. The food is insanely delicious; at first I am ginger with every bite, wondering what it all cost, and then I relax (could also be the bubbly), and just enjoy it. Enjoy being pampered. Charles isn't the world's wittiest conversationalist but he's not absolutely terrible, either. He talks a good deal about Vanna and Rupert because I know them, to make me feel comfortable, and he asks loads of questions about my work, about the script, about Kitty and Eli and my job, and about Mark. He's either genuinely interested or a very good actor.

Either way, it's nice.

'So you just introduced yourself,' he says, shaking his head. 'Amazing. And just gave him the script. Such a brave thing to do.'

'Seemed like I had to,' I mutter.

'No, it was very brave,' he pronounces. 'And you got your just reward, which Kitty finds very annoying, but I wouldn't let her get to you.'

I toast him back, well aware I am smiling foolishly, but it's just so nice to have someone in your corner.

'This has been such a lovely evening,' I tell him.

'Well, it's not over yet,' he says. 'Port? Cognac? Brandy? Something—'

'I can't.' I push the last delicious sliver of bitter chocolate and ginger tart around my plate. 'I've had way too much. Got to go to work tomorrow. Can you imagine what Swan would say if I were late again?'

'At least have coffee,' he says.

'Sure. Love some coffee.'

'And petits-fours.'

'Mmm,' I say. I should say no but I love petits-fours. Such

a fancy name for mini cakes. That's just what they are, isn't it? Tiny little cakes. Some part of me wonders why Charles chose such a heavy meal. Didn't he say to that bony cow back at Vanna's dinner party how much he admired self-discipline? And he made mean cracks about my weight. Yet now we're dating and he's plying me with chocolate tarts and roast potatoes . . .

Well, he's stopped being a bastard, so maybe this is all part of the deal. Accepting people for who they are.

But something's still niggling at me, because there's something wrong with this picture, something subtly wrong, but still wrong.

'I think I'll just have the coffee.'

'But you must try at least one,' he says, shoving the silver tray in my direction. All my favourites: brandy snaps, tiny profiteroles, little funny squares of lemon and white chocolate, praline wafers . . .

'I'm supposed to be watching my weight,' I say. He looks surprised.

'You are? But what's the point?'

I'm not sure I find that totally flattering. 'What do you mean?'

'You're not *hugely* fat,' he says, 'so why not just be who you are?'

'I thought you liked self-discipline,' I remind him.

'Oh. That. Well, that's fine for Priscilla,' he says. 'But you're Anna and I don't need to change anything about you.'

That response should strike me as truly romantic, and I'm not quite sure why it doesn't.

'Maybe just one,' I say, reaching for a brandy snap, because I've had quite enough of analysing things today. 'Thanks,' and I smile at him again.

Charles drops me off around eleven, refusing to let me get a taxi home – he says he has to 'escort' me. I spend the journey nervously wondering when he's going to invite himself up for 'coffee' in our comparative hovel, and wondering what I should do. But he doesn't. When he pulls up as close to our building as he can manage he says, 'May

I?' and then, when I nod, he kisses me chastely on the cheek.

'I had a wonderful time,' I tell him, giddy with relief. 'See you Saturday, up at Chester House, OK?'

'Absolutely,' he says warmly. 'Can't wait.' He grabs my large hand in his small one and kisses it.

'What's that for?'

'Oh, nothing,' he says. 'I just think you're the one for me.'

I smile back at him because I'm not quite sure what to say, and maybe he senses it, because he shifts gears, calls out, 'See you Saturday!' cheerfully, and drives off down Tottenham Court Road.

I walk in through the narrow corridor next to the unlit feminist bookshop and wonder what on earth that was all about. He sounded keen, didn't he? Really keen.

In fact, he sounded as if he wanted to marry me.

7

Janet and Lily are lying on the floor when I come in. There's a huge bottle of champagne empty between them; Cristal, so it must be one of Lily's many admirers with more money than taste. There are magazines everywhere, pages ripped out of them and scattered across the rug.

'What's this? Rebelling against the tyranny of the perfect body?' I ask. 'Seen the light? No longer willing to force impossible physical standards on the typical British size sixteen woman?'

'Oh, don't be pathetic, Anna,' Lily snaps. She raises her head, a cloud of expensively dyed hair in shades of platinum and butter spilling around her bony shoulders. 'I'm just going through some recent shoots with Janet. Trying to show her where she's going wrong.'

'Janet's not going wrong,' I protest.

'I am,' says Janet, gloomily. 'My bookings are really drying up.'

'She's getting calls for *catalogue* work,' Lily says scornfully.

'Whatever pays the bills, right?' I suggest.

'Wrong,' says Lily. 'If you get pigeonholed into that kind of thing you're done for. Photographers know it, agents know it, designers know it. You think that Versace or Dolce are going to want some girl strutting down their catwalk when her last job was for *UKfashions!*' She clicks her fingers. 'You're over. Done. Toast.'

'I turned them down,' Janet says nervously.

'Darling, of course you did,' says Lily. 'I should think so too.'

'Oh, this is so stupid,' I burst out. 'Janet, you're not going to make it onto the Paris catwalks anyway and I bet a few catalogue jobs wouldn't stop you finding work in the glossies. I saw Helena Christensen in a catalogue once.'

'I bet you didn't,' Lily says.

'I did.'

'Which one?'

I can't remember. 'BHS,' I lie.

'Well, even if you did, Helena's been done for years,' Lily says witheringly. I would be prepared to bet Helena Christensen makes more in a month than Lily does in a year, but I keep my mouth shut.

'Maybe Anna has a point,' Janet says timidly.

'Oh really?' says Lily, eyes narrowing. 'She does, does she? You're going to take modelling advice from someone who looks like *that*?' She gestures at me, standing there in my neat navy dress and fake pearls. Hey! This is one of my better outfits, actually.

'I could make Anna look all right,' says Janet, protectively.

'No you couldn't. Don't be ridiculous,' Lily says. 'No offence, Anna.'

I swallow. This is a bit beyond a joke. 'But it is offensive.'

'Excuse me?' she demands, looking up at me. 'I *said*, no offence.'

'I know you say that. Usually after you've said something really horrible and mean-spirited,' I reply. I know I've gone bright red, but I don't care.

'Actually you do do that a lot,' Janet mutters.

'I don't.'

'You just said, "Janet, you're twenty-eight, you have to work like a slave to get anybody to book you." And then you said, "No offence."'

Lily flicks her golden hair. 'You two are so thin-skinned. For heaven's sake, I'm only saying, Janet, how many covers has *Anna* booked recently? Who's the professional here, me or her?'

'You are,' says Janet, meekly. She looks at me apologetically. 'Sorry about what I said about making you over, you look great how you are.'

'That's OK,' I say, because Janet just seems so miserable. 'You're going to get me some new clothes before the dance at Chester House, right?'

'Right,' she says, perking up. 'I can do wonders for you. You just wait and see.'

'Janet,' Lily says severely. 'Am I wasting my time here?'

'No. No. I'm listening,' says Janet, placatingly.

'Now take Shalom's look in this shoot,' says Lily with a long-suffering air.

'I'm off to bed,' I say. Nobody notices, so I go to my closet cum bedroom, peel off my clothes, and I'm asleep in less than five minutes.

I'm getting ready to head out of the door when my mobile trills.

'Anna, where the hell are you?'

It's Kitty. I jump nervously.

'I was just heading down to Swan Lake.' That's his company.

'I don't think so,' Kitty hisses ominously. 'You must come into the office every morning before you go gallivanting off with Mr Swan. You need to get your instructions and make your report.'

Wow. I sound like 007 or something.

'Um, OK,' I say, placatingly. I can't get Kitty angry. 'I'll be right there.'

'I should *think* so,' she snaps, hanging up on me. Oh bugger. I race down the stairs, doing little sums in my head. What's the quickest way to the office? I look around for a taxi, then think better of it, the traffic's crawling. And there was a bomb scare at Covent Garden so the tube's out. I feel the panic rising. First I have to go and kiss Kitty's arse, and then get to Swan's, but that means I'll be late to him. Twice.

There's nothing else for it. I start to run. I make it in fifteen minutes, red-faced and sweating, and race up the stairs to our floor, where Claire's waiting for me, resplendent in a tiny red leather mini skirt and stack mules.

'Eli Roth around today?' I ask, nodding at the skirt, which is more of a belt with pretensions.

She shakes her head, disappointedly. 'But you better get in there,' she whispers. 'She's on the warpath.'

Great. I square my shoulders and knock on the door of Kitty's office.

'Come,' she barks.

I enter. Kitty's sitting behind her desk, stilettos tapping impatiently. She's wearing a scarlet Dolce & Gabbana suit with huge black buttons with the logo on them, enormous diamond studs, and a thick gold cuff bracelet. Obviously she's in a really bad mood. The more *Dynasty* the outfit, the angrier she is.

'I don't know why you would think you can go directly to Mark,' she says.

'But he doesn't like me to be late,' I say tremulously.

'Give me your report from the last meeting,' she snaps, ignoring this.

'Um, OK. Well, it was a story meeting. He gave Trish some notes and discussed the part with Greta.'

'And what else?' she demands, eyes narrowing.

'Nothing.'

'Are you holding out on me?'

'No.'

'Why does he *want* you,' she asks, bitterly. 'You! Of all people!'

'I think he thinks somebody senior would be wasted in those meetings, all I do is take notes,' I say, tactfully.

She nods. 'Yes, possibly. Well, anyway, that isn't enough work. I want you to continue with your reading. And I also want you to be Greta's assistant.'

I blink. 'What?'

'I brought Greta to this project,' Kitty says importantly. 'I cultivated Greta.' This is true. 'And I want Greta to be made to feel *special* by Red Crest Productions and especially by Kitty Simpson.'

I want to argue but I've got no time. It's already ten thirty. I shudder to think what Mark Swan will say when I show up.

'OK,' I say desperately, 'sure. Whatever you want. I'll be Greta's assistant.' This is just perfect, of course. Greta has a reputation as one of the most spoiled actresses in Hollywood. She may not dare to act up in front of Swan but I'm sure that'll be no problem with me.

'I've already told her that I've instructed you to attend to her every need.'

This is not my job, but I don't dare say so. 'OK.' Ten thirty-two.

'I better not hear any complaints,' Kitty says viciously. 'And each day as soon as Mark is done with you I want you back here in the office. You work for us,' she adds again.

'Of course,' I say. 'Um – thanks, Kitty.'

Why? Why am I thanking her for making me be this age-ing diva's dogsbody? I'm supposed to be a reader, up for promotion.

'You may go,' Kitty says graciously. I rush out of the door again, pulling out my mobile, tapping in the number as soon as I reach the street.

'Swan Lake. This is Michelle.'

'Oh, Michelle, hi, this is Anna Brown.'

'From the producers,' says Michelle. There's an ominous touch of triumph in her tone. 'He says you needn't bother to come in today.'

I look at my watch. 'But I was unavoidably delayed.'

'Whatever,' she says. 'He said to say if you called not to bother to come in.'

'I'm coming anyway,' I say.

'Nobody gets in this building unless Mark's expecting you,' she says meanly. 'So just accept it and maybe he'll let you back next meeting. Or maybe not.'

'I've got a good explanation.'

'They all say that,' Michelle says curtly. 'I have to go, goodbye now.' And she hangs up on me.

I'm only a few minutes away from Swan Lake anyhow, so I keep walking. What else am I going to do? Although I don't know what I'm going to say, how I'm going to explain to him . . . OK, here we are. Dean Street. And there it is.

161

And there they are. Swan. Greta. Trish, who sees me and shoots me a sympathetic look. A couple of other people I don't recognize, a pneumatic blonde and an andoyne young media type with a goatee. They're all piling into a couple of taxis. I rush forward.

'Mark,' I say. 'Sorry I'm late, but—'

He gives Trish a little push in the small of her back.

'I don't want you around today,' he says flatly. 'If you can't respect other people's time, Anna, then I've no use for you.'

Why do people only use other people's first names when they're cross with them or it's bad news? First Kitty, now him.

'But I've got a good excuse.'

'I'm not interested in excuses,' he says. 'I take this stuff seriously and I expect my colleagues to as well.'

He starts to climb into the taxi. I grab his arm.

'You've got to listen to me!' I protest.

'No I don't,' he says. 'See you.'

'Oh fine,' I snap, losing it. 'That's just perfect. You're absolutely right, don't bother giving Anna ten seconds to see if she's got a reasonable explanation. Ohhh no, just abuse your power totally and make other people feel terrible when it isn't even their fault.'

And I turn on my heel and walk away, heart pounding. I want to cry. That's it, then. That is the sum total of my Mark Swan adventure. Because he'll call Kitty and she'll be only too delighted to sack me so she can go and hang out with him herself.

'Anna.'

I turn. It's Swan.

'Look, I'm sorry I said that, OK?' I tell him tearfully. 'I've – I've had a really bad morning. Just please don't get me sacked because I need the money for rent and things.'

His face softens. 'Tell me what happened.'

'I was on my way to your offices and Kitty rang me,' I said, 'and she made me go into work and I had to leg it because of the tube, you know it was shut down, and she talked to me for a bit and then I had to run to your place . . . I would have got here in ages of time.'

Swan just stands there for a second. Then he holds open the taxi door.

'I'm sorry,' he says. 'You were right. I should have given you a chance to explain.'

I open my bag and fish out a tissue and blow my nose loudly, which isn't very sophisticated but unfortunately necessary.

'Please hop in,' Swan says. 'We're going over to the production designers to do some storyboarding, talk about sets.'

Gratefully I run to the taxi and clamber in the back. Trish and Greta scoot over, Greta scowling, while Swan gets in the front seat. He turns round, looks at me.

'We'll have a talk later. OK?'

'OK,' I mumble.

Greta looks disapprovingly at my red eyes. 'Stop making a scene,' she stage whispers. 'We don't need the Maestro distracted.'

Swan leads everybody into the production design offices, a nondescript building off Oxford Street.

'You guys go on up, I'll be right with you,' he says.

'You follow me, Anna,' Greta says majestically. 'I have some requirements.'

'Of course,' I say, obediently.

'Actually, I need to speak to Anna a second,' Swan says. 'She'll be right with you.'

Greta nods. 'Whatever you say,' she says, adoringly, but she narrows her eyes at me. I swallow a sigh. Greta is obviously Kitty's spiritual twin and this isn't going to be fun.

'Look,' Swan says, when the lift doors have hissed shut on the rest of them. 'I'm sorry about before. I acted like a total idiot.'

'That's OK,' I say. I'm not really used to important people admitting they were wrong. Or saying sorry.

'I thought, you know, we had a good talk last night. And,' he says, passing his hand over his hair, 'for some reason it really, really angered me when you were late again. I don't know why I took it like that. So personally.'

'That's all right.'

'I just don't think of you the same way I do the rest of them,' he says. 'I don't know why. You're different. Not so plastic. I think that's why I got so pissed off. I didn't want you to take it for granted.'

'My boss wants me to report to her every morning before I come to you,' I tell him.

He shakes his head. 'Unacceptable. I'll tell her, don't worry.'

I smile gratefully at him. 'Thanks.'

'Forgive me?' he asks.

I nod. I can't help smiling at him. He's so nice.

'OK. I promise not to be a slavedriver any more,' he says. 'Well, not to you, anyway. Still have to keep the actors in line.' And he winks at me.

Oh my goodness, he is *so* attractive. I look away.

'Well, I'd better be going upstairs,' I say, in a rather high-pitched voice.

'I'm just going to call my office, you can tell them I'll be right up.'

'You got it,' I say briskly.

Greta pats the empty seat at the table next to her as soon as I arrive.

'Sit here, Anna dear,' she purrs. 'Kitty's told me all about you.'

I can imagine. I grit my teeth. Kitty won't like it when Swan calls and tells her I have to go to him first, so I have to be very careful with Greta.

'If there's any way I can assist you,' I say humbly, 'any way at all . . . Kitty is very keen that you should have anything you want.'

Greta's eyes glint. 'Of course, dear. Got a pen?'

I fish around in my bag for one. Miraculously, I also have a yellow Post-It pad in there.

'What would you like?'

'First of all, I can't stand that filthy swill they serve for coffee in these places,' she says, 'so I want you to run out and find me somewhere that serves a proper cappuccino, and bring it back, of course not in a paper cup, I only drink from china. It's about self-respect.'

'Mmm,' I agree.

'And then my dry-cleaning has to be picked up daily. I'll want fresh flowers delivered to wherever we're working, and I need you to pop off to Harrods and pick up my Creme de la Mer order.'

I nod, writing everything down furiously.

'You can start with the coffee,' Greta says.

'OK,' I whisper.

Mark Swan walks into the room just as I get up. He raises his eyebrow.

'I'll be right back,' I say hastily. 'I'm just getting Greta's coffee. I'm her assistant,' I add, to his look. 'My boss wants to make sure she's taken care of.'

'Does she, by—'

I give him a pleading look.

'Does she,' Swan says, calming down. 'OK. But be back quickly. We're late already.'

'Thanks to you,' Greta says loudly to me.

'A decent cup of coffee is a necessity,' Swan says, judiciously. 'How does everybody take theirs?'

He goes round the table, taking orders. I dutifully write them all down, wondering how I'm going to manage to carry them all back.

'And you, Anna?'

'Me?'

He nods.

'Oh, I don't want one, thanks,' I say. 'I think carrying five is probably my limit. Anyway, I take my coffee plain so whatever they have here is fine. Sometimes I like Hazelnut Coffee-Mate, but . . .' I notice Greta glaring at me and realize I'm babbling. 'I'll be off, then,' I say brightly.

'You're just getting Greta's coffee,' Swan says. 'You're her assistant, is that right, Greta?'

Greta nods.

'Well then, somebody has to get Anna's coffee,' he explains. 'So it better be me. Anna can get what you need, Greta, and I'll just run out and take care of everybody else's order.'

Greta splutters. 'What? But that's ridiculous.'

165

'I intended Anna to learn from me,' Swan tells her, easily, but with a touch of steel in his tone. 'So if she's not here, there's not much point in me being here. We can run our errands together. Of course, business will have to wait until I get back, but I don't want to come between you and Kitty Simpson, Greta. Whatever you've worked out is fine with me.'

Greta swallows. 'Well, of course, I don't want to hold up our work *further*,' she says, looking meanly at me. But Swan is having none of it.

'You don't want Anna assisting you, then?'

Greta shakes her head.

'Good,' says Swan. 'And naturally I'm sure Kitty won't complain, as this is your wish. Right?'

'Kitty will be fine,' says Greta.

'Excellent,' says Swan, relaxing, and Greta lowers her eyes. It was a battle of the egos, and he won it, no problem.

I find chills are creeping all over my skin. I daren't look at him.

'Tell you what,' Swan says. 'Anna, if you feel like it, you could run those errands for Greta today, and join us again tomorrow.'

I look at him gratefully, delighted with the chance to placate Greta just a little. And to get out of there. I know I shouldn't think about it like this, but having him defend me, it's just so . . .

Well. It's sort of electric. And I mustn't think of it that way.

'Of course,' I say to Greta humbly. 'I'll take care of these for you and I'll see you tomorrow.'

The next week is hectic. I go to meetings with Mark Swan, that huge control freak, and I manage to survive them. I get there early, take notes, listen to what he says to the actors, the crew, watch how he slots in our pre-production around the film he's actually shooting. Swan gets me a pass to go on set, so I get to saunter past that security guard, who pretends he doesn't recognize me from before. I stand behind Swan on Hampstead Heath, watch him ride up on a crane or walk around with a megaphone, look at him coaxing performances out of the actors.

I can see what he's doing, take it in. He's a brilliant director. Implacable, but amazing at getting his actors and crew to do the exact best thing. You can see from the monitor, a shot you'd thought was perfect he'll redo, and then it'll be much better. I trail around after him like a little puppy, and he asks me questions, sharp ones, to make sure I've understood. And when I answer right he nods as if I'm a puppy who's learned to hold up its paw for a piece of cheese.

But I'll tell you something about this process. It's bloody boring.

I'm bored out of my mind. Who the hell wants to stand there in the drizzle, with a whipping breeze, watching a bunch of luvvies flub their lines or fake a passionate kiss? Who's really interested in hearing an assistant director go over a bunch of storyboards? I feel guilty, though, I know there are loads of people who'd kill for this chance, the way he says.

You know the types. Film geeks. People who like the kind of mind-numbingly dull and worthy movies that John finds appealing. People who actually watch *I Love Lucy* and like nothing better than those extra scenes on DVDs that go behind the scenes on the set of a movie. People who plunk down twenty quid for a big coffee-table book on the 'art' of the *Lord of the Rings* trilogy.

I am, I realize, not one of those people.

I don't understand. It's quite worrying, in fact. For years I've bitten the bullet of my low-salary, low-prestige job in the hope that one day I'd get my big break, get to make movies, and I'd be rich and fulfilled . . .

'Anna.'

I look up, clutching my notepad. It's Thursday morning, and we're standing outside on Wimbledon Common, in drizzling rain. The sky overhead is as grey as Barbara Bush's hair.

'Are you . . . yawning?' Swan says to me, eyes narrowing.

'Ahem, ahem,' I say, hastily turning it into a cough. 'No,

goodness. Absolutely not. Got a cold,' I say, trying to be perky.

Everybody else on the set is perky, and most of them don't even drink caffeine. It's disgusting really. Health nuts standing about in the rain and rhapsodizing over this possible shot and that possible shot and won't this be the *perfect* location for the dog-walking scene . . .

It makes it easier for me to be around Swan when I'm this bored. I can't look longingly after him because I'd get caught, and anyway I feel too self-conscious. So while he's standing there with Trish and the DPP, the guy who sets up the cameras and the look of the movie, I sort of hang back and concentrate on keeping warm.

'Mmm,' Swan says, eyes glinting.

Oh crap. Am I in trouble?

'I thought that the pond over there would add some great visuals,' I offer weakly. 'Maybe the dog could chase a duck and pull Elsie into the pond and then her nice dress is ruined.'

'That's funny,' Trish says. 'I like that.'

'And she'd get all pissed off, but the dog wouldn't care.'

'It could just lick her face.'

'And ruin her mascara,' I add, thinking of Winston at Vanna's. 'And then she has to go back to the vicarage looking like a total fool.'

'And Mrs Wilkins makes fun of her,' Trish says, getting into it. 'And she's seething. That's fucking great! You're brilliant, Anna.'

I smile at her gratefully and look over at Mark Swan, feeling rescued.

'And this relates to the wedding rehearsal scene how?' asks Swan.

'What?' I ask, nervously.

'We moved on from the dog-walking thing an hour ago. We decided this wasn't a suitable location,' Swan says drily. 'Wrong light. Remember?'

No.

'Oh yes,' I say. 'I remember,' I add confidently. 'The light wasn't any good.'

'Take five, everyone,' Swan says. 'Anna, why don't you just step over here with me a second?'

Oh, hell.

Don't show him, don't show him. Directors are like sharks. They can smell fear! I paste a suitably radiant smile onto my face, in the manner of an American cheerleader, and walk towards him as he heads for the privacy of a weeping willow.

'Anna,' Swan says.

'Yes. Can I help you?' I ask. 'It's going great, isn't it? The location shoots and everything.'

'How would you say you're doing?' Swan asks.

What kind of a question is that?

'I'd say I'm doing *fantastically* well,' I say firmly. 'I've not been late once!'

'That's true,' he concedes.

'I've taken *loads* of notes, I've watched you, and I've reported back to Red Crest and everybody's happy,' I say. Ha! I don't fold under questioning like some people I can think of, such as Sharon breaking down and blubbing when John accused her of flirting with every man in the office because she never did any work and didn't want to get sacked.

'Everybody except one person.'

'Greta's perfectly fine. I got her that Creme de la Mer she wanted,' I protest. 'She's been very cooperative with you. Maestro.'

'It's not nice to make fun of Greta.'

'I wasn't,' I lie.

'I'm not talking about her anyway. I'm talking about you. You look like me in a marketing meeting. Bored out of your skull.'

'Well, what do you expect?' I protest. 'Standing about here all day without even a fire to keep warm, looking at the same boring patch of grass. How can you do it?'

'Anna,' Swan says, gently. 'This is producing a film. This is pre-production. You know, checking out locations. It's part of it.'

'I have been paying attention, you know, mostly. I've been on time. I'm doing everything you want. I could try to pretend to be more interested if you like. I'm not trying to be bad. I'm really grateful to you, honestly.'

'It's OK,' he says. 'I'm not angry.'

'You're not?'

'No.'

I breathe out.

'I want you to come back to my house this afternoon,' he says. 'I want to talk to you about something.'

I shake my head. 'I can't. When we don't have pre-production they make me go back to the office.'

Kitty doesn't want me spending any more time with Mark Swan than is absolutely necessary. She watches my hours with him like a hawk. As soon as we're done with the day's chores, storyboarding, location scouting, rehearsals, script rewrites, I have to be in a tube station within five minutes and back in the office in ten. Which is fine with me. It keeps me away from Swan and his gorgeous eyes and his muscular chest. It takes my mind off watching how he controls everything, and everybody fawns all over him, and all the pretty young girls bat their eyelids at him . . .

He doesn't have a girlfriend, by the way. He told me one morning over coffee, when I caught a particularly obvious fling-herself-at-him from the pneumatic blonde from the production designers. Her name was Susan, and she was working on a storyboard. Swan likes to storyboard, that's when you get drawings of what every shot in the movie will look like, so you can plan it in advance. He's thorough.

'Oh, Mr Swan,' she kept saying breathily, 'this is such an honour.' And then she'd flutter her eyelashes at him and lean forward so he could see her humungous, surely fake, boobs in that low-cut top even better. *And* she'd put blusher on them to give herself even more cleavage. Plus, do you think a cartoonist needs to come to work in three-inch spike heels? No, neither did I.

I did actually roll my eyes and he caught me doing it.

'Coffee, anyone?' he asked, to cover his laugh, beating a hasty retreat to his office kitchen. Swan always made every-body's coffee since the showdown with Greta. He's not a prima donna, so nobody else dares to try it, not even her. Anyway, he comes back with a tray of coffee, and boobs-girl excuses herself to go to the loo and presumably slap on some more war paint. Her partner, a bloke, heads off for a quick fag and Swan draws me aside.

'Sorry about that,' I begin.

'Don't be.' He grins at me. Oh, he is so gorgeous. 'It's because of Misty.'

'Misty?'

'My girlfriend.'

I stiffen, I can't help it. Of course Mark Swan would have a girlfriend and of course she'd have a name like Misty. She's American, no doubt, a flawless Heather Locklear clone with bronzed skin, platinum hair, perfect, laser-whitened teeth, and a nice line in the beauty queen wave, the one where you only waggle your fingertips, because anything else is too vulgar for a girl. (You know the kind of woman. Never swears, doesn't drink, eyebrows are always shaped.)

'A model?' I ask cynically.

'An actress,' he says.

'And?' I can see there's more.

'And former cheerleader for the LA Lakers,' he admits.

I knew it.

'Anyway, I broke up with her last month, and there's a bit of . . .' he's too nice to say 'gold-digging', 'flirting going on,' he says.

'Why did you break up with Misty?'

'She was boring,' he says.

Ho-hum. I wonder how good-looking you'd have to be for Mark Swan not to find you boring?

'I'm sorry if it's a bit awkward,' he says, nodding towards the locked loo door.

'Oh, don't worry,' I say. Awkward? Why would it be? It's totally normal, a pretty girl flinging herself at a rich, gorgeous man. If I were her I'd do it too.

This is life, Anna, I tell myself. You've got to get over it.

'You could always date her,' I offer. 'She's, you know, she's very attractive.'

Swan looks towards the loo door with horror. 'The hell she is,' he says. 'What, her?'

I stare gloomily into my delicious coffee, to which he's added Hazelnut Coffee-Mate. He bought it especially because he knows I like it, even though he'd never be caught dead with anything so naff as Coffee-Mate himself. Swan doesn't consider boob-girl attractive, not with her waspish waist, nor her blond hair.

Hey, it's not so bad. At least I know where I stand.

I don't know why I'm thinking about a man I can't have anyway. This is not like me. I have Charles, and I try to concentrate on him. I go back to the office and tell a rapt Sharon all about Mark Swan's commitment-free status. Good luck to her, maybe she has a shot. I certainly don't.

'You don't need to worry about Red Crest,' Swan says, jerking me back to the present. 'Let me take care of that right now.'

'You don't understand,' I plead. I really don't want to go to his house. By myself? It's hard enough hanging around him with all these other people. Why his house? Why me? What if I stare at him too long and he catches me? 'You can't stop it, Kitty will be furious. You already called her about me coming straight to you in the mornings. She doesn't like me hanging out with the talent . . .' I trail off. Have I said too much?

He winks at me. 'I know the type, honey. Watch this.'

He flicks open his mobile, punches in a number.

'Kitty Simpson, please. Mark Swan. Oh, hi, Kitty,' he says. I can almost see her jumping to take the call. 'How are you doing?' He listens for a second. 'Yes, well, that's very kind of you. Very kind. Actually I'm calling about Anna. She's been talking to me all about your Oscar and your leadership on *Mother of the Bride*. Yes . . . she told me everything and I must say, I'm very impressed.'

I can't believe it. I can't stop the grin from spreading all over my face. Swan listens as Kitty gushes like an oil well.

'Mmm,' he says. 'Anna never stops talking about you, you're her heroine. I was wondering, can you send me a memo with your ideas on foreign marketing? Especially in

Italy? Since your Oscar was won there . . . I gather you have real expertise in that market and my people are stuck. How to get our English humour across . . . great, you can? That's perfect. I wanted to borrow Anna this afternoon for some grunt work, one of our runners is sick. Any chance? Oh, thanks. That's really useful. Yes, it will be good for her. So I'll look forward to getting your ideas, Kitty. Brilliant. OK, bye.'

He hangs up.

'I think you'll find you won't be in any trouble now,' he says.

'Yes. Thanks.' I look away, because it's just too much. It's so sexy, the way he can snap open a phone and take care of Kitty in five minutes. Swan is so self assured. I wonder if he ever had a klutzy moment in his life.

'You can go,' he says.

'What?'

Swan waves his rough-skinned hand, dismissing me. 'Get out of here. You're just an extra body on the set, you're worse than useless.' He looks at my crestfallen face. 'Just turn up at my place around half five. I'm having lunch with Rachel Weisz, can't get there before that.'

'Of course.' Rachel Weisz is bloody gorgeous. I try to remember, is she married to anybody? Please let her be married to somebody. They'd make a perfect power couple . . .

'You know the address.'

'Yes.' I shake my head, ashamed of myself. Why am I being so dog in the manger? Why shouldn't he go out with Rachel Weisz. I *have* a boyfriend. This is so bad. My throat thickens, and I swallow hard. I don't look at Swan, I feel panicky, as though he might catch me.

'Yeah, no problems,' I say quickly. 'I'm going to lunch with my boyfriend.'

'Ah, the millionaire,' Swan jokes.

'That's right,' I reply, tilting my chin up. 'The millionaire.' And I stomp off, away down the path, trying to tell myself that at least I'm going to get to be warm.

I call Charles from my mobile.

'Hi, what are you doing?'

'Me? Just pottering about. Writing. Planning a sequel to my novel.' I can hear the smile in his voice. 'How lovely to hear from you. Wasn't expecting to see you until Saturday, at Chester House.'

'I just wondered if you were free for lunch? I'm buying,' I add quickly. I should have enough money in my account for one lunch at least. 'Anywhere you like,' I say recklessly, hoping that it will not be a place with appetizers with unpronounceable names and fifteen-quid price tags.

'You certainly aren't,' Charles says, severely. 'I've never let a lady pay for lunch in my life.'

'I insist.'

'Nevertheless, I am paying,' says Charles, firmly. 'And of course I want to see you! I'll rustle up a table at the Savoy again, shall I? That's not too far from you.'

I look down at my wet jeans and sodden windbreaker. 'I just need to go home and change.'

'How about one, does that suit?'

He's so polite. He's such a sweetie. I don't know why I'm close to tears, still. And suddenly it just hurts to be wet and cold and poor and the sound of a lovely lunch in the Savoy served immaculately by career waiters, all warm and cosy, is perfect.

'That's wonderful,' I say gratefully. 'I'll see you there.'

Thank heavens there's a taxi, coming towards me with its lights on like a black and orange angel of mercy. I hail it and jump in, settling back into its lovely heated interior. Yes, I know it'll probably be an eighteen-quid ride, but I can afford it. I'm not paying for lunch.

I quickly shower and blow-dry my hair, then pull out my navy dress with the pumps again – I know Janet would yell at me, but I don't have time to experiment. I'm dreading her taking me shopping. Looking at myself in changing-room mirrors – ugh. But that's a nightmare for another day. Right now I will go with my shapeless but classic dress, my fake pearls and the matching shoes. Then I take off the fake pearls. In the Savoy they all wear real ones. Ho-hum.

I want to make myself up, but nobody's here to help me, so

I settle for foundation and the traditional dab of bronzer on both cheeks. This is about looking nice – well, as nice as possible, for Charles, nothing to do with the fact I'm going to Swan's place later. I twist my hair back in a neat bun. OK, at least I look businesslike now. Then I grab my bag and start walking.

Charles is waiting for me when I arrive. I breathe in, trying to relax. And it's not made too hard. There's that buzz that's always around him, the air of money. Here, it's reflected in the muted sounds of the dining room, a sort of very low-pitched hum of polite conversation. Rich people seem to murmur when they eat, ever noticed that? You never get a rowdy crowd if the median income is six figures and up.

I sit down when the waiter pulls out my chair.

'I took the liberty of ordering some champagne,' says Charles, and indeed there's a flute fizzing by my place. 'Hope you don't mind.'

'Not at all,' I say, taking a big slug. Then I remember where I am and sip it instead, delicately as I can.

'Oh, don't mind me,' Charles says. 'Knock it back, old girl. You look frazzled.'

Do I? I thought I'd done OK on the old face-repair job.

'It's been a rough morning,' I say.

'Why? What did they make you do?'

'Oh, nothing. It's not that so much as . . .' My voice trails off. 'Tell me about your day,' I suggest, brightly.

Halfway through lunch I escape to the sumptuously appointed loos to take a deep breath in and examine my face for treacherous mascara smears. But there's nothing wrong with it, apart from the usual.

I stare at myself in the mirror for a long moment.

What am I thinking? Here I am, having lunch with a lovely, kind man who jumped at the chance to take me out, at no notice, to one of the most expensive restaurants in London. I'm drinking champagne and eating fillet steak. I'm being spoiled rotten. Me, Anna, who's never had any attention in her life. And after lunch I'm going to go to a meeting with one of the most powerful directors in the world who apparently wants to mentor me in my career.

Three months ago I was just a grunt reader for Kitty Simpson, desperate to hang on to a spotty oik with halitosis.

So what the hell am I so unhappy about?

Some fat lady in Chanel enters the loo and glances disdainfully at my H&M shift dress. But I don't care. Charles is waiting patiently for me outside, with a pudding menu that includes chocolate soufflés and ganaches and homemade sorbets and things.

I take a deep breath.

I am *not* going to blow this because of some hopeless infatuation. I'm going to get real. My choices are Charles or nobody. And I *like* Charles.

I head back out to him, forcing myself to smile.

Mark Swan lives in the heart of Notting Hill, in a gorgeous Queen Anne house with a walled garden that puts Vanna's to shame. I've been there before one time, for a script meeting. We walked there from a local restaurant, after lunch, and I got to see the full gamut of women who wanted to attract his attention.

Everywhere he goes people want to be in bed with him. On the one hand you can understand it; a man that powerful, all the film people want to schmooze him. But women seem drawn to him no matter what. The average person wouldn't recognize Mark Swan if he sat next to them on a bus, but women preen whenever he walks into a room. Sometimes without even noticing it. Legs are crossed, toes are pointed, hair is tossed. I think I've seen enough lips tentatively licked this week to last me several lifetimes. There are the fingers laid casually on his jacket sleeve, the light laughter, eyes glancing his way, then away again, then back – all the little tricks. What they used to call 'feminine wiles'.

I don't have any of these, so I'm OK. It's just fun for me to observe Swan and his effect on the women all around him. It's like watching a rock star walk into a bar. I'm a student of pretty women, and seeing him set the cat amongst the pigeons is fascinating. In a purely anthropological sense.

I have just been taking care of myself, talking to Greta, making sure she's OK, reporting back to Kitty and Eli (mostly making stuff up – there's nothing really to report; what am I going to say, Pre-production is fine, wish you were here?) and trying to improve myself. Nothing fancy, just a little. I go for a jog most mornings, I'm eating salads and diet sandwiches, I switched to Diet Pepsi, and I'm down to one packet of Quavers a day. It's no big deal. I'm too busy trailing Swan to eat much anyway. He likes to keep me busy. As well as asking me testing questions on the day's notes, he likes to grill me on movie trivia, which I'm great at, as long as it's about a big budget Hollywood film shot no sooner than '84. He does ask me questions about dull arty films and seems delighted when I don't know any of the answers.

Apart from the fact I fancy him, I want to be just like him. I know it's impossible. I don't know how to direct and I don't want to. But just once, I'd like to inspire that kind of reaction in other people, I'd like them to drop their voices and whisper when I come into a room. Find me incredibly impressive, the way I think of him. But I just keep jogging and making notes and staying out of his way and trying to seem interested. That's what's going to get me through these couple of months, and that's what'll get me my promotion. Nothing else matters, right?

I square my shoulders as I turn in through Swan's cast-iron gate. I feel stronger than I did this morning. The champagne, the fact I'm not freezing. And Charles's patient attention.

I *do* love movies and I *do* want a good career. I don't need to dread seeing Mark Swan, just because he's handsome. This is a huge career opportunity. I'm glad I can be here.

OK, right. Here we go.

I march up his garden path (old-fashioned red bricks, lavender on the sides, hollyhocks and lupins) and ring the bell.

'Coming.' A muffled yell from inside and then Swan wrenches the door open. He's wearing black karate pants, tied at the waist, and . . . that's it.

His chest is bare.

Bare. And cut, as the Americans say. He's got hair all over his chest, but you can still see the muscles, the biceps, the what do they call those things on the chest? Ooh. I love muscular men. His chest hair looks like Sean Connery.

I take a step back, dry-mouthed. Stop staring. Stop. Staring!

'Oh, I'm so sorry,' says Swan. 'I was working out, lost track of the time. Come in, come in, I'll just get changed.'

Don't bother on my account.

Stop that!

'I'll make some coffee, shall I?' I ask. My voice has gone sort of hysterical and squeaky. I hastily beat a retreat to the kitchen and his sleek black coffee machine.

'Sorry about that,' Swan says, two seconds later. He's pulled on a pair of chinos and a large T-shirt. 'I got carried away. Love martial arts.' He shakes his head. 'Great stress-reliever. Of course I don't mean to scare innocent young women.'

'You didn't scare me. I mean that's fine.' I busy myself measuring out the coffee. It's scented with vanilla, smells heavenly. 'Were you breaking wood planks with your bare hands?' I joke.

'No, I'm doing bricks now,' Swan says.

'You're breaking bricks with your bare hands?' I ask, looking at him.

Swan shrugs. 'It's all technique. Don't look so impressed.'

'I'm not impressed!' I lie. Why? Why me? Will I ever be able to get that image out of my mind? And I was doing so well, with the nice lunch and all.

'Have a coffee,' I say, severely. 'Down to business.'

'Mmm, business,' he agrees, eyes glinting again. 'You're looking very . . . professional this afternoon, Anna.'

I stiffen. 'What's wrong with it?'

'Oh, nothing,' he says. 'Except that I feel like I'm about to get a rap on the knuckles with a ruler.'

'You probably need one,' I say haughtily.

'Of course you could do that sexy-secretary thing, you know, unpin that severe bun and shake your long hair loose. You could run your fingers through it,' he suggests.

178

Now I know he's laughing at me.

'It's not nice to make personal remarks,' I tell him. 'Don't you want to talk to me about something?'

'Sorry, I know. Only your boyfriend is allowed to flirt with you,' he says.

'That's right. Don't flirt with me,' I say, rather snappily. Then I blink. I mustn't snap at my mentor. But him being flirty with me is the absolute last thing I want. I don't need to be positively tortured.

Swan holds up his hands. 'OK, OK. Point taken. Let's talk about your role in pre-production. Learning the business.'

Finally.

'Yes?'

'Your heart's not in it,' he says. 'Why not?'

I start and my mug jerks, spilling coffee all over the table. 'Oh! I'm sorry.' I jump up and rip off a piece of kitchen towel, dabbing it up.

'Don't worry about that,' he says. 'Just answer my question.'

'I don't understand what you mean,' I say, shaking my head as though puzzled. 'You asked me this this morning. But it's not fair! I've been paying careful attention to the film-making process. I've got here early—'

'Yes.'

'I've been watching what you've said to the actors, watching you on set, listening to the script notes . . .'

'Yes,' he agrees, pleasantly enough.

'I have been paying attention, honest,' I plead. 'I can show you all my notes.'

Swan smiles lazily back at me, that relaxed self-confidence just oozing out of him. 'I know that,' he says. 'You're not stupid. You didn't want me to kick you off the pre-production.'

'No,' I say. 'It's fascinating.'

'It is, but obviously not to you.' Swan looks at me intently, as though I am a particularly interesting Japanese promo ad, or a still that needs analysis. 'You only come alive during the script meetings.'

I do love those meetings. Listening to the finished story take shape, watching Trish and Mark work through the characters, beef up the scenes and beats. I suppose it must have showed.

'Those are my favourites,' I admit.

'Why is that?' he asks.

'I love the story,' I say honestly. 'I just think the story is funny and poignant and near the knuckle and . . . I love it.'

'But you find discussion about sets tedious.'

'Oh, fuck it,' I say suddenly. I just don't feel like pretending any more. 'It's so bloody boring, I don't know how you do it. But I don't think you should sack me. I've done everything you said,' I argue, defensively. 'I've listened, I've learned, I've observed.'

'What about rehearsing?'

'That's boring too. I wouldn't do it for a million pounds,' I hear myself say dismissively. 'Standing around in the rain watching a bunch of overpaid actors say the same bloody line over and over. How hard can it be, eh? They talk about motivation in the scene – "Mark, what's my motivation here?"' I imitate Greta perfectly.

He laughs softly. 'What would you say?'

'Um, "Your bloody enormous pay check"?' I suggest.

Swan laughs again, really amused. 'You aren't impressed by the actors' craft?'

'I think they're a bunch of . . .' I stop. 'I expect some of them are really nice,' I say diplomatically. 'And not tossers at all.'

'Very well put,' he says, his face grave. 'So basically what we're saying here is that you only really like the story development stuff.'

'Well, yes. But that's the most important part,' I plead in my own defence.

'When you were bullshitting me this morning, when I caught you drifting off,' he says, 'no, no need to deny it. You made up that scene where Elsie's dog pulls her into the pond. Right off the top of your head.'

I think about an excuse, but it's clear he's caught me.

'Yes?' I ask warily.

'Well, like Trish said, that was good stuff.' He takes a drink of coffee. 'And that dialogue you wrote at our first meeting. That was good too.'

'Thanks,' I say, blushing.

'I asked you then if you'd thought about being a screen-writer.'

'Oh, yes. That was nice of you.'

'Well, have you?' he asks, looking me right in the eye.

I shrug. 'I don't know . . . I'm just a reader.'

'Could you do better than most of the scripts you read?'

'Oh, fuck yeah,' I say. 'Excuse me,' I add hastily. Nice girls don't say 'fuck' do they? Cheerleaders for the LA Lakers and Rachel Weisz.

'Listen, you're not terrible at producing,' he says. 'You could wind up with an OK career on the conceptual side. You found the right script for the right actress and you went after the right director. That's a big part of it. But all the grunt work of producing, locations, marketing, hiring crew, casting smaller parts, you hate all that stuff. So there's going to be a limit as to how well you do. Bringing a film in under budget depends on attention to detail a lot of the time.' He looks at me over the top of his coffee mug. 'What's the matter?'

I find I'm gazing at him adoringly again. The irony is, though, that the first time he's actually caught me I wasn't thinking how sexy and hot he is. I was just thinking, nobody has ever talked to me like this. Not as long as I've been in movies. He's taking me seriously. He's not telling me to stick with him and he'll make me a star. He's not offering me the moon. He's just listing strengths and weaknesses, as if I was a real movie person, as if I could have a career. It's almost as if he *respects* me.

I don't think I've ever been paid a greater compliment, not one that mattered.

'Nothing,' I say. Then I think better of it. 'No, it's not nothing. I was just thinking that –' I'm blushing but I plough on – 'this is really kind of you to do this.'

'What am I doing? I'm not doing anything,' he says, easily. 'Attached myself to a good script.'

'That's not true. We both know it,' I say. 'First of all you didn't get angry when I didn't recognize you. And then you let me give you the script. And you read it.'

He shrugs. 'It was ten minutes out of my day.'

'Yes, but big stars like you don't do that sort of thing. They only read what their agents send them. And that wasn't all, you made my bosses give me a chance to come and learn from you. You threatened to pull out of the deal unless they let me go.'

He grins. 'Like you say, I'm a big star. Got to get my own way.'

'And even though I turned up late the first time you didn't sack me. And now you're talking to me like this. Like you believe in me.'

'I do believe in you,' he says, dead serious. 'I saw in you something I haven't seen for a very long time.'

I look at him, asking the question.

'Passion,' he says. 'Passion. Love of movies. Love of stories. Enthusiasm. Most people have love of deals. Not you.'

'Thanks,' I say. I can barely whisper it out. I clear my throat, try to pull myself together. 'I mean, I want to thank you, so much, for all you're doing for me. I never had a chance till you came along.'

'You always had a chance,' he says. 'Think about this: I didn't just come along, you came to find me.'

There's a moment's pause. I'm staring into his eyes. I force myself to break the look, to wrench my eyes away, even though I wish I could stare at him like that forever.

'Doing something tonight? Got another date?'

'Not tonight,' I say, warily. Is he laughing at me?

'Would you like to go for a drink?' He spreads his huge, sunburned hands. 'No obligation. You're not going to get kicked back to Winning Productions if you turn me down. I want to discuss something with you, and I think better over beer than coffee.'

'Well . . .' It's only 6 p.m. 'I suppose that'd be OK,' I say, insouciantly. But my heart is leaping. What can I say? I can't help it.

He walks me out, down the road and round the corner where there's a small place called the Queen Adelaide. Probably one of the last pubs in London that hasn't been taken over, given the stripped-pine look and re-named something funky. The Queen Adelaide has a dark, burgundy fabric, musky-smelling interior thick with smoke and a slightly pungent reek, tables, chairs and benches of worn, dark wood, a dartboard and just a couple of outside tables in cast-iron; a grudging concession to London's cafe culture.

The bartender looks up when we come in.

'All right, Mark?' he says. 'Who's yer ladyfriend?'

'This is the lovely Anna,' Swan says. I stare at him suspiciously for signs of mockery but it doesn't look like there are any. 'Usual please, Mike,' he says, and the barman pours him a double rye whiskey and looks at me.

'Let me guess,' Swan says. 'Babycham.'

'Oh, fuck off,' I say. 'Half a cider is fine.'

'Living dangerously,' he comments as he slides some money across the pitted wood.

'I don't want to get too hammered,' I tell him. 'I might lose all my inhibitions and start giving you some home truths.'

He laughs, delighted. 'Isn't she great?'

'Yeah, I'm his pet seal,' I say grumpily. 'Available for weddings and bar mitzvahs.'

'Not like your usual type,' the barman comments. Swan scowls at him and he winks at me.

'What's that? Boobs like barrage balloons and a waist like a wasp?'

'She's got your number, mate,' the barman says, and I'm satisfied to see Swan flush.

'Let's sit over there,' Swan suggests, indicating a weather-worn bench with fairly manky-looking cushions tucked into a corner.

'Not outside?' The rain has stopped, and now it's a lovely sunny evening, warm and golden; even at sunset it's not chilly.

'I thought we might be more private in here,' Swan says.

'What do we need to be private for?'

'Some people recognize me, sometimes,' he admits, blushing. 'Film students and stuff. Actors. You get a lot of them round here.'

'Yes, it is a fairly pretentious area,' I concede. 'All right then, we'll sit over there to avoid your legions of fans.'

Swan slides into the corner, disappearing comfortably into the gloom, and takes a pull at his rye, instantly relaxing. I can see he feels safe here, protected. People don't bother him.

'I hope you don't think I'm pretentious.'

I don't say anything, just nurse my cider.

'It's OK,' he reassures me. 'You can say anything you like and I won't come down on you.'

'Anything I like?'

'Yep.' His eyes are sparkling.

'Sounds like a trap,' I say suspiciously. 'You'll lull me into revealing some of those home truths and then you'll call up Eli Roth and have me fired.'

'I wouldn't do that,' he says.

'Oh.'

'I'd call my agent and have her do it. I can't stand Eli,' he says, and grins.

'Very funny.'

'Seriously, you can trust me. We're breaking bread together. Or at least booze. That's got to be sacred.'

'All right,' I say, sipping my cider, which is flat. 'I do think you're a bit pretentious.'

'Oh yeah?'

'Yeah, but it goes with the territory. All directors are pretentious. They think the sun shines out of their arses.'

'Don't sit on the fence, Anna,' he says, stretching his hand out on the table. 'Tell me what you really think.'

'Well, it's true,' I say, hotly. 'What is this "A Rob Reiner Film" anyway? Why is it known as the director's film?'

'That's called the possessory credit.'

'I know what it's called.'

'The director is the one who's blamed if the film goes wrong, he's the one who makes all the decisions. I think it's fair.'

'You would,' I tell him. 'Directors are so unimportant, I'll never work out why they've got so much power.'

Swan goggles at me. 'You think we're unimportant?'

'Sure. Anyone can coax a performance out of an actor, they just want their egos stroked. And directors don't set the look of the film, the DPP does that.'

'So who's important then?' Swan ticks off the names on his huge fingers. 'The stars are self-indulgent tossers, the director's a meaningless appendage . . .'

'The writer,' I say triumphantly.

'The writer's the low guy on the totem pole.'

'I know that,' I say indignantly. 'But she shouldn't be. She's written the screenplay, made the story.'

'She,' he says, grinning. 'But a movie's more than a screenplay,' he adds, reasonably.

'I know that too. But the screenplay is the blueprint, isn't it? Film isn't anything more than a story told in pictures.'

'I'm responsible for the pictures.'

'Anybody could make the pictures,' I say hotly. 'But only the person who thinks of the story can make the story. Anybody could act the part, too. Just because one person's good in the role doesn't mean somebody else wouldn't have been just as good.'

Swan says, 'I admire your passion.'

'Oh, give over,' I say, scornfully. 'Now you really do sound like an LA executive.'

He laughs, a rich, deep belly laugh that seems to go on and on and sends half the pub staring in our direction. I nudge him. 'Cut that out,' I hiss.

'I'm sorry,' he says, wiping his eyes. 'I just think you're priceless. And fearless,' he adds, before I can take offence. 'Do you know how long it's been since anybody talked to me like that?'

'Well, you can't get me in trouble, you promised.'

'I wouldn't dream of it,' Swan says. He reaches across the table and takes my hand in his. I look at it, lying there, and for the first time in ages I'm not embarrassed of my too-large, too-utilitarian hands. Inside his large, thick fingers, they seem slender, feminine.

Mark Swan dwarfs me. I understand now, with a sudden shock of recognition, one reason why women react around him the way they do. He's *huge*. It's as if the masculinity of his personality has somehow manifested itself physically in his body. There are plenty of tall men around, but Swan is six four and built. His muscles are thick, he's barrel-chested, hairy – everywhere, even his eyebrows are thick and beetling – his hands look like they could crush up Coke cans without a second thought. He looks as though he's come forward in time from King Alfred's day when Viking longships were harrying the coasts.

Swan wears suits sometimes, nice ones which he has specially made, but on him they seem like a costume. There's nothing at all about him that *People* magazine would think of listing in its 'Fifty Most Beautiful' edition; he's a million miles away from Brad Pitt's smooth good looks, or Colin Firth's reserved manliness. No, he looks as if he's about ready to grab some poor wench by her long plaited hair and drag her off screaming to his cave, though you feel most of them wouldn't be screaming very hard.

I jerk my hand away.

'I think I've discovered your problem,' Swan says, apparently not noticing.

'Oh yeah?'

'Yeah. I couldn't work you out. You seemed to want this badly, your movie, and I thought you were fun,' he says lightly, and I feel a shiver all up my legs and back. 'So I thought I'd give you your break. And once I got you here, you obviously loved movies. But you were miles away.'

I blush.

'You thought you did a pretty good job of hiding it?' he asks, and winks at me. 'You forget, I'm around actors all day long and most of them are better than you.'

I blush some more. Whether it's being caught out or the wink I don't know, but I stare away, into my half of cider. Mark Swan at this proximity is very . . . disturbing, and the last thing I want to do is start giggling inanely like all the gorgeous groupies and hangers-on and adoring twenty-somethings he runs into every day.

'I'm doing my job,' I say, gruffly.

'Yes, but I told you, you'll never be good at it.'

I bristle. 'And why not?'

'Because you don't love it,' he says simply. 'I know you love movies, but you don't love producing and you never will.'

'So what's my vocation?' I demand. 'To buy tickets and sit there eating popcorn so men like you can be paid far too much for sitting in a chair and shouting "cut"?'

'You should be a writer,' he says, ignoring my hostility. 'You've got a great feel for story, you're obviously creative. You appreciate good scripts. You think the film is the story. You should write screenplays,' he says simply. 'Last time I mentioned it, it was just a suggestion. Now it's an order.'

I shiver with pleasure. Swan has just recited my secret dream back to me. The dream I've dismissed for about as long as I dismissed the idea of having a solvent boyfriend without acne. But that's happened, so why not this?

'An order?'

He nods. 'You want my help, right? After you started riffing on that dog-walking scene I knew for sure. She'd be a good writer. Not right away, but if she worked at it. You owe it to yourself to at least try, Anna.' He takes a long pull at his drink. 'This is pure selfishness on my part. 'I'm fed up with the dire scripts out there. Anything I can do to infuse new talent into the market, the more choice for me, right?'

I can't help it, I beam at him, and Swan leans back against the wall, his hazel eyes flickering over me.

'When you smile like that your face lights up,' he says. 'You should do it more often.'

'Thanks for the suggestion,' I say. 'It's – it's really nice of you.'

'If you'll take one more suggestion from a useless wanker of a director,' he says, 'don't go around repeating what you told me about directors all being a waste of space. It's not all that likely to get you hired.'

'I won't,' I say. He smiles at me, and I feel that bolt of electricity again and take a deep drink of cider to cover myself. I'm not going to do it, I'm not, I tell myself fiercely. Wouldn't that be a joke? Flirting with a man like Mark Swan, me, Anna Brown? Imagine how embarrassing. Just like Claire, pathetically dressing up and walking around giggling and flicking her hair for Eli Roth, while he doesn't know she's alive.

'I'm not promising you anything,' he says. 'For all I know you could be rubbish. But just in case you aren't, if you can actually produce a good screenplay . . .' he shrugs. 'I might be able to help you out.'

I don't dare ask what that would mean, but the adrenaline is coursing through my veins. Whatever it means, it's my ticket out. I look at him again, differently. This is *Mark Swan*, I tell myself. Probably the single most powerful man in the British film industry. He knows all the major Hollywood studio heads personally. Every super-agent in LA has been courting him for years, and there's not a star around who wouldn't love to work in one of his films.

If he's offering me his help, he's offering me the moon.

'Only if you're good,' he says, sternly. 'You might not be.'

'Thank you,' I say. 'Very much.'

'I haven't done anything yet. Maybe point you in the right direction.'

'I appreciate it. You're . . .' My voice trails off.

'Don't look at me like that,' he says gently. 'I prefer it when you're telling me to sod off than when you look awe-struck.'

'Awe-struck! I'm not awe-struck. You're just another overpaid loser as far as I'm concerned.'

'Better,' he says. 'Better.' And he reaches out one stubby finger and, grinning, tucks a strand of loose hair back behind my ear.

His touch is electric. Instantly, shamefully, I feel my nipples harden, my stomach liquefy, I want to squirm on my seat. My skin mottles.

I can't do this! I can't be like all those other girls. Mark Swan actually likes me. He wants to help me. I'm not going to blow it.

'I should go,' I say, as lightly as I can manage. 'Better get back. Still have a huge weekend read to do.'

'Isn't that what the weekend's for?'

'I have something on this weekend,' I say, thinking of Chester House.

'Your busy social life?' he says.

'Something like that.'

'You've hardly finished your drink.'

'It's almost seven,' I say pleadingly.

'Well, don't let me keep you from your cup of hot cocoa and your pile of dull scripts,' Swan says. 'See you Monday.'

Once I get safely home, I breathe out. That was OK, I think. I handled it fine. Didn't I? Mark Swan wants to help me be a screenwriter. And I didn't melt into a puddle of seething hormones at his feet and ruin all my chances.

I'm going to be Mark Swan's protégée. That's absolutely brilliant! I can't let anything stand in the way of that. The weekend read is indeed piled up on my bed, but I can't face it right now. I'm way too excited. I make myself a hot chocolate, if you can call it that, one of those diet things from Shapers. It's not the same as the full cream milk, four sugars and spoons of Nesquik I prefer, but it'll have to do. It's part of my health kick these days.

I peel off my clothes and take the mug to bed with me, where I fall asleep, trying to think of my career and not about the man who wants to help me with it.

8

'You promised,' Janet says.

'Well, I know, but I've changed my mind,' I tell her. It's Friday afternoon, I've reluctantly come away from the office, and we're standing outside Harvey Nichols. Just me and this incredibly sexy woman. Everybody is staring. We're like Beauty and the Beast.

Janet is sticking to her guns.

'Not good enough,' she says, spiritedly. 'You *promised* me I could make you over, Anna.'

'I'm not that superficial,' I say. Why are we doing this, except to make me go through changing-room trauma? I'll still emerge looking like a cross between a giraffe and a sack of potatoes. And Gonzo from the Muppets.

Janet snorts. 'Rubbish, you've just got a complex.'

'Well, you would have too,' I say defensively.

'You can't tell me you don't want to improve your look, deep down inside,' she says. 'What about all that running? What about all the stuff you've been eating?'

I blush. I was kind of hoping nobody had noticed. I leave at six in the morning and six at night and my flatmates are not supposed to be awake then, or notice when I get back.

Janet has, though. I feel aggrieved. It's very embarrassing, isn't it? Old lard-arse suddenly getting all Jane Fonda? I hope Lily hasn't picked up on it. She'd never stop teasing me.

'You threw out those chocolate Hob-Nobs,' she says relentlessly.

Yes, that was hard.

'And you've been eating Boots Shapers sandwiches and apples. And you switched to diet Pepsi.'

'I don't know why you have to snoop around in my food cupboard,' I say haughtily.

'Hey, it's Jay-Me's crib too,' she says. 'I know wazzup in the hizzouse. I think you've lost some weight already.'

'I have not,' I say. 'Don't be ridiculous.'

'At least three pounds and probably five,' she says judiciously. 'But most of that is water weight. Still, it's coming off.'

I shrug.

'You've been doing it because of that man,' she says.

I blush deeper. 'That's bullshit!' I say defensively.

Janet looks at me as if I'm insane. 'Yes you have. Copying his lifestyle? Modelling your millionaire mentor? Eli Roth, right?'

'Oh – yes. You mean like that. Eli. Well, sure.'

'And for love, of course.'

'What?' I protest.

'For Charles,' she says, nudging me. Oh, yeah, Charles, right. 'Huh? Huh? The big party? Now what's the point of taking the weight off and getting healthy if you don't dress to match?'

'Harvey Nichols, I can't afford that. Or the hairdressing.' My raise is looking more and more anaemic.

'My treat,' Janet says. 'Don't argue. I get staff discount here. Got connections,' she says airily. 'And the hairdressing is for free. Paolo's doing it as a favour to me. I spent ages setting that up. You can't let me down, I'll look really stupid.'

'OK,' I say glumly.

Honestly, what is the point? I'm wearing a black Gap T-shirt and 501s. Another don't-notice-me outfit. Janet is wearing a pair of sprayed-on white shorts, little tangerine leather sandals with kitten heels and sexy straps tied round her ankles, a white shirt knotted under her huge boobs à la Daisy Duke, and masses of jangly silver bangles. The thought of me in an outfit like that would turn the stomachs of strong men.

'Come on!' says Janet gaily. And she drags me through the revolving doors.

'I want to go home,' I mutter when we reach our floor. It's horrible. Rack after rack of clothes that cost the earth and fit skinny girls. So many clothes. Taunting you, really. Trying to get you into a changing room. And don't start me on changing rooms. They are just torture chambers, aren't they?

'Don't be stupid,' Janet says. 'OK, waist.' She whips out a little tape and comes at me with it.

'What the hell are you doing?'

'Measuring you,' she says. 'Fit is *everything*. Waist . . . boobs . . . 37, very nice . . . bum . . .'

'Keep your voice down,' I hiss.

'OK, let's try something,' she says, professionally. Janet moves through the racks going, 'This one . . . that one . . . this one . . .'

They all look unappetizing over her arm. Is she some kind of idiot savant? Like people who look at those Magic Eye things and instantly go, 'Hey, it's a boat?' All I ever see is stupid dots and then I get a headache.

'OK,' she says. 'Off to the changing rooms!'

Bloody hell. There's no escape.

And it's every bit as bad as I'd feared. I try not to look too hard, but you can't help it, can you? It's even worse than the bathroom, because there's a harsh overhead light. And my cellulite . . .

I'm never going on another run. Or eating another apple. The whole thing is meaningless.

'How are you doing in there?' Janet says. 'Let me see what you got, girl!'

I pull on one dress without looking at it. At least it's black. And unlike most times I try on stuff, it fits. No tugging to get it to cover my arms.

I step out.

'There,' Janet says. 'Not bad, for a start.'

'Don't be daft.'

'Look at yourself,' she says triumphantly, and whisks back the curtain.

I'm amazed. I'm not going to tell you I looked like Kate Winslet at the Oscars, but the dress gives my body a little something. A waist, to be exact. The fabric sort of paints one on by cinching in. The sleeves are three-quarter length, exposing my forearms in a strangely girly fashion, and there's a plunging neckline to show off a bit of cleavage.

I stand there open-mouthed. I feel . . . I feel like people wouldn't shout rude things at me any more. Not necessarily. And that's a huge plus!

'Skirt hides your tummy, see,' Janet says. 'Until you get it in shape,' she adds hastily. 'Won't be long with all that running. OK. Want to try the trousers? With the jacket. Not that one, the navy one.'

'All right,' I say grudgingly, but I go back inside the changing room, and this time I look at the outfit after I've put it on.

'Hi,' Paolo says forty minutes later. Janet hands her shopping bags to the coat check girl while I stand there, head down. This hairdresser's is not my sort of place. It smells costly. It looks costly. There's lots of chrome and sleek leather, and hairdressers with spiky, fashionable crops, and women all carrying little designer bags – Gucci pochettes, pale pink Chanel, Louis Vuitton in ice-cream summer colours.

I'm not used to this. I'm used to Supercuts.

'I see why you breeng her,' he says to Janet. 'She need 'elp.'

'Don't mind me,' I mutter, but Janet treads on my toe.

'Only you can rescue her, Paolo,' Janet breathes as I yelp. 'She's too far gone to even *think* of anybody else.'

He preens and runs his fingers through my hair.

'Very thick,' he says. 'Lanky . . . greasy . . . too heavy . . . no life.'

Is he talking about me?

'Revolting spleet ends,' he goes on. 'No shape at all. No colour. It ees rat.'

'Mouse,' Janet says.

'Rat, mouse . . . nasty,' he pronounces.

'Come on,' I say. 'Surely you know what mouse is? You must have seen it a million times.'

'Not,' says Paolo dramatically, 'in *my* salon.'

'Shut up,' hisses Janet.

'I can 'elp,' he says. 'Jay-Me, you just leave us. Two, maybe three hours.'

I can't believe somebody else actually called her Jay-Me – wait a second, did he say two to three hours?

'Uh . . . Janet . . .' I say weakly, but it's too late. She's given me that little waggle of her fingers and she's out the door.

'Now, sweetie,' says Paolo with an evil grin, marching me to a chair. 'You are totally in my hands. Yes?'

'Ready?' he asks.

Ready? I've been ready for the last three hours. The man is a maniac. He has me over in a little chair in the corner, facing a wall. He won't let me see myself in a mirror.

'I don' wan' any interruptions,' he said. 'You are canvas, *carissima*. Canvas does not tell artist what to paint, no?'

'No,' I say meekly. It can't be any worse than what I have now. He's been slashing with a razor, talking to himself, lecturing me. Everything is wrong about me, apparently. My hair is a total disaster. My jeans are all wrong.

'Wear lower waist,' he says. 'That way you not seem so *beeg*. Unnerstan? And what *ees* this?' he asks, grabbing my hand.

'My hand?'

'Not that. *Thees*,' he says, pointing at my nubby, ripped-off nails. 'Thas *deesgusting*.'

'Sorry,' I say meekly.

'Clara! Clara, *cara*,' he says, snapping his fingers. A young girl comes over, she can't be more than eighteen, and Paolo rattles something off to her in Italian. She swears. Obviously, I am a completely revolting specimen.

'She will geeve you manicure,' he says. 'You are friend of Jay-Me. One time, no charge. *Va bene*?'

'Yes. *Si*. Thanks . . .' I trail off lamely. Please not a manicure. I hate my fingers. They are all workmanlike and thick, not slender and delicate. I don't like anybody looking at them or fussing with them. But I can't very well say 'no, sod off', can I? Paolo is donating his valuable time here, not to mention several little plastic plates of foul-smelling, dark goopy colour that he's painting into my hair, foil segment by foil segment.

Other girls really enjoy this sort of thing, don't they? To me it's torture. I can't remember when I was last so bored. And I have to make conversation. Paolo finds out I'm in the film business and he wants to tell me all about his idea for a movie. Which involves a hairdresser and a gay beauty pageant and how the hairdresser wins. And I go on about how fabulous it sounds.

Finally, it's over, apparently. Paolo has wheeled my chair over to a mirror.

'Yes, I'm ready,' I tell him.

I prepare for the worst.

Paolo spins me round. 'Tah-dah!' he says.

For a second I can't quite believe it. Is that me?

My hair – it's gone. Most of it. What's left curls down an inch above my shoulders, and has a feathery, choppy way of swinging about my face. My nose actually looks smaller. I have a fringe, it softens my high forehead, and best of all I can tell that this cut will just fall into place. There aren't any special gimmicks required, no curlers for soft waves. I can wash it in the morning and it will still be there.

And then there's the colour.

I never realized colour could make such a difference. My hair, my mousy hair, has gone. Now there's a blonde there, not brassy like Claire, but shot through with silky highlighted strands in millions of different shades, from spun gold to copper, champagne, lemon, honey . . .

It hasn't just changed my hair. It's changed my whole face. My skin doesn't look so pasty any more. The thick dark cloud around it has gone.

I'm still not pretty. But I look normal now, almost normal. Just slightly on the plain side, sure. But not . . . not really ugly, apart from my nose.

'I don't know what to say,' I say, and I have tears in my eyes.

Paolo looks genuinely pleased.

'Yes, yes, I am a magician,' he says. 'You come back and next time you pay!'

'Oh, yes,' I say, and right now I feel I'd happily give him my firstborn daughter and half my kingdom.

Janet arrives to collect me, and after five minutes of air-kissing and mutual backslapping we get to pick up my bags and leave.

'When we get you home I'm going to pack for you,' she says. 'I know exactly what you must take. And I'll do your make-up. Are you going to wear the blue dress for the dancing or the green?'

'The green.'

'Better be the blue,' she says.

I smile gratefully at her. 'OK, the blue. Thanks, Janet. Thanks a lot.'

'All it takes is confidence,' she says. 'And the right fit.'

Well, that's not true. But I do feel an awful lot better. I'm almost looking forward to the party now. Just wait till Charles sees me! I fall into a lovely reverie where he's so dumbstruck he proposes on the stroke of midnight in front of the whole crowd of Hoorays, and all the beautiful women who were only after him for his money are all crying and wailing that it isn't fair. Only in my fantasy Charles obviously has a few muscles. And I'm a bit thinner and I've had a nose job. Charles is also a bit taller, and dark, without a goatee. In fact, he looks a lot like . . .

Stop that.

'Here we are,' Janet says, paying the taxi. 'Let's go and get changed. You're practically ready anyway, but I've got loads to do.'

We walk together up to the flat and just to put me in an even better mood, I'm not winded. I could really get into this fitness stuff. My new haircut makes me want to pull my jogging pants on and go out right now for a run round

Covent Garden. But I suppose that might mess it up . . .

We walk through the door, and Janet pulls out a case of hers. 'Right. You'll take this . . . and that . . .'

She's packing me as efficiently as an air hostess and has the bag zipped in two minutes flat. It normally takes me at least half a day.

'Where did you learn to do that?'

'Oh, you know. Packing for shoots. All round the world,' Janet says airily, then her face falls. 'Course, been a long time since I did any international gigs.'

The door bursts open and Lily appears, talking into her mobile phone. 'Yes, well, must dash, kiss-kiss, darling, ciao-ciao,' she says, flicking it off. 'Oh my God!' she squeals. 'Look at *you*! That's too funny.'

'What's funny?' Janet demands. 'She looks fantastic.'

'Well, let's not go too far,' laughs Lily.

She's wearing a tiny pink sundress with a cute little matching pink cardigan and pearly buttons, the palest pink sandals, and not much else. She's been to the hairdresser too. Her platinum blond mane spills expensively down her back.

'But,' she adds, 'it's definitely a *great* improvement, although what wouldn't have been, eh, Anna?'

Instantly I feel all my lovely happiness seep out of me.

'Fuck off, Lily,' snaps Janet. 'She looks great. You really do,' she adds to me.

'Oh, you do, you do,' says Lily gaily. 'You'll be the belle of the ball. I must run to the bathroom and repair my face,' she says, scrutinizing herself in the mirror over the mantelpiece. Her radiant beauty stares back out at her. 'I look simply *vile*.'

She flounces off into the bathroom, locking it.

'Sod her,' says Janet stoutly. 'Don't let her ruin everything for you.'

'I'm not. I know I – look better,' I say. But it's no good. I'm still the Ugly Sister. It's still a joke to have me turn up with these two in tow.

'You know, Lily doesn't have a boyfriend,' Janet says.

'What about Claude?'

'Her sugar daddy? Come on,' Janet says disparagingly. 'If

198

she's so hot how come she can't get herself someone her own age with pots of money? They just hear her talk for five minutes and it's all over. Anyway, better go tart myself up. You take off those clothes,' she adds sternly. 'I've put your travel outfit on your bed.'

'Thanks, Janet,' I say, and impulsively give her a hug. She's actually turning into a really good friend.

'And then I'll do your make-up,' she adds.

I chuck my jeans and T-shirt in the laundry and change into Janet's choice for the car: flat-fronted, low-cut charcoal grey trousers and a silk-looking silver shirt. And some slides. I turn in front of my mirror and try to get some of the euphoria back.

OK, I'm still not pretty. But Charles is going to be knocked sideways. And everybody in the office will be really surprised. It'll be great for my career. Like Anna Brown is coming out of her shell. And Mark Swan will be . . .

Mark Swan won't care, obviously. Except he'll think I'm blooming. Professionally.

'Ready?' asks Janet, sticking her head round the door. She's wearing a gorgeous lemon-yellow halter-neck dress that shows off her boobs and butt and a single canary diamond on a chain, with a pair of black strappy high-heeled sandals.

'That's very J-Lo,' I say.

'You think?' she asks, pleased. 'Now sit still, this won't take a second,' and then she's all over me, dabbing at my face with little sponges and pencils and brushes and what looks like a pot of lip gloss that's a scary, blood-red colour.

'There,' she says. And when I look, it's wonderful. All very neutral, but painted-on cheekbones and smoky eyes, and red, wet-looking lips, more make-up than I'd ever have dared. It really doesn't look that bad, apart from my nose. I feel . . . what's the phrase the Americans use? Pulled-together. That's it.

'OK!' says Lily, emerging from the bathroom looking even lovelier, in palest-pink lipstick and with glitter on her cheeks. 'Got the map? Let's go!'

* * *

It's seven thirty by the time we're slowed down, looking for the turning into Chester House. Everybody's in a filthy mood. Janet because she's done most of the driving, Lily because she's not in a limo, and me because I'm starving.

Losing weight sucks.

Normally, I quite enjoy long journeys in the car. I put on my music and every time I stop at a garage it's straight into the shop for some supplies. Huge family-sized packets of Quavers, chewy mints, a Strawberry Mivvi and a packet of Rolos, or equivalent. And I never feel guilty about it either. I mean, it's garage food, it doesn't count, does it? It's on the road.

This time, when we pulled into a garage, Lily came back with a huge packet of salt'n' vinegar Discos and a Milky Way for me, but as soon as we pulled onto the motorway Janet chucked them out of the window.

My stomach's rumbling embarrassingly. I can't wait for dinner. Suddenly the procurement of food has taken precedence over not wanting to do this.

'You must have missed it,' Lily says petulantly. 'Go back. Go back to that pub!'

'I bloody haven't missed anything.'

'Well, you must have.'

'There haven't been any turnings!'

'What's that?' I say.

Up ahead there's a mini traffic jam – Land Rovers and Jags trailing back along this tiny little B road with its thick hedges and overhanging trees.

'They're turning in,' says Lily, relieved. 'This is it! Can't you hang back a bit, Janet?'

'Jay-Me,' corrects Janet. 'And why would I do that?'

Lily pouts. 'Well, we're only driving an old Renault,' she says. 'I don't want anybody to see.'

'Oh, fuck off,' I say, and thankfully, we're bumper to bumper, and everybody is trailing in to the left.

'Ooh, look at that,' says Lily, sounding excited. We all look up. There are two pillars at the side of the road, enormous, thick, old grey stone, and two rampant lions perched on top of them, hunched over shields.

'Ooh,' says Janet.

'Very clichéd,' I say, pretending not to be impressed.

'Look at this drive,' Janet says, and indeed we're through the gate now and onto a wide, bumpy road, and the cars are streaming down it. On either side rolling grass, gentle hills, copses of oak trees dotted here and there. You can instantly imagine horses and carriages trundling down here for similar parties a couple of hundred years ago. It's sort of like heaven. Or Cinderella's castle or something.

'He can't own this,' Lily says. She's sounding uncertain.

'Apparently he does,' I tell her.

'My God,' she snaps. 'Why you?'

'Look at *that*,' breathes Janet in awe.

And we do. There's the house. If you can call it that. It's more like an enormous mansion that you go to visit on school trips. A vast edifice of the same grey stone as the pillars, complete with spikes and balustrades and statues on the top, like St Peter's in Rome. It's got ivy trailing all over it. It's magnificent.

'How many bedrooms?' asks Lily, sounding expert. 'Twenty? Thirty?'

'Probably more,' says Janet.

'This *has* to be National Trust,' says Lily.

'I've got no idea,' I tell her.

'It isn't,' say Janet triumphantly. 'I looked it up on the internet. It's all his.'

No wonder all those girls come after him, I think. And catching sight of Lily looking at me through narrow lids I have to agree with her. Why me?

'It's only been a few dates,' I say humbly.

'Exactly,' says Lily instantly. 'You could hardly call it a relationship, could you?'

'Forget it,' says Janet coldly. 'You're not going after him. I think it's destiny,' she adds to me.

I don't know what to think. I mean, I like Charles. I just never saw a future with him. Let's face it, I never saw *bed* with him, so how was I going to get into wedding bells? But at the risk of losing sympathy, I have to say, right now I can't

201

think straight. All I can see is this huge. Enormous. Incredible. House.

'I'm glad we did that makeover,' Janet murmurs to herself.

We pull into a wide, circular driveway at the front of the house, with ushers showing all the cars where to park.

I wind my window down for the uniformed man checking the names.

'Anna Brown and guests,' I say.

Lily shoves her head past me. 'That's Ms Lillian Venus and Ms Janet Meeks.'

'AKA Jay-Me,' says Janet, as I blush scarlet.

'Very good, madam,' the man says serenely. He beckons to one of the parking ushers.

'This is Miss Brown's car,' he says, significantly. 'If you'll just follow him, madam,' he adds to Janet.

We are being conducted straight through the rows and rows of gleaming cars, our wheels crunching gravel, trundling past everything from Porsches to Ferraris, with lots of Jeeps and the odd souped-up Volvo. The usher beckons us to a halt.

'Right here, madam,' he says.

'Look at this,' says Janet. We have been allocated a parking spot right by the front porch.

The usher springs into action, opening our doors and taking our little cases, but I insist on grabbing them back from him. I already feel incredibly self-conscious.

'Anna,' says a familiar voice. I turn round to see Charles standing there. He stares at me for a few seconds, blinking. 'Good Lord,' he says. And pauses. 'Good Lord.'

I smile encouragingly, but does he have to look quite this shocked? He recovers, reaches towards me, gives me a peck on the cheek, beaming. 'Good of you to come, good of you to come.'

'We wouldn't have missed it,' says Lily huskily, shoving herself forward and thrusting her hand at him, 'for the world. I'm Lily Venus.'

Charles kisses her hand, and she gives a practised, little-girl giggle.

Lily Venus? Her real name's Frutt. Doesn't she realize that

if she's going to use a pseudonym it shouldn't sound like a porn star's?

'Delighted,' he says, staring at her awkwardly as Lily licks her lips. 'Ahm, delighted.'

'And this is my other roommate, Jan – er – Jay-Me,' I say.

'All right!' says Janet in a friendly voice. 'How's it hanging?'

'Yes. Marvellous,' says Charles faintly. 'Shall I show you to your rooms?'

The entrance hall is thronged with people in evening gowns and penguin suits, all braying and trying to kiss Charles, but he threads his way rather expertly through the crowd and leads us up a wide, stone staircase, lined with solemn-looking oil portraits of his ancestors.

'Here we are,' he says eventually, after we've been through three corridors and I'm lost. 'The William Suite. I hope that's acceptable?'

'Why is it called that?' asks Lily, peering in at the sumptuous interior.

'Oh, you know. William the Third liked it. Stayed here quite often with one of my predecessors. It's a bit fusty, I'm afraid,' he adds apologetically. 'Anyway, I'll leave you ladies to change. Cocktails before dinner downstairs. Just follow the crowd. Anna,' he adds, in a low, intent voice as Janet and Lily race into the room in delight. 'You look absolutely sensational.'

'Thanks,' I say, going pink with pleasure.

Charles gives a stiff little bow and withdraws.

I don't look sensational, but it's nice to hear it. Very nice. And everybody's sucking up to Charles madly, and yet I'm his date.

'Bloody hell,' says Lily, looking around as soon as the door shuts. 'I could get used to this.'

It's like being allowed to walk inside one of those roped-off room exhibits in a palace or a stately home. Everything in here is probably a priceless antique. The walls are hung in a rich yellow damask, there's a vast four-poster with intricate carvings on its dark legs, and an opulent daybed; the thread-bare carpet looks Persian, the chairs look Louis XIV . . .

Janet is staring out of the huge, lead-panelled windows, and even though it's twilight, you can still see the orchards spread out below us because they've been lit with flaming torches.

'Imagine, Anna,' she says, disbelievingly. 'All this could be yours.'

We get changed with lightning speed, Janet and Lily because they can't wait to unleash themselves on some unsuspecting duke or other major landowner, and me because if I don't eat soon I'm going to start on wads of loo roll just to stop the gnawing pains in my stomach.

Lily is wearing a slimline strappy number in palest gold silk, with teetering matching sandals, and diamond drop earrings some past boyfriend bought her, her hair up in a chic French twist.

Janet has chosen a burnished copper gown in crushed satin, a full ball gown look, with a corseted waist that throws out her boobs and bum. She's let her raven hair fall loosely and she looks stunning, very Catherine Zeta-Jones in *Zorro*.

I'm wearing the blue dress Janet picked out. It has a silky look to it, a full-ish skirt, and a boat neck over three-quarter sleeves. It shows a bit of cleavage and covers my bum and my legs. I've also got on some real pearls which Janet lent me – she wouldn't let me go with fake ones.

I look like an ordinary girl with a big nose. But let me tell you something, that's a huge improvement! It even feels strange to wear colour. I've got enough black in my wardrobe to run a funeral parlour.

I resolve to try not to look too much at the other two. Anyway, if we don't get downstairs soon I may just eat them.

Cocktails are finished, and people are being ushered into the dining hall. Thank God. If I drank on a stomach this empty I'd be puking my guts out before nine.

Charles is there, waiting and hovering.

'Ah, ladies,' he says. 'You all look absolutely incredible.'

'Fuck me, who's this?' asks a man standing next to him.

Despite the upper-class accent and the black tie he looks like a huge oaf. 'Charles, you lucky sod. You're not seeing both of 'em, are you?'

'Shut up, William,' says Charles, loudly. 'Please excuse my friend,' he says to us.

'Sure,' says Lily, smiling at him. She grinned when he said 'both of them' instead of 'all three of them'.

'Jay-Me and Lily, this is William Lyons,' Charles says. 'And Anna is my girlfriend,' he adds, with a touch of proprietary pride.

'We're not seeing anyone,' says Lily.

And suddenly there is a throng of black-tie men all around us, some shaking my hand perfunctorily, others ignoring me altogether as they try to get closer to Lily and Janet. Charles and I are shoved to one side, but not before I've heard one of the Yahoos say to his mate, 'Fancy a bit of rough?'

'Mmm, yeah,' says the other one, in a low voice. 'Good for a couple of tumbles, at least.'

They laugh and try to get through the little crowd of hopefuls. And I suddenly feel a burst of anger. I actually feel protective even of Lily. How dare they talk about her like that? Just because she isn't some horse-faced gentleman's daughter called Camilla or Prudence. Poor Lily, she's got no idea, I think instantly. She'll have a rough time of it looking for a husband in this crowd.

I push my way back to them. 'Excuse me.'

The crowd melts – they have to, I'm a woman – and I thread my arm through both of the girls'.

'I think we should go in to dinner now,' I say.

'Let me go!' hisses Janet, outraged. 'You've got yours!'

Some of the men hear her and chuckle to themselves, which makes me even angrier. I drag the two of them into the hall. It's easy. I've got thirty pounds on both of them.

'She's right,' whispers Lily to Janet. 'Always leave them wanting more,' she says, like an expert.

I examine the seating chart for their places. 'You're on four and you're on nine,' I say to them. 'And look, find me after dinner. I really need to have a word with you.'

'No you don't,' Janet says to me reassuringly as she walks away. 'I've done all I can. You're on your own now, Anna.'

Dinner is delicious. Luckily for me there's no bread basket, because I would have ripped through the whole thing right away. As it is, my stomach keeps rumbling and I have to keep coughing to cover myself.

'That sounds nasty, old thing,' says Charles solicitously. 'Have some champagne.'

We're sitting together on table one. It's the most enormous dining hall with an iron-clad fireplace big enough to take an entire tree, and a huge vaulted ceiling. I can imagine long medieval trestle tables with benches in here, but today there are round tables, covered in white linen, with chairs set around them equally covered in linen. Little fairy lights and chiffon drapery have been pinned to the walls, giving the whole room a sort of Snow Queen feel. There are tall crystal vases crammed with flowers and gold-painted grapes scattered around the bases.

'Just a little party,' Charles says modestly, at my open mouth. 'You know, sometimes I really push the boat out. I was thinking about something special for my birthday. Maybe you could help? Plan it with me. You know. As the hostess.'

'I . . . maybe,' I say. 'This is amazing, Charles.'

'Well, you look divine,' he says. 'The dress, and the . . . the . . .' He waves his hand at me and I feel a little thrill of pleasure. He's not so bad, Charles, is he? He honestly seems really nice.

Then he puts his hand gently on my upper thigh and squeezes it through the blue folds of my dress.

I half choke on the champagne. Urghh . . .

'Caviar, ma'am?' asks a waiter.

'Oh! Yes, thanks,' I say, jumping, so Charles is forced to dislodge his hand. What's wrong with me? He was only putting his hand on my thigh. He's sweet, isn't he? Free from any obvious deformities? No perversions? Enormous mansion, millions of pounds?

So why do I feel like bolting as fast as I can leg it?

The waiter heaps a huge mound of glistening grey pearls on my plate, and everybody tucks in, mixing in chopped egg and other things from little bowls.

'Sevruga,' Charles says. 'I don't care for Beluga, I think it's a bit bitter . . .'

It tastes good. Fantastic, even. But a can of dog food would probably taste good at this point. Fortunately, mixing and scooping and eating pre-empts too much conversation, and then the middle-aged woman in black taffeta next to Charles is shouting in his ear, and I can relax again.

The main course is unremarkable, roast pheasant with stuffing and mashed potatoes, and I'm just wolfing mine down when the man on my other side speaks.

'Hello,' he says. 'I'm Ed Dawson.'

'Anna Brown,' I say, shaking his hand. He's about my age, maybe a year or two older, rather nondescript, with light brown hair and hazelnut eyes. A bit skinny for my taste but a nice smile.

'So I gather Charlie's lucky enough to have you as a girlfriend?'

'I – well, we've only just started . . .' I begin, but become aware Charles is listening in. 'Yes,' I say lamely.

'That's fantastic,' he says warmly. 'I'm his cousin. We've been hoping for ages he'll find someone nice. He usually dates such slappers.'

'Um, well. We – we have fun,' I say.

'I was just wondering,' he says, awkwardly. 'The young ladies you came in with . . .'

'Yes?' I ask. Defensively.

His face falls. 'Nothing. Nothing. Obviously they have boyfriends . . .?'

'Actually, no,' I say, 'but they're both very nice girls.'

'Of course,' says Ed. 'They look . . . *very* nice.'

'You mustn't judge them by their looks,' I say hotly. 'Just because they're beautiful, they're not people's toys.'

'Gosh, no,' he says humbly. 'Of course not.'

'They've got their own brains and . . . careers and things.'

'What do they do?'

'They're . . . models,' I say reluctantly, 'but they take it very seriously and . . . manage their own money.'

'That's excellent,' he says. 'Tell me about the blonde girl'.

'That's Lily,' I say. 'Lily Frutt – er, Venus.'

'Lily Frutt-Venus,' he says, reverentially.

'Just Venus,' I correct myself. 'And the brunette is called Janet, but she likes to be called Jay-Me.'

'Why's that?' he asks.

'Just because,' I say, and then Charles leans over to ask me something, and Ed Dawson turns to his other side.

Dinner proceeds just fine after that. I have maybe just a tiny bit too much champagne, but I do justice to everything – the green apple sorbet, the plate of cheeses, and the petits-fours with coffee and brandy. Charles doesn't put his hand on my knee again, just makes small talk. I'm really starting to enjoy myself, actually, the champagne and brandy combining to make me *incredibly* mellow.

'Have you come to a decision on the book?' he asks.

'Huh?'

'My book,' he reminds me. 'Anna's been reading my book so her company can make it into a film,' he tells the battleaxe on his other side, proudly.

Oh help. I asked John for his report too, just to be certain, I mean it's more his type of thing. Theoretically. Only it wasn't. 'Total bullshit,' I think it began. 'Sophomorphic approximation of Proust without any of the fun! Takes himself completely seriously and is completely awful . . . no plot . . . no characterization . . . melodramatic description . . . no potential whatsoever.'

'Oh yes,' I say, trying not to slur my words. 'I've been meaning to talk to you about the book.'

'Is it absolutely divine?' asks the battleaxe loudly.

'Anna's reading my book!' Charles announces to the entire table. And now they're all looking at me. I should just tell the truth. But obviously I am not going to.

'Well,' I say carefully. 'My reader thought your manuscript was very different, Charles. He even said it was unique.'

This is the worst book I have ever read, went the conclusion.

'Really,' says Charles, beaming.

'It stood out of the crowd for him.'

In two years as a reader, I have never been subjected to such drivel . . .

'He thought it was amazingly intricate.'

Muddied, over-complex and pretentious . . .

'So?' Charles says eagerly. 'Are you putting it into production? I might consider directing, you know.'

My mouth goes dry. Suddenly I remember how Vanna handled it.

'I don't think it's right for film,' I say carefully, 'because it's so complex, so varied. Film would butcher it!'

'I see,' says Charles, and his face falls.

'Oh, it's too good for the pictures,' says the old boot. 'Isn't that so, Anna?'

'Absolutely!' I say brightly. 'Too good for them!'

But Charles is swigging from his brandy and now I feel horrible. All mean and nasty, when he's tried to be so nice to me.

'Charles,' I whisper, but he shakes his head.

'Come on, everybody,' he says loudly as the waiters whisk away the plates. 'Shall we dance?'

The dancing is pretty awful. The room is glorious, of course, hung with golden silk drapery and featuring a band doing live Scottish reels, which I am flung into without knowing any of them. And then there's disco, which finally gets Janet and Lily onto the dance floor. Charles dances the upper-class shuffle with me. It involves putting your arm round a girl's waist, then lifting one foot and putting it down in exactly the same spot, and repeating with the other foot. Sort of marching in place. With the occasional turn. When we finally break for a while, I think it must be 3 a.m but it's only been an hour.

'Bugger,' says Charles to me. 'Must go and check on people. Host, you know. Is that OK?'

'Oh, of course,' I say, gratefully. 'I'll be at the bar.'

I rush off to get myself some more champagne (where

would evenings like this be without booze?) and find Ed standing there at the bar, looking longingly after Lily who is dancing surrounded by a crowd of admirers.

'You could always ask her to dance,' I say.

'What's the point,' he responds, gloomily.

Lily glances our way and waves at me. A second later she's broken off from the throng of men and is teetering towards me on her high heels.

'What shall I do?' asks Ed, panicked.

'Champers!' says Lily loudly, snapping her fingers at the barman. She grabs the crystal flute without thanking him. 'Anna,' she says ingratiatingly, 'wonderful party. All the boys tell me Charles is very taken with you.'

'He is,' agrees Ed.

'He thinks you might even be the one, apparently.'

After the book thing, I doubt it. 'Don't be silly. I've only just started—'

'No, it's true. You see, Charles is tired of pretty girls,' Lily explains to Ed, with a sigh.

'Then why is he going out with Anna?' responds Ed, chivalrously.

Lily gives her patented breathy giggle. 'You're so funny. And you are?'

'Ed Dawson,' he says eagerly. 'Lovely to meet you. Lily, isn't it?'

Janet bounces up to us in a froth of bronze silk.

'Top party,' she says, breathlessly.

'Have some champers!' Lily says brightly.

'Just water,' Janet gasps. She is a bit red-faced. 'I was doing my J-Lo salsa and I'm knackered. Wazzup?' she says to Ed.

'Excuse me?'

'Don't be so affected, Janet,' snaps Lily.

'I'm Ed,' he says, recovering, as Janet pumps his hand. 'And you must be the lovely Jay-Me.'

'Yeah,' she says, going redder with pleasure. 'Right on!'

'So, Ed,' starts Lily.

Here it comes. The patented Lily Frutt Inquisition. She's going to find out everything about him. Or at least everything that can be written on a balance sheet, which I guess is all that matters to Lily.

'Dawson,' she says. 'Same name as Charles. Coincidence?'

'Not entirely,' he says, diffidently. 'I'm his cousin.'

'Older branch of the family?'

'No, younger,' he says. 'We're quite distant. Second cousins.'

'And what do you do?'

'I work on a farm,' he says.

Lily pulls a tiny face. 'How interesting,' she says distantly. 'And I suppose that must be hard if you have a big place to run? Like this one?'

'Oh no,' says Ed. 'I live in a small flat in Bath.'

'You own it?' she asks brightly.

'Rent,' he says. 'And it's daylight robbery. Of course Bath is a very pretty town. Very interesting. Some of the history is absolutely fascinating—'

'Excuse me,' says Lily, cutting him off with a cold smile. 'I must get back to my dance partners. Nice to meet you, anyway.'

And she sets down her crystal flute and marches off instantly. That's actually not the worst I've seen her. Once we were in a club and a guy comes up and asks her to dance. She says, 'What kind of car do you drive?' He says, 'Ford Fiesta.' She turned her back on him.

Ed's been let off easy, but he doesn't seem to appreciate it. He stares miserably into his champagne.

Janet and I exchange looks.

'I've always wanted to go to Bath,' Janet says warmly.

'Really?' Ed asks.

'Would you like to sit down?' she says, kindly. 'And tell me about . . . farming?'

Some of the humiliation clears from Ed's face. 'We could go and find some chairs,' he says. 'And they're serving sorbet in the next room. Charles has an ice-cream parlour set up.'

'I'd love a sorbet,' says Janet.

'That's a jolly nice frock you're wearing,' he says.

She smiles at him. She seems to like him. I'm relieved, and suddenly exhausted. All I want to do is go to bed.

'Excuse me,' I say to them, but they're deep in conversation already and ignore me. I scope out the room. Oh yeah, there's Charles . . . surrounded by girls. It's like Lily in reverse, only somehow I don't think it's his body they're after. I walk up to him.

'Hi,' I wave.

He sees me through the crowd and extricates himself. He looks hunted.

'Anna,' he says with relief. 'This is Anna,' he says to the crowd. 'My girlfriend,' he adds firmly.

Damn. This is ridiculous. I like Charles but . . .

I know I should be sensible. I'm a girl without many romantic prospects. I've managed to claw my way up the beauty ladder to only three or four rungs below normal. And here's a nice guy with a huge fortune and he's actually interested in me.

But I look at the goatee and the platform shoes and I just know . . .

There's no spark. At all. I just can't spend the rest of my life with this man.

I'll tell him, I promise myself. I'll tell him tomorrow.

'Lovely to see you, darling,' I say just as loudly, kissing him on the cheek. 'You know, I'm . . .' I resist the temptation to say 'knackered', 'rather tired. I think I might pop off to bed.'

'OK,' says Charles gloomily.

'I've been having the most wonderful time,' I hasten to reassure him. 'I'll see you in the morning.'

I wake up not quite sure where I am. There's the familiar sound of Lily snoring like a train and the unfamiliar sound of twittering. It's birds. Outside.

I blink and clear my head. Right, I'm at Chester House. Blearily I haul myself out of bed. No chance of waking the two Sleeping Beauties, they're both crashed out in their party frocks and full make-up. I can feel the beginning of a nasty headache crunching around my temples but at least I'm in my nightie. They're going to suffer when they get up.

As I step into the bath, this thought cheers me up. But only a little. I've got to break up with Charles today.

I look regretfully around me at the beautiful house (the bathroom has antique Chinese wallpaper), the furniture, the carved four-poster, the view outside the window – in the daylight even more special: apples trees covered in green apples, lush grass, terraced lawns, a large pond, even some deer – actual deer – grazing in the background. I must be mad.

But I just can't do it to him.

Poor sod. He's been mucked about by women after his great country pile all his life. I can't do the same thing, can I? Go on dating him, just because he's loaded? We've nothing in common, except that we're both part of life's rejects club.

I put on make-up, quickly, check myself – the hair still looks pretty good, Paolo really is a magician – and unzip my case. Janet has packed everything, bless her. The morning outfit is the black dress I first tried on. With a pair of stack-heeled slides. Add the bronzer on my pale skin and I look sort of OK. Definite result.

I've just managed to zip myself up in the back when there's a quiet knock on our door. I tiptoe over (though nothing short of a nuke is going to wake those two right now) and open it a crack.

It's Charles.

'Are you decent?' he hisses.

I nod. 'The other two are asleep.'

He fidgets a bit. 'Can you come out for a walk?'

'A walk? What time is it?'

'Half eight,' he says. He looks totally despondent. 'I understand if you don't want to.'

'No, no,' I say hastily. 'I'd love to come for a walk. Obviously.'

Charles leads me down a flight of back stairs.

'We won't go through the kitchen,' he says. 'Bit of a zoo in there. Lots of people never even went to bed.'

'I expect they're having breakfast,' I say hopefully. I could kill for a good breakfast. A bacon sandwich. And scrambled eggs. And maybe some grilled mushrooms and tomatoes . . .

'Want a Husky?' he asks, handing me a coat. 'And some wellies? You don't want to ruin those lovely shoes . . .'

I suit up obediently. How long will this take before I can steer him back to the important subject of breakfast?

'We'll just slip out here,' Charles says, opening a little side door. And now we're outside. It really is a beautiful day, already warm, the dew starting to evaporate. He leads me along gravel paths lined with lavender bushes, down the terraced steps of his lawn, past huge stone urns covered with trailing roses.

'Nobody can see us now,' he says breathlessly.

'That's great,' I say nervously. He isn't going to make a move, is he? Expecting a good morning quickie?

'Anna,' he says, and suddenly he just looks so down I want to hug him, 'my life is shit.'

I glance behind us at the grey stone mansion.

'Um . . .'

'You don't need to lie about the book,' he says plaintively. 'Vanna got drunk last night and told me I had no talent.'

'I – well –'

'I thought it was amazing,' he says. 'I thought it was a masterpiece. It took me ages to write,' he adds defensively.

'You had great discipline,' I say weakly. 'Just because she – we – didn't care for it . . .'

He shakes his head. 'It's rubbish. Like everything else I do. You don't know what it's like,' he says, unhappily. 'Everything I try goes bad. I went into the stock market and lost half a million. I bought a racehorse and it never won. I tried to be a lawyer once . . .' shakes his head. 'Stupid exams. And now this. It was my way out, it was everything. I'm nothing, am I? I'm just a useless failure. And people are all so horrible to me. All the bloody time.'

'You know, when I met you,' I say carefully, 'you were a bit . . . stand-offish and stuck-up. But you've got a lot better since then,' I add hurriedly.

'It's only because people laugh at me,' he says pathetically.

'If I don't stand up for myself they all laugh. So I have to say that it's literature. And you know, then they can't laugh so much. I don't give them the chance.'

'So . . . you're striking first,' I ask. 'In the stuck-up stakes?'

He looks at me, smiles weakly, and nods.

'And girls?' I ask relentlessly. 'You were quite mean about how I looked.'

'Same thing,' he says. 'If you look like me. All those pretty girls. At first I believed them,' he says bitterly. 'But they were always just laughing. I can't help it if I'm a bit short. And my hair's going.' He turns to me and takes my hand. 'You're so different from them, Anna. You didn't know about Chester House. You've got your own life. A real career,' he adds, admiringly. 'More than I've got. I want you to help me find something to do – just to help me,' he adds, sadly. 'If I didn't have you I wouldn't have anything.'

I open my mouth. Now would be the time. To bravely break up with him.

'You're the one good thing in my life,' he says, and to my horror breaks into sobs.

Pants!

'Don't – don't cry,' I say, pleadingly, fishing around for a tissue. There's a disgusting old hanky in one of the pockets of the Husky. 'Here.'

He blows his nose loudly and offers it back.

'Um, you keep it,' I say.

'I'm so lucky to have you,' he goes on. 'If you left me I don't know what I'd do.'

What can I say?

'Don't worry,' I tell him with a forced smile. 'I'm not going anywhere.'

Back in the kitchen there are indeed loads of people, and to my amazement two of them are Lily and Janet. They've pulled themselves together in record time. Lily is wearing sexy white jeans and a clingy pink sweater, and Janet has on a red skirt and matching silky top that look wonderful with her skin tone. They are both nursing Alka-Seltzers. And sitting with blokes!

OK. The fog lifts. Nothing but men would persuade them to get up this early. Janet is curled up with Ed on a cushioned window seat, and Lily's sitting on the lap of someone I haven't met yet. He's younger than her usual type. Very handsome, dark and intense. And there are Vanna and Rupert; I didn't see them last night.

'Anna,' Vanna says, looking absolutely thrilled. 'You two lovebirds been out for a romantic tryst?'

'Yes,' says Charles smugly. 'In the orchard.'

Everybody goes 'Oooh'. My skin prickles with embarrassment.

'Love the new haircut,' says Rupert loudly. 'You look a much better filly like that. Have you lost weight? Bloody good job. Ow, Vanna! Mind your damn foot, it's on my toe!'

'Vanna, have you met my flatmates?'

'We've been introduced,' Vanna says politely, which means she doesn't think much of them. Vanna's quite high-powered. She doesn't value modelling as a career.

'I hope you slept well, Anna,' says Lily, warmly. 'We don't want you too tired on the way back. Anna's a top movie executive,' she announces to the room. 'She needs to be fresh! For her business.'

I blink. Apparently I have travelled into another dimension, where Lily is nice and sings my praises to everybody.

And then it dawns on me. Charles still has his arm round me. Vanna and Janet are beaming at him. A few of the Sloaney girls curled round their coffee cups are casting death stares in my direction.

They all have me married off already. I might as well be registered at Harrods and popping in to Vera Wang for a fitting.

'Who's this?' I ask Lily, desperate to change the subject.

'I'm Henry,' he says, extending a firm hand. 'Nice to meet you.'

'Henry's going to be my boyfriend,' Lily purrs.

'Only if you're very lucky,' Henry says sternly. 'And play your cards right. And know how to iron a shirt properly.'

Everybody laughs while I just blink. Nobody talks to Lily

like that, even for a joke! But she doesn't seem to mind. She's tracing her fingers across his chest.

'Henry's in property,' she says. 'It's a very exciting field.'

'Deathly dull, actually,' Henry says. 'But a family tradition.' He shrugs.

'That's Henry Marsh,' Lily tells me, with a sly nod.

Oh, I get it. Henry Marsh. Must be of Marsh and Strutter, one of the biggest estate agents in the country. You see them everywhere. Trust Lily to pick up a property mogul for herself. But she really seems taken with him, at first glance. Not just stringing him along, like she does her older, richer sugar daddies.

'Henry Marsh and Lily Venus,' says Ed, squeezing Janet's hand. 'Men are from Marsh, women are from Venus.'

'Oh, her real name's not Venus,' Henry says stoutly. 'That was a load of old bollocks. It's Frutt.'

Lily nods meekly. Bloody hell, I really like Henry.

'Want something to eat?' Charles asks me.

'Oh! Yes,' I say. 'I'd like bacon and eggs, please, and toast and marmalade, or jam will do . . .'

I catch Janet's eye.

Damn.

'On second thoughts,' I say morosely, 'maybe just some cereal's fine.'

On the drive back, I don't have to say much, luckily. Janet does ask me some questions about Charles, but all she really wants to do is talk about Ed. And Lily is equally taken with Henry. All that's required of me is to drive home, and wonder how the hell I get out of this.

'He's bound to be worth masses.'

'I think Bath's a really beautiful place. He wants me to go for the weekend.'

'They've got offices everywhere. They're *the* people to see in Chelsea.'

'He says he loves milking cows. He knows how to do it even though they use machines now.'

'Henry's probably got lots of properties himself.'

'Ed likes rugby – do you think I could learn rugby? He said he'd take me to a game.'

'I bet he owns, like, huge lofts in Soho and strings of buildings.'

'Ed said I was better looking than J-Lo. And he said he likes the name Janet.'

'I bet he's got a place even bigger than Chester House.'

'Ed said he'd like to take me to the theatre. He's coming down to London.'

'He can't stay in our flat,' snaps Lily, momentarily distracted. 'There's no room.'

'He says he's got friends,' says Janet.

'I don't know how you can go out with him, Janet,' says Lily airily. 'He's obviously totally skint.'

'Money isn't everything,' says Janet, protectively.

'If you can't find a man that has any,' shrugs Lily.

'You really can be a complete cow, Lily,' I snap. 'Just shut up. Ed's very nice. So what if he's not rich?'

'Darrrrling,' says Lily in a patronizing drawl. 'I'm just *saying*. Janet could do *so* much better. And money matters, you know. Sometimes even the prettiest models don't make it all the way . . .' She sighs with satisfaction, inspecting her reflection in the rear-view mirror. 'Though I've been lucky. Janet has to look to her future!'

'Janet's career's doing just fine,' I say.

Janet stares at her lap. 'I don't care,' she says mulishly. 'I like him. He's interesting. He talks about interesting things.'

'Like farming?'

'I like animals,' says Janet. 'Anyway, I'm going to see him again. On Tuesday, actually. We're going out to dinner.'

'Yeah, at McDonald's,' Lily says.

'At least he's not ninety, like Claude Ranier,' I say. 'He's young and . . .' I can't very well say good-looking. 'Pleasant.'

'Well, pleasant doesn't pay the rent.' Lily examines her nails as I join the M25. 'Now Henry, he's young and gorgeous. As well as loaded.'

'And deaf, obviously,' I say. 'Or haven't you shown him that side of you yet?'

Lily smirks. 'He thinks I'm ideal girlfriend material, actually.'

Does he? I wonder. Part of me wants to tell her what all those Yahoos were saying, before dinner. Warn her. But that would make Janet feel bad too, and anyway, just now I don't feel tremendously sympathetic.

'When are you seeing him again?'

'Probably Saturday,' says Lily. 'If I decide to let him take me out.'

'He seemed to have a pretty good handle on you,' I say. 'Maybe you'd better not try your normal routine.'

'What do you mean?' she asks innocently.

'Oh, you know. Keeping him waiting in the lobby for twenty minutes while you read *Marie Claire*,' I say. 'Vetoing all his restaurant suggestions. Sending back whatever you order. Taking the flowers he brings and dumping them into the bin right in front of him.'

I have seen Lily do all these things, and the older guys she dates usually don't say a word. Unless it's to apologize.

'I don't think he's going to take it,' I warn her.

'He'll take whatever I give him,' Lily says, shrugging. 'And he'll like it.'

I remember what Henry said about her name.

'Whatever, Lily *Frutt*,' I retort.

'Just shut the hell up,' she says. 'And keep driving. And before you take relationship advice from *her*, Janet, I do hope you've noticed she bagged herself the richest guy at the party.'

'I know,' says Janet loyally. 'I think it's fantastic.'

I sigh. I wish I did.

9

I have a lot of time to think about that over the next couple of weeks. Mark Swan is called to LA to edit his last movie, and I'm left back at the office, in charge of the care and feeding of Trish Evans. Kitty, delighted to have me under her beady eye once more, debriefs me totally and then swoops in to massage Greta's ego, and all her co-stars' too. She leaves the office almost every day at noon to have a power lunch with some actor or agent, trading fulsome compliments over Pellegrino and rocket Parmesan salads, leaving me trawling through scripts once again and writing up her memos to Eli Roth (apparently these are too 'sensitive' for Claire to do, but my theory is that Kitty just wants to put me in my place).

I don't care. On the first Thursday a courier package arrives for me from Hollywood. I rip it open, while John hovers around my desk.

It's from Swan. Scriptwriting software.

'*Final Draft*?' he asks. 'What does Trish need that for? She's already written her script.'

Yeah, but I haven't, and this stuff costs a bomb.

'Oh well, you know directors,' I say non-committally, stuffing the FedEx envelope in the waste bin so he won't see it's addressed to me.

There's a note in with the package, unsigned. It says, 'Make it good.'

I hug the package to myself, letting the emotions wash over me. Thankfully, he's away. There's no danger of me being tempted by crazy thoughts of him while I'm with

Charles. But he hasn't forgotten me. He's sent me *Final Draft*.

The image of him in the pub rushes back, like a movie, his face etched on my brain.

I look down at the box. It seems more romantic than any bunch of roses or box of truffles . . .

Of course, he didn't mean it that way. But he does mean me to be a writer. I look round at the office and suddenly think how small and petty it all is. And I glance at the pile of useless scripts on my desk.

I can do better. I'm sure of it. Mark Swan believes in me, and so do I.

Let's see. I start playing with ideas almost immediately. A comedy, definitely. I like those best. Something cheap to make. High-concept. But cheap, and unusual. How about a ghost? I like ghosts . . .

Sharon slinks over to me. 'Back at last?' she purrs.

'Yep,' I say, pretending to be very busy reading one of the crap scripts. Why can't she go away and let me think about my script? I can feel all the ideas simmering at the back of my brain, bursting to come out. All at once, I feel engaged, excited, in a way I haven't done in years. I can't be bothered with Sharon's office politics now!

'You know, I've been waiting for your call,' she says.

'Oh?' I pretend not to know what she's talking about.

'Mark Swan. Whether he has a new girlfriend yet.'

'Oh, that. Well, I didn't ask him.'

'And why not?' she demands, eyes narrowing into little slits. 'That's very selfish of you, Anna. I do think you might have done me that favour.'

'I feel awkward asking him personal questions,' I say. 'But you've no chance,' I tell her.

'And why not?'

'Everybody wants him.'

'I'm a little more than *everybody*,' she retorts, with supreme self-confidence. She lowers her voice so nobody can hear us, but looks me right in the eye. 'Anyone would think you wanted him for yourself!'

'Actresses want him,' I tell her. 'Models. Film students.

Just ordinary girls. They fling themselves at him wherever he goes.'

'So?' she snaps.

'So what does he want another reasonably pretty girl from the film business chucking herself in his path for? Look, Sharon, I haven't seen him off-hours – much, anyway – but he doesn't live with anybody. And he never mentions anybody.'

'Maybe he's gay.'

I laugh, the idea is so ridiculous. 'He's as straight as a ruler.'

'Then you can't say he wouldn't want me,' Sharon hisses. She pulls back her slender shoulders and shoves her not-very-impressive tits in my direction. 'Who wouldn't want to get with *this*?'

'Mark Swan for one,' I say, giggling. She's so ridiculous!

'I could be a model if I wanted,' snarls Sharon furiously. 'You're quite simply jealous. You want to sabotage another woman's chances with the perfect man.'

'And why is he the perfect man?' I say, ignoring her shots. 'You don't even know him. You've never *met* the guy, for heaven's sake. You don't even think he's good-looking, you said so, before. The only thing about him that appeals to you is that he's rich and powerful.'

Sharon laughs bitterly. 'Oh yes, and you're such a bloody saint! Going out with that book guy, and why? Because of his sex appeal?' she says, sarcastically. 'Can't resist those pint-sized good looks?'

'Shut up,' I snap. 'You do know you're a total cow?'

'I thought so,' she says, triumphantly. 'Oh, let me see, what could it be about Charles Dawson that makes you want to go out with him? It couldn't possibly be all that money and the big manor house, could it?'

I shudder inwardly. She sounds just like Lily.

'Or maybe it's because a girl like you could never get a man like Mark Swan,' she says, with a mean glint in her eye.

'Well, nor could a girl like you,' I tell her. 'And you couldn't get a man like Charles Dawson either. He's a great person and he wouldn't touch you with a ten-foot barge-pole.'

'He'd never get the chance,' she says. But it's a lie and we both know it.

'Goodbye, Sharon,' I say, picking up my script and ostentatiously opening it in front of my face.

She makes me think, though. I don't particularly want to, but I have to face it. Lily said it before – not that Sharon would ever have an original thought in her head anyway, but I know the idea's out there. That I'm dating Charles because he's loaded. And it stings. All the more because I'm not being fair to the guy.

I look at the cool cardboard box of software on my desk. I have a career going now, I have my dreams. I don't want to use Charles Dawson just because I'm afraid of winding up alone. I'm going to have to break up with him. I'm going to have to actually go through with it this time. And it makes me sad.

'What's the matter with you?' Lily says, as I walk through our front door and flop down morosely on the couch. 'Your brilliant career taken a dive?'

'No.' I glance inside my bag at the *Final Draft* software box. I'm going to boot it up on my laptop tonight. Start writing something, and not something about bloody love and romance. My funny ghost story. I'm not in the mood for any more suffering. There's enough of that in real life.

'Charles?' she says.

'Bugger off, Lily,' I tell her.

'He's dumped you?' she asks with fake concern. 'Oh well,' she adds. 'I thought it wouldn't last. You two are just so different.'

'He hasn't dumped me.'

'Oh,' she says, disappointed.

I ignore her. I go to my tiny desk, slide the software into my laptop. I've already got the idea. A ghost story . . . a caper . . .

I flex my fingers, and start typing.

It feels really good. I write the first scene, then the next, then the next. I stop to read it over, love it, and write some more. It's as if I can't stop!

By the time Janet comes out of the bathroom, her slim body wrapped in a gorgeous cashmere dressing gown, I've written fifteen pages. I look enviously at the dressing gown! Models do get the best perks.

'Hey, girl,' she says. 'Wazzup?'

'Not bloody much,' I say, gloomily. 'Janet, what should I wear to . . .' I look over at Lily. 'To an important dinner with someone I respect?'

'Hmm,' she says. 'Businesslike?'

'Not really. I just don't want to be too sexy.'

'Ho ho,' Lily says. 'Too sexy! That's a good one.'

I try to ignore her.

'Conservative,' Janet pronounces, 'but feminine, flattering. Long skirt, fitted round the bust . . .'

'Would you go shopping with me again?' I ask diffidently. 'I need this one outfit.' I think of work and Mark Swan. 'And maybe a couple more casual. For later.'

'Love to,' she says instantly. 'When do you need it by?'

'Tonight or tomorrow night . . .'

'I can't make it till the weekend but I'll pull you out something nice, anyway,' she promises. 'Something from Joseph. What you really need, Anna, is for me to show you what to buy, so you can get it yourself. Once you know what to do, you can look stylish all the time.'

Lily snorts.

'No need to mock,' Janet says severely. 'Just think of Chloe Sevigny.'

'Anna's not going to look like you,' Lily says. 'Don't you think you should stop filling her head with nonsense, Janet? It's not kind,' she adds, piously.

I look at Janet, uncertain.

'I know that,' she says stoutly. 'But Anna can look stylish. Anybody can look stylish. You've got some good features,' she promises me. 'We just have to bring them out on your body, the same way we did with your hair.'

I smile at her. Why not, eh? She did do amazing things with me last time.

The phone rings and Lily dives on it.

'Hello, Lily Venus's residence,' she says. 'Oh *hi*, Charles, *darling*, how are you? Oh, that's *wonderful*. Anna?' Yes, she's here, she says reluctantly. 'It's for you,' she tells me, her tone surly.

'Hello, sweetie,' says Charles. 'How was your day?'

A fresh wave of guilt breaks over me. 'Oh, fine. Look, do you think we could go out for dinner? I'd like to talk to you,' I say, picking my words carefully so Lily doesn't read anything into them.

'Love to,' he says. 'How about tomorrow? That new place, Vespacci's? New Bond Street. Great reviews.'

'That'd be perfect,' I say, and I feel a great sense of relief because I know I'm doing the right thing. Of course he'll be sad, but he'll thank me for it later.

On the other hand I'm sad right now, and I wonder if I'll be thanking myself . . .

'Hello,' says Charles, pecking me on the cheek and then stepping back admiringly. 'You look absolutely marvellous.'

'Thanks,' I say. I've dressed really carefully. You'd never know how depressed I'm feeling. Somehow, breaking up with Charles merits the best I (and Janet) have to offer in the old style stakes. I'm wearing my most chic dressy dress, a pale yellow silk thing with an A-line skirt that comes just below the knees, a little cream, lacy cardigan, ivory heels, and I'm back to Chanel No. 5. I imagine Vespacci's is hideously expensive, and I don't want to disappoint him.

I'm going to dump the guy. At least I can try to make him look good. Right?

'You look fabulous yourself,' I lie.

Actually, he doesn't look bad, for Charles. He's skinny as a rake, but at least he has a good tailor.

The restaurant is perfect. I'd never heard of it, but obviously a lot of very rich people know it; you could blind yourself just from the glitter of gems sparkling in the candlelight. The waiters are immaculately dressed and very discreet, they seem to sort of glide between the tables like ghosts. There's plenty of space between each table, a good deal of mahogany and burgundy, leather, and absolutely wonderful food. Charles orders smoked salmon and a cheese soufflé, and I get a salad and some roast beef.

'Let's have champagne,' he says. 'Goes with everything.'

'We're not celebrating, are we?' I ask, anxiously.

'Every day with you is a celebration, Anna,' he says, making a little bow with his head. Oh, this is a disaster. How am I supposed to break up with him? Please don't say things like that. And the waiters. Whisking things away, setting things down . . .

I've had absolutely no practice at being the dumper. I am a lifelong dump-ee. I try to think of the ways various exes have done it to me before. It started with Robby Caldwell in fourth form. I think his parting shot was 'You're fat and ugly'. And then there was Pete Villa in college. He told me he needed to find himself in the arms of another man. Kevin Feathers said we should start seeing other people. 'Not including ourselves,' I remember him adding, in case I hadn't got the point. And Brian, my latest and lamest, something along the lines of fresh stimulus and real positivity. Oh yes. Plus, that looks were important to him.

I don't think any of these will do. Charles doesn't deserve the unwanted open-heart surgery without anaesthesia that men apparently think it's OK to practise on women.

'I'd rather have the house red,' I say to the waiter. 'Just a glass. To go with the beef,' I explain to Charles, who's looking a bit crestfallen.

'I suppose I'll have some Pouilly-Fuisse,' he says.

Quick, change the subject. 'I'm really enjoying being back in the office,' I lie brightly. 'Away from Mark Swan for a bit. Getting lots of work done. I thought I might write a script.'

'Oh yes?' Charles says, warmly. 'You're a star, Anna. You have to give me some career advice. I'm bloody useless, can't do anything.'

That takes care of the starter course; I rabbit on inanely about Kitty and John and Sharon, between telling Charles he isn't useless at all, he just hasn't found the right job for himself yet. The wine also helps. I am convinced that without alcohol the entire planet would come crashing to a halt. I take big, healthy gulps, get another glass, and almost start enjoying myself. Charles is so genuinely pleased for me. I really like the guy. It's such a shame I'd rather chew my own foot off and swallow it whole than think of his hands on my boobs.

But nothing lasts forever, and eventually we are halfway through our main course, the waiters are all melting quietly towards other tables, the natural beeswax candle is flickering low and warm, and I still haven't said a thing. And Charles is going on about some new hedges he's having planted at Chester House, and how difficult it is to find craftsmen who know how to lay a drystone wall . . .

I take one more big, fortifying gulp of wine and plunge right into it.

'Yes,' I say, 'I do understand about the hedges.'

'They're such a terribly important microclimate.'

'Yes . . . look, Charles, we need to talk.'

He looks perplexed. 'We are talking, aren't we?'

'About us, I mean. I think you're a fantastic person, but I don't think we're right for each other.'

He looks amazed, as though I've just remarked I really work for the CIA.

'Do you mean romantically?'

'Yes,' I say gently.

'But why?' he asks, bewildered. 'We get on OK, don't we?'

'Oh, absolutely,' I say, agreeing with him. 'I think you're great.'

'I know I was a bit bloody at first,' he says, apologetically, 'but I thought you'd forgiven me.'

'Oh. I have. Totally.'

'Then what is it?'

Arrrgh! I thought nothing could be worse than being dumped, cruelly, by selfish fuckwits, but obviously I was wrong. There must be a tenth circle of hell where all you get to do is break up with poor, innocent men twenty-four hours a day.

'I think you're a wonderful friend and I really enjoy your company,' I say. 'But there just isn't that spark between us.'

'Oh,' he says, sounding relieved. '*That*. I know *that*. That's not a problem.'

'Um . . . how exactly isn't it a problem?'

'Oh, you know,' Charles says airily, eating some more soufflé. 'That goes away. Everyone says so. You have passion for a few years, then it wears off, and you're left with friendship. Long-term success is about getting on with the other person. Seeing eye to eye.'

'I think I need that passion, though.'

'You don't,' he says, sighing heavily. 'You need to be part of a family. To not be lonely. To spend time with a person you like. Otherwise, life's bloody awful. Passion? What's passion? I wanted passion with all those other women,' he says, a glint in his eye. 'But it didn't last. Soon I was looking for any way to get them out of my life.'

'I don't know,' I say uncertainly.

What he says makes sense. It sounds completely reasonable. So why do I still feel like I want to run away?

'How many girls do you know who have followed passion, only to be completely miserable within six months?'

I think of Janet, chasing Gino. And others. 'Maybe one or two,' I concede.

'And old married couples. How passionate are they?'

Well, I can't visualize Mum and Dad rattling the bedposts. Thank God. But I do know other older couples who are still walking around hand in hand. And some who can't bloody stand each other.

None of them appear to have any passion at all, as such.

'Not very,' I say. 'But don't you think people should at least start out with passion? Then they mellow as they get older.'

'Yes, ideally,' Charles concedes. 'But look, we're just skipping that stage. I don't fancy you,' he says earnestly.

229

'Oh. Well, thanks.' That explains a lot. All the delicious food, for one thing. What was it he said about my dieting? 'What would be the point?' I suddenly understand why it rankled when he told me he liked me for my personality – it meant he *didn't* like me for my looks.

I don't want much. I know what I look like. Logically I shouldn't expect anything in that department. But the fact is, the thought that Charles is completely indifferent to me *does* hurt.

On the other hand, if I insist on waiting for a man who actually wants me, I'll be waiting forever, won't I? Waiting alone. Charles sees the look on my face, tries to decipher it.

'But you know, I could fancy you,' he offers. 'It's possible. You look very elegant,' he adds politely. 'It wouldn't be difficult to *learn* to fancy you, Anna. The more we get to know each other, I mean. You look so much better since you had your hair changed. That evening at the ball, you almost looked pretty.'

He's so sincere it's hard to be cross.

'And I know you don't fancy me,' he says. 'Who would?'

'Lots of girls,' I lie.

'But, you know. Dark room, couple of bottles of champagne, we'd be OK,' he says heartily. 'I bet you'd be a great mother. And I could be a good husband. Supporting your career. Providing you with nice things.'

'I don't care about that.'

'I know,' he says, smiling. 'It's one of the main reasons I like you so much. But we could be so happy together. Close friends. We could have a nice family and lots of money. I can give you anything you want.'

I look down at my glass of wine. For the life of me I can't think of a good reason to say no. It may be a little out there, but he's sounding awfully practical. I can't stop myself thinking of the seagrass flooring and gorgeous oil paintings in his flat. Imagine Lily or Sharon, say, coming to an engagement party there. I can just see it now. They'd be furious!

I'm grinning foolishly at the idea.

'And you say you want to write films? I can support you while you do that,' he offers earnestly. 'You'd never have to be a struggling writer. You can live in comfort with me for as long as it takes for you to be successful.'

I smile at him. What a gorgeous thing to say. He's not asking me to give the writing up – he knows exactly what to say.

'It would be a terrific wedding,' he says, encouraged.

I blink. 'Wedding?'

'Of course wedding,' he says. 'Girls love weddings,' he adds, with the air of an expert. 'Imagine ours. You could plan it, anything you wanted. Of course, it'd have to be at Chester House,' he adds hastily. 'No eloping, things must be done properly. But you can pick the flowers, any gown you like. Vera Wang's very popular, isn't she? Perhaps your friends could be the bridesmaids. Janice and Lila.'

'Janet and Lily.'

'Right,' he says, making a dismissive little gesture with his hands. I beam at him. Ha! I introduce Charles to two younger models and he doesn't even register their names. Brides-maids, yes, they *could* be bridesmaids. Janet smiling and happy for me and Lily, aggravated beyond belief, scowling her way down the aisle. Course, she'd most likely try to spill red wine on my Vera Wang, so maybe best not to ask her.

I drift off for a few seconds into fantasy – a warm stone village church, its porch wreathed in flowers, white and yellow, I think, a huge cream-coloured marquee on the lawn, the chink of champagne glasses, as yet faceless bridesmaids in moss-green empire-waisted dresses, and me (in Vera Wang – Vera would be there personally, of course, with the safety pins and the needle and thread in case I needed a last-minute touch-up) looking all stately, because even in my fantasy I am not transformed into a delicate, elfin little bride.

Ooh. It *is* a nice picture. We'd have really good food, too, not manky trays of coronation chicken and boring cold whole salmon that nobody eats. And maybe I could invite Brian and his new girlfriend. Oh yeah. I'd *love* to do that. In fact I could track down *all* my ex-boyfriends and invite them, with their dates, and then kiss the air on the side of their cheeks dismissively . . .

I never thought I'd get married. Not unless it was to some utter loser in the Brian mode who'd offer me a shitty registry office ceremony where you don't even get to wear a dress, and afterwards everybody goes to some crap restaurant for a meal. I hate those civil weddings. What the hell's the point, apart from making sure your man's liable for a divorce settlement if he ever dumps you? It's about as romantic as watching a plumber fix the bog. And, naturally, it was what I assumed I'd be getting, if some poor desperate fool ever decided to hitch his star to mine.

'Are you for real?' I ask. 'You want to marry me?'

'I'm not joking, Anna. I'm serious about you. You're the one,' he says, reaching out and taking my hand. 'You're clever, you're ambitious, you don't want me for my money, you're nice, you're interesting.'

'Charles, I – I don't know what to say.' I'm flattered, but no. No thank you. It's an honour, but I can't. That's what I should say, and yet I don't. I don't want to say any of those things, because I absolutely don't want to wind up alone, and he's offering me a way that I don't have to.

'Then don't say anything,' Charles suggests. 'Think about it. It's all I've done since I met you. You probably need more than one night.'

'I can't make any promises,' I tell him.

Charles shrugs and polishes off the last of his soufflé. 'I'm not worried,' he says. 'Because, like I said, you're very clever.'

'You're sweet, but what's that got to do with it?'

'Everything,' he says confidently. 'You'll mull it over, you know. And you'll realize that I'm right. We're trained to think of love and marriage as decisions to make with your heart, but that's silly. They're too important for that. They're decisions we should make with our heads.'

I look at him, longingly.

'You take your time,' he pleads. 'Sleep on it.'

It's been so bad, you know. Being lonely all this time. Dating losers that treated me like rubbish, just so I had someone to go out with. Wanting and hoping to look different, and waking up every morning in the same tall awkward body with the same big-nosed face. Having everybody think I could only get a man like Brian – and then he'd dump me. Sure, I made a joke of it, because you have to, don't you? You have to laugh at pain or you'd never make it through your workday. I thought that maybe now, when my career's going better, I'd stop caring about romance. I'd just give up on it and find fulfilment in my job.

But for some reason that hasn't worked at all. I still want love. And the wine and gloom in the restaurant are softening Charles. Like he said, a couple of bottles of champagne, turn down the lights . . . I'm sure I could cope. Plus, I like him, he wants a family, and he wants to give me a life of luxury. That song from *Four Weddings* starts playing in my head: the one about getting married and not being lonely any more.

I want that. Not to be lonely any more. And I can actually have it. I look at Charles Dawson and I feel a huge wave of gratitude, real gratitude, and I think, gosh, I *love* you, I love you for rescuing me.

'No need for that,' I hear myself say. 'I'd love to marry you, Charles. Thanks very much.' And I reach across the table and kiss him on the lips.

'Oh God,' he says, eyes blinking owlishly. 'Oh God. Really? That's – that's *fantastic*. We're going to be *so happy*.'

'I know we are,' I agree, and take another huge gulp of wine.

Charles pays the bill and then pours me into a taxi.

'D'you want to go home?' he asks.

I shake my head. If we're going to be married I'm going to have to sleep with him, so I might as well get it over with, I think. If Charles can be all logical about it, so can I.

'Let's go back to your place,' I suggest, lowering my voice.

He smiles at me. 'Sure.' And kisses my hand reassuringly. The taxi interior seems to be swaying a bit, so I lean my head against the window and watch the jewel-like raindrops that are glittering against the glass. Anything not to think about it. Charles seems to know how I'm feeling. He backs off, removing his hand and looking out of the other window, and I'm grateful that he's not trying to push it.

When we get to his building he over-tips the driver and offers me his arm to go inside, which could be out of chivalry but could also be a cunning plan just to keep me upright. I stagger inside the flat and he takes me straight into his bedroom, which is just as I expected: cosy, with cream linens and bookshelves, and not a bit sexy.

'I'll just go and freshen up,' I say brightly. He's got a little ensuite bathroom with a tub and no shower, and I root around frantically in the medicine cabinet, desperate for some KY jelly. Only there isn't any, so I have to use water. I sluice out my mouth with spearmint Listerine and that's my ablutions taken care of. I peel off my clothes, grit my teeth and open the door, trying not to be too frightened. It's only sex, after all. I've done it before. Never liked it, but so what? You don't like going to the dentist and you have to do that, as well.

'Coming, ready or not,' I say brightly, stepping forward seductively, but the room's pitch dark and I catch my foot on the end of his rug, which slips loose from its moorings and slides forward two inches, sending me tumbling to the ground.

'Bloody hell, are you all right?' asks Charles, from some-where ahead of me.

'Fine,' I say, feeling like a total prat.

'Let me get the light,' he says.

'Oh no, don't do that,' I beg. Heaven forbid. I don't know which is less appetizing, the thought of being seen naked, my love handles offered up for inspection without so much as a filmy, forgiving negligee to cover them, or the idea of having to gaze on Charles as nature intended. He's so short, especially compared to me. And I feel sure he's got a really bony chest, I mean he's got to, being as skinny as he is . . .

'No problem,' Charles says, sounding relieved. 'Come straight ahead. Yes ... yes ... there you go,' as I stumble forward and reach the soft goose-feather duvet.

Not skinny, I tell myself, slim. Not short, just ... compact. I want to ask if we could make a start on those two bottles of champagne now, but I don't dare say anything. It would sound too mean. Besides, I'm half slaughtered as it is, I don't want to actually pass out on him. I don't think he'd attribute it to a frenzy of passion.

I crawl into the bed. Charles has pulled back the sheets and he reaches for me, his bony (and yes, it is bony) arm snaking round my plump waist.

'Don't worry,' he says, reassuringly. I would like to think he was psychic, but it's probably actually just because I'm as stiff as a board with nerves. 'It'll be OK.'

And he does his best, fiddling here and stroking there, and it's not as bad as I'd expected. At least, it's not painful. But it is kind of nasty and embarrassing and the only thing I want to do is get it over with. At least it doesn't last long. Charles doesn't say much, just grunts a bit, and manages to finish in about two minutes, so even though they're two really, really long minutes, I tell myself that this won't be a problem, that I can handle this.

After he's done he turns me round so we're spooning and kisses the top of my shoulder, and then he falls asleep right away. He snores lightly and I think I'll never be able to get to sleep, but it doesn't take long before my body un-tenses and I've slipped off myself.

'Wakey-wakey.'

I blink. I'm not sure where I am. And then I remember. Charles's face is looming above me, like Richard Gere's over Julia Roberts' in *Pretty Woman*. With some marginal differences, such as that Charles looks sod-all like Richard Gere. Sadly.

Oh no, he doesn't want a morning quickie, does he? Involuntarily I pull the goose-feather duvet round me, but Charles is already fully dressed (grey suit, white shirt, cufflinks, unfortunate pink tie) and not making any aggressive moves towards my naked body.

'What time is it?'

'Nine fifteen,' he says.

'Oh bloody hell, I'm late.' I jump out of bed and race to his little ensuite bathroom, shutting the door with lightning speed so he can't get a good look at my nude flesh in the morning light.

'Couldn't you take the morning off?' he asks, standing outside and knocking gently on the door. I open it a crack, so he can see just my face.

'I can't, I'm really sorry. I have to get back home to get some new clothes and everything,' I wail. 'It's going to be half ten by the time I get in, at the earliest. Kitty'll kill me!'

'Say you're meeting a writer,' he suggests. 'It's true, at least technically.'

I pause. 'That might be worth a try.'

'I want to go shopping with you,' Charles says.

I blink. 'Shopping? What for?'

'Well, we are getting married,' Charles says reasonably. 'Don't you think we should get you a ring?'

'Afternoon,' I say, happily, tucking a strand of hair behind my ear. 'Beautiful day, don't you think? Not too hot, just perfect,' I add, waving my left hand insouciantly around to indicate the perfection of the weather. Then I curl my hand round my coffee cup and start to drum my fingers idly on my desk.

It's 2 p.m. and I'm just back in the office after lunch with Charles. I told Kitty I was doing some scouting work for Mark Swan. I'll have to spend an afternoon typing up totally fictitious location reports, but it's completely worth it. Charles spent all morning with me in jewellery shops I'd normally be frightened to even go into, and now I'm flinging my hands around like a TV evangelist just hoping somebody will ask me about the fantastic rock on my finger . . .

'What the fuck is that?' asks Sharon, suddenly.

She noticed! I think, gleefully. Well, she probably noticed five minutes ago and has been battling with herself not to ask me, not to give me the satisfaction, and now she's finally given in.

How could she not notice?

'This?' I ask casually. 'Oh, this is my engagement ring.'

'No it bloody isn't,' snaps Sharon. 'That's not *real* . . .'

'Whatever you say.'

'My God,' she says, her face draining of blood. 'It *is* real?' She snatches up my hand, gasping. 'No way, Anna. No way!'

'What's all this?' Kitty asks, emerging from her office. I try to extricate my hand from Sharon's, but it's too late. Kitty has descended on both of us like a hawk, scowling thunderously.

'What's going on?' she asks, fake sweetly. 'Anna, you have my copy of the location reports?'

'Not just yet.'

'What's this?' she asks, taking my hand from Sharon. 'Pretty. Is it Butler and Wilson?'

My ring is gorgeous. A glittery princess-cut ruby, pigeon's blood red and translucent, four carats, with two trillion-cut diamonds on each side.

'No,' I mutter. 'It's real.'

Kitty's groomed eyebrow lifts. 'What? Real?'

'Yes, Charles and I got it at Garrard's,' I say.

'Charles is your fiancé now?' asks Sharon, setting her teeth.

'Who's Charles?'

'The book guy,' Sharon informs her. 'Turns out he's Anna's boyfriend.'

'You can't do favours for your personal friends, Anna,' says Kitty severely. Her body has gone all rigid.

'I don't know why he'd bother, he doesn't need the money,' Sharon informs her. 'Got a stately home and everything.' She sighs. 'Well done, Anna,' she adds grudgingly. 'You really scored there.'

'I scored because Charles is a very nice person,' I say, looking at my ring.

'It's nice, if a little ostentatious,' sniffs Kitty. Apparently only she is allowed to flash huge diamond rings around the office. 'Congratulations.' Then a light comes into her eyes.

'Shall I tell Mark you don't want to work with him any more?'

'What?' I ask, panicking. 'Why?'

'Well, you won't need to now, will you, dear?' she says. 'You'll be far too busy planning a wedding, and then there'll be babies and who knows if you'll even come back to work.' Kitty's eyes are beady little slits. She sounds as sexist as my grandad.

'I'll still be working, Kitty,' I say, as firmly as I dare. 'I want to make movies.'

She gives me a saccharin smile. 'Perhaps you can get your husband to bankroll some.'

I pretend I haven't heard.

'Get me those location reports within the hour,' she says crisply, and flounces off.

Her words leave a touch of ice in the warmth of my day. I can see it now. Kitty will go to Eli Roth, try to make it seem like I'm marrying money and dropping out of movies. But I'm not going to, and I'm not about to let her push me out, either.

'It's a lovely ring,' says Sharon, ingratiatingly. 'You've always had such taste, Anna. I can't wait to see what your wedding dress is going to be like. I am going to be invited, aren't I?'

I look at her eager face and can't help it.

'Oh, sure,' I say.

'Brilliant. Thanks, Anna,' she says, gratefully. 'Weddings are the perfect place to meet guys. Maybe I could even find one like yours,' she says, wistfully.

I thought Charles was meant to be a midget and the only man I could get, I want to say, but don't. It's been a lovely day, I tell myself firmly, and I don't want anything to spoil it.

I'm finally alone at my desk. Thank heavens. I open up *Final Draft*. I want to get to the end of act one.

My mood lifts. I'm crackling with energy, I'm thrilled. I set to work, start typing furiously, one eye on Kitty's door. No way am I letting her see this.

* * *

Charles is obviously going hard for Boyfriend of the Year award. He has two dozen blood-red roses delivered to my desk in the afternoon, provoking long drawn-out sighs of sheer envy from most of the girls in the office (the news had spread everywhere thirty seconds after Sharon left my desk), and he calls to say he's booked us in to the Ivy for dinner because it's nice and close to my work. And asks if 'Jane and Lucy' can join us. I decline on their behalf.

'We'll have to tell everyone,' he says enthusiastically. 'I thought we'd start with your friends. After your parents, of course.'

Mum and Dad! With a guilty start, I realize I had forgotten about them. Of course I must tell them.

'What about your parents?'

'Both dead,' he says. 'If they were alive I wouldn't have Chester House.'

'Right.'

'But I've got you now,' he says, delightedly. 'You'll be my family.'

'Course I will,' I say, loyally.

'I want to put the announcement in the papers. And we must have a party. Vanna will be thrilled.'

'She will,' I agree.

'Maybe you two can plan it together,' Charles says. 'Vanna's wonderful at parties. We want to make it really big with loads of guests,' he says. 'All your friends.'

'Right,' I repeat. All my friends? Who are they? Janet and Vanna, basically. Lily and Sharon, possibly. Claire Edwards. And, um, that's it.

Oh, except Mark Swan. A chill runs through me at the thought of inviting him, somehow. I don't know why. Lily and Janet will be all over him, yeah, but he's used to that.

No, I don't know why, but I definitely don't want him there. Probably because it's not a good idea to mix business and personal. I'm a serious professional now. I'm a screenwriter. Or I will be one, soon. No need to mess things up for myself.

Yeah. That must be it.

'So we'll go and see your parents on Sunday, for a late lunch. About eleven start.'

'Sorry?'

'Your parents,' Charles says patiently. 'And it's "what", darling, not "sorry".'

'Oh. Yes. Absolutely,' I say, trying to sound more enthusiastic.

'You're OK with that, aren't you?' he asks, guardedly.

'Oh sure,' I say. 'It all seems to be moving a bit fast, that's all. But it's great,' I add hurriedly. 'You put the announcement in the papers, then.'

'I will. Do warn your parents, they'll want to cut it out and keep it,' Charles says. 'Day after tomorrow. Toodle-pip.'

'Toodle-pip,' I say, dutifully. Toodle-pip? I'm going to have to get used to that. I have a sudden vision of myself in ten years, a great strapping country matron with a ruddy, weatherbeaten face, clad in a Husky and wellies, tramping through my fields to check on my horses, or something. And I'll say things like 'down the hatch' and 'chin chin' and stuff.

That's pretty depressing, so I look down at my ruby and diamonds to cheer myself up. It works. They're so bright and sparkly it's like wearing a firework.

The phone rings again. I expect Charles, but it's Vanna. She doesn't actually say any words but I can tell it's her by the high-pitched squeal of delight that starts at a note only dogs can hear and rises from there.

'Daaaarling,' Vanna shrieks when the squealing has tailed off. 'You're brilliant! Incredible! Fantastic! Spectacular! Amazing!'

'You make me sound like I've just acquired super-powers,' I say.

'But you have,' Vanna cries. 'X-ray vision is nothing compared to the power to obtain vast estates with a single syllable! "Yes," ' she sighs. ' "I do." Well, that's three syllables but who's counting? I'm so *proud* of you, Anna. I knew you had it in you. All those years and you were just waiting for the right chap. I bet you're glad I lied to you about that dinner party now.'

240

'So you admit it,' I say darkly, pouncing.

'A little white lie,' Vanna says airily. 'And look where it got you! I'm coming over.'

'I'm still at work.'

'I'll pick you up. It's almost five thirty now.'

I glance towards Kitty's office. She's in there with Eli Roth, the two of them probably discussing my made-up location reports. Normally I hang around the office till six or later, but with my luck, Roth will come out here in his sharp suit and ask me too many penetrating questions and I'll get caught. Mark Swan says it's driving him nuts that he's shut out of the *Mother of the Bride* decisions.

I wonder what Swan's doing in LA right now.

Anyway, never mind about that.

'Yes, come over,' I hiss to Vanna. 'In fact be as quick as you can.'

'I'm just down the block, darling,' she coos. Her offices are in Covent Garden. 'Be there in a jiff.'

Vanna arrives ten minutes later in a cloud of glory. Or it could be Chanel No. 19. It's hard to tell, frankly. She struts out of the elevator looking amazing in something tight and black, possibly Azzedine Alaia, complete with a devastatingly chic Prada mock-croc clutch, outrageous fifties style open-toed stack stilettos, and a huge pair of wraparound glasses. Despite her tiny size – even with the three-inch heels the girl is barely five six – people make way, drawing back from her path. She's so formidable she makes Kitty look like Anthea Turner.

I quickly gather up a few scripts for cover but, possibly scenting a rival, Kitty has emerged from her office. Bugger. And Eli Roth's right behind us.

'OK, let's go,' I say hurriedly, but it's too late. Kitty has sauntered over to my desk and is staring coldly at Vanna as though she's a half-dead bird some particularly nasty cat has dragged in.

Vanna, however, does not flinch. She returns Kitty's stare with an equally icy gaze. I look at her enviously. How I'd love to be able to face people down like that!

'Can I help you?' Kitty asks her.

'I don't think so,' Vanna says easily. 'I have a meeting with Anna.'

'Oh really,' says Kitty, giving a little laugh for Eli Roth's benefit. I try not to look at him, but my peripheral vision sees him standing there, tanned lips set in a disapproving line. 'Another of her friends being entertained during office hours?'

'I am a friend of Anna's, yes,' says Vanna smoothly. 'Who isn't in London?' I splutter, but manage to turn it into a little cough. 'But I'm seeing her on business now.'

'What are you guys meeting about?' asks Roth, friendly enough. But not underneath the plastic smile.

'Books,' says Vanna. 'She's looking for new properties to adapt. Right?'

'Right,' I say meekly.

'Oh dear,' say Kitty, superciliously. 'Another unpublished author. You really must stop this, Anna, it's a waste of company time, and I do hope you weren't thinking of leaving the office for this, or charging coffees to petty cash.'

'I'm not an unpublished author,' Vanna says. 'I'm not an author at all. I'm the editorial director of Artemis Books. Of course, if Winning isn't interested in discussing any of our titles—'

'It's Red Crest now,' says Eli Roth, moving forward and effectively elbowing Kitty to one side. She stands there blushing as he hurries to repair the damage, smiling at Vanna with a full-wattage LA beam. 'You must be Vanessa Cabot.'

'That's me,' Vanna says coolly.

'I make it my business to know the names of all the players,' Roth says, smugly. 'So you're going to give Anna first look at some of your stuff?'

'Of course. I heard she was working with Mark Swan.'

'News travels fast,' Roth says. He slaps me heartily on the shoulder. 'Well done, Anna. Best get off to your meeting.' He looks over at Kitty rather crossly, and she, in turn, shoots me a very fleeting but very nasty death stare. I shiver, but Vanna is smiling and dragging me away. Thank God.

'You could have been a bit nicer,' I protest.

Vanna gestures impatiently. 'Darling, no I couldn't. She was challenging me and therefore you. She has to do that, of course, because she's so scared of you.'

I laugh out loud. Luckily I haven't yet had a sip of my fat-free, taste-free cappuccino. It would have gone everywhere.

'You're so blind sometimes,' Vanna sighs. 'Imagine if you were – what's her name?'

'Kitty.'

'Kitty. You're a woman of a certain age in an industry that only values youth. You're paranoid about showing even a wrinkle, so you have to go in for surgery and Botox. Your company just got bought out, and the only person with a viable script going is a young girl who works for you, so you steal the credit, but the young girl attaches a major director all by herself, and is his preferred partner from the producing team. What has this Kitty done? Nothing. And she knows her boss probably knows it.'

'She told Roth that she found it.'

Vanna snorts. 'And you don't think he knows the truth? A shark like that?'

'He can't. He's never given me any credit for it.'

'Doesn't mean he gives Kitty credit for it, honey,' Vanna says cynically. 'He knows. He's not dumb. The only reason he hasn't started firing people yet is he's figuring out the most cost-effective way to do it.'

'But Kitty won an Oscar.'

'What for?'

'Best Foreign Film.'

'Doesn't count,' says Vanna dismissively – and somewhat accurately. 'He won't fire you, because you've found him something, got it off the ground. But he won't give you much pay or credit either, unless you stand up for yourself.'

'But why not?'

'Because he doesn't have to,' Vanna says simply. 'Do you remember when we were children and they said "I want doesn't get"?'

'Um, yes?'

'In real life "I want" does get,' she says.

'Wow,' I say. I'm really impressed. I wish I were like Vanna, so petite yet ferocious, elegant yet deadly, like a gorgeous Siamese cat.

'But who cares about bloody business?' Vanna exclaims. 'Let me see the rock! Ooh. A ruby.' She pulls my hand closer to her eyes, scrutinizing the stones carefully. 'Darling, that is stunning. Rubies are getting very rare, you can't find the good clear ones for love nor money . . . very attractive,' she pronounces. 'And tasteful.'

'He does have good taste,' I concede. 'But don't you think we're going to look kind of silly?'

'Silly? What do you mean, the height difference? Don't be ridiculous,' Vanna scoffs. 'Nobody ever thinks millions of pounds are silly. You are going to be mistress of almost a hundred acres of prime English countryside, a bloody huge manor house that everyone I know would give their eye teeth for, including me, a lovely flat in London—'

'It is a lovely flat.'

'And entrée into society with a capital S.'

'I don't care about Society. With or without a capital S.'

'I know, but you will,' Vanna says comfortably. 'Everybody does. They pretend not to, of course, but they do really. Just like people want knighthoods and never admit it. Now Charles doesn't have an actual title, but he's even better. His family is so old . . . I wonder if I have an old copy of *Burke's Landed Gentry* anywhere.'

'Vanna, you're being ridiculous.'

She shrugs elegantly. 'But darling, I'm just so happy for you! Everybody will be green with envy. Everybody. All those years of people saying cruel –' she pauses, not wanting to give anything away, like I didn't know already – 'of people wondering why you weren't married yet and now this! You couldn't have *done* any better. Unless you'd bagged Prince Andrew, of course. I wonder . . .' she muses, then shakes her head. 'No, I suppose it's too late now. Anna's off the market!'

I smile weakly at her. 'It is kind of amazing. But I don't want to marry Charles for his money, you know.'

'Of course not,' says Vanna immediately. 'It can only be true love. You'll laugh at me now, but at that first dinner party I didn't think the two of you were attracted to each other.'

'We weren't.' I blush. 'He grew on me,' I lie. 'And he's nice, once you get to know him.'

'And he treats you well? Not tight with his money?'

'Oh no, he's very generous,' I say, looking at my huge sparkler and thinking of all the delicious meals and taxis and flowers.

'And he makes you feel good?'

'Yes.'

'Supportive of your career?'

'That too,' I say. 'Even when I had to let him down about the book.' I swallow, thinking of how devastated he was by that, thinking of him crying in the orchard at Chester House. How small I felt. And now I'm going to try to write, is Charles being resentful? Not even slightly. He wants to support me while I labour over dialogue and third act breaks and inciting incidents.

'So he's the perfect boyfriend,' Vanna says, triumphantly.

'I suppose so,' I say.

Vanna shakes her head. 'You should be jumping up and down, Anna. Don't you understand? You've just won the Lottery! And so has he, of course,' she adds loyally.

'Yes, I know. I – I'm really thrilled,' I tell her. I suddenly feel overcome with tiredness and have to smother a yawn.

'You go home and get your beauty sleep,' Vanna says, dropping a note on the table. 'You'll need your strength. We have a huge engagement party to plan. Charles wants to make a really big deal out of it.'

I smile weakly. 'Sounds terrific.'

Vanna drops me home before rushing back to leafy Barnes to break the news to Rupert (I can just imagine him at the engagement party, can't you? Red-faced and florid, probably braying to all and sundry how lucky I am to have been taken off the shelf) and I walk up Tottenham Court Road back to the flat. It's early evening and I feel drained, but I keep glancing at my ring as if for strength.

There's a ring on that finger. A bloody expensive one, too. Not the diamond chip a good-size salt crystal could put to shame that I'd been expecting from some loser, some day, if I got lucky. And even though I'm not looking forward to Rupert Cabot pointing it out, I am in fact off the shelf. As I thread my way past drunk students on their way to an eighties night at the Astoria and a wino passed out in front of the off-licence, it strikes me that I won't have to do this any more. Walk back to Lily and Janet's, I mean. I'll be sharing a bedroom and some expensive Persian rugs with Charles in Eaton Square splendour. No more Pot Noodles for supper either. It's the Ritz all the way . . .

And suddenly all I really want is to get back to our flat above a shop, our cheerful, coffee-stained IKEA sofa, and my aggravatingly gorgeous, bathroom-hogging room-mates. I pick up my pace, almost jogging, and sigh with relief as I reach the little alley next to the Moon Goddess bookstore, enter our doorway and step into our ancient, coffin-sized lift.

But of course this is only nerves, isn't it? Fear of change. Which is utterly ridiculous, for years now my main fear has been staying the same.

I paste a big, hearty smile on my face as I fling open the door. They're both there; Janet is flicking through *Heat* with her headphones on, and Lily is drawling on the phone to someone or other. I can hear a few 'dah-lings' and 'sweethearts', though that doesn't necessarily mean it's a bloke.

'Hi!' I say, gamely. 'Big news!'

They ignore me.

'Big news!' I repeat enthusiastically, in the manner of an American cheerleader.

'Do you mind?' Lily hisses.

Janet looks over at me, fingers on her headphones. 'Waiting for tonight,' she warbles tunelessly. 'When you would be here in my arms – woah-oh . . .'

'You shut up too!' Lily spits. 'Sorry, sweetie,' she coos into the phone. 'Can't get a second's peace . . .'

I make exuberant hand gestures, giving them a little wave

and waggling my fingers like a Playboy bunny but they still don't notice.

'I got engaged!' I shout. Loudly.

That does the trick (subtlety has never been Lily's strong suit). She quickly says, 'Call you later, baby, *ciao bella*,' and hangs up. 'What?'

Janet slides the headphones off her ears. 'What?'

'I've. Got. Engaged.' I thrust my hand towards them waggling it wildly, and as luck would have it, thrust it into one of the last rays of the setting sun blazing dustily through our windows. Little bursts of light sparkle all over the room.

'Oh. My. God,' says Janet.

'No. Way,' says Lily.

We must stop speaking in Morse code!

'Yes way, actually,' I say, proudly. 'Charles proposed. And I accepted!'

Janet screams. 'Aaaargh! Aaaaargh!' She leaps up from her chair and jumps up and down like a mad yet gorgeous baboon. 'Aaaaaaaaaaaarrrghhhhhh!'

'Ssh,' I say, frightened that next door will hear and call the police.

'I don't believe it,' Lily says faintly. 'Chester House. Eaton Square.' She looks sick. 'Chester House,' she says again. Then she gives herself a little shake. 'Congratulations,' she manages, with an effort. It's almost as though she and I were facing off in the finals of *Pop Idol* and I had somehow bizarrely won. 'Let me see the ring,' she says with a forced smile. I have to give it to her, though, she's at least making an effort.

I thrust it forward a bit further.

'Oh my,' says Janet, excitedly. 'Bling bling!'

'That *is* nice,' says Lily, seizing on it. She holds it up to her eyes. 'Princess-cut ruby, about four carats, translucent, two trillion-cut diamonds, two carats apiece, cut is very good, colour's a D, clarity . . . hmm . . . SI1 to SI2,' she rattles off.

'Lily knows her jewellery,' says Janet wisely. 'She's been given so much of it.'

Nothing as nice as this, though. She knows it and I know it. I can see that Lily thinks this situation is so mad as to be psychedelic. She, the most radiantly lovely, blonde, tiny elfin model, has been upstaged by a great strapping size fourteen with a big nose.

'What does that mean?' I ask innocently.

Lily swallows drily. 'Retail? About thirty grand,' she says.

Thirty thousand pounds? I feel sick. Charles didn't let me see the price tags when we were choosing the ring. I want to rip it off my finger and lock it in a safe somewhere.

'After the marriage that'll be yours to keep,' she says. 'Very portable asset,' she adds longingly.

'What are you talking about? They won't be getting divorced,' says Janet loyally. 'I think it's wonderful. Congratulations, hon,' she says, kissing me on the cheek.

'Thanks, Jay-Me,' I say, squeezing her trim waist.

'Yes, well done,' says Lily. She forces another smile. 'Well done indeed,' she adds, turning away quickly. She really has gone very pallid. 'I'm very tired, I think I'll be off to bed. Congrats,' she repeats faintly.

I'm touched, at least she's being civil.

'I expect you'll have one of these soon,' I say. 'From Henry.'

'Yes,' Lily says, shaking herself again. 'No doubt. And he's loaded,' she adds, more to herself than me. But her heart doesn't seem to be in it. 'Goodnight, guys,' she says, and goes into her room, shutting the door quietly.

Janet hugs me again, oblivious to Lily and her moods. 'This is *so* brilliant, Anna. Tell me everything, you've got to tell me everything. Did he go down on one knee? Was it totally romantic? Was it, you know, like the great love of your life and everything?'

No. No. And no.

'It was more just a conversation, but it was really nice,' I say, sounding defensive, but I can't seem to stop myself. 'Anyway, I'm actually pretty tired too so I'm going to turn in as well. See you tomorrow.'

'OK,' says Janet, uncertainly. 'Fantastic news!' she calls after my retreating back.

I shut the door to my closet-sized bedroom and feel my racing heartbeat slow down. Thank God that's over; nobody else left to tell, besides Mum and Dad. I pick up Mr Bear, my imaginatively named teddy, and hug him close to my chest as I crawl on top of my duvet, clothes still on, ready to pass out. Just before I sleep I look in the direction of my ring for some sparkly reassurance, but it's gloomy in here and I can't see it glitter.

There is one more person I have to mention it to, I think as I drift off.

Mark Swan.

10

I wake up in the morning feeling a bit disorientated. Maybe it's because I'm now cold, with no covers on. Or because I fell asleep in my clothes even though I wasn't drunk. Hastily I pull them off and stuff them in my canvas laundry bag, and put on my ratty brown towelling dressing gown so nobody realizes what a total slattern I am.

'Morning,' says Lily brightly as I emerge. 'It's the bride-to-be!'

I start, then it comes back to me.

'You can have the shower now,' she adds generously. 'Janet left already. A shoot. Though who knows who's booking her,' she adds with a touch of contempt.

'*Vogue*, I expect,' I say loyally.

'Yeah, right,' says Lily, contemptuously. 'Anyway, you can use my Aveda shampoo if you want,' she offers, with very uncharacteristic generosity.

I goggle. 'Are you sure?'

'Of course I'm sure. You know you're always free to use any of my stuff,' she says airily. 'What's mine is yours.'

'Are you feeling OK?' I ask, gingerly. Lily marks the level of her cosmetics on her bottles with wipe-off magic marker and has threatened legal action over my occasional sneaking of a squirt or two of one of her expensive potions. I got my own back, though. Diluted them with water. And once I decanted her entire bottle of Perlier bubble bath and replaced it with Safeway's own and she never knew the difference.

'Of course,' she says briskly. 'You know me, share and

share alike.' She smiles at me winsomely. 'You have to keep up your bridal beauty,' she says, smiling.

My bridal what?

'He only just proposed. I'm not getting married just yet,' I say. Obviously Lily has slept on it and is terrified that what's mine might not be hers. But she needn't have worried. She can come and spend a year in one of Charles's many guest bedrooms for all I care. I suddenly wish with a pang that she'd do just that. I want to clutch on to everything familiar, even Lily.

'Oh, you know Charles. I'm sure he'll be setting the date very soon,' Lily says.

Actually, now she mentions it, so am I.

'Have you told your parents?'

No. Let me do that now. I dial their number, just wanting to get it over with, and feel a huge sense of relief when I get the answer machine.

'Hi, Mum and Dad, it's me,' I say redundantly. 'Anna. Anyway, I've wonderful news, I'm getting married. His name's Charles Dawson and he's a . . .' what? 'a writer. I don't want to tell you this on the answer machine,' I lie, 'but he's keen on putting the announcement in the papers, so I had to be sure you'd hear it from me. He's very nice, you'll like him. Call you later, bye!'

I hang up feeling as if a sack of potatoes has just been lifted off my shoulders. I couldn't face the conversation. What? Who? How long have you been going out with him? Why haven't we met him? Can he support me in the style to which I have become accustomed? Well, that last one would have been pretty easy even for Brian. I've become accustomed to thinking of store-made sandwiches as the height of luxury.

Still, I can't put it off forever. I make a mental note to ring them properly tonight. Otherwise there's the risk that when I take Charles round to see them they'll simper and fawn with joy when they find out how much dosh he's got. I can just see it. Must be stopped. I must prepare them for the fact that, against all the laws of probability (except in my dad's eyes) I, Anna Brown, am going to marry money.

'Let's go shopping for wedding dresses!' Lily suggests. 'I know a *fabulous* boutique – invitation only, but not for you and me, of course.'

I start. 'I can't. Got work.'

Lily pouts. 'Haven't you resigned yet?'

'No, and I'm not going to.' I think about my ghost comedy. I can't wait to get back to work on it. Get into the second act.

'Well, it's your choice,' she says reluctantly. 'If you will insist on slaving away for peanuts when there's no need to.'

'Women need to work, Lily,' I tell her severely. 'We need our own independence and careers.'

'Good God, why?' she asks.

I think about giving her a long and passionate answer then decide it's not worth it and head off to the shower instead. Thank God I *do* have work to think about, or I'd be going nuts. Oh yes, I am a professional woman.

And, just as an aside. Not that it's really important. Mark Swan is back in England today. I'm leaving the office to work with him again. And it's such a great *professional* opportunity.

I hail a taxi. Why not? I am about to be a woman of substance, after all.

Not my own substance, true. But maybe that can change with work on my movie.

'Morning, Anna,' says Michelle coolly. 'Good to see you again.'

I doubt it. I hadn't seen much of Michelle, because Swan had mostly had me with him on set, or at some hotel, or over at his place. And I don't think she was crying too much about it, somehow.

Michelle is young and casually dressed, but I don't let that fool me. She's super-efficient, slavishly devoted to Swan and, of course, great-looking.

Today she is resplendent in daring jeans and a sprayed-on T-shirt over her neat little bud breasts that says, 'I'm only wearing BLACK till they make something DARKER!'

'Is he about?' I say, foolishly, looking round the office. 'Did he have a good holiday?'

'Going to Hollywood to battle the studios is hardly a holiday,' corrects Michelle. 'Coffee?'

'No thanks,' I say, erring on the side of caution.

'Suit yourself,' says Michelle, icily. 'He's in a meeting.' She nods at the closed door.

'How long will he be?'

She looks at me with disdain. 'As long as it takes, I suppose. You can't rush genius.'

I want to say he's only a director but don't dare. I cast around for something to read, but there's only this week's *Variety* and I read that ages ago.

'What's that?' Michelle asks suddenly.

She's pointing at my ring.

'My engagement ring,' I say, wishing to twist it out of sight. Why? It's a perfectly lovely ring. 'Here, have a look,' I tell her boldly, thrusting it towards her. Because, obviously, I'm not ashamed of it.

'Ooh, it is stunning,' she says, in a markedly more friendly voice. Then she pauses and a shadow crosses her pretty face. 'Who . . . who's it from?'

'His name's Charles and he's really nice,' I say. My standard answer.

'Oh!' she says, brightening. 'Well, that's great! Congratulations. True love makes the world go round, don't it? Sure you don't want that coffee? We've got herbal tea and all. Even some PG Tips,' she offers, generously. I instantly get the impression that she would no longer be spitting in it.

'That's OK.'

'Let me buzz Mark for you,' she says, smiling full wattage at me. It's bizarre, as if she thought I was an axe murderer but I have just been revealed to be actually Mother Teresa.

'Isn't he in a meeting?'

'Oh, that don't matter, not with news like this,' Michelle says airily. She goes back to her desk and presses the buzzer before I can stop her. 'Mark?'

'What is it?' asks his disembodied voice, a bit tetchily.

'Anna Brown's here.'

'Yeah?' He sounds pleased to hear it. 'Great, I'll be right with her.'

'And she's got engaged!' says Michelle, loudly.

'What?'

'*Engaged*,' Michelle repeats. 'To be married. You should come and see the ring!'

There's a pause and I find I'm blushing.

'That's great,' Swan says, politely. 'As soon as I've finished my meeting I'll be out there to admire the ring.'

'Oh-kay,' singsongs Michelle.

'And no more interruptions, please,' he says, hanging up.

'You didn't need to do that,' I tell her, but Michelle shakes her head.

'Wonderful news like that, he's gonna want to know straight away,' she says. 'You madly in love?'

'Oh . . . yeah, sure.'

'Tall, dark and handsome is he?'

'Not tall . . . I like the way he looks,' I lie. 'We make the perfect . . . he's a real gentleman,' I say.

'A real new man?'

'He always pays for dinner,' I defend him.

'What's his career? In the film business?'

I look helplessly at the door. Please, rescue me. 'He's a writer.'

'Talented writer,' says Michelle, with satisfaction. 'You going to have lots of kiddies?'

'I expect so.'

'He's loaded, ain't he?'

'How do you know?'

She points at the ring.

'Well, yes. But I didn't marry him for that.'

'Course not,' she says indignantly. 'Country gent, is he?'

'You can't tell that from the ring.'

'I knew it,' she says with satisfaction. 'Don't know, just seems your type. Country gent. Getting away from it all.'

But it's not, I want to tell her. Not me at all. I'm London. I'm crowded tubes and air conditioned offices. I'm Starbucks and *Loot* and drooling over over-priced flats. And most of all

I'm walking around Soho, dreaming about writing movies and joining the Groucho Club.

I'm Mark Swan's world, at least in my dreams. But now I'm marrying a land baron, they expect me to instantly give it all up and settle down to sensible skirts, Labradors and walled kitchen gardens. Michelle smiles at me encouragingly.

'You're so lucky,' she says.

'Mmm,' I reply. And then the door opens. I jump to my feet. Swan is shaking hands with another man in a suit. He's American, you can tell by the deep tan and the loose, unstructured suit, informal yet breathtakingly expensive. He looks like Eli Roth.

'Thanks,' he says to Swan. 'Good to see you. I hope you'll consider the offer, Mark.'

'I'll consider it,' Swan says in a tone that adds he's not promising anything.

'Good day, miss,' says the guy to Michelle. She simpers.

'See you soon, Mr Giallo,' she says.

Wait a minute. Not . . . not Frank Giallo? My God. It *is* Frank Giallo. I recognize his face now, it took up the whole third page of last week's *Variety*. He's the new president of Artemis Studios, took over from Eleanor Marshall last week. He's one of the most powerful men in Hollywood. After David Geffen and Steven Spielberg, this bloke is *it*.

And here he is, courting Mark Swan.

I feel slightly faint. Swan catches my eye. He glances at me neutrally, his eyes flickering to my ring, then sees the expressions of awe and sheer terror crossing my face, and his body relaxes slightly, as though he's thinking better of something.

'Hold on a second, Frank,' Swan says, stopping him. Giallo pauses, looks back at him eagerly, as though he's hoping Swan will give instant consent to whatever he just offered him. 'Here's somebody I want you to meet. Anna Brown, this is Frank Giallo.'

'How do you do . . . sir,' I say, dry-mouthed.

'Sir!' says Giallo, pumping my hand and chuckling. 'Love those English accents. Too cute. You can call me Frank, honey. Anybody recommended by Mark is on first-name terms.'

Michelle scowls at this but I think I'm the only one who notices.

'Anna's a talented producer, but she's going to start writing scripts,' Swan says, grinning.

'That right?' says Giallo to me, but only a squeak came from me. 'Are you any good?'

'If she is,' Swan says, 'I'm going to messenger her script over to you and you're going to read it.'

'Are you attached, Mark?' asks Giallo, cannily.

'Don't pass it down to some shitty vice-president either,' says Swan, not answering the question.

'Oh, that's OK,' I say, because my face has gone so red it makes a becoming match for my ruby. 'You don't need to do that, Mr . . . Frank.'

'Mr Frank!' says Giallo, delighted. 'Listen, sugar, a word of advice. If you have powerful friends and they pull strings for you, don't say no. That's Hollywood. Don't bother with that Brit reserve. I'll be reading the first thing of yours I see because Mark Swan recommended you. If you've impressed him,' he shrugs, 'that impresses me. See?'

'Yes,' I manage.

'Backers don't come more powerful than him,' he tells me.

I can't even look up at Swan. 'So now what do you say?' Giallo asks.

'I – thank you,' I say.

Giallo grins. 'She's learning, Mark. It's been good to meet you, Anna,' Giallo says, to let me know he remembers my name. And then he wishes us a good day, steps into the corridor, and I watch the elevator swallow him up.

I'm left standing there, with Swan and Michelle, staring after him, open-mouthed and slightly hyperventilating.

'I can't believe you just did that,' I say.

'Why not?' Swan asks easily. He reaches round and grabs his coat where it's laid over a chair. 'Ready? We're meeting Trish again this morning.'

'Oh.' I swallow hard, try to get my act together. 'Yes.'

He doesn't pay any attention to me all morning. Not to my ring. Not to my haircut. Trish sits there and complains as he demands more changes, and he ignores her, just tells her why he wants things done. And she gives in. I sit there taking notes for Kitty and Eli, but I'm doing it on autopilot. I'm miles away. Trying to process what just happened.

I mean, I knew he was big. And popular. But power . . . I don't think I'd realized exactly how much power he had. Or thought he would flex those muscles for me. Oh, on one level, yeah, but I thought it was just nice, encouraging talk. I don't think I ever expected him to do something like that for me.

In thirty seconds, he just turned my whole life upside down. Now screenwriting is more than a dream, now it's something I have a chance at. A studio head's going to read it. People can spend ten, fifteen years trying to break into Hollywood and never get a chance like that.

And Mark Swan got it for me in ten seconds.

I try to concentrate and be professional, but it's all a bit pointless. I keep staring at him. He towers over Trish, it's ridiculous. She's willowy and blonde with pale English skin. Tiny against him. I wonder what she would look like if he kissed her. I wonder what—

'Anna.'

'Yes?' I ask, guiltily.

'Anything to add?' Swan says.

'No. No, I think you nailed it,' I say. Luckily he nods and doesn't ask for anything more.

'That was good work,' he says to Trish, and I can see her smiling, blossoming under the sun of his praise. 'Lunch?'

'Can't,' she says. 'Got lunch with Peter.'

'Who's Peter?'

'My boyfriend,' Trish says, happily. 'Peter's a lawyer. But he's not a complete git or anything.'

'That's certainly a recommendation,' says Swan, smiling.

'Oh, eff off,' says Trish, airily. 'Rather go out with him than you and your ugly mug, quite frankly.'

'Understandable,' says Swan. 'Oh well.'

'You go with Anna, have a good time.'

'Anna. Yes.' Swan turns to look at me, without much

enthusiasm. 'Sure you can't come?' he asks her again, a bit more pleadingly this time.

'No. Thanks,' says Trish. 'See you guys, yeah?'

'Bye, Trish,' I say as she exits. Swan turns to look at me and I let my eyes slide to the floor.

'Well,' he says, after a pause. 'I suppose we'd better go out to lunch.'

'OK,' I say, feeling hunted. I was so longing to see him, to tell him about my script, to tell him what I've done with his gift. But now the moment is here I just want to run away.

I don't want to talk about Charles and me with him.

'How's Edgardo's?'

'I don't know it.'

'Little tapas place in Holland Park. Not very fancy,' says Swan. 'I suppose you're only eating the best now, huh?' he adds, nodding at my ring.

'Who, me?' I shake my head. 'McDonald's is my idea of fancy. You know me.'

'Not all that well, perhaps,' he says. What does he mean by that? 'Edgardo's it is.'

'Fine,' I say. 'As long as you don't order off the menu for me,' I add with a touch of my old spirit. 'Why do men always do that and think it's sexy? It's not. It's aggravating.'

Swan looks at me and his mouth twitches very slightly. 'I promise you can order for yourself.'

'All right then,' I say. 'Let's go.'

The sooner we get to the place, the sooner I can get out of there. I wish Trish had come. I don't want to eat with him by myself.

He makes me nervous.

Edgardo's is, as promised, not very fancy. It has small tables covered with oil-proof laminated cloths in a cheesy French check, plain white walls, and ancient-looking menus. The tables are all crowded together, the waiters look harassed, and the place is heaving. I love it immediately.

The delicious scent of Spanish cooking is everywhere, little glass carafes of cheap red and white wines are knocking about, everybody is stuffing their face and having a good time, and the conversation is punctuated with laughter and sizzling. Plus, nobody looks at Swan.

We thread our way carefully past people's elbows and coats draped on the back of chairs and find one of the last empty tables.

'Nobody knows you here?' I ask as we sit down.

'The waiters,' he says. 'Nobody else. I don't like to be bothered while I'm eating. Or have photographers outside ready to take a picture of me and stick it in the papers.'

'How can you stop that?'

'Simple maths,' he says. 'A snapper's not going to hang about a cheap and cheerful tapas place waiting for the possibility of getting a shot of Mark Swan. He's going to be outside Quaglino's on Friday night, so he can get a shot of at least four or five celebs. People whine about privacy, but they don't do anything to actually protect themselves. Unless you're Madonna, you don't have to be in the spotlight unless you want to.'

'People think you're cultivating a deliberate air of mystery,' I say. 'Mark Swan, reclusive superdirector. Mark Swan, Hollywood's secretive superstar. Mark Swan . . .' I realize I'm getting a bit lyrical and trail off, blushing. 'You know.'

'Apparently not,' he says, drily.

'Most people think it's a strategy on your part.'

'To get famous by avoiding publicity?' He snorts. 'That makes a lot of sense. My only strategy is to avoid being fucking bothered all the time.'

A waiter swoops down on us, grinning, bearing four or five platters of little hot, delicious things.

'Señor Mark,' he says. 'Ow good to see you! Here you go.' He lays them out for us, steaming plates of crepes, little grilled sardines, some herby sausage things, some pitted, marinated olives, and fried jalapeno peppers stuffed with cheese.

I scowl at Swan.

'*Gracias,*' he says to the waiter.

'I get the rest of it. And your wine. Señorita,' he nods to me. The stress of this meal is mounting by the second.

'I thought I told you I hate people ordering for me,' I snap. 'God, you're such a egomaniac control freak, Mark Swan!'

He picks up an olive, admires its glossy black skin, and puts it in his mouth.

'Aren't you even going to answer me?' I demand, furiously.

'These aren't for you, sugar britches,' he says. 'These are for me.'

Oh. I sit back in my seat, feeling like a total idiot.

'Your paella, señor,' says the waiter, arriving with more food. 'Your cold meats . . . your fried artichokes . . .'

The little dishes are placed down in a seemingly never-ending stream. 'And miss, for you?' he asks.

I shake my head. 'I'll share some of his.'

'Very good,' the waiter says, putting down a large carafe of red and withdrawing.

'Sorry,' I mutter.

'Who said you could share mine?' Swan asks lightly, picking up a sardine and crunching it in his mouth. He pours a glass of wine for himself. 'I wouldn't want to intrude on your culinary decision-making.'

'You love making me sweat, don't you?'

'Mmm,' he says, winking at me. 'You make it so easy.'

He's so gorgeous. So tanned and huge and self-confident. And the way he's mocking me, so lightly . . .

I swallow hard against the overwhelming wash of desire that rips through me. Don't be bloody stupid, Anna, I tell myself. I reach for the little dishes, helping myself to food, pouring the wine, anything to distract myself.

'That's what I like to see,' Swan says. 'A girl with a good appetite.'

I glance down. My plate is heaped with sausages, pan-cakes, rice, olives, fish, cheeses, all in a big pile as if I was a starving refugee let loose at an all-you-can-eat buffet. I imagine the girls Swan's been seeing in LA. All blonde Heather Locklear clones whose idea of lunch is two lettuce leaves, no dressing. He's probably looking at me right now and going 'No wonder she's such a fat knacker', I think miserably.

Well, too late now.

'Thank you for what you did,' I say. 'In the office today. And you know, thank you for the *Final Draft.*'

He nods, acknowledging this gracefully. 'Always happy to help out a friend.'

'I won't let you down,' I say.

'You're still planning to use it, then?'

I blink. 'What do you mean?'

'You're going to try and write? Not quit your job?'

'Bloody hell,' I say. 'Why does everybody keep asking me that? Why would *you* ask me that?'

'Hey.' He holds up his hands. 'Just checking you're not being pushed into something you might not want to do any more.'

'Why wouldn't I want to write? You said I had talent. I'm almost halfway through it already,' I tell him proudly. 'And you told Frank Giallo all about me. He said he'd read my script.'

'Don't bet on it,' Swan says, demolishing a stuffed mushroom.

'What do you mean?' I ask. 'He said he would.'

'Don't you know, in Hollywood nobody ever says no? They say yes even when they mean no. Giallo would say anything to please me.' He shrugs. 'Most likely, he forgot about you in five minutes.'

'Right.' I feel very small. The man is such a good friend to me. Look what he's doing for me. And all I can do is go around being hostile and prickly.

'Can you do it?' Swan asks. 'You think you can write a good screenplay?'

'Oh, absolutely.' I smile at him, back on solid ground. Movies. My career. 'It's a great premise.'

'Tell me,' Swan says. Not a request. His dark eyes are holding mine, assessingly. As if he is about to weigh my talent in the balance. The man who can make me or break me.

I realize he holds my future in the palms of his hands.

'It's not a very arty film,' I begin. 'It's just a little comedy really and—'

He holds up a hand. 'Stop. You don't do it that way.'

'What?'

'You don't begin with an excuse. You have a story to tell, tell it. Don't start out by saying how bad it is. Or how good it is. Just tell it.'

'OK,' I say. Suppressing a wild urge to say, 'Yes, sir.' He's speaking with such authority, it's insane.

'And,' he adds, with a grin, 'especially don't start out saying what cross it is.'

'What?'

' "This is *Die Hard* meets *Rambo*," ' he quotes. ' "This is *Gladiator* meets *Pretty Woman*." '

'That one sounds quite good, actually.'

'Doesn't it? Might have to do that one myself.' We smile at each other and I feel relieved. Despite all the teasing, and the massive help, he's seemed a bit cold. And I've been so ungrateful and snappy. I love it when he smiles at me. It feels like it used to. Because I value his friendship.

'Go on,' he orders.

'All right.' I swallow nervously, but there's nowhere to bolt to. 'It's called *Mrs Watkins*.'

'Shit title.'

'I know, I have to think of something better. Anyway, it's about these two antique dealers who go around ripping off old ladies and housewives and buying their antiques for nothing. And one day they buy this old Welsh dresser from this old lady. And they take it away but it's haunted. And the ghost is making their lives miserable so they try to return it. But the old lady's moved, and they can't sell it or dump it so they have to drive round the country looking for her . . .' I daren't look at him. I stare at one of my delicious mini-pancakes, even though I'm far too nervous to eat it.

'And hilarity ensues?' he asks, deadpan.

I sigh. He hates it. I might have known. 'Yes, hilarity ensues,' I admit, feeling about as hilarious as Eeyore on a particularly gloomy day.

He doesn't say anything, so I push my food around my plate, glumly.

Swan reaches out and puts a calloused fingertip under my chin, tilting my head up so I have to look at him.

'Anna, that's not bad,' he says.

I start. 'Really?'

He nods. 'It's an old-fashioned caper. It's fresh, it's funny – at least as a premise. You have a long way to go before anybody buys it, but I'd be at least interested, based on the pitch.'

'Thank you,' I say.

'Don't thank me,' he says, sipping his wine. 'Just finish the script. Make me proud.'

My throat thickens. I so badly want to make him proud.

'I will.'

'That's if you've got time, of course.'

'What?'

'Well,' says Swan, looking me over. 'You've got to spy on me for Winning.'

'Red Crest.'

'Whatever. And of course you'll be spending a lot of time with your new interest.'

'My what?' I ask, blinking.

Swan raises a craggy eyebrow. 'Your fiancé. Don't tell me you've forgotten about him already?'

'Oh.' I start. 'Of course not.'

'It's a very nice ring,' he says cordially.

'Thank you.'

'When were you going to tell me about this?'

'It only happened last week,' I say, defensively. 'I was going to tell you today. But Michelle beat me to it.'

'So she did.'

There's a long pause. I eat a giant olive to fill it. It's absolutely delicious, but it tastes like ashes in my mouth. I don't want to discuss Charles with him. Because Charles and me is too private, I tell myself.

'I hope you'll be very happy,' he says.

'Do you want to know who it is?'

'I already know who it is,' he says. 'Charles Dawson.'

I blink. 'How do you know Charles's name?'

264

'I have my sources,' he says.

There's another pause. He disapproves, I think nervously. Maybe he's like Lily and Kitty and thinks I'm going to drop my career.

'So you know about Chester House and all that,' I gabble. 'But it's not going to affect me at all.'

'Of course it's going to affect you.'

'I'm still going to work. I still want to write movies,' I protest. Why do I suddenly feel like crying? I blink it back, furiously. 'I have my own dreams,' I insist.

'I'm glad to hear it.'

'Aren't you going to congratulate me?' I ask, and hope I'm not sounding too whiny and shrill.

'No,' Swan says. 'I'm going to wish you joy. That's from a Jane Austen adaptation I directed in college. They used to wish the women joy. Only the men were congratulated, because they had been lucky enough to win the hand of the woman.'

I try not to stare at him. 'That's very . . . chivalrous.'

'I'm old fashioned in many ways,' Swan says, and looks away from me.

'Many . . .' I can't stop myself, it's like worrying a tooth. 'Many people have been, you know, congratulating me because Charles is well off.'

'A bit more than well off, I believe.'

'Yes,' I admit.

'What does that have to do with anything?' he says, and looks at me, and now his eyes are narrowed.

'Nothing,' I say hastily. 'That's what I'm trying to tell you. Nothing. I'm not going to become a rich country lady and not bother with the scriptwriting.'

'So, you're not marrying the guy for his money?'

'I would never do that!' I say, angrily. And this time I really am angry. My eyes are flashing. I can't stand it that Swan would say that to me. I don't care when other people suggest it, but him? 'Never! Never.'

There's another horrible pause, then he says, 'I believe you, Anna.' And signals for the waiter, who is at our table in

265

less than five seconds with a bill. Swan doesn't even look at it, just hands over two fifty-pound notes, which the waiter pockets and disappears as fast as he'd come. Nothing on that menu cost more than four quid.

Swan stands up, grabbing his coat.

'I'm beginning to see why you get such great service,' I joke. He smiles down at me, but it's distant. Withdrawn.

'I've got to get over to the editing suite,' he says as we emerge from the restaurant. 'Can I drop you at Red Crest?'

'No need,' I tell him. 'I'll make my own way. It's just round the corner.'

Swan nods and hails a taxi. He drives off without looking back.

When I get back to the office I fling myself into my work. Which is to say I type up some bullshit about our 'working lunch'. That gives me an hour or two to work on my script. It's the only thing that eases the stresss. The second act is flowing now. My fingers hammer over the keyboard, and when I finally take the floppy disk out I exhale. At least that's something done today.

Next I bite the bullet and call my parents. They're delighted, Dad blustering on about how lucky Charles is, and my mother giving me the third degree about his money; honestly, she's worse than Lily. I make an appointment to take Charles round tonight. Best to get it over with.

'Anna.' It's Claire, looming over me.

'Here it is,' I sigh, holding out the ring for her inspection.

'Yes, it's really nice,' she says, awkwardly. 'Really nice. But actually, can you go upstairs?'

'Upstairs?' I stare at her. 'What for? I have to work,' I say severely, minimizing my Free Cell before she can get a look at my computer screen.

'Eli Roth wants to see you,' Claire says, apologetically.

I feel a shiver of apprehension. 'Why?'

Poor Claire is practically hopping from foot to foot. She looks as if she's desperate to tell me something, but she doesn't open her mouth. And then I see Kitty behind her, standing in the open door of her front office, hovering. Giving me what can only be described as an evil smirk.

My heart thuds. 'Claire, what is this?' I whisper.

She shakes her head miserably. Obviously terrified to say anything with Kitty watching. This is not good.

'OK,' I say, nervously. 'I'll be right there.' Is there anything I need to take off my computer? I wrote my script on a floppy, thank God, I think, as I slip the disk into my pocket. What else? Any emails saying John is a pretentious idiot or Mike is a sexist wanker . . . or . . .

'I'm so sorry,' Claire mutters, 'but you have to go up right now and I have to escort you. Otherwise . . .' She jerks her head behind her and I see two security guards standing in the corridor, looking over in my direction. And, as I turn my head about the office, I realize that everybody has stopped what they're doing and is staring at me.

You don't need a degree in office politics to figure this one out.

'I'm getting fired?'

'Please, Anna,' pleads Claire, unhappily.

'I'm getting fired,' I say wonderingly. A strange sense of calm comes over me, which is jolly surprising, as I would have thought I'd be crying buckets right now and begging for my job.

'Are you coming?' Claire asks.

'No, I don't think so,' I say. I stand up and pull my bag out from under the desk.

'But you have to,' says Claire, almost crying.

'No I don't,' I tell her. 'They're going to sack me for something. Whatever it is, it isn't fair. I'm the only one in this whole office who's done anything worthwhile for ages. And I don't think I'm going to change their minds. I mean, I can't do a better job than this. I found a great script and a great director. If that's not enough then no amount of talking from me's gonna change a damn thing.'

'But what shall I tell Mr Roth?' wails Claire.

'Tell him he can stick his lecture up his arse,' I suggest. 'You have my home number. No hard feelings, OK?'

'OK,' snuffles Claire, looking at me admiringly. As, I notice, are all the people working within range. But nobody says 'Well done' or 'You tell 'em' or even 'Goodbye, Anna'. They all want to keep their jobs. I pick up my handbag and head for the elevators. There's nothing I absolutely need to take with me, thank God, and if the tech guys unearth that long email about fancying Rufus Sewell, well, they can just enjoy it.

'Where the hell do you think you're going?'

It's Kitty. Like a vampire, she has moved at the speed of light, yet silently, from her position in her office doorway to block my path to the elevators. There's a light of real malice in her eyes. So much for my being her right-hand woman and her go-to girl. But somehow I'm not in the least surprised.

'I'm going home,' I say simply. 'See you.'

'You can't get out of this, Anna,' she pronounces, and she's speaking very loudly and clearly so everybody in the office will get it. 'You're going to be terminated whether you pretend to be off sick or not.'

'I never said I was sick. I said I was going home,' I tell her.

'You always were a slacker,' she hisses. 'Aren't you even curious as to *why* you're being fired?'

'Not really, no,' I say. 'It's some line of bullshit you and Eli Roth have concocted so you can steal the credit for *Mother of the Bride*.'

'That's my project. I found the script,' says Kitty. 'And you agreed to that publicly,' she reminds me.

'Because you told me I had to.'

'Oh, you're just a fantasist,' she says, scornfully. 'You're being fired,' she adds loudly, 'because you gave me a set of fabricated location reports yesterday. You weren't at those locations!'

I suppose she must have a spy who reported seeing me with Charles.

'That's true, I made them up,' I say. 'But taking a half day sick is hardly a sacking offence when I'm the sole person in this office to have come up with a viable project in the last six months.'

'You had very little to do with this film. Really nothing.'

'I found the script and the director,' I say.

'You provided a bit of coverage, and Mark Swan would have attached anyway. He did it for the script, not for you. He was already on our list to submit to. In fact, the way you circumvented his agent was *most* unprofessional,' Kitty snarls. 'And as for the rest of your so-called contributions, you haven't given us any usable information on the pre-production process. Eli and I are very disappointed in you.'

'You're just jealous,' I tell her. 'Jealous I can find good scripts, jealous Mark Swan likes me, and now you're jealous I'm getting married,' I say, suddenly understanding that it's true. 'Women are meant to help each other in business,' I tell her. 'But you're just one of those sad old cows that can't stand to have a rival. You only liked me when I was invisible and put-upon. As soon as I started to get somewhere you wanted to get me out of here.'

'Don't think Mark Swan will protect you!' snaps Kitty. 'He's under contract now. We've got him,' she adds, viciously. 'He can't just walk away to save the hide of the sixty-foot woman. Yes, that's right, get lost,' she shrieks, as I punch the button and step into the elevator. 'Don't you have to go and ravage Tokyo or something?'

'Don't you have to go and get some Botox shots?' I retort. 'Those wrinkles aren't going to surgically eliminate them-selves. Bye, Kitty. Enjoy being an old maid,' I say loudly, and have the satisfaction of watching the look on her face before the doors mercifully hiss shut and I'm riding safely down to the lobby.

I make it out into the sunny streets of Soho before I break down. Mostly because my devastation is mixed with shame. I threw the same kinds of insults at Kitty that everybody throws at me, insults to do with her face and body and lack of a man, when really I should only have insulted her bitter, selfish, vengeful personality. But I instinctively knew what would hurt, and I used it. Why not? Those insults really hurt me, after all.

Attack of the sixty-foot woman. That was actually a pretty

creative one, I tell myself, trying to laugh at it. To take away the sting.

I decide to take the tube home, because then I'll be crammed in with millions of people at lunch hour, and I won't be able to cry.

Lily's sitting on the couch when I get in, brushing her long waterfall of platinum hair with one hand, and holding the phone to her ear with the other.

'Yes,' she says, laughing. 'Absolutely, darling!' She looks over at me with disappointment. 'Actually, she just walked in,' she says. 'Do you want to talk to her? Really? Oh, I suppose if I *must* let you go,' she concedes, giggling flirtatiously. Then she holds out the receiver to me, giggling snapped off as if someone had flicked a switch. 'It's for you,' she says crossly. 'It's Charles.'

What's she doing giggling with Charles? I stare at her suspiciously, but Lily simply puts her head upside down and starts to brush her roots.

'Hello, darling,' he says. 'Having a nice day?'

'Not really,' I say. 'I just got fired.'

The brush stops dead in Lily's hand. She's listening, agog.

'No!' says Charles, outraged. 'What are they, morons?'

He's so supportive, I think. I should be grateful. No, I *am* grateful.

'That's very nice,' I say. 'We've got dinner with my parents tonight, if you're free.'

'Free? Absolutely,' he says, gleefully. 'What time?'

'Eight.'

'I'll pick you up at six. Five thirty, as you're not working, and it's rush hour.'

'I don't want to go to dinner with my parents,' I cry. 'I want to stay home and get drunk.'

'That's not good for you, darling.'

'Who sodding cares?' I say, then feel bad. It's not his fault, is it? 'I mean, you're right,' I say, dully.

'Does this mean that you won't be working with that Mark Swan chappie any more?' says Charles.

Oh.

'Hello, are you still there?'

'Yes. Yes, I'm here,' I say faintly. Of course, it does mean that, doesn't it? I should ring him. Right now. But he's not in the office, he's holed up in an editing suite.

It doesn't mean I'll never see him again. I try to calm myself from the sick, panicky feeling coursing through me. He said I was his friend. Of course we'll see each other in the future.

'You should keep up your links with him,' says Charles, judiciously. 'Why don't you invite him to the engagement party, darling?'

'Um . . .'

'I'd love to meet him.'

I swallow. Of course he would. And why not? I'm going to wind up meeting all of Charles's friends, sooner or later.

'Sure,' I hear myself say. 'I'll ask him.'

'I wouldn't worry,' Charles says. 'You'll have a new job in no time. And if you don't, well, you don't bloody need one, to be quite honest,' he adds, with a touch of pride.

'Thanks . . . darling.'

'Pick you up at five thirty, then,' Charles says, merrily.

We hang up.

'Got fired?' asks Lily brightly, knotting her slender plucked eyebrows together in fake concern.

'Yes,' I mutter.

'What for? Embezzlement? Sleeping with the boss? Drugs?' asks Lily, excitedly. 'Oh well, hardly matters now, does it?'

'Of course it matters,' I say, wanting to cry. 'I want to be able to pay my own bills.'

'Don't you understand, Anna?' she says, exasperated. 'You don't need that. You've *won*. He's *nuts* about you,' she says, in a voice that says she finds his attitude inexplicable. 'Your trouble is you want to work as well. You want *everything*,' she says, resentfully. 'And you've already got the only thing that matters.'

Somehow I don't think she means love.

'Once Henry and I are together I'll probably retire from modelling,' Lily says airily. 'Not that I'll have to or anything.

271

I just won't want to do it any more. I'll be concentrating on my marriage.'

'I have to jump in the shower,' I say. 'Family dinner tonight.'

'I heard. You don't think your parents will put him off, do you?'

'No.'

'Because, let's face it, they're not exactly top drawer. But you're probably right,' she adds with a sigh. 'He probably doesn't care. It's as though he's made up his mind what he wants. And you're it.' She looks at me accusingly, but I can't help her. I can't explain it either.

Charles arrives in the Rolls bang on time and we head off to Surrey. He doesn't ask me about being fired, which is great, because I don't want to talk about it. All he wants to do is discuss wedding plans.

'. . . and I know a marvellous chappie who can get orange blossom, even out of season. You must have orange blossom, darling, all the Victorians did. It smells heavenly . . .

'The cake . . . I hope you don't want anything exotic . . . I was thinking of traditional, maybe six tiers, white icing with lots of decorations. They can put motifs from the family crest in. And two flavours, tiers of white chocolate with fresh raspberry sauce and then lemon with pureed strawberries. And what do you think about an ice cream? I love lemon. You can get it so rarely, it's always sorbet, but it makes a fabulous ice cream . . .

'The church is St Mark's in the Fields in Greenhampton, it's sixteenth century, very pretty, and you'll love our vicar. He's so nice and gives such short sermons . . .'

He goes on and on. I give him the occasional 'Right' and 'Uh-huh'. I mean, I *want* to care. I know I will do, later. It's just that weddings are apparently really complicated when you get down to it. You don't think about it, but there's a hell of a lot of stuff to organize.

'Look, um, darling,' I say eventually as we're pulling in to my parents' village. 'Maybe we should go for something a bit simpler.'

He looks at me in horror. 'What? Simpler? Why?'

Why indeed?

'It's just that I've been sacked so I'll be looking for another job,' I extemporize. 'And that's going to take absolutely ages, you know. And I have to finish off my script. I don't know if I could do justice to this kind of really fancy wedding.'

'Oh,' he says, blowing out his breath in relief. 'So you've no objection to a nice wedding in principle?'

I think those weddings on the beach with just your closest relatives are nice, but I know what he means, so I say brightly, 'Absolutely not! It's just that there's no time to plan one.'

'Don't worry about that, sweetie,' Charles says reassuringly. 'Bunty and Crispin used a fabulous wedding consultant for their wedding. I'll just hire her. Flora Maxton.'

Flora Maxton? I've read about her in *Tatler*. She's the *sine qua non* of society wedding planners. She works on a percentage of the budget, which has to *start* at seventy-five grand.

'But Charles!' I protest, horrified. 'How much are you thinking of spending?'

'It takes a fair bit to put on a decent show,' he says. 'And I'm only going to get married once. I want it to be perfect. Starting our lives together in style,' he says. 'Is this your road?'

'Yes. Third on the left.'

'You don't have to worry about anything,' he says. 'You can leave the whole thing up to me.'

'I . . . OK.'

'Your career's important to you, so it's important to me,' he says supportively. 'Is this your parents'?'

I look up at our house. It's a three-bedroom semi with pebble-dash and concrete, a neat little garden lawn and an amusing gnome baring his bottom. I have always been fiercely ashamed of the gnome baring his bottom, but now I stare at Charles accusingly, just daring him to feel the same way.

'Yep. Home sweet home,' I say challengingly.

His jaw's set firm; not even a sniffle of a sneer.

'Lovely,' he says, bravely.

Mum and Dad react just as I'd expected. Dad, who's tall and strong, looks a bit askance at Charles's height – he's barely taller than Mum – but I glare at him and he backs off. And Mum is fawning even before she's figured out what he's worth. I can see the relief written all over her face – it's true, Anna has finally got somebody! Very flattering.

My mother's petite, elfin beauty is not reflected in her cooking. She serves roast lamb with traditional British overdone vegetables, boiled to green mush – Mum has missed out entirely on the UK's culinary revolution, Jamie Oliver and she are complete strangers – and Charles makes polite conversation. He compliments the cooking. He compliments the lawn. He compliments the gnome baring his bottom. And then he asks for a second helping of mushy broccoli. He's being a complete star, no doubt about it.

'We're thinking of having the wedding at Chester House, if you give your permission, of course,' he says, tactfully. 'The bride is usually married from her parents' house, and that would be *wonderful*, but we do have rather a lot of guests. It's purely logistical . . .'

'What's Chester House?' Dad asks.

'That's Charles's house in the country,' I say.

'Our family seat,' says Charles, modestly. 'It's quite pretty, for a wedding.'

My mother pauses, her eyes fairly glazing over with pleasure, the way my cat's used to do when you scratched its head.

'But I thought you lived in London, Charles?' she asks, a bit too eagerly.

'I've got a flat there. Comes in useful,' Charles says.

'It's in Eaton Square, Mum,' I say, as she's about to ask herself.

'Ooh,' says my mother, going almost catatonic. Even Dad nods briskly, as though Charles's finances have now been cleared as up to snuff.

After I nudge Mum warningly in the ribs she shuts up until Charles excuses himself to go to the loo (where he will be confronted with a loo-roll cover in the shape of a Spanish lady playing castanets). Then the dam breaks, and I am overwhelmed with teary-eyed hugs and loudly whispered congratulations, and Dad saying loyally that it's Charles who is the lucky one. But secretly he's obviously every bit as relieved as Mum. By the time we manage to get out, and Charles is driving me back in the darkness, I'm well aware exactly how happy I have just made my beloved parents. And Charles isn't even being a dick about their middle-classness or anything, he's saying how great they are, and can he send a limo to ferry them to the engagement party . . .

My mother said, 'You're the luckiest girl in the whole world!'

And I know I am. There's no doubt about it. I'm really lucky.

Definitely.

11

I wake up late. I'd switched off my alarm clock last night, and by the time I surface it's half nine. Light from the living room's dusty windows streams in through my open door, and I lie in bed, blinking and groggy, not quite sure where I am. Charles's place? No. Here. I glance at the clock, and, horrified, swing my feet out of bed. Fuck it, I'm late for—

And then I remember.

There is no work. I've been fired. For fabricated location reports.

I groan miserably and head into the kitchen to make myself some coffee. There are a few bagels (mine) and boxes of Special K and Slimfast bars (theirs) but I don't feel like eating. It's chilly outside, and I just wrap my hands round the steaming mug and wonder what to do.

'Morning.'

Janet emerges from her bedroom, looking sensational in a pair of boxers and a little camisole, yawning, her hair all sexily mussy. Blimey. I don't look that good even after three hours of professional primping.

'Hi,' she says. Then she gives her head a little shake and focuses on me. 'Anna.'

'That's me,' I admit glumly.

'Lily told me,' she says. 'I'm so sorry.'

'That's OK,' I say, even though it isn't. 'I expect I'll get another job,' and I swallow hard to down the lump in my throat. I don't have any idea if I'll get another job. A wave of fear crashes over me. I found the script and attached Swan, but I can't prove that, can I? And no doubt Kitty's been busy on the phones, making my name mud. Plus, it's not like I'm twenty-three, is it? I'm thirty-two. Too old to start at the bottom again. If I go in as a secretary I'll stay a secretary.

'I'm sure you will. You've had a brilliant career,' Janet lies, kindly. 'And you've got that director bloke. He'll help you.'

I look at her. 'Janet, that's brilliant! Of course he will. He'll help me.' I gulp another quick swig of coffee and pick up the phone.

'Mark Swan's office,' Michelle says. I glance at Janet, who's giving me the thumbs up.

'Hi, Michelle, it's Anna.'

'Hello,' she says guardedly. 'How was your lunch?'

Lunch? Oh. With Mark. 'Fine,' I tell her.

'And how's your fiancé?' she asks, a bit accusingly.

'Also fine. Look, is he there?'

'Hold on,' she says, resentfully. There's a brief pause, and then Swan's voice on the line. Rough, impatient.

'Anna? What's up?'

'I got fired yesterday,' I say.

'Really? Why?'

'Over nothing. Over making up some location reports for Kitty. I was out . . .' I trail off.

'Yes?'

'Out shopping for a ring.'

'I see.'

'It's still not fair,' I say tearfully. 'And she knows it. I found the script, I found you – nobody does anything at that shitty company.'

'So why don't you complain to the boss?'

'Eli Roth? He was in on it. He wanted me fired.' I pause. 'Come to think of it, I don't know why. Kitty was jealous, but Eli . . .'

'Eli didn't care about you,' Swan says, shortly. 'He cared about me. I've signed on for the movie now. I can't quit without damaging my reputation, maybe lawsuits. So because I insisted on having you around, he thought it'd be a great time to flex his muscles. Crack the whip on the maverick director.'

'What are you going to do about it?' I ask, momentarily distracted.

'Something. I haven't figured out what yet. But it will cost him.'

'Can you get me my job back?'

'Did you make up the reports?'

'Yes.'

'Then no, not really. You're right, it was only an excuse, but you gave them one.'

My cheeks flame. 'You've got power,' I accuse.

'Ah, but I only use it for good,' Swan says.

'Very funny.' I breathe in. 'But I don't need a job, do I?'

'I gather not.'

'No, I don't mean Charles,' I say, blushing. 'I mean, my script's almost done. You'll read it and you'll help me sell it, right?'

'I'll read it,' he says. 'But I very much doubt it'll be ready to show.'

'Why?' I demand. 'I thought you said I had talent as a writer.'

'I think you might. But this is just a first draft, Anna. It'll take a lot of work to get that script ready to show anybody. And no, I'm not going to help you sell it.'

'Why?' I demand. I find my voice has gone all high-pitched and shrill. And I'm close to tears. 'Why won't you help me? Everybody listens to you!'

'Because you've got to make it on your own,' he says. 'I thought you knew the film business better than to think your first attempt at a script is going to be perfect.'

'I thought you'd be supportive,' I snap.

'I am being supportive,' he says. 'You just can't see it.'

I swallow hard. 'OK,' I say, feeling defeated. 'OK. Well, thanks for all the help you've been, and everything.'

'Got to go,' Swan says, sounding distracted. 'Important call from the studio. Talk later, OK?'

'No problem,' I say, but I'm listening to a dial tone.

Janet is watching me from the couch. 'So he wasn't much help then?'

I burst into tears and she comes over and hugs me. 'He's so powerful,' I sob into her shoulder. 'And he could have helped me . . . but he wouldn't. He said I wasn't good enough . . . now I'm not working there any more it's like he doesn't believe in me any more . . .'

'I'm sure you're good enough,' Janet says loyally. 'I bet your script's absolutely fantastic.'

'He won't help me get an agent. And I need one to sell my script. I don't want to rely on Charles, you know?'

'I know,' she says, sympathetically.

'It's over for me. I'll never get an agent. Nobody will take me on.'

'Why wouldn't they? Isn't your script good?'

'You don't understand,' I wail. 'The script's great, but I'm thirty-two, I don't even look the part. These days writers have to be young. And chic. Stylish. They all look just like you,' I say despairingly. 'I don't look polished. Not even with this lovely haircut.'

Janet pulls back, hands me a Kleenex. I blow obediently.

'I've got a plan,' she says.

Ominous words. I look at her warily.

'What, just because I look like this I've got to be thick?' Janet asks defensively. 'That's not fair either, you know.'

'I know. I'm sorry.'

'My career's not going perfectly,' Janet says, 'so I do stuff. I call bookers and magazine editors, I send out headshots. I go to the right clubs, let them know that Jay-Me's where it's at!'

'And how's that working out?'

She frowns slightly. 'Well . . . early days yet. Anyway, you must do the same thing. You'll be applying for jobs, writing letters. And you need to look the part.'

'No chance of that, is there?' I say glumly.

'On the contrary,' Janet says triumphantly. 'There's every chance. Have you forgotten how I made you look for the ball?'

'No, and that was very kind of you,' I say, 'but this is real life. You can't hide everything in a big cloud of taffeta when you're trudging up Dean Street looking for an agent.'

'Huh!' says Janet scornfully. 'Do you think it only works for evening dresses? I can fit you out for everything.'

I sniffle. 'Everything?'

'You just dress wrong,' Janet says. 'I can make you look sharp.' She snaps her fingers. 'You already have the hair.'

It's true. My hair still looks good from the fabulous cut Paolo gave me. And I remember how pleased I was that night at Chester House. I did look kind of OK.

'But clothes are so expensive,' I say. I *want* to believe it, but . . . 'Have you ever been into Miss Sixty? It's called that because the cheapest thing in it is sixty quid.'

Janet raises one beautifully tapered eyebrow. 'Anna,' she says patiently, 'how much have you got in your bank account?'

I think. 'About one and a half grand.'

'Then let's spend it.'

I gawp at her. 'Are you insane?'

'Are you?' she retorts. 'Maybe you've forgotten but you *are* marrying one of the richest men in England.'

We both glance down at my ring.

'You want to get that new job, don't you?' Janet presses. 'You want to look like a film-biz person?'

I nod.

'Then don't argue with Jay-Me,' she insists. 'Follow me!'

Fuck it. Why not, eh? I've got nothing better to do.

We take a taxi to New Bond Street, Janet's personal Mecca. Crammed to the gills with scary designer clothes and shoe shops, as well as those expensive knick-knack places where a scented candle costs ten pounds. She's as excited as a child at Disneyland, jumping up and down and going 'Ooh' and 'Aaah' all over the place, keeping up a one-woman monologue.

'That'd be perfect on you . . . Ideal for covering the upper arms . . . V-necked cashmere, perfect . . . Leather skirts with a panel, brushed stretch cotton, low-risers . . . Red's very smart with cream. It's the new navy . . .'

'But Janet,' I wail. 'Look at these prices.' I pick up a tiny scrap of cashmere pretending to be a sweater. For whom? Barbie? 'This is three hundred and eighty. And those shoes are five ninety-five.'

'They're Armani,' she says.

'But at this rate I'll only be able to buy three things.'

'Hmmm,' Janet acknowledges. 'Don't you have a marital credit card or something?'

'No I do not,' I say stiffly. 'And I don't want to spend Charles's money anyway.'

'Are you mental?' Janet asks.

'I don't want to be a kept woman.'

'You're not, you're a bride-to-be. It's so romantic.' Janet sighs. 'And remember, with all his worldly goods he thee endows. And that's quite a lot of worldly goods.'

I shake my head, mutinously.

'Oh bloody hell, all right,' says Janet crossly. 'No Voyage, then. No Armani. No Donna.' She sighs. 'No Chloe.'

'I get the picture.' Gloom settles over me again. 'So it's no good?'

'I didn't say that,' Janet says, brightening. 'We can get you one or two nice pieces anywhere. I'm thinking H&M, I'm thinking Zara, I'm thinking Banana Republic. You just have to know what to look for.'

'And what are we looking for, exactly?' I ask.

Janet nods disparagingly at my favourite pair of black 501s with my comfy big black sweater from M&S. It hides my massive boobs perfectly.

'Not that,' she says, firmly. 'Ah. Here we are. Banana Republic. First stop. Now,' she says, pausing in front of the revolving doors, 'you have to agree to put yourself entirely in my hands.'

'Whatever you say, Trinny,' I tell her.

Janet tosses her sleek black hair. 'Those two? Amateurs,' she says, dismissively. 'Ready?'

I may have mentioned before that I absolutely hate shopping. And I hated it this time too. Struggling in and out of clothes, trying not to look at my naked thighs in the mirror, trying to tell myself it isn't all a colossal waste of time. Plus, it's embarrassing getting changed with Janet standing right in the cubicle watching me. I know she's a girl and she's my mate. But she's also an incredibly slim, beautiful girl with perfect skin, luminous eyes, and nothing jiggling anywhere.

I don't look in the mirrors. I don't really have a chance to. As soon as I pull anything on, Janet says, 'Yes, OK,' and snatches it from me, or, 'No. Gross!' and removes it from the cubicle. In fact I'm bending over and pulling on and struggling in and out of clothes so much it's quite a workout. I get dragged from store to store, getting more red-faced and tousle-haired each time, until I just can't take it any more.

'Yes, OK,' says Janet for the millionth time. 'We'll take those.'

I unzip yet another pair of flat-front camel-coloured trousers and hand them wearily to her. We're standing in the changing room at Zara, and it's already one fifteen and I'm sick of trying not to look in mirrors.

'Please,' I beg. 'Haven't we got enough now?'

Janet considers.

'Water,' I croak. 'I need water.'

'Well,' she concedes. 'The bags are a *bit* heavy, I suppose.'

Heavy? Arnold Schwarzenegger would have trouble hauling them around. Janet couldn't possibly carry them, so yours truly, being a strapping lass, has to strain and grunt her way into each successive shop like a pit pony hauling a cart full of coal.

'I'll just pay for these,' she says, holding her hand out for my Visa. 'Meet you out front. We'd better get a taxi,' she adds judiciously.

I flag one down. I can't really afford it, not any more. Who knows how anaemic my bank account is right now? But on the other hand I also can't fight my way into the tube. I wait for Janet and look at the huge bags groaning with clothes, laid out on the taxi floor. The clothes are all folded and look sort of boring and innocuous. And there are *tons* of them. The hopeful feeling I had this morning is starting to evaporate. I have no idea why I let myself be talked into this.

'There.' Janet climbs in and shuts the door.

'Where to, darlin'?' asks the driver, looking her over appreciatively. Despite the fact I've already told him where to. I sigh. How I wish, how I just wish I could look like Janet, if only for five minutes.

I glance down at my ring for reassurance but it doesn't help. Charles is offering me a life raft, but he doesn't fancy me, does he?

I know I'm being selfish and demanding, but I just want someone to *want* me. Not simply like me. How would that feel? I haven't a clue . . .

'What's the matter?' Janet asks, concerned. 'Why are you crying?'

'I'm not crying,' I say firmly, wiping away a tear. 'It's just dusty in here.'

'It's your job, isn't it?' she says.

Oh. Yeah. My job. Forgotten about that.

'Don't worry so much,' she says, seeing my expression. 'We'll go home and dump the clothes and then you'll see the results.'

'Can't we go for some lunch?' I plead. My stomach rumbles embarrassingly and I cough to cover it, but that never works, does it?

'I'll pop out for a sandwich,' says Janet firmly. 'No time for *lunch*.'

Twenty minutes later we're home. I'm sitting on my bed, gloomily, wrapped in my ratty towelling dressing gown. I've had a shower and washed my hair, and now Janet's blow-drying it. My new haul of clothes is laid out carefully around the room, and Janet has my make-up bag next to her.

'If you're going to cheer me up with a makeover why can't we use your cosmetics? You've got all the good stuff.'

'No. This isn't a makeover,' Janet says, authoritatively. 'This is about changing your life. We have to use what you've got so you can do the same.'

'Don't get me wrong, this is really sweet of you,' and I give her arm a little squeeze to show that it is, 'and it's nice to get your hair and make-up done and get some new frocks but it doesn't exactly change your life.'

'Look.' Janet puts down the hair-dryer. 'I know I'm not clever like you, Anna. Or ambitious,' she says. 'And I'm not focused like Lily is. That's why you two are so successful and I'm not.' Her voice sounds thick.

'Janet, you're very successful. You're a model!'

'Yeah, me and thousands of other girls,' Janet says. 'Anyway, this isn't about that.' She shakes her head as though to drive the thought away. 'It's about you and what I'm telling you is that I am an expert about some things. I'm an expert about clothes.'

'I know,' I say meekly. 'I didn't mean anything by it.'

'You meant it,' Janet says. 'But that's because you don't understand. I know how you feel about how you look, but it's all in your head. If you could just see that.'

'Please don't tell me beauty comes from the inside.' I sigh. 'You know, that with a bit of self-confidence I could be mistaken for Britney Spears.'

'I'm not saying that. You're never going to be petite or, you know, pretty. In a conventional way,' she adds hastily.

I sigh. 'You're not making me feel much better.'

'But you don't have to be pretty to be sexy, Anna. You've got a sort of fire about you. And you have great skin, and you're tall, and you have lovely eyes and hair.'

'And Gonzo's nose.'

'Your nose is *distinguished*.'

'I'm going to get it fixed.' I say fiercely. 'Soon as I've married. I know I said I wouldn't spend his money but I will for that.'

'Don't do that,' Janet says in horror. 'You'll take all the character out of your face.'

'I don't *want* character. I want to be *pretty*.'

'Why?' Janet asks. I look at her sharply but she doesn't seem to be joking. 'Why do you want that? Every other girl is like that. You look different.'

'In a bad way.'

'Only you think that.'

'I look like a bloke,' I say despairingly. 'All huge and tall and strong.'

Janet laughs. 'You look nothing like a bloke. You're more feminine than you think. Let me show you, OK?'

She continues to blow-dry and I just sit there. I'd love to go and get a cup of tea and put on Oprah and stop all this rubbish, but I know I can't. Janet's being so nice. I just have to take it.

'There you go,' she says.

I'm sitting in front of the mirror in Janet's bedroom. Janet has a kidney-shaped dressing table with a framed picture of J-Lo on it and a mercilessly lit mirror with lightbulbs all around it, like in theatre dressing rooms. I wince when I see my skin. It emphasizes every line, every large pore.

'Don't worry, everyone looks like that before foundation,' says Janet. 'OK. Look. Are you looking?'

'I'm looking, I'm looking.' She's laying the contents of my make-up bag out in front of me as though she's a surgeon and these are her implements.

'Sponges.'

'Check.'

'Foundation.'

'Check.'

'Stop saying check and pay attention!'

'Sorry,' I say meekly. 'I'm grateful, really I am.'

'Bronzing powder. Where's your blusher?'

'I use the bronzer.'

'That's good, but you'll need something for winter. A pink. Throw out these browns, they do nothing for your skin tone. Here.' She retrieves a small plastic square with a rose colour I'd never dare to try and gives it to me. 'You can keep this.'

'Maybelline? Don't you have all those fancy brands?'

'Maybelline are excellent,' Janet says. 'I read books. *Don't Go to the Cosmetics Counter Without Me*. You can get it in the States. It gives the chemical composition of the stuff. Mostly it's just packaging.'

'Oh,' I say. I had no idea she took it so seriously.

'So, blusher, mascara, lip gloss, eyeliner. Lose the black, too heavy for you. Go with a light brown. Picks out the blue in your eyes.'

'Wouldn't blue do that?'

'Blue? Hell, no,' Janet says, severely. 'Do you want to look like Banarama from the eighties? Well, *do* you?'

'No.'

'First we start with moisturizer. It should be non-comedogenic and contain a sunscreen.'

'You what?'

'Doesn't give you spots,' Janet says, dabbing some on. 'Oil of Olay is fine. OK, so wait until it's sunk in. Good.'

Good? My face is as shiny as Rudolph's nose.

'Next, concealer. This goes under foundation not over it. That gives a more even look. Here you go,' she says, dabbing my long-lasting Max Factor under my eyes and gently rubbing it in with a sponge. All of a sudden I do look a bit less tired. 'And then foundation, you sweep up and away, up and away, blend it in at the hairline.'

She gives a practical demonstration. I watch her in the mirror. OK, well, yes, I can see my face is smoothing out a bit. Fair enough.

'Next do your blusher.' She considers. 'There are a few techniques but you're a bit of a beginner, so tell you what, we'll just go with fail-safe. You dab some on the apples of the cheeks, here, and sweep a touch under the jawline – and brush lightly on the chin – and I'd go on top of the forehead for you. There,' she says, satisfied. 'You following, Anna?'

'Yes,' I say. 'I am, actually.'

It's really quite impressive. I turn my face from side to side. My cheekbones sort of stand out more, and my face looks slimmer. And I have this rosy glow, and all my eyeshadow bags have gone.

'Don't overdo it. Just a light hand for day. OK? Less is more. And the eyes,' she says, moving on. 'You've got great eyes, don't overdo them. Just this light brown shadow or a nude, one shade, just on the lid. Like this, super-simple. And you've quite dark lashes. I wouldn't ever wear mascara in the day. For evening just a coat on the upper lashes only, like this . . .' She swipes. 'Makes them look even bigger. See?'

'Yes. Gosh, you're really good at this,' I say.

'Liner – you don't need any if you're using mascara, it'd be too much. So that just leaves lips. Yours are very sexy,' she says. Is that actually a note of envy in her voice? 'Very full. For day, what you do is basically nothing. You dab on a tiny bit of Vaseline or a clear gloss at the most. OK?'

She demonstrates. Suddenly I actually notice my lips. They *are* quite big and soft. Who'd have thought it? With the clear gloss they look as though I've been licking them.

'For night you can use a nude colour, make sure it tends more to pinks than browns. And *absolutely no lip liner*. Nothing looks worse than a girl whose make-up's worn off and she's got a ring of liner on her lips and nothing else. That's total porn star,' Janet says.

'I'll remember.'

'Then perfume. You want a signature perfume. Mine was designed by J-Lo. Do you want some? It's the shiznit,' she promises.

'Um, I'm sure it is the shiznit,' I say, backing away. 'But I'm not a scent sort of girl.'

'Bullshit,' says Janet. She reaches behind me and before I can stop her, sprays me with something from a white flowery bottle.

'Get off – what is that?' I sniff. It doesn't smell that bad, actually. Sort of light and floral.

'Anais Anais,' says Janet.

'That stuff?' I protest. 'That's so uncool.'

'It's a gorgeous perfume. Totally floral. Very feminine. And that's what you are, Anna, you are very feminine. Don't be such a snob,' Janet says firmly. 'Do you think Charles will know what the smell is? No, he'll just think Anna smells like a lovely bunch of flowers.'

'OK,' I say humbly. I keep staring at myself in the mirror. Is that me? Truly? I look like a new person. And there isn't even any shading around my nose. I've got big eyes and rosy cheeks and glossy, sexy lips, and OK, I'm still not pretty, or

even normal, but I don't think I look ugly either. I look –
what's the word – striking.

Yeah. Striking.

I turn my head around.

'Not bad, huh?' demands Janet with a certain amount of
pride.

'It's really great,' I tell her. 'Thanks.'

'You haven't seen the clothes yet,' she says, excitedly.
'Come on!'

I get up and go with her, refusing to feel bad. I know I have
to look at my body now. No hiding from it. But at least there
won't be changing room lights overhead, and I know how to
do my make-up now. I can distract from my body with my
face! And Janet's being incredible, no way do I want to rain
on her parade.

'All right,' I say, forcing some enthusiasm. 'Let's go!'

Janet looks at me shrewdly. 'It's amazing,' she says.

'What?' I ask guiltily.

'I make you over twice, you look incredible both times,
and you *still* don't trust me. It's like you don't think Jay-Me
knows wazzup.'

'I know you know what's up.'

'Then what's that weepy look on your face for?'

'I'm just thinking you can't make a silk purse out of a sow's
ear.'

'Man,' Janet says. 'You really do have problems. You're not
a sow's ear. How can you walk around thinking of yourself
like this?'

'Because other people think of me like that,' I mutter.

'What other people?'

'Boys. Men. Everyone, really.'

'Oh, excuse me,' Janet says with heavy sarcasm, 'and you
the girl marrying one of the most eligible bachelors in
England?'

I want to burst right out and tell her. Charles doesn't fancy
me. Nobody fancies me. But I keep quiet. Somehow, I know
it wouldn't be fair to him to say anything. But I know what
he thinks, don't I? What he said about my diet and my

jogging. That there was no point, basically. I'm sure he'd think there was no point to this either. I know Charles. He's the kind of bloke who settles for his woman looking modest and presentable. I'm sure he pictures me in, ooh, I don't know, a tweed skirt and sensible brown brogues. Old lady clothes.

'I suppose,' I say. 'But love is blind, you know that.'

Janet purses her lips. 'Girlfriend, I think *you're* blind. OK. First thing, try on this,' she flings me one of the Zara flat-fronted trousers, 'and this.' A very ordinary-looking cream jumper from H&M.

Reluctantly I shrug off the dressing gown and step into the trousers. They fit very well, at least they're comfortable. But the sweater . . .

'I can't.' I look at it despairingly. 'Why did you pick this? It's all fitted and it's got a V neck.'

'Because it's fitted and it's got a V neck.'

'But that will show my boobs,' I point out.

'Exactly,' says Janet, triumphantly. 'Now put it on.'

I pull it over my head. I'm still turned away from the mirror, she isn't going to reveal the full horror to me yet.

'Now we accessorize,' says Janet. 'Put on this,' she chucks me the chunky leather wristwatch she bought me, 'and this,' the gorgeous Coach handbag she made me buy, and you have no idea how much it was, 'and slip your feet into these.'

I stare at them. 'What are they? I didn't buy those.'

'Shoes. Pied à terre. I bought them, they're a present.'

'But . . .' I struggle with myself. 'Janet, that's so lovely of you, and everything, but I can't possibly wear those. Do you still have the receipt? I'll exchange them for something really gorgeous, I promise.'

'You'll wear these,' Janet says.

'But those are *heels*,' I explain. 'Kitten heels.'

'Yeah. So?'

'Um, I'm five eleven,' I say. 'If you hadn't noticed.'

'So what? I'm five six.'

'Bit of a difference.'

'Anna,' Janet says patiently. 'You are not going to learn to love yourself if you keep trying to be Lily.'

I swallow hard.

'You're not going to shrink, you know. There's no point always wearing flats and slumping your shoulders like you do, trying to be smaller. Flats don't do anything for you. No, don't start crying,' Janet says. She rushes forward with a Kleenex. 'If you bloody well ruin that make-up job I'll kill you.'

'Sorry,' I sniffle.

'These are only an inch, maybe less, but what they do for your legs is . . . well.' She steps back and looks at me, satisfied. 'Turn round and look in the mirror. You'll see.'

I turn round. And look.

And blink.

Who is that? I mean, I know it's me. But it's not. It's somebody else. I don't have a tall, confident pose, and shapely hips, and attractive arms and delicate wrists – the watch makes my useful, thick wrists look small in comparison. And the low-slung pants fit me perfectly, and they shave off inches everywhere. And the heels . . . the heels are sort of forcing me to arch my toes, throw out my bottom and stand up straighter, which makes me look about eight pounds lighter. The top is very fitted. It shows off my cleavage, but it's thick cotton, and it doesn't look slutty, just . . . womanly. OK, not delicate, but womanly. And polished.

I gasp in delight.

'I can't believe it. I look great.'

'I know you can't believe it,' Janet says. 'That was always your problem, Anna.'

I examine myself again. It's true, it's absolutely true what she told me. I still look like me, but instead of hefty, I'm seeing myself as strong; instead of gigantic, I'm seeing myself as tall; instead of thick-set, I'm seeing myself as curvy. I *do* have a waist. The flat front of the trousers is hiding my tummy, you don't even notice it. And my boobs look impressively large instead of slutty or like Daisy the cow. They fit the rest of my body. The whole thing just . . . *fits.*

'Don't just stand there preening,' says Janet triumphantly. 'Take that lot off. We've more to get to. Lots more.'

It's incredible. Almost everything she's picked, I love. The boring-looking clothes come alive when they're on my body. The neutral colours are soft and chic and they all blend with each other; it's an autumnal fantasy of black, cream, red, and beige. Janet has chosen piping and details, but no prints, lots of block colours.

'You're not ready for patterns,' she says. 'I want to keep it simple.'

'Whatever you say,' I say eagerly.

'Everything mixes. Try that burgundy leather skirt with the knee-high boots. And the fishnets.'

'I can't wear . . . OK. I'll try them.'

'And the black sweater, the cashmere. The white cotton will work too. Or that white V-neck tee with the brown bomber jacket.'

It's actually fun. It's so much fun. I thought she'd lost it when she showed me the fishnets, but the amazing thing is, when there are only a couple of inches visible, between the top of the calf and the end of the skirt, they don't look tarty, they look pulled-together, a bit Manhattan, even. They show off my legs, which aren't bad, in a sexy way. Wow. I look so film-business.

I could ask for an agent anywhere dressed like this. I could walk right into William Morris and . . .

OK, let's not get carried away. But I'm amazed at how it makes me feel. The clothes aren't light and flippy and sprayed-on like the things Lily wears, but they still look feminine. I feel an intense surge of confidence. Is it possible?

'I told you,' Janet says, as if she can read my mind. 'I told you. This can change your life. I know you think it's only clothes, and it's really superficial, but when you look the best you can, it helps. It just helps.'

'Thanks, Janet,' I say, giving her a hug. 'Thanks so much. You don't know what this means to me.'

'You're welcome,' says Janet, tossing her gorgeous hair.

'You get . . . um . . . mad props,' I say, and she laughs.

I manage to kill a whole afternoon playing with my clothes. Janet has to go out to visit her agent, and I just stay in the flat, peeling off one outfit and putting on another. Mixing them up. Trying on all my new, cheap accessories. Everything looks great.

I know I'm still not beautiful, but I actually like how I look. Maybe it's not all nonsense, maybe there is something to that idea about character and so forth. Anyway, I feel a lot more cheerful.

I put my last (white, silky) shirt back on the rail in my tiny bedroom (no room for a wardrobe) and pick up my address book. I ring a couple of numbers, production houses I've heard of, bigger places that might have something. Once again I check out my reflection in the mirror, but nothing's changed. I still look pretty good. Good enough to walk into any of those places, I tell myself.

I walk determinedly over to my desk, boot up my laptop, and start writing. The script pours out of me. I feel so confident, now, like nothing can stop me. I write and write until my fingers get cramp, and finally, that's it, the end of the third act. I'm done.

Screw Kitty and Eli. And screw Mark bloody Swan, too, I tell myself mutinously. I don't need any of them. I'm a screenwriter now. I can do this!

I pick up the phone. I only hesitate for a moment, then I dial the first number.

The phone buzzes just as I'm on my way out of the door.

'Brown,' I say snappily. Wow, this new me is something else! It's incredible what a little change in how you look can do for you.

'Anna? Is that you?'

'Hi, Charles. Darling,' I add.

'Darling.' I can hear the pleasure in his voice. 'Are you quite sure you don't have any extra guests?'

'For what?'

'The engagement party,' he says patiently. 'It's tonight, remember. At Vanna's.'

'Oh. Oh, yes.' I'd forgotten all about it. 'I know,' I lie. 'No, just Janet and Lily.'

'You never gave me the address of your friend Mr Swan.'

I feel perverse relief. 'Oh, no need to ask him,' I say. 'As we're not working together any more.'

'But I found it out,' Charles continues. 'And I asked him.'

I'm silent.

'Does it matter?'

'No. That's OK,' I say. 'I'm sure he doesn't want to come anyway.'

'Oh, he does, don't worry about that. He accepted right away.'

Did he? 'That's great,' I say. 'I can't wait to introduce you. What time is it?'

'Seven for seven thirty, but we should be there early. Say I pick you up at six thirty?'

'Sounds good.' I've got that green dress I bought with Janet first time around, the one I didn't wear to the ball. Well, who cares what Mark Swan does, I think. I've got my green dress. I'm going to look *fantastic*.

I look at my watch. My first interview's in twenty minutes and I've got another one half an hour later. And then one more tomorrow morning. It was amazing how easy it was to get people to see me. I just called a couple of the big agencies, and mentioned my script, and how I've been working with Mark Swan, and they all asked me to come in!

This is easy. What was I concerned about? I'll have representation by tomorrow, and then my agents can start selling the script to Hollywood. I wonder how much I can get for it?

'Gotta go. Don't want to be late. I'll see you tonight, sweetie,' I tell him.

I will not think about Mark Swan. I will not think about Mark Swan. I repeat this in my head like a mantra as I turn into Soho Square. Of course, this method repeats the name Mark Swan, so it doesn't work all that well. I try just clearing my mind. That doesn't work either.

He thinks I'm not good enough. He won't back up my

script. All that encouraging my hopes and dreams, and now he's pulling away . . .

And he didn't seem all that upset about me getting sacked. In fact, he didn't seem even slightly upset.

It hurts. It does. More than it should. But it's a blessing in disguise, I think to myself. I was, just possibly – so ridiculous – just starting to develop the *tiniest* crush on him. And I'm going to be married. So it's a good job that I've been sacked and he's not helping me any more, I tell myself. After tonight I won't have to see him again, and I'll be safe enough tonight, won't I? I've got my own fiancé.

OK. The Gryphon Agency, Inc. Here we are. I push through the revolving doors, a new spring in my step, mentally rehearsing the pitch for my script. Gryphon has a very cool lobby, or cruel lobby, depending on how you look at it; the walls are thick strips of chocolate brown leather upholstery, punctuated with an equally thick strip of mirror. Hundreds of Anna Browns gaze right back at me but, thankfully, they still look quite cool. I've paired the leather kitten-heel boots with some flat-front, low-rise black trousers and a slightly off-the-shoulder burgundy top, added the wristwatch and some dangly earrings and a chunky garnet ring I stole from Janet's jewellery box – well, she's done so much for me already she's not going to mind my nicking the finishing touch for an hour. I wear this on my right hand, to go with the watch, and not to clash with my engagement ring. My bag is that new black Coach one, slung casually over my shoulder as though I'm used to looking stylish.

The receptionist is a black girl in a black dress, sitting behind a kidney-shaped desk made entirely of smoky glass. She's gorgeous, but this time I'm not intimidated.

'Hi.' I smile at her confidently. 'Anna Brown, here for an interview with Paul Fallon?'

She glances at her sheet. 'Oh yeah, he's expecting you. You can go right in,' she says.

Wow. This is brilliant. I thought for sure she'd tell me to have a seat and I'd be stuck staring at my fingernails for half an hour. Agent interviews make you wait for ages, normally.

They're the only occasion your time is valued less than it is at your doctor's.

'Thanks,' I say, striding into the office.

Where's Fallon at? He must have heard about me, I think. Heard on the grapevine how good I was on story during the *Mother of the Bride* meetings. I bet he can't *wait* to read my script.

'Paul about?' I ask one of the black-clad, John Lennon lookalikes who seem to be everywhere. One of them sullenly jerks his thumb towards a corner office. I head over there, knock on the door.

'Come,' he says.

I open the door, step inside. Everything in here is very futuristic, lots of chrome and glass and more mirrors. Fallon has posters for *Bladerunner* and *Minority Report* on his walls.

'Anna Brown,' he says warmly, getting up out from behind his desk to greet me. He's got that Manhattan beat-poet thing down pat: black turtleneck sweater, wire-rimmed glasses, expensive yet ill-fitting black slacks. Although not too many beat poets could afford that gold Rolex. 'So good to finally meet you,' he says.

'And you,' I say. 'I'm a big fan,' I lie. We smile Cheshire cat smiles at each other and he gestures me to take a seat. Man! This is so easy. I'll be represented here by the end of the day.

'So, I've heard all about your great work on *Mother of the Bride*,' he says pleasantly.

'You have?' I beam.

'Sure. London watering holes are buzzing,' he says.

'Well,' I say confidently, exhaling and stretching out my legs. 'It was a great script. I knew as soon as I found it that I could—'

'How did you attach Mark Swan?' he asks, cutting me off and looking at me intently.

'I found him and asked him to read the script.'

'Just like that?' he asks, sounding disappointed. 'Didn't you have some kind of special connection with him?'

'No, quite the opposite, actually. I bumped into him in a newsagents and I didn't even recognize him at first. It was a

really funny story, because I'd originally gone up to his set
to—'

'So you didn't know him from before? Film school?'

'I didn't go to film school.'

'And you weren't friends?'

'No,' I say. 'But I did get him to read the script. I think I
found a good story.' I smile confidently at him and pull my
script out of my bag. 'In fact, Mr Fallon, that's what got me
interested in writing my own scripts, because I was good at
story. Mark pointed that out, and what I've written is a
comedy. It's the story of—'

'Anna, let me put my cards on the table,' says Fallon briskly.
'I'm sure you're a great writer, but scripts aren't exactly in
short supply, and our slate of writers is full. Unless you can
bring something spectacular to the table?' He looks at me
expectantly. What does he want me to do, a backflip onto my
hands, burp the alphabet?

'But I've written a really funny comedy. Isn't that
spectacular?'

He chuckles. 'That's a bit naive, honey, don't you think?
You use what you've got in this business.'

I blink. Are the new clothes so good that he's proposition-
ing me? What is this, finally the casting couch?

'Mark Swan,' he explains impatiently. 'You have to use
your relationship with him. We hear that you're friends with
him.'

'You hear from where?'

'You've been seen around,' he says. 'Now, obviously, we
here at Gryphon *love* Mark and we *adore* his vision. Do you
have influence with him? Can you attach him to this script?'
He jabs a manicured finger at my neatly bound script but
makes no move to take it.

I sigh. 'No, Mark does his own thing.'

'But if you *did* have influence with him, you could use it?'
Fallon persists. 'We could get you some kind of a deal with
Mark attached.'

'But the script's good enough by itself!' I say mulishly. 'You
should at least try it out.'

'With Mark Swan, sure. Without him . . .' Fallon spreads his hands. 'It's been nice to meet you.'

I stand up and walk out. I don't know how I have the guts to do it, really. The old Anna would have cried, would have begged and pleaded. And yes, I feel upset, but I'll be damned if I'm going to let this bastard see that.

I get outside, half choking on fury and hurt. Oh well, I think. My cheeks are burning, but I'm not going to let it affect me, I'm going to shrug it off. I've another interview coming up, after all. They weren't the only people to say they wanted to see me.

A sudden, nasty thought hits me. I walk into the little park in Soho Square, perch myself on the edge of one of the benches, avoiding the sleeping student sunning her perfectly flat abdomen, and pull out my mobile. The other agency's name is Westin, their offices are over in West Kensington.

'Hi there. Can I speak to . . .' What was his name? 'Richard Hatherley, please. This is Anna Brown, his four p.m. Yeah, thanks.'

Hatherley comes on fast. 'Anna,' he says. 'Not calling to cancel on me, I hope?'

'Oh, I don't think so,' I tell him. 'I just want to be sure we're on the same page. You are interviewing me about my script, aren't you?'

'Absolutely,' he says, enthusiastically. 'That's exactly what we're interested in, Anna. Can't wait to read it.'

I exhale, letting all the air out of me.

'Phew,' I exclaim. 'For a moment I thought you might only be interested because you thought I had some influence with Mark Swan.'

There's a long pause.

'You mean you *don't* have any influence with Mark Swan?'

My heart sinks. 'No, none. I've had a falling out with him, actually.'

'A falling out? How do you "fall out" with an A-list director?' Hatherley demands. 'He's the only reason we wanted to speak to you. How do you "fall out" with someone

298

like that?' The jovial, fatherly tone has disappeared, and now he's snapping at me. It reminds me vastly of Kitty.

'Very simple,' I say, seeing red. 'You tell him to fuck off. Like this. FUCK OFF!' I yell, then press the red button. It's not the same as slamming the receiver down, though, is it?

OK. I think it's safe to say they won't be reading my script.

The hum of conversation in the park dips for a few seconds as everybody stares at me. I lift my head, cheeks flaming, and stuff my cell phone back in my bag. Then I haul myself to my feet and walk off down Greek Street. Trying to avoid anything cool. It doesn't work. I walk past the entrances to production companies, hip bistros, private clubs, record companies, the London outposts of the major studios. In short, everything I've ever wanted in my career. And now, after the only bit of success I've ever had, what have I really got out of it?

Nothing.

When I was in college and full of dreams, do you know what I wanted? To be a millionaire by the time I was thirty. Vice-president of a Hollywood studio.

But instead, the reality of it is that I'm thirty-two and an unemployed script reader.

My script feels thin and pathetic inside my handbag. I'd pinned all my hopes on it. But Mark won't help me submit it, and agents aren't interested unless he's attached.

I thought they wanted to see my work. But they only wanted to get to him.

Despite my new, cool clothes, my eyes fill with tears, so I fish out a bit of loo roll stuffed into my chic new handbag and dab them away. I can't afford to cry, not now. For one thing, it'll ruin my fabulous make-up job. For another, I have to get myself over to Vanna's for the engagement party. I can't let Charles see me all upset at that.

And for a third, Mark Swan will be there.

I shudder. No way am I going to let Swan feel any pity for me. If I want to cry, I can do it tomorrow.

12

Vanna's place is already packed by the time I get there. I was hoping to have a quiet chat with her, maybe even a glass of wine, before Charles arrived, but I guess there's no chance of that. Cars are stacked three deep in her glorious gravel driveway, there are blokes outside rigging outdoor fairy lights into the trees and bushes, there are caterers coming and going with crates of champagne and trays of delicious-looking food.

I thread my way into the hallway past them.

'Vanna? Vanna?'

Rupert comes lumbering up to me, followed by Winston. I reach down with a pre-emptive ear scratch before he can ruin my make-up again.

'Upstairs, slapping on the old warpaint,' he says. 'Who are you? We haven't been introduced.'

'Rupert.' I wave frantically at him. 'It's me, Anna.'

'Anna. Anna?' he says, astonished. He comes closer, peers at me. 'Anna! By God, it is, too. I didn't recognize you.'

'Oh,' I say, feeling pleased.

'You look *terrific*,' he says, sounding amazed. 'I say, did you have a nose job?'

I grit my teeth. 'No, Rupert, I did not have a nose job.'

'Because it looks a bit smaller,' he says charmingly. 'Although now I get up close I see that's just an optical illusion. And you look thinner! You look normal now,' he tells me, graciously.

'*Thank* you, Rupert.' I sigh. The sarcasm is lost on him.

'Let's have a look at you,' he says, turning me round.

'Good Lord! That's a very flattering outfit. Course, I always prefer ladies in dresses. You should wear dresses more often, Anna!'

'I'll remember that.'

'That way, people won't assume you're a lesbian,' he says.

'Top tip, Rupert,' I say. 'Thanks.' I look round desperately for Charles, but he hasn't got here yet. A mature person would handle this whole situation with zen-like calm found deep from within. On the other hand, I'm going to handle it with alcohol. 'Pass the champagne, could you?'

I'm sipping away as Vanna descends the stairs and appears. She stops dead in the entranceway to the kitchen, blinks.

'Anna? Is that you?'

'Hi, hon,' I say, going over to give her a hug, but she holds me back so she can take me in.

'I don't believe it,' she says. 'I don't believe it.'

I don't know whether I should be flattered or annoyed by all this incredulity.

'I didn't look *that* bad before,' I mutter. 'Did I?'

'No, it's not that,' she lies blatantly. 'But now you look *delicious*, darling!'

'Let's not go overboard,' Rupert guffaws. Vanna shoots him a look before which a Cossack would quail and he beats a hasty retreat to the morning room.

'He's an oaf,' she says with unusual candour, 'but you do look wonderful, Anna. Have you had some . . .' Her voice trails off. I lift an eyebrow.

'*Work* done?' she hisses.

'Nope.'

'No Botox? Collagen lip injections?'

I shake my head.

'Not even a little contouring on your . . . Oh. No. Really, well, that's fabulous. So this is all you,' Vanna says, warmly. 'And now we can see what you should look like. You should have done this years ago. Did you use a personal shopper?'

'Sort of.' I explain to her about Janet.

'The model I met? The brunette or the blonde?'

'The brunette.'

'You know what I think of models, darling.'

'They are my friends, please be nice to them.'

'I'm always *nice* to people, even people I can't stand,' Vanna says. 'Hostess with the mostest, you know. But I do think I'll have to re-assess with your friend Janet. She's done a simply marvellous job.'

'She has.' I can't deny it. Under all the sick disappointment of my job interviews, and even though I'm dreading seeing Swan, I also want to know how he'll react when he sees me. He's never seen me in anything better than baggy jeans and shapeless T-shirts. Hey, at least I'll be looking good when I have to face him. I wonder if he'll say anything, or will he—

'Charles! Angel,' Vanna says. I go bright red and spin round to see Charles standing there, looking as pleased as Punch, clutching two huge bouquets of roses. 'Darling, your fiancé's here.'

'Yes,' I say, guiltily. 'Hi, Charles.'

'Ladies,' says Charles. He's wearing his stack heels and a dark suit, goatee beard neatly trimmed. 'These are for you, Vanna, to thank you for throwing the party. 'Fraid your house will be total bedlam.' He hands her a vast bouquet of yellow and pink roses and twiggy branches that's almost as tall as he is. 'And Anna, my dear, these are for you.' He passes me an equally ginormous bouquet of crimson-red roses. 'To thank you again for agreeing to become my bride,' he says.

'I didn't bring you anything,' I say, as Vanna beckons her housekeeper over to whisk away the flowers and stick them in water. 'I'm sorry . . . sweetheart.'

'You're giving me the ultimate gift,' Charles says, bowing stiffly. 'Your hand in marriage.'

I hug him and reach for my champagne glass again. The caterers have put on their uniforms, the first guests are drifting in through the door, and the party's about to start.

'Let's go and stand next to Vanna and welcome everybody in together,' I suggest. Charles smiles up at me gratefully. I thread my arm ostentatiously through his. Dear Charles, he's so sweet. I am absolutely not going to be ashamed to be seen at his side.

'Do you like my new dress?' I ask him. 'My make-up? I had it all done specially for tonight.'

'Oh, yes,' he says, giving me a cursory glance up and down. 'You look very nice, darling.'

I tell myself it doesn't matter. Men don't have an eye for these things, everyone knows that.

Mark Swan doesn't show up.

It's a fabulous party. One of Vanna's best. Everybody seems to be having an incredible time. There are three different kinds of champagne, including pink (I love pink champagne, even though it's probably very common of me), there are little designery snacks – chips served in miniature newspaper cones, dim sum, tiny tartlets of various styles, beef carpaccio speared through with a toothpick – that must be the upper-class version of sausage on a stick – all sorts of stuff. There's a string quartet out in the garden, and all the trees are strung with fairy lights, and candles in glass jars are strategically dotted around the lawn. Vanna's catered a huge buffet with every kind of delicious food imaginable. I don't seem to have much appetite, though. I'm picking away at smoked chicken, sitting at the top table with Charles and Vanna and Rupert, trying to remember to laugh at all his jokes, and generally watching everybody else have a fabulous time.

Is it just me, or is it really dull showing everybody your engagement ring every five seconds? It's like going to university and being asked over and over again what A levels you did. I'm so sick of extending my hand and simpering. I know, I know, I must sound like a total grouch. And I *want* to be better at this. Happier. I don't know what's wrong with me.

I do enjoy Lily and Janet, though. They've turned up with Ed and Henry. Janet's all over Ed while Lily is trying to do her princess act on Henry. And Ed, who's rather scruffily dressed, is very solicitous, constantly going back and forth fetching things – glasses of champagne, napkins, slices of watermelon. Mostly for Lily, as Henry doesn't seem to be asking how high when she says jump. Janet's beaming soppily at Ed, and Lily keeps tossing her blond hair and pretending to be all stiff and cold, but she actually looks quite happy. Happier, now I come to think of it, than I've seen her for ages, maybe since I've known her.

'Darling.' I turn to Charles. 'Would you excuse me just a moment? I'd like to go and have a chat with Lily and Janet.'

'Of course, poppet,' he says, turning back to Rupert and his boring anecdote. I slip away.

'Hi, guys,' I say.

'Anna, congrats,' says Henry, shaking my hand warmly.

'Charles is a lucky man,' Ed says. 'Let's have a look then. Ahm, yes,' he says, examining the ring. 'Pretty.'

'I wouldn't need something that big,' Janet says eagerly. I wince for her, but Ed doesn't seem to have noticed.

'Oh yes, I prefer smaller stones, myself,' he says.

'I bet you do,' mutters Lily with a warning look at Janet, but Janet's oblivious.

'But you know Charles,' Henry says cheerfully. 'He never spends one pound where ten will do, does he?'

Lily sighs, then hastily covers it with a delicate little cough.

'Tell you what, Dawson, let's go and talk to him, leave the girls by themselves for a few minutes,' says Henry to Ed.

'Righto,' says Ed. 'Be back in a sec.'

'Don't be too long,' purrs Lily.

Ed looks at her, blushes. 'I won't,' he says.

'Oh, bloody hell, Anna,' Lily exclaims as soon as they're out of earshot. 'This party is ridiculous. Just ridiculous! How much did he spend?'

'I doubt Vanna let him pay, but I'm sure he tried.'

'He's lovely,' says Janet, loyally. 'And you look awesome,' she says with triumph. 'Anna B in the hizzouse! Lettin' 'em all know wazzup!' She does a hip-hop hand motion.

'Oh, cut that out, Janet,' says Lily.

'Don't be a hater,' Janet says, hurt.

'I do look great, thanks to you.'

'He was knocked off his feet, wasn't he? Blown away?'

'Um, absolutely.'

'I knew it,' Janet says in triumph.

'But never mind about me,' I say, wanting to do anything to change the subject. 'What about you girls? You look like you're in love.'

Janet sighs blissfully. 'I totally am.'

'And the whole world knows it,' Lily tells her, shortly. 'You're not going to get whatever costume jewellery Ed can afford acting like that.'

'Am I being too keen?' Janet asks humbly.

'I've seen limpets less clingy than you,' Lily says witheringly. 'But it doesn't matter if you ruin your chances with Ed, because you really can't go out with him. He's too poor.'

'No he's not,' say Janet and I together. And Janet's bristling, mild-mannered Janet looks as if she might turn into Bruce Banner and rip Lily's head off.

'Tell me about Henry,' I say, to ease the tension. Lily shrugs.

'He's all right,' she says.

'You look like you're quite keen on him.'

'I'm not keen on men, they're keen on me.'

'But you have to admit, he's not bad looking.'

Lily relaxes slightly, sighs. 'He is gorgeous,' she admits.

'Nice to talk to?'

'I like him,' she concedes. 'Even though he's very arrogant. He wouldn't fetch my drink,' she pouts.

'That's because you changed your mind three times. He got you whisky, you said you wanted vodka. He got you vodka, you said you'd decided on cognac.'

'A woman has a right to change her mind,' Lily insists, tossing her hair. 'Anyway, Ed got it for me.'

'He's very chivalrous,' Janet says adoringly. 'Couldn't move fast enough, could he?'

'It's such a fantastic party,' Lily congratulates me again. 'You're sooo lucky,' she says.

'Yeah, I know,' I say. 'I'm really lucky.' I look at Janet staring adoringly after Ed and suddenly feel a painful clench around my heart. But I know I have to shake this off. Charles is a good man who wants to take care of me for the rest of my life, and if I hold out for some stupid, romantic notion of true love . . .

It's just that I like myself in my new clothes. I think I look pretty damn good. And I wish my man felt the same.

Charles gives me a little wave and I wave back.

'True love,' Lily says. She nudges me. 'I did tell you, Anna, it's as easy to love a rich man as a poor one. Glad you made the right decision!'

'I keep telling you, it's not like that.'

'Oh, don't worry,' Lily says infuriatingly. 'I'm not going to spill the beans.'

And then I see him. Standing in the kitchen doorway, looming over the garden. Scanning it, a large present wrapped up in his hand. Looking for someone. Looking for me.

'I think I'll just go and circulate,' I tell them. 'Your boys are heading back this way.'

Distracted by Ed and Henry, the girls don't even give a backwards glance as I walk away from them towards the kitchen, my kitten heels crunching on the gravel.

He came. Three hours late, but he came.

Swan is standing there, hovering, looking, I'm glad to see, extremely uncomfortable. I square my shoulders, shake out my hair. The green dress Janet picked is ruched at the bodice, it looks rather Georgian, empire-wasted, draping skirt, and my make-up is the same as it was this afternoon; I just freshened it. And I'm wearing Anais Anais, and I have a sweet little evening bag in white clutch silk . . .

But why would he notice? Charles didn't.

I take a deep breath. I'm only doing this to take his gift and be polite, I tell myself. And after tonight I probably won't see the guy again.

'Hello, Mark,' I say, in what I hope is a suitably patronizing tone. 'How good of you to come,' I add coldly.

'Anna,' he says. He stands there clutching his gift and looks me up and down. 'Anna,' he repeats.

'That's my name, don't wear it out,' I joke.

'You look stunning,' he says.

I melt just slightly but try to freeze myself back over. Swan looks pretty stunning himself. In honour of the occasion, he has decided to wear a suit. Totally out of character, but he

307

looks good in it. It's dark, charcoal grey, with a pale blue shirt. It fits him perfectly; with his body, it has to be bespoke.

'Thanks,' I mutter. I catch sight of our reflection in Vanna's glass door. Tall though I am, he dwarfs me.

'This is for you,' he says, awkwardly, shoving the gift at me. It's quite solid.

'Thank you, you needn't have bothered,' I say neutrally. *You certainly didn't bother to help submit my script* hangs in the air.

'I can see you're consumed with curiosity as to what it is,' Swan says drily, 'so let me put you out of your misery. It's a Sony Vaio.'

'The souped-up laptop?'

He nods. 'You can load the *Final Draft* I gave you right onto it and start rewriting your movie.'

'How do you know it needs rewriting?'

'How do you know it doesn't?' he asks, patiently.

I pull myself together.

'Very kind of you,' I say dully. 'You shouldn't have. It's far too generous a gift.'

Swan shrugs. 'I've got a lot of money. As, I suppose, do you.'

'I don't have anything,' I tell him. 'I just got fired. Remember?'

Swan says nothing but makes a gesture with his head at the party raging expensively all around us.

'This is my friend's house,' I tell him. 'And yes, Charles is rich, but so what, that's not my money.'

He grins. 'Wives all say that, until the divorce courts.'

Oh! He's so arrogant, standing there, looking down at me. The awkwardness has gone and he's back to his old self-confident self. I think of the interview today. 'If you can bring us Mark Swan,' and the phone call, same thing, and Frank Giallo's sly look at him, 'Are you attached?'

It's Swan, Swan, Swan. It's all about him, isn't it?

'Speaking from experience?' I snap, seeing red. 'Bitter because you have to pay her alimony?'

'Not at all,' he says, and his tone's pleasant enough, but a shutter comes down over his eyes. 'I was glad to give Maryann what she asked for. In fact I gave her more.'

'Well, aren't you Mr Perfect,' I retort. 'I expect meanness to ex-wives makes for bad PR. I know you probably find it a wild and crazy idea, but some of us want to make our own money.'

'Sure,' he says, aggravatingly patronizing.

I thrust the gift box back at him. 'Maybe you should take this back.'

'Why? Are you still sulking because I wouldn't help you submit your script?'

'I'll never be able to get an agent now,' I cry. 'People always want you. They always want to get to you!' I'm glaring at him. 'I'm worse off than I was before, I've no job, and now I'll never get one. Nor an agent. I can't make it on my own, I'll always be seen just as a short cut to *you*. And you won't even *help* me.'

Swan looks at me. 'Anna,' he says. 'You need to think about this. You're just being pathetic.'

'Pathetic?' I demand furiously. How dare he?

Something snaps inside me. Mark Swan, standing there, looking so gorgeous, so disdainful. I step forward, hand raised to slap him round the face, but his own hand darts out, lightning quick, and catches mine. My wrist seems tiny in his hand; it's stopped as dead as though I'd tried to smash it into a force field.

'You shouldn't resort to physical violence,' he says softly, and he's looking at me like . . . like . . .

We're close, his chest is inches from mine, his head, his weatherbeaten, craggy face, and that sexy, kind, laughing mouth . . .

I feel my own lips part, involuntarily. My heart is racing. His eyes are flickering down over my face . . .

'Don't look at me like that,' I manage.

'But I want to,' he says, still softly.

My heart flips over. Thudding. I sway on my feet a little, I feel almost faint. He lets me go and I take a step back. It can't mean what I think it does. Can it?

'Darling.'

I give a little shriek and spin round. 'Charles! You scared me.'

'Sorry, poppet,' he says. 'I missed you. She has to mingle, I suppose,' he says jovially to Swan. 'Can't keep her all to myself.' He looks at me expectantly.

'Charles, this is Mark Swan,' I mutter. 'He was my former . . . colleague. Mark, this is Charles Dawson, my fiancé.'

'Congratulations,' says Swan neutrally to Charles. I see him stand there, drinking Charles in. And Charles is looking up at him, rather hostile.

'Thank you,' he says. 'You must be very upset that Anna's been fired, having had the good fortune to work with her.'

Dear Charles. Trying to protect me. He slips a slender arm round my waist possessively.

'I think Anna can do bigger and better things,' Swan says, politely. He pauses, then adds, 'And how long have you known Anna?'

Charles smiles. 'Whirlwind romance, wasn't it, darling?'

'Yes,' I agree miserably.

'A couple of months,' he says.

'That's very nice, to have passion for someone like that,' says Swan, and looks me right in the eye. I feel dizzy. I swallow.

Don't be so stupid, says the voice in my head. *This isn't Brian, this is Mark Swan. He could have anybody he wants.*

'Indeed,' Charles agrees, smiling trustingly at me. 'Anna's the one for me, aren't you, darling?'

Both men look at me. 'Oh, yes,' I agree. 'I definitely am.'

There's a pause.

'And you are for me, too,' I add hastily. Charles smiles in relief.

'It was very good to meet you, Charles,' says Swan. 'You're a lucky man. I have to go. Goodbye, Anna.'

He puts his gift box down on the kitchen table, turns round, and walks out, without looking back at me.

Charles pulls me closer. 'The party's getting late,' he whispers. 'Want to go back to my place?'

So not.

'Sure,' I say, turning to him and smiling weakly. 'No problem.'

Afterwards, sitting up in bed, Charles has a bony arm round my shoulder.

'Thought it went brilliantly, didn't you?' he says. 'Everybody was there. And all so pleased for us, darling. Vanna was thrilled to bits, and Rupert said ... Well ...' He colours a little. 'Anyway, Rupert thought it was a great idea, says we're well suited. And your girlfriends getting on so well with Ed and Henry. I was the cause of that,' he says, rather proudly. 'They met at my ball! Charles Dawson, the lurve doctor,' he says, in a fake American accent.

I force a smile. 'It's all very nice.'

'We must get on with planning the wedding.'

'I know, but I told you, I'm just a bit busy trying to get a new job, and—'

'Come down tomorrow,' he says, turning to me and pleading. 'Just one morning won't hurt. I want to show you my ideas. If you sign off on them, I'll do all the planning, just like I promised. Vanna can help me.'

'I don't know.'

'I've been working on some things to show you,' he says hopefully. 'Putting some scrap books together and stuff, got some sample menus in, that sort of thing.'

I can't say no to those puppy-dog eyes.

'All right,' I say. I try to force some enthusiasm. 'Sounds like fun.'

He's as good as his word. We wake up at seven to Charles's old-fashioned alarm clock with a little buzzing metal bell on top, go into the kitchen for a quick breakfast – toast and Marmite, Charles has marmalade, with coffee, delicious actually, he has all the modern gadgets – then he takes a shower, drives me home.

'We're not going to be long, are we?' he asks, nervously checking his watch, as I let myself into my flat. 'I want to get on the M1 before all the traffic starts.'

311

'Quick as I can, I promise,' I say. 'I just need a quick shower and a change.'

Lily's sitting on the sofa, curled up like a particularly lissom and blond cat. She stirs, stretching her body beautifully as Charles comes in.

'Hello, you two,' she purrs. 'Back from your lovers' tryst? Anna, I'm jealous.' She smiles.

I know she is.

'Come and sit next to me, Charles,' she suggests, patting the couch beside her. 'Anna won't be a mo'.'

Charles looks hunted, but I nod at him.

'I'm jumping in the shower,' I say, but Lily is oblivious, she's bending all her charm on my fiancé.

'So, tell me about Henry,' she suggests. 'His family's very rich, aren't they? The Marshes?'

'Oh, ya,' says Charles, and I watch Lily relax and beam with approval. 'Harriet and Fred Marsh are absolutely rolling in it, but you needn't worry, it hasn't spoiled Henry a bit.'

'Oh no,' breathes Lily, as I grab my towel and clothes. 'It doesn't worry me. I don't object . . .'

It ought to annoy me, as I step into the shower and quickly baste myself, soaping all over just to feel clean, to shake the nasty feeling of last night away. But somehow it doesn't. All I do is sigh and think, poor Lily.

But with Charles hovering, there's no time to pore over Lily and her many mental problems. I towel off, duck into my closet-cum-bedroom, and choose another Janet-approved outfit. Red A-line skirt, cream short-sleeved, scoop-neck cotton jumper with red trim, camel-colour slide shoes, camel bag. Yep; still looks fabulous. A quick blast of the make-up and I'm done. I check myself out. Apart from my wet hair, I look wonderful.

'Tah-dah,' I say, emerging from my bedroom and doing a twirl for Charles, who's just sitting there gazing at Lily.

'Oh. Right,' he says, wrenching his gaze away. 'You're ready,' he notes, his eyes barely glancing at my outfit. 'We'll be off then.'

* * *

I'm bloody glad when we finally pull in to Chester House's long, bumpy drive. For one thing I need the loo and I need to stretch my legs. For another, Charles has spent almost the entire journey alternating between in-depth postmortems of the engagement party and in-depth predictions for the wedding. Guest lists, seating arrangements, different types of music . . . I do my part, though. I'm hiding behind a huge pair of Ray-Bans so he can't see the total lack of interest in my eyes, and I sort of nod and go 'Mmm' and inject just enough questions to show that I'm listening. For the most part, I just watch the glorious English countryside slip by. And concentrate on not thinking about Mark Swan.

I'll admit it. It's something of a losing battle.

I keep replaying that moment in my head. His hand, catching my wrist. The firm set of his jaw, his sexy mouth, the way he looked at me, so intensely. The hostility to Charles.

But no. No. I mean, he already had every opportunity to say something to me . . .

And anyway, why would he? I mean really?

Get over yourself, Anna, I tell myself firmly as Charles parks the car. It's not going to happen.

'Here we are,' Charles says.

I look up at the massive, glorious, old warm stones of Chester House. The kind of dream place I'd always imagined buying if I won the Lottery. Twice. Once might not have been enough to afford it.

'Not a bad place to throw a wedding,' Charles says, proudly. 'Is it?'

I look it over, drinking it in. 'Nope,' I say. 'Not bad at all.'

We're served tea and homemade scones by Mrs Milchen, Charles's venerable housekeeper, and delicious they are too.

'It's very nice to see the young lady, Mr Charles,' she says warmly. 'Very nice to see you indeed, miss.'

'Please call me Anna,' I say, horribly embarrassed.

'Very good, Miss Anna,' she says, beaming. 'I always said it were about time Mr Charles got himself a young lady, didn't I, Mr Charles?'

'You did, Mrs Milchen,' says Charles, jovially. Mrs Milchen

313

hugs him round the shoulders as I smile faintly back. Charles's butler and dodgy valet – blimey, the guy looks as if he eats puppies raw for breakfast – seem equally pleased. And he's so relaxed around them, they obviously like him.

He's such a nice bloke, I think. I completely understand now why he was so nasty the first night, at Vanna's. Women just using him. No self-esteem. Hey, I can relate. Sometimes I think women are a bit hard on men. We assume it's fine to use them and blow them off; after all, they're only after one thing, right? But contrary to popular belief, they do have feelings. And Charles's were trodden on pretty well.

He squeezes my hand and gives me that loving, grateful look, and I feel something constricting my heart like a band.

Pull yourself together, Anna.

'Let's go through into the blue drawing room – I've got some stuff in there.'

'The blue drawing room?' I joke. 'Come on, like there's more than one?'

'There are three,' Charles says seriously. 'And two morning rooms.'

'Right.'

'And a parlour.'

'OK.'

'And a music room.'

'I get the picture.'

'I can take you to see the two libraries if you want, and then there are the studies – you haven't explored the place properly, have you, darling? Let's do that now.'

'No, no,' I shake my head. 'Let's go through the wedding stuff. I really want to see what you've come up with,' I lie.

'OK,' he says, beaming. We adjourn into the said blue drawing room, which is indeed painted in a gorgeous Wedgwood blue with cream trim and sports blue-and-white based Persian rugs. On an antique oak coffee table there are about twelve huge folders, laid open. I can see charts and lists, magazine pictures, glossy brochures.

'Charles!' I exclaim. 'How much planning have you done?'

'Oh, just the odd bit here and there,' he says, modestly.

'Shall we start with the menu ideas? Or the flowers? Or shall we pick a marquee? I got six quotes. See which design you like best. Here,' he says, passing me a brochure richly illustrated with full-page photos of dream-like silky tents.

I sit beside him on the robin's egg damask couch and try to pick and choose. And actually you can get kind of lost in it. It's the fairytale, he's offering me everything. A horse and carriage. Masses of yellow and white blooms everywhere. Rivers of champagne. A tent that would do credit to the Arabian nights, with a hardwood dance floor. A DJ and a live band and a string quartet. Any designer wedding gown I like, a tiara from Basia Zarzycka, hand-made silk wedding shoes embroidered with gold thread . . .

It will give Charles and Di a run for their money. Or Madonna and Guy, come to that. It's just about everything I've ever dreamed of in a wedding.

Except the groom.

I try to get away after that, but he insists I stay for lunch at least. I endure more delighted fussing and clucking from Mrs Milchen, get told off for not eating enough, and then finally manage to persuade Charles to run me to the station.

'I can drive you back, sweetheart, it's no trouble,' he assures me for the millionth time.

'No, that's fine, really. I – I like the train. And you need to do more planning,' I tell him.

'There is that,' he concedes. 'It's quite the operation!'

'I know.'

'Call me soon, OK?'

'You got it,' I tell him, kissing him on the cheek and hastily heading off to find my platform. I have to get away from him. I have to do some serious thinking.

I buy a couple of magazines in the station W.H. Smith's. *Cosmo* and *Company* and . . . *Heat*. I pick up the magazine, hold it in my fingers.

I can't help it. The memory of Mark Swan, reaching out and grabbing my copy. Me lecturing him about his fags. The way he looked at me when he took the script.

The way I felt when he called me, in Kitty's office.

The way I looked at him in 47 Park Street, when I couldn't concentrate because I was staring at his mouth instead of listening to him talk.

The pub. My hand in his. My hand looking . . . delicate, in his.

My engagement party, and all I wanted in the whole world was . . .

. . . to have him kiss me . . .

'Love.'

I blink. The woman at the till is scowling at me, the people behind me shuffling their feet.

'Sorry, I was miles away,' I mutter. I fork over some money and run for the shelter of the train.

It's absolutely no good, I think to myself. I'm in love with Mark Swan. I need to see him, I need to tell him . . . and I need to tell Charles.

I shudder at the thought. It makes me want to cry. Poor, poor Charles. I'd rather chew my own arm off than do this, but I still have to do it. I can't marry him. I know I agreed with all the logical, practical reasons, and I know he wants to marry me more than anything in the world, and that he's a lovely person and he doesn't deserve to be hurt . . .

But I still have to hurt him.

I'm going to do it today. I can't marry him, because I want to marry Mark Swan, and that's all there is to it. Yes, maybe there are people who can do the sensible thing, marry the suitable man, forget about passion, and be perfectly happy. I know history, that used to be just about every rich man and woman in England before this century, didn't it?

But maybe I'm not sensible. Or maybe I'm just plain nuts. But it doesn't matter, it doesn't matter, I think, half hysterically. I just love Mark, I want Mark.

I jump in a cab and give the driver Mark's office address. Even though I've only been there once, I still remember exactly where it is. My heart's thudding, pounding as he inches his way maddeningly through the traffic. I check myself in the rear-view mirror about a hundred times until he asks me to get out of the way, like I'm a nutter. Which I probably am. But at least I'm a nutter with a great make-up job whose hair has dried.

I don't know what I'm going to say. I don't know what I should do, how I should act. I only know I want him and I'm going to risk everything to tell him.

I burst through the front doors of Swan Lake Features. The burly man on reception tries to challenge me, but I quell him with a look.

'I'm Anna Brown from *Mother of the Bride*,' I say imperiously. 'And I want to speak to Mark.'

'You got an appointment?' he asks, insistently.

'No. Yes,' I change hastily, seeing his face. 'A standing invite, Mark said to drop by anytime.' I subtly spread my fingers over his desk to steady myself. To shore up my trembling knees.

'Wait a second,' he says severely, and turns away from me, mumbling into his phone.

'Michelle says you can go up,' he grunts, obviously quite disappointed.

I'm not, though. I jump into the elevator and ride it up to the main floor, breathing deeply, trying to calm myself. He's got staff here, he meets people. There's Michelle. I don't want to show him up by bursting in on some business, like a hysterical lovesick baboon.

The elevator doors hiss open.

Michelle's sitting there, wearing more black, this time a drop-dead gorgeous scoop-neck, body-hugging short number.

'Anna,' she says coolly. 'Bit of a surprise to see you here, and what was that *Mother of the Bride* stuff, did they reinstate you?'

'Oh, that. No, that was just a ploy to get past your security guard.'

'And why would you wanna do that?' she demands.

'I have to see Mark,' I say. My eyes are sparkling. 'I – I just have to see him.'

Michelle pauses. 'Is it about business?'

'No, it's definitely not about business,' I say. 'I need to tell him something, something personal. Is he here?'

Michelle's eyes narrow and she looks at me, consideringly.

'No, he's not here but he should be back soon,' she says. 'Do you want to wait?'

'Oh. Yes. Thank you,' I say gratefully. I sit down and pick up a magazine.

'This . . . this about your love life?' Michelle asks casually.

I smile at her, trying to warm her up to me. 'Yes, it is, actually.'

'Oh,' she says, examining her nails. 'Well,' she says, giving me a smile that doesn't quite reach her eyes, 'I'm sure he'll be delighted to hear how everything's going with your fiancé and all.'

'As a matter of fact—'

'You know, because we're so happy together,' she says, looking me dead in the eye.

'What?'

'Because we're so happy. Since we started going out.'

My mouth falls open. 'What? You and Mark?'

'That's right,' says Michelle, triumphantly. 'Me and Mark.'

'But you weren't going out.'

'Not till he asked me, no,' says Michelle. 'But we've loads in common. I want to be a director too. And we work out together. Mark loves women with good bodies,' she says with heavy emphasis. 'Slim and fit, you know. That's when he asked me, when we were in the gym one day. I think he'd only waited because I work for him. We've been close friends for a while now. You know Mark,' she says, shooting me a significant look. 'He loves women, he always gets close to the women he works with.'

I put the magazine down. I'm trembling. I try to steady myself. But I daren't speak because my voice is going to wobble and she'll know.

'He told me he really liked you,' Michelle says. 'You and I'll have to go out, Anna. Get to know each other. Maybe swap a few wedding ideas.'

'He asked you to . . . to marry him?'

'Not yet,' Michelle says. 'But I think it's coming. Fingers crossed, eh?'

Oh yeah. Fingers crossed.

'Tell you what,' I manage. 'I'd better be going. Just, just give him my love and . . . and ask him to let me know where I can send the wedding invitation.'

'Just send it to Notting Hill,' Michelle says, smiling at me like a crocodile. 'It's really nice of you. We'd *love* to come.'

13

I don't know how I manage to get home, but I do it somehow.
I get on the tube like a zombie. I feel so bereft, and so stupid.
Of course I misread him. What was I thinking? Of course
Mark Swan would want a girl like Michelle, she's so slender,
so pretty, young and hard-bodied, interested in sodding
directing.

At least, I tell myself, I haven't ruined anything. I haven't
said anything to Charles *or* Mark. I can still get married, still
have a companion, still live in Chester House . . .

I open our door. And stop dead.

Janet is sprawled over the sofa, shaking with sobs. There's
a huge bottle of gin right next to her. She looks as if she's had
a quarter of it already. The smell of the alcohol hits me like a
punch in the face, mixing with my hangover, but I ignore it.
I rush right over and put my arms round her.

'What happened?'

'Ish my agent,' she says. 'He shays he doeshn't want me any
more.' And she dissolves into a fresh round of heaving sobs.

'Stay there,' I say. I pick up the gin bottle and empty it
down the sink. Then I run downstairs and out to the Boots
across the road, and pick up loads of bottles of freshly
squeezed orange juice. Janet loves freshly squeezed orange
juice. And I get some more Shapers sandwiches and some
fizzy mineral water with elderflower and some bubble bath
(lavender – she needs to be soothed). Then I race back up to
our flat. On the way up the stairs I notice I'm taking them
two at a time, and I'm not even out of breath. Am I getting
fitter? And why am I thinking about myself?

Janet's exactly where I left her. I put some ice in a tall glass and pour the elderflower water over it.

'Drink this,' I say.

Janet takes a sip then pushes it away. 'It's sweet, I can't drink thish,' she wails. 'I'm *fat*, thash why I've got sacked. Fat an' old an' no one wantsh me!'

'It's diet,' I say. 'No calories. Look!' I shove the bottle under her nose. 'Now drink!'

Obediently she chugs it down. I force her to drink four huge glasses of water until she rebels and says she won't have any more.

'Orange juice then,' I plead, unscrewing the cap and waving the plastic bottle temptingly under her nose.

'Definitely calories,' she says.

'Just a few and full of vitamin C. Which makes the skin smooth and the – the eyes bright,' I say. 'It's been proven by scientific studies.'

'*Hash* it?' gulps Janet.

'Yes, and it also cures wrinkles . . . and . . . freckles,' I say. Janet hates her minute sprinkling of freckles.

It works. She drinks almost the entire bottle of juice, I have the rest, and then she manages to eat two low-cal prawn sandwiches. Eventually she seems to calm down a little bit, the water works, she's not so drunk. But she still seems incredibly miserable.

'OK,' I say. 'Tell me what happened.'

'I called Marcel about this gig,' she says tearfully. 'First thing this morning. I was all ready. I'm never late, always professional, you know? Not like some girls.'

'Who was it for?'

'Harpers and Queen,' she says. 'It was my first big national shoot for a while. But when I called in, Marcel said they'd decided to go with another girl. Laura Boynton. D'you know her?' she asks, pathetically. 'She's the one with that lean athletic look and the cropped brown bob. She's nineteen. She's so hot right now.'

'That's only one shoot,' I say sympathetically.

'I tried to say that. To put a positive spin on it,' says Janet with forced brightness. 'Make lemonade out of lemons. I was modelling my millionaire mentor. That's J-Lo. I don't know her but I feel I do spiritually. What would J-Lo do? That's the motto of the Jay-Me Crew.'

'Ahm, yes.'

'But Marcel said, "We have to talk, Jay-Me," ' she sobs. 'And then he tells me that I'm a very beautiful girl, but the market's got different needs right now . . . and . . . some other stuff . . . and he couldn't be my agent any more.'

'So what did you say?'

'I said maybe somebody else at the agency could handle me. And he said he was speaking for the agency. He told me they've got a new policy and they only represent girls over twenty-three if they're exceptional.'

'Oh, Janet,' I say, hugging her.

'He said, "To be honest, you've been living on borrowed time, Jay-Me. Maybe you should think about retiring. Or try one of the specialist agencies for older women," ' she wails. 'It was so *humiliating*.'

'But . . . couldn't you do that? There are model agencies for girls over thirty, aren't there? And you're not even thirty.'

'I will be soon,' says Janet darkly. 'Thirty. Imagine.'

'That's not eighty.'

'Might as well be. Take up with one of those agencies? I'll be modelling mumsy skirts and high neck blouses for the *UKfashions*! catalogue,' Janet says, breaking down completely.

I pass her some Kleenex.

'OK, look,' I say, when there's a ten-second break in her crying. 'Look, Janet. You are beautiful.'

'I'm not.'

'Yes, you are,' I said. 'Incredibly beautiful. If you don't fit into the model mould any more, so what?'

'I'll have to live on the streets,' Janet says. 'I'll starve. There aren't any charities to help old models,' she adds plaintively.

'I don't think Help the Aged starts at twenty-eight, no.'

'I suppose I could get married. D'you think Ed would still

be interested?' she asks hopefully. 'Even if I'm not a model any more?'

'I'm sure he would, but I wouldn't bring up marriage,' I say hastily.

'Why not?' she demands.

'Erm, because you haven't actually dated for a month yet,' I suggest.

'Oh yeah,' she says, dejectedly. Her shoulders slump again.

'But why does that have to be your only option?' I ask. 'You're a clever girl. You've got personality. Why is that the only thing you could do?'

Janet sighs. 'Never done anything else.'

'Do you have any qualifications?'

'Oh. Those,' she says, as though retrieving a distant memory. 'Yes. I've nine GCSEs and two A-levels.'

'That's great!' I say enthusiastically. 'What are the A-levels in?'

'Pottery and art history.'

'OK,' I say, carefully. Obviously the career as an English teacher or international merchant banker is out the window. 'Are you good at making pots?'

'No,' says Janet, tears trickling out of her eyes again. 'I only got a C. I kept failing my practical, my pots came out all wrong, and I tried to say they were conceptual pieces but they just marked them as wonky.'

'OK,' I say. 'Well, we'll think of something. You can start a whole new career. I bet you can be really successful and make loads of money.'

'D'you really think so?'

'Absolutely,' I lie. 'How much money do you have right now?'

'I don't like to look at my bank statements,' she says.

'Maybe we should start. Why don't you get them?'

Janet gets up a bit unsteadily, goes into her room, and comes back with a sheaf of unopened Barclays statements.

'Which is the newest one?'

'That one.'

'OK, open it and see what it says.'

Nervously, she rips it open. 'Oh, that's not so bad,' she says. 'Five thousand and thirty-eight pounds and sixty-two pence.'

I breathe out. 'There you go. You'll be fine while we find you a new career.'

'What does OD mean?' she asks.

'Overdrawn,' I say. 'Does it say OD by that figure?'

'Yes,' she says, miserably.

'So you're actually overdrawn by five grand,' I say carefully. 'Well, not to worry. We can fix it.'

'How?' Janet says.

I have no idea! How can she pay off a five-thousand-pound overdraft with no job and no prospects? But she's looking at me hopefully with those big brown eyes. I've got to do something . . .

'We can call the bank,' I suggest. 'Get you on a payment plan. And I can lend you a bit of money. To start.'

'Don't be silly, Anna,' says Janet kindly. 'You haven't got any money. You spent it all on those clothes.'

'I had a promotion, remember? I've still got a bit left.'

'The rent's due next week,' Janet says, also miserably. 'I thought something was wrong when I went to the ATM machine last week and it ate my card.'

'It did what?' I ask in horror. 'How have you been managing?'

'These!' Janet says brightly. She fishes her purse out of her bag and shows me a dazzling array of credit cards in gold and platinum. 'They're brilliant!'

'Janet,' I say. 'You must have . . . at least fifteen.'

'Oh yes, but I'm very responsible,' she says. 'I do open their letters.'

'That's a good start,' I say encouragingly.

'And I'm not behind on any of them,' she says proudly. 'Look. I'll show you.' She goes to her room and returns with another huge pile of envelopes. But, as she said, all of them have been neatly split open. I read through a few.

'But almost all of these are maxed out,' I say.

'Not these three, they're new,' she says, displaying a gold

visa and two platinum Mastercards. 'There's loads of space on these ones.'

'But on this one,' I tap an Arsenal FC card, and Janet hates football, 'you owe three and a half grand.'

'But I only have to pay twenty-nine pounds,' Janet says brightly. 'See? It's easy. You only pay a bit.'

My headache is starting to come back.

'So you're making minimum payments on all the cards,' I say, 'and when each card maxes out you just get new ones?'

'They send me new ones because I'm not behind,' Janet says proudly.

'And you pay the minimums . . . out of the bank?'

'Well, it ate my card,' Janet pouts, 'so now what I do is I get cash advances from the new cards and pay off the old ones with them! I've loads of space. My limit's six thousand on this one.'

As it sinks in, for one wild moment I feel glad. Isn't that awful? But I had been beating myself up because Janet and Lily always go to the best hairdressers, the most expensive restaurants, the chic little boutiques, and I couldn't afford any of them. And now I realize, neither could Janet. I wasn't so totally out of it before.

But then I feel really mean for thinking that. Poor Janet, she's clueless. She's got no idea what she's got herself into.

'Um, can I have that calculator?' I ask. She passes it over, and I tap in the figures.

'OK,' I say. 'And now can I have your purse?'

She passes it over. 'What are you doing?'

I take it into the kitchen. 'Nothing,' I say, fishing around in the drawer for the scissors. OK, there they are.

'Nooo!' Janet shrieks. She rushes into the kitchen as I start cutting up all her credit cards, arms flailing at me. But I fight her off. 'What are you doing?' she shrieks. 'I need those!'

'I'm leaving you this one,' I say. I hand her back the gold Visa with the six grand limit.

'What the hell did you do that for?' Janet yells. 'Are you out of your mind?'

'Janet,' I say, 'You owe these credit card people twenty-one thousand pounds.'

She blinks.

'Your monthly minimum payments come to eight hundred pounds, give or take,' I say.

'So, I'll have paid them all off in two years,' Janet says defensively.

'No. That's just the interest. Your balances won't actually go down. You could pay eight hundred pounds a month for your whole life and you'd still owe the same amount.'

'But – but that's robbery,' Janet says faintly. 'They're just thieves.'

'And with the bank overdraft, that's twenty-six grand. And you've rent of seven hundred a month, so you need to find one and a half grand every month after taxes and that doesn't include any bills or food,' I say.

'I don't . . . I can't . . .' Janet says. Her eyes are tearing up. 'I don't know how this can be happening. Oh my God, Anna. What the hell am I going to do?'

'I don't know yet,' I tell her. 'But I'll figure something out.'

What the hell. At least this is distracting me from my own pain. I put my arms round Janet's slim body and give her a huge hug.

I do what I can. I call some debt services and negotiate a lowered interest rate, and get Janet on a payment plan. Same thing with her bank. They won't give her any more overdraft, but they have stopped threatening legal action.

'It's a start,' I say.

'No it's not,' she says, about to dissolve.

'It is,' I insist.

'What about the rent?' she asks unhappily. 'I can't pay! You know what Lily's like.'

Yes, I do. 'Maybe you could move out,' I suggest. 'Get somewhere cheaper.'

'But I wouldn't be with you two,' Janet says. 'You're my friends. And this is Zone One.'

'I don't think you can live in Zone One any more,' I say

327

gently. 'You've got to look for somewhere cheaper. Maybe a share in Zone Four. We can get a copy of *Loot* . . .'

'Oh my God,' says Janet, starting to cry again. 'I've been dumped because I'm too old and too fat, and now I'll never make any money again and I'll have to live in Neasden!'

'We need to find you a job,' I say. 'Let me just think, OK? And maybe you could borrow some money from your parents. For a couple of months.'

'I had a quarrel with my mum,' Janet says. 'She didn't like Gino.'

'See? She sounds like an excellent person,' I say. 'Why don't you give her a call?'

'I'll look like a stupid failure,' Janet says. 'My dad told me I shouldn't be a model. And now he'll be right.'

'Just call them,' I say. 'And don't worry about anything. We'll fix it all.'

'You know, you're really amazing, Anna,' says Janet wistfully. 'I wish I was like you.'

She wishes she was like me? That's good for a laugh.

'But you're gorgeous,' I say, self-consciously.

'You look pretty good too, these days,' Janet says. 'And you're so clever and funny and everybody likes you.'

'Well, everybody likes you, too. And I know you've done a great job with me. I know I look all right now,' I say gratefully. 'But still, I'd only look like you in my dreams.'

'You don't understand,' Janet says. 'If you're clever and make a lot of money you can make yourself beautiful. But if you're beautiful and not clever you can't really do anything.'

Lily comes home at 3 p.m. after a successful *Company* shoot, in a good mood because she'd been picked for solo shots out of the group. I'd like to say she's sympathetic and understanding. But I'd also like to be able to do the splits. Neither seems particularly likely right now.

'But I don't understand,' she says, screwing up her face in mock concern. 'Why did Marcel say that? Oh, he only mentioned your age? Not your weight problem? Older women agency, I suppose you could *try*,' she says. 'But there are lots of girls going for those spots, aren't there? I don't think you can rely on them accepting you, to be quite honest. Maybe if you went on a fast. For two weeks. And have you considered plastic surgery?'

'Lily, pack it in,' I say quietly.

'Oh, you wouldn't understand the modelling world, would she, Janet? It's all about beauty. Types of beauty,' Lily says, taking a quick look at her porcelain skin in our living-room mirror. 'You have to work *very* hard if you want to keep your bloom. Janet really hasn't, so she's got to expect this sort of thing.'

'I'm warning you,' I say.

'It's partly your fault, Anna,' says Lily to me severely. 'You've changed Janet. You've been making her eat chocolate and crisps. And I've caught you both drinking alcohol. You've pushed her along this path to ruin.'

'You're a loony,' I say.

'Oh really?' Lily demands. 'What about her men? You've got her going out with this Ed loser.'

'He's not a loser!' Janet says fiercely.

'He doesn't have any money,' Lily says. 'And now you've got no job, who's going to look after you? Huh?'

'Maybe she can look after herself.'

'Nobody will hire her, Anna,' Lily says patiently. 'It's too late for her. You should let me find her somebody to help. Since I'm seeing Henry now, maybe I can hook you up with Claude Ranier,' she says. 'He might still be interested.'

Janet looks pale.

'She's not interested,' I say flatly.

'She doesn't have that many options,' says Lily. 'She's not like you, marrying a millionaire. The rent costs money, maintenance costs money. You can't keep up the right look without certain things,' she says, examining her manicure. 'A beautiful girl is like a thoroughbred racehorse. Her upkeep costs money.'

'It does,' Janet agrees, looking guiltily at me. 'That's why I had to buy those things. Be seen in the right places. You know.'

'Image is everything,' pronounces Lily.

'Janet's going to be moving out,' I say to Lily. 'She's had some financial troubles and she can't make the rent.'

'What?' Lily snaps. 'You can't do that! It's due next week. You have to give notice.'

'I didn't know,' Janet says humbly. 'I'll look for somewhere else right away.'

'That's not good enough!' Lily screeches. 'You *owe* me that money!'

'I'll pay you back,' Janet pleads. 'I'm – I'm going to call my parents.'

'I can't wait around for you,' Lily says nastily. 'This mess is your own fault and you're just going to have to pay me, that's all.'

'But I've no money. The machine ate my card.'

Hearing this, Lily pauses for a second, going rather pale. Then her eyes narrow.

'Well,' she says. 'You can pay me in assets, can't you?'

'Assets?' I repeat.

'Yes. Assets.' Lily marches into Janet's room and flings open her wardrobe. 'Look at all these clothes! You've got your Dolce, your Chloe, your Voyage, your Armani.'

'Leave those alone!' Janet says in horror.

'Lily, be reasonable,' I plead. 'We're friends. Be supportive . . .'

'Supportive?' Lily shrieks. 'What are these?' she demands, holding up the most exquisite, delicate pair of pale green high heels decorated with dark green ivy-leaf straps. 'These are bloody Patrick Cox, that's what. And she's got eight pairs of Manolos! Why should *my* debts go begging?'

'They won't fit you,' I say hastily. 'Janet's a different size to you.'

'I'll say,' agrees Lily spitefully. 'What size feet?'

'Five and a half,' Janet says.

Lily tosses her hair. 'I can't wear those big clod-hoppers, I'm a four. Narrow,' she adds.

Janet breathes a sigh of relief.

'But that's not the only thing you've got,' Lily says. 'What about accessories? Let's look at your handbags. Louis Vuitton. I quite like that one. Fendi baguette. So yesterday. Chanel . . . Chanel . . . you don't have any, I'll take this Coach one and the Louis Vuitton and I'll have this Gucci one and the Doone and Bourke . . . and these Kate Spades,' she adds triumphantly, loading her arms with all Janet's handbags.

Janet starts to cry again.

'Don't be so wet,' says Lily. 'I've left you the Fendi baguette, though personally I wouldn't be seen dead with it. And I'll have that DKNY you're using right now,' she says, reaching for it.

I smack her hand.

'Ow!' Lily yelps.

'Leave those alone,' I say. 'You can put them all back.'

'These are in lieu of rent,' Lily snaps. 'She owes me a month's notice.'

'That's fine,' I say. 'I'll pay her rent for next month and you can consider this your month's notice.'

'Anna, you can't,' says Janet tremulously. 'She can have the Kate Spades.'

'I'll expect it on the first,' Lily says coolly.

'No need to wait that long,' I say, reaching for my coat. 'I'm taking Janet out to dinner and then I'll come back with both our rent money. And it's one month's notice from me too. I'm leaving as well.'

'Fine, see if I care,' Lily says, flinging Janet's bags back in her wardrobe in a heap. 'You're both just a couple of losers and I can't *wait* to get rid of you. Of course you've landed on your feet, Anna, but I wouldn't count on it lasting with Charles,' she hisses. 'One of these days he's going to wake up and see sense!'

I take Janet round to Bella Pasta.

'I can't possibly,' she protests. 'All the complex carbs! Pasta's refined flour.'

'You can eat now if you're not going to be a model any more.'

Janet looks hopeful. 'You think? But won't I just explode?'

'Not from the odd spag bol, no,' I promise her.

'I do like pasta,' she admits.

'You'll eat like an almost normal person,' I say, 'and perhaps you'll go up to a size ten. You'll still be skinny. However little you eat you'll never look like Lily.'

'Why was she so horrible?' Janet asks.

'I don't know,' I say. I order a couple of Diet Cokes and a grilled chicken salad for me. 'But it's a chance for you. A fresh start. And maybe you should think about your wardrobe, you know. You could sell a few things on Ebay and raise quite a bit. That Fendi baguette, I bet you'd get a couple of hundred for it.'

'It *is* so yesterday,' Janet says solemnly. 'I could sell that one.'

'And maybe a few of last season's shoes? You wouldn't wear them anyway.'

'Yes,' Janet says thoughtfully. 'That is a good idea.'

'And you'll call your parents,' I say. 'And we'll get you a job.' Suddenly I have a brilliant thought. 'Ooh. Janet. You were quite good at the history of pottery, weren't you?'

'Yes,' she said. 'I just couldn't make the pots.'

'And the art history?'

'I got an A in that one,' she says rather proudly. 'A grade A A-level. Me!'

'I know,' I say. 'Maybe you can't do any art yourself, but I bet you could work in it. What about a museum guide or something?'

'I think you need a degree,' she says.

'Yeah, maybe. Well, something in art. And after you call your parents you should call Ed.'

'But I don't want to tell him,' she says pathetically. 'He won't fancy me any more.'

'Of course he will,' I say.

'Do you think Lily's right?' she asks, all worried. 'I mean, it does look like he doesn't have any money.'

'I thought you said you didn't care about that.'

'I don't, I still really like him,' she says.

'You don't need a man to take care of you, you know,' I say gently. 'You can actually make some money of your own.'

'Do you like working?' Janet asks suspiciously.
'I did,' I sigh.

When we get home Lily has stormed off out for the evening, leaving a huge pile of bills on the coffee table with what we owe marked out in big red letters. I write out a cheque and leave it on her bed. I write 'Janet and Anna's last month's rent' on the memo section, just in case she tries anything funny.

I really don't understand that girl. Why is she so nasty? You would think Janet had slapped her round the face or something. Oh well. One more month and I'll never have to see her again.

'I think I'm going to bed early,' Janet says weepily. 'If that's all right with you.'

For a second I panic. I want to stop her; after all, if I'm looking after Janet I won't have to think about myself.

'That's fine,' I say. 'You get some rest.'

I sit by myself on the couch and try to take stock.

OK, I'm fired. I've spent all the money I have in the world. After the clothes, that rent cheque wiped me out. I can't get another job because Mark Swan's shadow is looming over me everywhere I go. And I'm in love with a man who's already got a girlfriend, a girlfriend who's much younger and prettier than me.

And I'm engaged to a multi-millionaire who's devoted to me.

I look down at my ring. It glitters beautifully in the twilight.

Then I pick up the phone and call Chester House.

'Miss Anna, how nice to hear from you,' Mrs Milchen says. 'But Mr Charles ain't here. He went back to London this afternoon, miss. He's in the flat, I reckon.'

'Thanks, Mrs Milchen,' I say.

I pick up my bag and ride our coffin-lift down to the street, hail a taxi. I get inside and settle comfortably into the seat, wondering what the hell I'm doing. No job, no flat, and now no man? I must be certifiable.

But I know I'm not. I know what I have to do. Yes, Swan

doesn't want me, I must have been kidding myself to ever have thought *that* was a possibility. And yes, Charles has almost everything a girl could wish for: he's rich, he's generous, he's kind . . .

But I don't love him. And knowing now that I can love someone, really love them, I'd rather be alone than settle.

But it's still hard. I feel the tears rising, try and fail to choke them back. I'm about to pull the trigger on my whole way of life. A couple of weeks ago I was in the film business, marrying into luxury, living in a flat with two friends. Now I'm about to go home to my parents, man-less, job-less, and money-less. And what have I got? New make-up, hair and a pile of clothes.

But the thing is, as superficial as it sounds, I quite like myself in the new clothes, with the new look. I don't feel unfeminine. I think maybe that that's why I've got the strength to go and see Charles now, even without any other options. I owe it to him to pull the plug before it gets any worse, before he spends another penny on me.

Ugh. I really hope he's home.

I pay the taxi, climb out onto the pavement. It's a cool autumn night and I press the bell. I shiver, but mostly, I think, not from the cold.

'Hello?' comes Charles's disembodied voice.

'Charles, it's me,' I say, teeth chattering. 'Can I come up?'

There's a pause, and instantly I know he knows.

'No problem,' he says dully. 'Come up.'

'And I suppose I can't change your mind?' Charles asks, stiffly, when I've finished. He's sitting next to me on his beautiful nineteenth-century chaise-longue, and he hasn't burst into tears or anything, but I can see the pain etched deep all over his face.

I feel sick. People give me pain, I don't give it to other people. Especially not nice ones who have never done me any harm, and only tried to help me.

'I'm so sorry.'

'You know, this other chap isn't available.'

I nod.

'You might end up with nobody.'

'I know.'

'And you'd rather be with nobody than with me, I see,' says Charles, and now his eyes do fill with tears, but he makes a manful effort to blink them back and I pretend I haven't noticed.

'It's not like that,' I say desperately. 'It's just that I thought I could do without being in love, and I can't.'

'I see.' Meaning he doesn't.

'Charles, women have always used you,' I say passionately. 'And you're better than that. You don't deserve to be used for your money. *Or* your good nature, which is what I'd have been doing if I'd stayed.'

'You've every right to hold out for better,' Charles says bravely.

'It's not better, it's just different,' I say. 'Look, has it ever occurred to you that you deserve better too?'

'I don't want better, I want you,' he says.

'No you don't,' I tell him. 'And that's the problem. You don't find me at all attractive. You didn't notice when I bought all new clothes and had my hair cut.'

'Was that the problem?' he asks, stricken. 'That I didn't pay enough attention?'

'No, it absolutely wasn't. You'd have paid attention if you'd fancied me.'

'But what does that matter, when I loved your personality?'

'Charles, you didn't,' I cry. 'You loved the idea of being married, having a wife who wasn't after your money, having a family life. We've nothing in common! You know what your biggest compliment to me was? That I was a good listener. You were just lonely. But there's someone out there for you, someone really nice, who you can also love. Not like me.'

He takes this in but doesn't respond for a moment.

'So what will you do?' he asks, eventually.

I shrug. 'No idea. Go back to Mum and Dad's for a bit. I have to, I've no money.'

'I can give you some money.'

I squeeze his hand. 'Thanks, Charles, but no. Although that's really sweet of you.'

'Sweet,' he says bitterly. 'I don't want to be bloody *sweet*.'

'I can find you someone,' I promise.

'But they'll still be after my money,' he says.

'Not the right girl. You'll . . . you'll just know.'

'And what about you, Anna?' he asks. 'What about you finding the right man?'

I sigh. 'I already did, but he's taken.'

Charles says, 'I'm going to get us both a drink.' He gets up, goes to fetch some of those glorious cut-glass tumblers, and hands me a giant whisky. I prefer mine with Coke, but any port in a storm, right?

'I'm sorry,' I say, after I've knocked back a giant swallow. I think of Janet, and Lily's viciousness, and Kitty, and my job interviews, and Charles's folders full of wedding ideas, and my eyes brim over. 'I'm just really sorry.'

He says, 'You have to tell him.'

'What?'

'This Swan chappie. You have to tell him.'

'Don't be silly, I'll never say anything. He's got Michelle. She's much prettier than I am.'

'That doesn't matter,' he says. 'You look fine.'

'You never fancied me,' I point out.

'Not as such,' Charles admits. 'But he did. I saw it in his eyes. Don't you think I can tell when another chap's after my woman?'

That makes me feel awful again. 'I'm so . . .'

'I know, I know,' he says briskly. I can see he's still hurting, but he's pulled himself together. 'Not your fault. You never wanted to be with me and I talked you into it. I don't regret that, you know,' he says fiercely. 'I still think we could have been happy. And I had to ask.'

'I understand.'

'Which is exactly why you need to see him,' he says. 'Can I give you some advice? As a friend.'

'You are my friend,' I say. 'My dear, good friend.'

'I hope I am,' he says. 'You risk only a little . . .' he pauses. 'Perhaps a little embarrassment, shall we say, if you ask him. And at the very least you'll have the satisfaction of knowing you tried. I had to ask you, Anna. I had to try.'

I can't help it, I start crying again.

'Be brave,' he says. 'Tell him.'

I sleep late the next morning. Exhausted and drained, I finally stagger into the living room at about eleven, to find Janet, in much better shape than before. She's got a copy of *Loot* open on the table, she's washed her hair and put on her best white jeans and she looks absolutely sensational.

Better than that, she looks *happy*.

'I called my parents,' she said. 'They're a bit tight right now, but they said they could lend me a thousand. That's a start, isn't it? And I called a few of those agencies.'

'Which ones?'

'The older ones. The catalogue ones. Don't tell Lily,' she adds, shyly. 'She'll only laugh. But I took my portfolio in to the Elegance agency and they said they could get me work starting tomorrow. They said I had a "perfect look" and I could make loads.'

'That's great,' I say, with a sigh of relief. 'That's great, Janet.'

'I mean, a job's a job, right?'

'That's right.'

'And Ed was really nice to me. Even though I got sacked. I even told him about modelling for the catalogues and he didn't laugh or anything. Actually, he said he was coming straight round. We're going out to lunch.'

'See? I knew he'd be like that,' I tell her.

'I was wondering,' she says diffidently. 'D'you think I could borrow just another fifty quid? To pay for lunch. I mean, he doesn't have any money, does he? I'll pay you back as soon as my mum's cheque gets here.'

'Sure,' I say, forking it over, although I shudder to think what I'm down to now. But as long as I've got enough to get me back to Mum and Dad's, I'll be OK, I tell myself.

The door rattles and Lily walks in.

'Oh, it's you two,' she says nastily. 'What are you doing, Janet? Looking for a place in Bermondsey?'

'You've got your rent cheque,' I say.

She tosses her hair. 'I guess so. I hope you realize I can start showing your rooms.'

The buzzer sounds and Janet leaps to answer it.

'Come on up,' she says. 'It's him,' she says to us.

'Oh yes,' says Lily, 'the boyfriend with no cash.'

'Can't you behave yourself even for ten minutes?' I demand. 'If you're going to sit here and say sarcastic things, Janet will just leave.'

'Don't worry,' Lily pouts. 'I don't embarrass people, even if they *are* stabbing me in the back.'

Janet looks dubious, but she doesn't have time to do anything about it because there's a quiet knock on the door and there he is.

'Hi,' Janet says, smiling hugely. 'Come on in.'

Ed enters, passing a hand nervously through his floppy Hugh-Grant-type hair. He's wearing cords and a dark blue shirt which is a bit frayed, old fashioned enough to have cufflinks, though. His wardrobe could use a bit of updating and so could his hair. But he has a nice smile. That ought to count for something, unless of course you're Lily.

'Wow, you look awesome, babes,' he says to Janet, giving her a huge kiss on the cheek. 'Brought you some flowers,' he adds, producing some wilting chrysanthemums wrapped in bright orange paper and Sellotape. They probably cost him all of three quid at the local deli.

'They're adorable,' says Janet, smiling and taking them. 'Thanks so much.'

'Hello, Anna,' he says. 'You look marvellous too. How nice to see you. And . . . Lily,' he adds, staring at her. He goes pink in the face.

Lily is wearing a string-mesh vest with a nude silk lining that makes her look like a slave girl in chains, coupled with a pair of the shortest, tightest crotch-hugging white mini shorts this side of an arrest for indecent exposure. Her long, flaxen hair is flowing loosely down her back, and her skin, with her endless legs, is perfectly golden like warm toast. She's wearing a pair of stacked white slides with a sexy little chain anklet with a tiny bell on it.

'Hi,' she says, without interest.

'So you'll be seeing Henry soon?' Ed asks her. 'He's a decent chap.'

'Very,' Lily agrees.

Ed swallows drily and looks away at Janet. You can see he's struggling manfully not to gape at Lily.

'How do you know Henry?' Janet asks.

'Met him through Charles,' Ed says. 'Very brave man. Went into the Marines for a couple of years. Rumours of gallantry in Bosnia. Good egg.'

'Yes,' Lily says sweetly. 'As good as gold,' she adds, looking at me significantly. 'Actually, he's coming by this evening too.'

'Great!' says Ed. 'Maybe we should stick around. We could all go out together.'

'Do you know,' Lily says coldly, 'I'm not sure Henry and I would be quite your speed, Ed.'

'No,' says Janet hastily. 'Let's do that some other time. You and me can go out tonight, just us. OK?'

'Oh. Yah. Sure,' says Ed, rebuffed. 'Whatever you say.'

'There are lots of great restaurants round here that are *very* reasonable,' Janet says earnestly. 'There's Mr Chow's Sun Fun Palace, there's Pizza Nation, there's Freddy's Fried Chicken. Anna lent me fifty quid, so we can go wherever you want, really!'

Lily snorts. Ed is looking at Janet as if she's started speaking Swahili.

'Um, of course,' he says. 'If you really want to.'

'Or we can find somewhere cheaper,' says Janet, a bit desperately. 'I don't care.'

Lily's humming under her breath. I realize it's 'Hey, Big Spender'. Luckily, Ed seems totally oblivious.

'OK, you two, have fun!' I say brightly, pushing them out of the door. 'I'll put the flowers in some water for you.'

I wait until I've heard the lift creaking its way downstairs then round on Lily.

'I don't know why you have to be so mean.'

Lily widens her cornflower-blue eyes. 'Mean? I'm just trying to help Janet. She may be over the hill and have weight challenges,' she says delicately, 'but she could still do better.'

'Bloody hell—'

The buzzer goes again. 'Maybe she's dumped him already. Best thing for everybody,' Lily says gaily. 'Hello? Oh, hi, darling,' she purrs, her voice dropping a couple of octaves. 'Yes, I'm all ready. Come on up.' She hangs up and looks at me. 'Can't you make yourself scarce, Anna? You're going to cramp my style, quite frankly.'

'Oh yes,' I say sweetly. 'I'll make myself as scarce as you did when Brian used to come round, shall I?'

'God, you're a cow,' she says. 'Oh well, it doesn't matter. We'll be off out soon. To Claridges or Nobu or the Ritz, I expect. And then most likely drinks at the Met Bar or something. Don't wait up.'

'I wasn't planning to,' I say.

Another knock on the door and Henry arrives. Lily shoots me an exultant look as he walks in. He's wearing an immaculate suit and carrying a bunch of dark red roses.

'Hello, gorgeous,' he says to Lily.

'Hi, Henry,' Lily purrs. 'Flowers? For me? How adorable.' She buries her tiny nose in the glossy dark leaves and gives me a significant look. 'Let me just go and stick them in some water.'

Henry comes over to me and offers me a kiss on the cheek. 'Anna, great to see you,' he says. 'How did you like the party? Charles never stopped going on about you.'

'Yes, well. He's a great guy,' I say guiltily.

'Darling,' says Lily, coming back from the kitchen, 'give me five minutes to repair my face. Kiss kiss.'

She vanishes into the bathroom again and Henry sits down on the sofa.

'Tea? Coffee?'

'Nothing, thanks,' he says. 'We should be off any second. Where's your other friend? Janet, isn't it?'

'That's right,' I tell him. 'She just left on a date, actually. With Ed.'

'Another great bloke,' says Henry. 'They should have come out with us.'

I don't mention Lily's veto. 'That would have been nice.'

'Course, it would have had to be somewhere fairly plain,' Henry said. 'And I don't think Ed's used to that.'

'He doesn't like exotic food?'

'No, I mean, price-wise. I know a bunch of great places to eat around here, but they're not exactly first-class. Not what Ed Dawson's used to.'

'Oh?' I ask, moving a bit closer to the edge of my seat. The water's still running in the bathroom.

'He rarely goes to any restaurants where there are prices on the woman's menu, if you know what I mean,' says Henry, smiling. 'Got more cash than most Third World nations, that lad.'

'Ed?' I ask, just to be sure. 'But isn't he a younger cousin? And rents his place?'

'Oh, ycah, no family money to speak of,' Henry says. 'But he's a stockbroker. Rather a brilliant one. Made his first million while he was still at school. Retired when he was thirty, bought a farm, potters about on it for fun. And he's renting that flat because his manor house is being rewired.'

I check to see if he's joking, but he isn't.

'But his clothes,' I protest.

'Oh, yes.' Henry grins. 'Scarecrow chic? Doesn't surprise me. He has absolutely no idea how to dress. I think he just doesn't care. Ed's a very astute art collector, but apparently doesn't know the name of a single good tailor. Maybe Janet can sort him out a bit. He could use the help.'

'Really,' I say, trying not to grin back. 'How fascinating.'

'Now I have to be more discerning,' Henry says. 'Find good suits on sale, have them altered, wear them until they fall to pieces. I've had this for three years,' he adds, touching the crisp fabric of his suit.

'So you're not quite as flush as Ed?' I ask, then blush. 'I'm sorry, obviously that's none of my business.'

'God, no,' says Henry. 'Poor as the proverbial church mouse. If not poorer. At least they aren't months behind with their bills.'

'But don't you work for your family's estate agency?'

'I do,' he says. 'But I'm not very good at it. Don't make all that much. I'm thinking about trying to do something else. I have to stay there, though, just to pay off my credit card bills.' He sighs. 'Taking quite a while, too.'

'But your family . . .'

'They're just cousins,' he says. 'Father was in the army, retired on a pretty measly pension. Took everything he had to pay my school bills. Oh well. I'll have to find something proper to do for a career eventually,' he says. 'I suppose I can't really go on like this at my age.'

'What do you like doing?'

'Music,' he says, blushing slightly. 'I used to play the cello a bit. Of course, that doesn't pay the bills either. I tell you, I'm quite pleasantly surprised a girl like Lily would go out with me.'

'Ahm. Yes.'

'Most girls in her position seem to look for something a bit more substantial.'

Lily sticks her head out of the door. 'Ready!' she says brightly. 'Where are we off to?'

'D'you like Moroccan?' Henry asks.

'Love it,' Lily says, flouncing out of the bathroom and pirouetting for his approval. 'Where did you have in mind? Momo's?'

'I was thinking more of Dhelirious,' he says. 'It's a funky little dive off Earl's Court. Great atmosphere, great food, no liquor licence so bring your own booze. I thought we could stop at an off-licence, pick up something robust, make a night of it.'

'A little dive?' Lily asks, her voice strained. Then she gives a sexy little wriggle and beams at him. 'But that's so *adventurous* of you! Don't you think it's *adventurous* of him, Anna?'

'Absolutely,' I say. 'In fact, the whole evening should be a real adventure for you, Lily.'

'Let's go,' Henry says to her.

'There'll be no problem finding a taxi,' Lily asks. 'Will there?'

'Tube's quicker anyway,' says Henry.

'The tube,' Lily says, faintly. 'How delightful! I haven't been on public transport for years.'

'Bye,' I say, waving them off merrily. 'Have a great time, you two.'

They walk out, Henry sliding his thumb and forefinger into the small of her back, and Lily giving a pleasurable little shudder at the touch. Henry's quite different to all the rich, ugly old goats she goes out with. I just wonder what the fireworks are going to be like when she finds out just how different.

I pick up the receiver and call my parents. Oh great, another slap in the face; they're on holdiay for a week. How could I have forgotten the annual trip to Greece? They rent that tiny townhouse with the roof terrace and don't leave their sunbeds except to go to the beach, and then come back every year with red, peeling skin and a stomach infection. I could get the spare key off Mrs Watley, the next door neighbour, but that would involve having to sit in her doily-filled 'front parlour' and have my life dissected for forty-five minutes. And I can't be bothered. So I'll have to gut it out here for another week. What the hell, I've paid the damn rent.

I'm all alone in the flat. I make a cup of coffee and try to think calmly about what to do. First, there's housekeeping. I call Vanna, explain I can't marry Charles, don't mention Swan – no need to make even more of a fool of myself. She's wonderfully sympathetic, offers to come round (no thanks), offers to send back all the engagement gifts (yes please).

'Don't worry about a thing, darling,' she says. 'I'll take care of them.'

'Thanks, hon,' I say. 'I'm really grateful. And I'm sorry you spent all that money.'

'Don't be! It was a great party and anyway, I've got pots of it. Speaking of which, would you like some? Just to tide you over.'

Oh dear, I'm going to cry again. I grab a tissue and blow my nose. 'That's OK,' I say.

'Well, consider coming to work for me. I can always use an extra assistant. Good pay, no coffee making. I'll teach you the business, give you some books to read, promote you,' Vanna offers.

'Wow. That's – that's really kind of you.'

'So you'll come? You're not without resources, Anna. You've got friends.'

I swallow down the lump in my throat. 'Darling, Vanna. But no, I don't think so. At least, not yet.'

'Why not? Don't let pride get in the way.'

'Oh, it's not that. It's just that I don't want to work in publishing,' I tell her. 'I – I want to be a screenwriter.'

'Well, if you change your mind . . .'

I hang up on her feeling really grateful. OK, I may not have a place to live, a job, or a man, but at least I've got friends. Vanna, and Janet. And Charles.

His words from last night come back to me, but I shove them away.

I can't tell Mark Swan. I can't face him. I don't want to see him ever again. Stop that! Think about something else. Think about your non-existent career.

It's time to see if I can hack it or not.

On my own.

Even though I feel kind of stupid and pretentious, like who am I kidding, who do I think I am, I refuse to let it bother me. I go into the bathroom and wash my hair, blow-dry it and put my make-up on. I change into one of Janet's more comfortable choices, the black Zara pants with the crisp white polo shirt and the cute little black heels. Even though I'm not going anywhere today, or seeing anybody, this isn't about that. It's about looking as good as I possibly can *for me*.

My heart's broken and I have no money and no prospects, but there's still this little kernel of something, somewhere inside me.

I've got my friends. And I believe in myself.

I open up my ancient laptop, boot it up, slot in the floppy disk. A couple of clicks and a script page appears on my screen. I take a deep breath, and type: FADE IN.

14

The atmosphere in the flat has been poisonous lately. Janet's been moving out, and ever since Lily discovered how stinking rich Ed is, she's been too jealous to speak to her. A procession of new, and thankfully vile, potential room-mates are forever traipsing round our rooms, I'm trying to rewrite and not think about Mark (yeah, right), and now Henry and Lily have stormed back in, slamming doors and fighting, and concentration's just impossible.

Lily is shouting. She's wearing a tiny, red, flippy dress with spaghetti straps, matching heels and a flimsy little chiffon scarf, and she really does look beautiful when she's angry.

'Remind me again,' she bellows. 'Why the fuck am I dating you?'

I try to hide myself behind the *Evening Standard*.

Henry follows her into the flat, and closes the door.

'I have no idea,' he says, quite calmly. 'But don't feel under any obligation to continue.'

Lily smoulders at him.

'Excuse us,' he says apologetically to me.

'Oh, don't mind me,' I say, pretending to be busy reading. I wouldn't miss this for anything.

'Why would he mind you? When he doesn't even mind me?' Lily spits. 'I arrive at the bloody restaurant and he's leaving!'

'You were thirty-five minutes late.'

'Traffic,' Lily snaps. 'Ever heard of it?'

Henry sighs. 'I told you, if you were late you should call my mobile. You didn't call.'

'I forgot,' says Lily, tossing her hair.

'We can do it some other time,' Henry says.

'You should have waited for me!' Lily screeches, stamping her foot. 'You're a sexist pig!'

'Oh, really?' Henry demands coolly. 'This is the third time this week you've turned up late for a date. Are you in the habit of doing this with every man you've ever dated?'

Absolutely.

'Of course not,' spits Lily. 'And anyway, I'm worth it.'

'Not to me, darling,' says Henry. 'Goodbye.'

Lily stands rooted to the spot as he turns and walks towards the door. Then she springs, like a cat, grabbing him by the shoulders and spinning him round.

'Do you think you can talk to me like that?' she demands. 'I'm Lily Venus! I'm a top model and men are begging to go out with me! Who the hell do you think you are?'

'Somebody who's not interested in a spoilt little girl playing head games, and doing it badly,' Henry says. He looks Lily over, not in the usual way men do, but more sort of considerately.

'It's a shame,' he says. 'Because underneath all the hostility and childishness I think there might be an interesting person. On the other hand, I'm certainly not going to bother to stick around through the tantrums to find out.'

'I can't believe I *slept* with you!' Lily shrieks at him. 'Do you even know how many guys would kill for the chance to be with me?'

'Not all that many,' Henry says drily, 'once they'd had you. You're half frigid. And you're not going to enjoy yourself until you drop all these sad little walls you've built up.'

'I'm the best thing you've ever had,' Lily hisses.

'You're not even in the top fifty per cent,' says Henry. 'I'm amazed I actually stuck it out this long, in fact. Histrionics and tantrums don't do it for me.'

Lily pauses, trying to think of a good riposte. You can almost see the hamster wheels turning in her brain.

'You're poor,' she says finally, with deadly venom. This is about the worst insult Lily can muster. 'You come in here with your good suits and your fun little restaurants, but you don't make anything, you couldn't even *afford* the Ritz!'

'Of course I couldn't,' says Henry. 'When did you last check the prices there?'

'You don't have any money!' Lily accuses. 'Ed's got more money than you do!'

'He's got more money than the Bank of England,' says Henry. 'So what?'

'A scrub like you can't afford a girl like me,' says Lily, devastatingly.

Henry shakes his head. 'Lily,' he says, 'you're never going to get a man until you stop talking like a high-priced hooker.'

Ouch.

'I could get anyone I wanted tomorrow,' Lily snaps.

Henry walks over to her and kisses her lightly on the mouth. Lily is completely shocked, too shocked to do anything. It's very sexy, actually.

'Well,' Henry says, licking the taste of her off his lips, 'good luck with that, baby.'

And then he really does turn and walk out of the door.

Lily and I both stay completely still, listening to his footsteps disappear down the stairs.

'He'll be back,' says Lily.

'He won't, you know,' I say. 'Maybe you should go after him.'

'Me? Run after some poor chump with holes in his shoes?' Lily laughs savagely. 'I don't think so, darling.'

'You seem pretty upset,' I comment.

'I'm not,' Lily says. 'Well, I am, but only because he was so mean and nasty. Coming from a man with no loot. What a total waste of time.'

She storms into the kitchen and I hear the fridge opening. Lily pops a bottle of champagne. She thinks no real model should ever be without at least one bottle in the fridge, and I must admit it's come in handy in emergencies.

'Want some?' she asks. 'I'm celebrating getting rid of that loser.'

'Henry's not a loser,' I say. 'Just because he wouldn't stand for it.'

'Wouldn't stand for it?' she splutters, shoving a glass of Bollinger into my hand. 'Puh-leese, I'm a few minutes late, *fashionably* late, and, you know . . .'

'And what?'

'I send back a few dishes, I order a few extra things, that sort of thing,' Lily says, flicking her hair again. 'He doesn't understand. I'm a *diva*. Frankly, it was an incredible concession just to see him, once I worked out he doesn't have anything.' She laughs bitterly. 'Think of it! Me, getting dumped by a poor person!'

'Diva is just a politically correct word for a spoiled bitch, Lily,' I say. 'Clearly you really like the guy. Why don't you just call him and apologize and start over?'

'Date that sexist pig? He called me a hooker!'

'You called yourself a hooker, you said he couldn't afford you.'

'God, you're such a prig, Anna,' says Lily. 'It's obvious what I mean. I intend to marry rich, not marry some broke chancer.'

'And how is that not hooking?'

'It's one man. It's *marriage*. It's been done that way since . . . since forever,' she adds for emphasis.

'Trophy wives are just hookers with one john, if you ask me.'

'Well, I didn't,' snaps Lily. 'Aren't you supposed to be giving me support in my decision?'

'It looked like his decision.'

'I would have finished it in a couple of days anyway,' says Lily firmly. 'I can't date a man who expects me to eat in nasty little dives.'

'Look,' I say to her. 'Chasing rich guys hasn't worked out for you, has it? You've gone from one rich guy to the next, never staying with anybody more than a month, and none of them proposed. Plus, you're still here, in this flat. And you finally find a man you like, someone who won't put up with your shit, which is exactly what you need—'

'I need a man who'll give me my own space and respect my routine.'

'A pushover?'

Lily shrugs. 'I'm a princess, take me or leave me.'

'He left you.'

'So what?' she says. 'The only reason I haven't hooked up with a man properly is because *I* wasn't ready to. As soon as I set my mind to it –' she snaps her elegant, fire-engine red talons – 'I'll have anybody I like reeled in. Not like you, letting Charles get away. I can be married within three months.'

'Suit yourself.'

It's funny, but looking at her, so beautiful, all angry and sulky, I just feel pity.

'I'm going to watch *Big Brother*,' she says, flicking the television on. 'I don't want to talk about this with you any more.'

'Fine with me,' I say, taking another sip of champagne.

Janet comes back in the next morning after a catalogue shoot, and I tell her the whole Henry story.

'I hope he's not too upset,' she says.

'He didn't seem to give a damn.'

'Ed will be sorry. He thinks Henry's wonderful.'

'How's it going with you two?'

'Oh.' Janet flushes. 'Actually, really great. Can you believe it? He's asked me to meet his parents next week.'

'That's fantastic,' I say, beaming. 'Do you think it might be serious, then?'

'I wouldn't mind,' she says, shyly. 'He offered to pay off all my debts. I was so surprised when I found out he had money.'

'What did you say?'

'I told him no.' She looks sideways at me. 'Was that stupid? Lily said I was being a moron.'

'No. I think that's cool.'

'I mean, I've got a few of those catalogue shoots now. And it is paying.' Janet looks determined. 'I like Ed, but I want to do it myself. Whatever Lily says, I quite like it.'

'I think that's wonderful.'

'But I don't want to do the catalogues, not forever. It's not because I'm proud, not just that, anyway. I remembered what you said about art.'

'Don't give up a paying job for a pipe dream,' I say nervously. She's only just started paying off those debts.

'Oh, no,' says Janet. 'I wouldn't. But it's not just a pipe dream. Ed had a suggestion, he said I could go to work in a gallery. I could sell art. You know, for one of the big dealers.' She blushes again. 'He said I was so pretty that everybody would want to buy from me.'

I stare at her. 'But that's brilliant.'

'You think so?'

'You'd be great at it,' I say. 'He's right! How could any of those rich collectors resist you?'

'Ed knows some people. He said he'd get me a few interviews.'

'So will you move into his place?'

Janet shakes her head, shyly. 'I found another flat-share. In Camberwell. I thought I should have a place of my own, you know, until we get married. It's cheap, it's not as nice as this, but it is mine.' She looks at me nervously. 'Do you think I'm being mental like Lily says?'

'Never mind about Lily, she's just jealous.' I hug Janet. 'I'm really proud of you.'

Janet scribbles her new number on a piece of paper for me. 'And what about you? Where are you going?'

I sigh. 'I'm moving back with my parents. Till I can get another job. Or . . .'

She looks at me quizzically.

'Or, you know, until my script sells,' I say. I feel stupid, but there it is. I said it.

Janet hugs me back. 'I think that's brilliant and I bet you'll be a wonderful writer.'

'Thanks. You've always believed in me,' I say. I can't believe it, but I'm crying. Over moving away from a model!

'Everyone who knows you believes in you, Anna,' Janet says. 'You just needed to believe in yourself.'

Two days later I'm out of there. My mum drives the Ford Fiesta all the way to London, to pick me up.

'You can't come on the train,' she says firmly when I try to object. 'All your things, darling. And anyway, you need looking after.'

The sad thing is you can fit all my essential things into two small suitcases. One laptop, one case full of lovely new Janet-approved clothes, and one case full of books and scripts. Lined up by the door they don't seem to amount to much for thirty-two years. The rest of my stuff is just papers, prints, pre-Janet clothes that I suppose I'll be donating to Oxfam and other such rubbish.

Lily is sitting on the couch staring at her fingernails, looking impossibly beautiful and impossibly bored at the same time.

'Will you be OK?' I ask her at the last minute, since she doesn't seem to want to volunteer any Janet-style tearful farewells. Janet had an interview at the Arnsdale Gallery this morning, and went off red-eyed and red-nosed after thirty minutes of weepy hugs.

'Oh, don't flatter yourself,' Lily snaps, flicking through the latest *In Style*. 'I've already got people lining up for your room. And anyway, I won't be staying here long. I've decided to get married.'

'Oh, you and Henry are back together? That's great,' I tell her.

'I wouldn't get back with that loser if you paid me,' Lily snaps. 'No, I'm just going to find some very *lucky* man who can look after me properly and marry him. I've had enough of being by myself and sharing with . . . people,' she says, making the last word sound really insulting.

'Well, I'm off then,' I say after a pause.

'See ya,' says Lily. She doesn't bother to get up. 'You might as well leave me your phone number.'

I'm tempted to ask why.

'Just in case you've forgotten something you need,' she adds, reading my mind. I scrawl it down on the yellow Post-It pad by the phone, pick up my cases, and lug myself downstairs, not trusting the lift.

Mum has the good sense not to say anything as we pull into the Soho traffic. She pretends to be busy looking at the road as I lean my head against the window, a fat tear trickling down my cheek. It feels like everything's over. My lame attempt at a career, my friendships, and especially my love life. Although can something be over when it never really even started?

When I get home darling Dad has made me tea and Marmite toast. My parents have cleaned out my old room, put fresh flowers in a milk jug by my bed. He tells me I did exactly the right thing (though Mum does look a bit disconsolate) and I have to wait for it to be exactly right. It's an incredible weight off my mind that they don't seem to care as much as I'd feared. And however I feel right now (like my heart's been crushed into powder) I don't want to upset them, so I put a brave face on it, as much as I can. I unpack all my nice clothes, put my scripts on a bookshelf, take out my laptop. And then there's a choice. I can either start rewriting, like Mark Swan suggested. Not that I wanted to listen. Or I can listen to Mum's plans to fix me up with Kevin Nealey, the recently paroled thug ('Such a nice boy, Anna, it's not true what they say about him') who used to live next door and occasionally returns to get his laundry done or scrounge some more drug money.

I start reworking my script. After all, there's nothing else to do.

I settle back into my old life. I'm writing every day, going for walks around the village with our ancient dog Rover (a most unlikely name, as he's as fat as a beer barrel and, prior to my coming home to lick my many wounds, has gone on about four walks a year) just to clear my head. Walking is supposed to do that. Clear your head. Funny, mine's just as muddled, wrapped up in a blanket of cotton wool and pain. There seems to be nothing to say. Yeah, I'm writing, and it's not going badly, but, I don't know, that's hardly a real job, is it? And I need one. I don't have Charles any more. Thinking about Mark is so painful that it makes me want to die, but the bottom line is, I'm not going to. The sun will rise every morning, and unless I do something drastic, it'll find me in my old attic bedroom when I'm forty.

I'm making plans to move back to London. I'll take the advice I gave to Janet, borrow some money from my parents, put a security deposit on somewhere cheap, and then find a flatshare near a tube.

I can't chase my dream any more. I need a job, an actual job that pays. Otherwise I'll never own a flat of my own. On my old salary it would have taken me about fifty years to save up enough for the down payment on a Zone 4 studio. Hey, maybe I should become an estate agent . . .

It's all incredibly depressing and I only have one remedy against it. Working on my script. I've grown up now, the hard way. I know I'm not going to get Mark Swan to help me, I know Frank Giallo isn't gagging to read it. In fact, the odds against it getting noticed and read are tremendous. But I keep writing it.

It's the only dream I have left.

'Anna,' my mother says anxiously when I come back from my lunchtime stroll round the park with a huffing Rover, who glares at me reproachfully as I slip his harness off, 'you've had a phone call. There's a message on the answer phone from somebody called Jamie, but it sounds like a girl.'

'Jay-Me,' I say, excited. Oh man, it's good to hear from Janet! I wonder if she and Ed have set a date yet. Even though my life is shit, I'm so glad hers isn't. She's such a dear heart, she can't help being pretty, can she?

'She seems a bit upset,' says my mother.

My heart somersaults. When Mum says somebody is 'a bit upset' there has usually been screaming and wailing you could hear in Scotland. I rush to the answer machine, press the button.

'Anna,' sobs Janet's disembodied voice. 'Ed . . . left . . . Lily . . . aaaahhh haaahhhh . . .'

She's crying, really bawling, so bad I can't make it out. I run upstairs, retrieve her mobile number. Nothing but voicemail. So then I try our flat.

'Hello?' It's Lily.

'Hi,' I say. 'What's going on? Is Janet OK? I need to speak to her. She called me, she was in a bad way.'

'Oh, Janet,' says Lily, with more ice in her tone than usual. 'She's left. Moved out. You can get her on her cell phone.'

'Already? The month's not up.'

'She wanted to leave right away and I thought it would be best,' Lily says, nastily. 'She's got her deposit back, so she can hardly complain.'

I have a sinking feeling.

'Lily,' I say slowly, 'what happened?'

'Oh, nothing.' I can almost see Lily tossing her hair. 'Just Janet thinking she owns people.'

'Tell me,' I say.

'She thinks she owns Ed after four dates,' Lily snaps. 'They didn't exactly have a relationship. They hadn't pledged their undying love.'

'Ed broke up with her?'

'That's right,' Lily says. 'He met somebody else, and as it turns out, Janet couldn't be objective about it. She's very immature.'

Poor Janet, oh poor, poor Janet. I know she truly loved Ed, thought he was the one. Her rescuer. Just when her self-esteem was recovering from the collapse of the modelling. And Ed was so nice, and then he just dumps her?

'Do you have any idea who the new girl is?' I ask.

'Well, of course I do,' Lily snaps. 'It's me.'

I blink. 'What?'

'You heard me.'

'But you were going out with Henry,' I say, stupidly. 'You liked Henry. I thought you wanted to get back together with him.'

'Henry's not the right man for me,' Lily says, airily. 'I told you, Anna, he doesn't have a bean. I can't go for that. Plus, he was cold and controlling.'

'You mean he wouldn't take your crap,' I say icily, because it's beginning to sink in.

'Ed's different,' Lily triumphs. 'He understands me and my schedule. He knows how to treat a lady. And he's *far* richer than Charles, by the way,' she adds spitefully.

Oh hell. I can just see it. Ed, so nice, diffident, just starting to really like Janet. But he always had that thing for Lily. Staring at her like a dog slavering over a steak. He's obviously forgotten how she dismissed him when she thought he was poor. Either that, or he's so blinded in lust he doesn't want to see it.

'And Henry?'

'He'll get over it.'

'What did he say?'

'To me, nothing. After he stormed out that night he just stopped calling, which is the best way for all concerned,' Lily says coldly. If there's a touch of regret there, I don't hear it. 'But he harassed Ed.'

'For going after you?'

Lily laughs. 'He says he doesn't care what I do, but that Ed shouldn't have left Janet. He actually told Ed it was beneath him and that he'd be miserable – we'd both be miserable. But that's not true, we're blissful!'

'You don't find Ed at all attractive, Lil. You wanted Henry.'

'He had his chance and he blew it,' says Lily.

'How could you do this to Janet?'

'Oh, please. Don't lecture me,' Lily says. 'She has no right to block our happiness. She hasn't even known him a month, she doesn't own him—'

'Lily,' I say, 'you've got problems. I have to go.'

'When will you be leaving?' she persists. 'I really want you and the rest of your stuff out before the end of the month. I think Ed might move in with me while we choose a new house together, there are some nice townhouses in Mayfair I was looking at online . . .'

'Later,' I say flatly. 'Goodbye.'

I hang up and dial Janet's cell phone again, but I still get her voicemail. Perhaps that's just as well.

I look at the clock by my bedside: 2p.m. I don't want to think about any of these disasters, mine, Janet's, Henry's. But I have to. I can't let things stand like this. Hastily I pack up a little overnight case.

'Can I borrow fifty quid?' I ask Mum. 'I have to go to London.'

* * *

I catch the high-speed train into Paddington, and get a cab to the flat. I had wondered if Lily would have changed the locks, but she hasn't – maybe she just hasn't got around to it. Nobody is there, thank God. I let myself in with a spare set of keys I forgot to give back to Lily and take a shower, but don't bother to unpack. I'm not going to stick around long enough to make it worthwhile.

Then I call Janet at her new number.

'Hi, how's it at home?' she asks, without enthusiasm. God, she sounds so down. As if she's talking through a blanket of fog.

'Never mind about that. I need to see you.'

'Oh, God, Anna,' she says. 'I just can't – I don't think I can take it right now. Going over it all again . . .'

'It's not that. I need a place to stay. Just for a week while I sort myself out.'

'But you're with your parents.'

'Not any more,' I tell her. 'I need to be in London. Job-hunting. I'm desperate. Please?'

Janet is momentarily shocked out of her depression. I know she didn't expect a cry for help.

'Sure,' she says. 'It's pretty grotty,' she warns me. 'One bed but I've got a sleeping bag.'

'Luxury,' I tell her. I scribble down the address, call another cab, and I'm over there in half an hour, knocking on the door of 2B in one of those ex-local authority blocks.

Janet opens the door. She looks wrecked. Her lovely olive skin has gone sallow, her eyes are red and puffy from crying, and she's got shadows under her eyes the size of watermelon slices. She's wearing her favourite white jeans and T-shirt, but they're hanging off her; she looks as if she hasn't eaten in a week.

'What happened?' she asks.

'I can't face living at home any more,' I tell her, part truthfully.

'But why don't you just go and stay with Charles?' Janet

asks. 'You're still friends, right? Not that you're not welcome here, but Eaton Square, you know.'

'Yes,' I sigh. 'It's very nice. But not right now. And I'm about to have a huge row with Lily and I need your help.'

'Why are you going to have a huge row with Lily?'

'Because she's stolen some of my money,' I lie, easily enough. Definitely getting better at that. 'She won't give back my deposit. And I need back-up from you.'

'No way,' Janet recoils. 'I can't go and see her again. Ever. Besides, he might be there.'

'Ed?'

'Yes, Ed.' Janet starts to snuffle again. 'He's been calling here . . . my cell phone.'

'Oh?' I perk up. 'What's he been saying?'

'I don't take his messages,' Janet says, drawing herself up and trying to look dignified. 'I don't want to hear it – oh, you weren't the one for me, I'm really sorry, we should both move on, whatever. Maybe he wants to be friends,' she says bitterly. 'And I don't want that either, so what's the point?'

'But you can't let Lily steal my money,' I plead. 'I need my deposit back. And you can say I didn't wreck my room.'

'Is that what she's saying?' asks Janet, with a flicker of interest. 'She's such a cow.'

'And Ed's not in London,' I lie. 'He's staying in the country with his parents. For the week. Lily told me. Please, let me call her and arrange something, and you back me up.'

'OK,' she says grudgingly.

'I really need that money,' I tell her. 'You are going to help me, aren't you?'

'I honestly don't want to,' Janet says, miserably.

'But I'm so poor,' I wail.

'Are you sure he's not there?'

'Positive. Look, only you were there when I moved in, only you can back me up about the condition of my room.'

Janet wavers. 'Well, let me just go and have a shower and put something on,' she says.

I call Lily when Janet's safely in the shower.

'Hi, it's me.'

'What do you want?' Lily snaps.

'Nice to talk to you, too.'

'Don't be so bloody juvenile, Anna.'

'I need to come round and return the spare keys. I took them home with me by mistake.'

'Oh. All right, I suppose.'

'Is Ed with you?' I ask.

'Yes,' she says. 'Why?'

I glance at the bathroom door, but Janet is still in there, washing her hair. Thankfully. I suppose she couldn't bear the idea of Lily seeing her looking so dreadful.

'Well, I should say hello to both of you. You know, as a couple.'

'Why?' she asks, suspiciously.

I try for innocent. 'Lily, Ed and Charles are cousins. They're close. We're going to be seeing a lot of each other if this works out for you. Charles and I are thinking of getting back together,' I extemporize. 'You want to be accepted into the family, don't you?'

She hesitates. 'Of course, well, yes. Naturally. Getting back together, huh? That's very nice.'

'So will he be there with you later? About half an hour? I'd like to say hi. Maybe Charles will be with me, I'm not sure.'

'Oh, yes. Charles is such a sweetie,' says Lily smoothly. 'Of course, *do* come over, I think a family get-together would be lovely.'

'Dawsons and future Dawsons,' I say.

'Future Dawsons,' she gloats. 'Oh yes! We'll wait in for you. Ciao.'

Janet nukes her hair and doesn't bother much with make-up, but that's OK. She's used concealer and bronzer to make herself look more healthy, and she's such an expert that she manages to look fresh and sexy with next to nothing – all the dark circles gone, a couple of blue drops in her eyes, and bingo! She's almost as good as new.

I can't say the same for myself, but that doesn't really matter any more.

We take the tube down to Tottenham Court Road and get

out and walk. Janet wants to bottle out, but I keep telling her I need her as my witness, to get the money back.

'Aren't you going to press the buzzer?' she asks, when we walk past the feminist bookshop into our narrow building.

I shake my head. 'Still got the keys.' Anyway, I can't risk Ed answering the door. Janet would definitely run away then.

We step out of the elevator and I knock on the door.

Lily opens it. I shove Janet inside.

'What the hell are you doing here?' Lily sneers at Janet. She's wearing a slinky little black dress with a matching chiffon cardigan, looks like Ghost, and is made up to the nth degree. Ed, on the other hand, is perching nervously on the end of our couch in a Hugo Boss suit, the type Eli Roth favours. It looks bloody awful on him. I've seen startled racehorses that look more comfortable than Ed does now.

When he sees Janet, it's hard to say which of them looks more horrified. Janet gives a strangled cry and makes for the door. But I'm blocking it. There's no way she's getting past me.

'What is this?' screeches Lily. 'What the hell's going on?'

'You set me up,' says Janet. 'Let me out!' she implores.

'Nobody's going anywhere,' I shout. 'Ed,' I say. 'What the hell are you doing? You broke up with Janet in order to date a woman who doesn't give a toss about you and is only interested in your money.'

'Please don't,' says Janet, starting to cry. I knew she would do that, but it still freaks me out. I stand my ground all the same. I mean, I've done it now, haven't I?

'How dare you,' shrieks Lily. 'I love Ed for himself! Get out!'

'That's not true,' I say, looking at Ed. 'She wanted nothing to do with you back when she thought you were poor, and she tried to talk Janet out of dating you because we all thought you had no money. But Janet stayed true.'

'Rubbish!' squeals Lily. 'That's complete . . .' she wants to say 'bollocks' but settles for 'nonsense', on the grounds of its being more ladylike, perhaps.

'How could you dump somebody that cared about you like that?' I demand. 'Ed, how could you?'

Janet is crying now, properly. Ed shoots to his feet. He looks like a man in an agony of indecision, but when Janet starts to sob he can't help himself any more. He bounds over to her, grabbing her hands.

'Please,' he says desperately. 'Janet, don't, don't . . .'

'It's all a bloody stunt, she'll get over it,' Lily snarls. 'Let her go.'

'I can't,' Ed says, miserably. 'I'm sorry, Lily. I – I love her.'

'Excuse me?' Lily demands, glacially.

'I tried to call you,' Ed says to the sobbing Janet. 'I tried to find you. I was so sorry, right away . . . it was just a moment's madness . . . when you caught us kissing . . .'

'Madness?' Janet sobs.

'Yes,' he says. 'Lily had . . . her cardigan slipped and I . . .'

'Slipped, right,' I snort.

'You're actually falling for this?' Lily yells. 'I can't believe you!'

'I never wanted to go out with you,' Ed says to her. 'I – I kissed you – I felt . . . obligated. You and I aren't a great fit,' he says, deferentially. 'I mean, I liked my old jumpers.'

'I liked them too,' Janet sniffs.

'Those old rags!' Lily yells, her eyes bright with tears of her own. 'You were a damn wreck. You need me, Ed! You wanted me!'

'I wanted you,' he admits. 'But needed you . . .'

He turns Janet's tear-streaked face towards him and covers it with kisses, kissing down the salty tracks on her cheeks.

'Well,' says Lily. Hot tears of rage and shame are now spilling out of her beautiful eyes. 'I hope you're happy, Anna, ruining my life like that.'

'I haven't ruined your life at all,' I say to her. 'You're in love with Henry.'

'I couldn't care less about him,' she says.

'That's bullshit and you know it. You're in love with him, Lily. You're just afraid. That's why you keep dating only men with money. That's why you were so horrible to Janet when her modelling career went wrong. Because you think it's going to happen to you, soon.'

'No it isn't!' she shrieks. 'Shut up!'

'And you don't think you can do anything else,' I continue, relentlessly. 'You don't have any faith that you can use your brains. You think that as soon as your beauty's gone it's all over for you. That's why you're looking so desperately for rich guys you don't even care about. Because you think your looks are all you're worth.'

'Liar!' Lily half screams. And then she bursts into proper tears, sobbing, great heaving, wrenching sobs that make Janet's waterworks look totally unimpressive. Janet comes over to hug her, horrified, and Lily clutches her, not that she can probably see who it is through the flood. I pass over the box of Kleenex.

Ed says quietly, 'Excuse me,' and withdraws into the hallway. I hope he isn't about to bail.

'You – you don't know what it's like,' Lily shudders, looking at me with red eyes and red nose. 'You've got no idea what it's like to think about your looks every single day.'

I smile faintly. 'Actually, you'd be surprised.'

'Every line,' Lily says. 'Every hair, every tiny wrinkle . . .' She shudders in fright. 'I use all the products but it doesn't help. Not really. My face is still changing, I count the months till my next birthday,' she sobs. 'It's easy for you, you've got a degree,' she says. 'I don't even own this flat, I just sub-let it.'

'You're clever,' I tell her.

'I never even bothered with university, I was modelling,' she sobs. 'And it didn't work out. Not really . . . never the big jobs, the big money. I could have bought a flat, I could have had a car . . .'

'You still can,' I say. 'You can do all that stuff.'

'With what? I've got no qualifications,' she says. 'GCSEs in English and French and that's it. All I know is modelling and fashion.'

'So go into modelling. Executive side. Become a scout. Start an agency,' I suggest. 'You know what it takes, you know the photographers, the ad executives. You keep a Rolodex for schmoozing, use it to make them customers instead.'

Lily's sobs calm down just a little.

'Do – do you really think I could?' she asks.

'You'd be great at it,' Janet encourages her. 'You could always tell which girls would make it and which wouldn't. You knew I wouldn't.'

Lily looks at her shoes.

'Sorry for being such a bitch,' she says. It's almost inaudible, but it'll do.

'That's OK,' Janet says, hugging her. Janet is such a sweetie. I hope she marries Ed and has sixteen babies. She's such a mother.

Ed walks back into the room.

'I just called Henry,' he says. 'He's on his way over.'

'Oh God,' Lily says, panicking. 'No. I can't see him. I can't. What can I say to him?'

'Sorry?' suggests Ed. 'That's what I said to him. Took it quite well. Good chap, Henry.'

'You know,' I tell Lily, 'you've got to start by getting some courage up. You've got to see Henry and tell him the truth. Just apologize and explain why you did it.'

'Do you think he'll take me back?' she asks, in a small voice.

'Honestly, I have no idea,' I tell her. 'But you'll feel better about it anyway.'

'I'll do that,' she says.

'OK,' I say. 'I'm off.'

Janet says, 'Thanks so much. I don't know how I can thank you.'

Lily squeezes my arm. 'Yes, thanks, Anna,' she says. 'You made me face up to some things,' she says, starting to cry again.

'It'll be OK,' I say. And for her, I think it will.

Ed and Janet leave, but Lily wants me to stay, so I sit with her curled on the couch and hug her while she cries.

'I can't see Henry,' she keeps saying. 'I look awful . . .'

'That doesn't matter, not if he loves you. And if he doesn't, then you don't want him.'

'And what about you?'

'What?'

'What about you?' Lily asks, blowing her nose loudly on a piece of tissue.

'I don't understand.'

'Oh yes you do,' says Lily, her reddened, still-lovely eyes narrowing shrewdly. 'You broke up with Charles over that Mark Swan man, and he still has no idea you want to go out with him.'

'That's completely different,' I say. 'He's got someone else, Henry doesn't.'

'Is he married?'

'Not yet.'

'Then you should still tell him.'

I shake my head. 'Lil, you don't know, honestly. I can't just tell him "Please leave your girlfriend, I fancy you".'

'Why not?' she demands. 'You've all this great advice for other people, but you don't listen to any of it yourself.'

'You should see his girlfriend.'

'I don't know if you've noticed, Anna,' says Lily patiently, 'but since Janet made you over you've been looking different. Better. You know, men could fancy you,' she says, encouragingly. I can't even be cross; Lily isn't going to change her whole personality in five minutes, and at least now she means well. 'And like you said, if the man's in love, pretty or not doesn't really matter.'

'Yeah, well. To a certain extent. I've still got a Gonzo nose.'

'You call it Gonzo, I call it Roman and distinctive.'

'I wish you had called it that,' I say wryly.

'I bet Mark Swan does too. I bet he was interested in you,' Lily says, ignoring me. 'On some level you know he was, or it wouldn't hurt. Women always know.'

I look at her with new eyes.

'I'm not as stupid as I look,' Lily says.

'You're not stupid, you're . . . you're canny,' I tell her. 'But you know, he had a long time to say something to me. And he never did.'

'All that time he thought you were going out with Charles,' Lily points out.

That's true.

'Don't you think it's at least possible that you didn't say anything because you thought you weren't pretty enough, and he didn't say anything because he thought he wasn't rich enough?'

'I—'

'Just call him,' Lily says, shoving the phone at me. 'You know you have to do it, anyway.'

'Why do I have to do it? I don't have to do anything.'

'You do,' she says. 'Otherwise you'll spend the rest of your life wondering.'

There's a knock on the door.

'Henry,' says Lily, bolting to her feet.

'I have to go,' I say.

'Stay,' she pleads.

'You'll be just fine. Hi, Henry,' I say, opening the door before she can respond. He's got a wild look in his eyes, and glances from me to Lily. 'I was just leaving,' I reassure him. 'See you guys later.'

'Anna,' Lily says. 'Thanks.'

'You're welcome.'

'And call him,' she adds.

I don't say anything. I just walk out of the door and close it on Henry striding over to Lily and folding her into his arms as she starts to cry again.

She's right. Of course she's right. I mean, what do I have to lose? Nothing important. Hope. Self-esteem. Face. The ability to hold my head up in public.

But still. I will always, always wonder.

I walk down the street. I have no idea where I'm going. Our flat is out, and I don't want to go back to Janet's if I can help it. Not tonight, she'll want to have a heart to heart with Ed long into the night. And I've no desire to play gooseberry.

I could go to Charles's place. I know he'd let me stay. But I don't want to reopen the wound.

I head down Tottenham Court Road, past Borders, walking towards Leicester Square, and I find myself turning right, heading into Soho.

My heart's thumping, but I keep going.

I'm going to Swan Lake productions. If I'm going to do this it's got to be right. A phone call won't do. Whether he says yes or no, I have to see him. Even if Michelle's there. In fact, especially if Michelle's there. I don't want to go behind her back, I'm going to tell her straight out that I'm in love with her fiancé and I have to tell him and I'm going to let him make the choice.

I almost want to smile. If this goes wrong, and let's face it, it's going to, it will be the single most awful, embarrassing moment of my life. It'll be worse than that bloody school dance.

But somehow it doesn't matter. I feel so much lighter, so much freer. This will finish it, one way or the other.

I enter Swan Lake. The security guard is snoring behind his desk, so that's one fewer thing to worry about. I slip quietly into the lift, ride up to Mark's floor, palms sweating, feeling sick. Calm down, I tell myself. Maybe the security guard was a good omen and Michelle isn't even here! Maybe I won't have to do any confronting at all.

The lift doors hiss open.

'What the fuck do you want?' snaps Michelle. 'You're not on the list. Did that moron downstairs let you up again? I'm going to get him fired.'

She's wearing a simple white T-shirt and jeans, and even with my carefully chosen, smart little outfit (shirt dress, jean jacket, kitten heels, chunky amber necklace) and my killer haircut and pro make-up she knocks me for six.

'I snuck up,' I say. My voice sounds tinny and small, as if it's coming from far away. Something to do with being terrified. 'I want to see Mark.'

'Not without an appointment,' Michelle says stiffly.

'I want to see him because I'm in love with him and I want to tell him that,' I blurt out.

She laughs, a fierce, hostile laugh. 'I *knew* it,' she says. 'Knew you were just like all the others. And you with a fiancé!'

I show her my bare left hand. 'I broke up with him.'

'Over Mark? He's never given you the slightest reason,' she says bitchily. 'I would know. He tells me everything.'

'I'm being honest with you. I need to tell him. I know you're going out with him. But I want him to break up with you and be with me,' I say.

She tosses her head. 'You can't, he's not here. Now go away.'

I stay put. 'I can hear him inside the office. He's on the phone. And I'm going to tell him. I still have his cell phone number and his address. And his email address. You can't stop me.'

'Do you realize how stupid you sound?' she demands. 'Just leave! Don't make me call security. I won't tell him how you embarrassed yourself, it happens to lots of women around Mark.'

I look at her, long and hard. Something in her voice. It's . . . it's *panicky*. She keeps glancing at Mark's inner office.

And then it hits me, in an overwhelming burst of sheer joy.

'And you're one of them! Aren't you?'

She doesn't look at me.

'You're one of them, Michelle. He's not going out with you, he never was. I should have known.'

'What the hell do you know about it?' Michelle asks, her young face crumpling. Her eyes fill with tears. 'He could have gone out with me! He would have, eventually. If it hadn't been for *you*,' she hisses.

'What do you mean?'

'Why can't you just go away?' she asks, and then she starts crying. 'You don't know what it's like, sitting here day after day. Booking lunch places for all the little girlfriends, and then he meets you, and you don't even want him, you're with somebody else.'

I fish out a Kleenex from my bag.

'I'm sorry,' I say.

'You had to take two men,' she says, sobbing. 'You couldn't leave any for anybody else, even people that love them and would never get engaged to somebody else and . . .'

I walk awkwardly towards her and rub her on the back. Is that how she sees me, some sort of femme fatale?

'I honestly love him,' I tell her. 'He probably won't care any more, but I have to tell him, I'm sorry.'

'I know,' she mutters.

The door opens and Swan is standing there. He looks at me with my arm round Michelle.

'Anna,' he says. 'What the hell's going on?'

Michelle looks at me, her expression hunted.

'Michelle just found out her . . . auntie died,' I say.

'That's right,' mumbles Michelle.

Swan looks at the two of us again. 'OK,' he says. 'Why don't you take the rest of the day off, Michelle. I can get the phones.'

Michelle nods, grabs her bag and exits without looking back. Poor girl. I know just how she feels.

'You'd better come in,' he says to me.

I walk in behind him, sit down on one of his couches. That helps, makes me feel like I'm not going to faint.

Swan sits down opposite me. I wait for him to say something. He doesn't. I glance over towards his desk, and see something familiar.

It's the boxed Sony Vaio he bought me as an engagement gift, still in its wrapping paper.

'You got it back then,' I say.

He nods.

'I split up with Charles,' I try.

'So I gathered,' Swan says drily. 'From your friend Vanna.'

'Aren't you going to say anything?' I ask, desperately, when he doesn't add anything. 'Aren't you going to ask me how, or why?'

'Why should I?' he demands. 'You expect to just walk back in here, Anna, and have me fawning all over you? Why should I care what you do with your life?'

I stand up. 'You're . . . you're right. I'm sorry, this was a huge mistake.'

So that's it then. I turn on my heels, don't want to cry in front of him.

A large hand comes down on my shoulder. Swan spins me back round, spins me to face him. Forces me to look up at him, to feel him close to me, towering over me.

'Where the hell do you think you're going?' he asks. 'You're not walking out on me again, Anna Brown.'

'Mark . . .' I say. I can hardly breathe.

'You were going to marry someone else,' he says. 'Somebody you didn't even love. And then you finally break up with him, and what happens? You don't even call. What was it, Anna? Still throwing that tantrum because I wouldn't help you hamstring your own career? Your script wasn't ready. You weren't ready.'

'I thought you didn't care about me,' I mutter.

'Didn't care? Damn, you're frustrating,' he says, shaking his craggy head. 'I'm the one who started you writing. I'm the one who got you to do what you really loved. You'd think you would have trusted me, but no. Because I wasn't going to send you out there with something half-baked, I don't care.'

I look at him, not daring to hope. Trying not to hope. In case I get crushed.

'You only see what you want to see,' he says, bitterly. He lets me go. 'And when you break up, and I think maybe . . .'

'Maybe what?'

'But you don't call. You don't even give me the courtesy of a phone call. Nothing. I guess I meant that little to you. As soon as I wouldn't help your so-called career, you didn't care about me any more.'

'It wasn't that,' I say.

'Sure it was,' he says.

I think of Michelle. Poor Michelle. I can't tell him the truth.

'I didn't think you'd want me,' I mutter.

He pauses. 'What?'

'I didn't think you'd want me.'

Swan blinks. The rage is gone from his face. He looks . . . he looks *astonished*.

'Excuse me?' he demands. 'Why the hell would you think that?'

I make a gesture at my face, at my nose. 'Men don't fancy me.'

'The hell they don't,' Swan says.

'But my nose . . .'

'Is beautiful,' he says. 'Like the rest of you.'

Now it's my turn to stare. 'Give over,' I say scornfully. 'I'm tall and strong and—'

'You're not tall from where I'm standing. And anyway, what's wrong with tall? I'm tall. Should I be offended?' Swan asks, with just a touch of humour. 'And stop fishing for compliments.'

'What?'

'Oh, you know how lovely you are,' he says. 'Those breasts . . .'

I blush.

'Those legs. Those eyes. Your hair. You look even prettier since you started showing it off.'

'I didn't think you'd noticed.'

'I notice everything.'

'But I'm not thin . . .'

'So what?' he says. 'I can't stand that anorexic look. You're gorgeous.'

'Maybe if I got a nose job . . .'

'Over my dead body,' he says. 'You've got an interesting face, not like all those two-dimensional bimbos.'

'You fancied me?' I ask. Pathetic, I know. But I can't help myself. 'You're not joking?'

'Are you delusional?' he asks. He moves towards me, stands right next to me again. Looking me over, head to toe. A sweeping look, full of intent. 'I hate the word fancy,' he says softly. 'I don't fancy you, I *want* you.'

I go all liquid. Instantly.

'And not just for five minutes, either,' he says. 'I liked how you looked, but I didn't want you because of that. I felt for you because of . . . I don't know. Your sense of humour. Your fearlessness. Your intelligence.'

'Really?' I whisper.

'Because you're just Anna,' he says. 'You're not like anyone else. You're . . . Anna.'

'I'm in love with you,' I say.

'I know that,' he says. 'But I thought you didn't. Or you knew it, but didn't care. Is that why you came here today?'

I nod.

'Not to ask me for help with your career?'

I shake my head. 'I don't want your help. I'm rewriting. I want to make it on my own.'

He chuckles. 'Not a chance, baby. I'm going to pull strings like you wouldn't believe.' He looks down at me, and gently kisses me on the mouth. It's a very light, very sexy kiss. His lips just brushing against mine, just a feathery, electric touch. A wave of wanting crashes all through me.

'Will you go out with me?' I whisper.

Swan gathers me into his arms, pulls me close to him. My weight is nothing to him, nothing at all.

'I'd much rather stay in with you,' he says. 'What are your plans?'

'For when?'

'For the next sixty years,' he says. 'We can start with that.'

And then he kisses me again.